P9-CPW-539

FRACTALS EVERYWHERE

David McFadzean
270-4190

FRACTALS EVERYWHERE

Michael Barnsley

School of Mathematics
Georgia Institute of Technology
Atlanta, Georgia

and

Iterated Systems, Inc.
Atlanta, Georgia

ACADEMIC PRESS, INC.
Harcourt Brace Jovanovich, Publishers
Boston San Diego New York
Berkeley London Sydney
Tokyo Toronto

Copyright © 1988 by Academic Press, Inc.
All rights reserved.
No part of this publication may be reproduced or transmitted in any form or by any means, electronic or mechanical, including photocopy, recording, or any information storage and retrieval system, without permission in writing from the publisher.

Figure credits and other acknowledgments appear at the end of the book.

ACADEMIC PRESS, INC.
1250 Sixth Avenue, San Diego, CA 92101

United Kingdom Edition published by
ACADEMIC PRESS INC. (LONDON) LTD.
24-28 Oval Road, London NW1 7DX

Library of Congress Cataloging-in-Publication Data

Barnsley, M. F. (Michael Fielding), 1946–
Fractals everywhere/Michael Barnsley.
p. cm.
Bibliography: p.
Includes index.
ISBN 0-12-079062-9
1. Fractals. I. Title.
QA614.86.B37 1988
516—dc19

88-12644
CIP

Printed in the United States of America
88 89 90 91 9 8 7 6 5 4 3 2 1

I dedicate this book to my wife
Diane Colette Barnsley

Contents

Acknowledgments

I acknowledge and thank many people for their help with this book. In particular I thank Alan Sloan, who has unceasingly encouraged me, who wrote the first *Collage* software, and who so clearly envisioned the application of iterated function systems to image compression and communications, that he founded a company named *Iterated Systems Incorporated*. Edward Vrscay, who taught the first course in deterministic fractal geometry at Georgia Tech, shared his ideas about how the course could be taught, and suggested some subjects for inclusion in this text. Steven Demko, who collaborated with me on the discovery of iterated function systems, made early detailed proposals on how the subject could be presented to students and scientists, and provided comments on several chapters. Andrew Harrington and Jeffrey Geronimo, who discovered with me orthogonal polynomials on Julia sets. My collaborations with them over five years formed for me the foundation on which iterated function systems are built. Watch for more papers from us!

Les Karlovitz, who encouraged and supported my research over the last nine years, obtained the time for me to write this book and provided specific help, advice, and direction. His words can be found in some of the sentences in the text. Gunter Meyer, who has encouraged and supported my research over the last nine years. He has often given me good advice. Robert Kasriel, who taught me some topology over the last two years, corrected and rewrote my proof of Theorem 2.7.1 and contributed other help and warm encouragement. Nathanial Chafee, who read and corrected Chapter 2 and early drafts of Chapters 3 and 4. His apt constructive comments have increased substantially the precision of the writing. John Elton, who taught me some ergodic theory, continues to collaborate on exciting research into iterated function systems, and helped me with many parts of the book. Daniel Bessis and Pierre Moussa, who are filled with the wonder and mystery of science, and taught me to look for mathematical events which are so astonishing that they may be called miracles. Research work with Bessis and Moussa at Saclay during 1978, on the Diophantine Moment Problem and Ising Models, was the seed which grew into this book. Warren Strahle, who provided some of his experimental research results for inclusion in Chapter 6.

Graduate students John Herndon, Doug Hardin, Peter Massopust, Laurie Reuter, Arnaud Jacquin, and François Malassenet, who have contributed in many ways to this book. They helped me to discover and develop some of the ideas. Els Withers and Paul Blanchard, who supported the writing of this book from the start and suggested some good ideas which are used. The research papers by Withers on iterated function systems are deep. Edwina Barnsley, my mother, whose house was always full of flowers. Her encouragement and love helped me to write this book. Thomas Stelson, Helena Wisniewski, Craig Fields, and James Yorke who, early on, supported the development of applications of iterated function systems. Many of the pictures in this text were produced in part using software and hardware in the DARPA/GTRC funded Computergraphical Mathematics Laboratory within the School of Mathematics at Georgia Institute of Technology.

George Cain, James Herod, William Green, Vince Ervin, Jamie Goode, Jim Osborne, Roger Johnson, Li Shi Luo, Evans Harrell, Ron Shonkwiler, and James Walker who contributed by reading and correcting parts of the text, and discussing research. Thomas Morley, who contributed many hours of discussion of research and never asks for any return. William Ames who encouraged me to write this book and introduced me to Academic Press. Annette Rohrs, who typed the first drafts of Chapters 2, 3, and 4. William Kammerer, who introduced me to EXP, the technical word processor on which the manuscript of this book was written, and who has warmly supported this project. Alice Peters, whose energy and enthusiastic support helped me to write the book. She and her production team, especially Iris Kramer and Ezra C. Holston, turned a manuscript into a beautiful book for you.

This book owes its deepest debt to Alan Barnsley, my father, who wrote novels and poems under the *nom-de-plume* Gabriel Fielding. I learnt from him care for precision, love of detail, enthusiasm for life, and an endless amazement at all that God has made.

Michael Barnsley

Introduction 1

Fractal geometry will make you see everything differently. There is danger in reading further. You risk the loss of your childhood vision of clouds, forests, galaxies, leaves, feathers, flowers, rocks, mountains, torrents of water, carpets, bricks, and much else besides. Never again will your interpretation of these things be quite the same.

The observation by Mandelbrot [Mand 1982] of the existence of a "Geometry of Nature" has led us to think in a new scientific way about the edges of clouds, the profiles of the tops of forests on the horizon, and the intricate moving arrangement of the feathers on the wings of a bird as it flies. Geometry is concerned with making our spatial intuitions objective. Classical geometry provides a first approximation to the structure of physical objects; it is the language which we use to communicate the designs of technological products, and, very approximately, the forms of natural creations. Fractal geometry is an extension of classical geometry. It can be used to make precise models of physical structures from ferns to galaxies. Fractal geometry is a new language. Once you can speak it, you can describe the shape of a cloud as precisely as an architect can describe a house.

This book is based on a course called "Fractal Geometry" which has been taught in the School of Mathematics at Georgia Institute of Technology for two years. The course is open to all students who have completed two years of calculus. It attracts both undergraduate and graduate students from many disciplines, including mathematics, biology, chemistry, physics, psychology, mechanical engineering, electrical engineering, aerospace engineering, computer science, and geophysical science. The delight of the students with the course is reflected in the fact there is now a second course entitled "Fractal Measure Theory." The courses provide a compelling vehicle for teaching beautiful mathematics to a wide range of students.

Here is how the course in Fractal Geometry is taught. The core is Chapter Two, Chapter Three, Sections 1 to 5 of Chapter Four, and Sections 1 to 3 of Chapter Five. This is followed by a collection of delightful special topics, chosen from Chapters Six, Seven, and Eight. The course is taught in thirty one-hour lectures.

Chapter Two introduces the basic topological ideas that are needed to describe subsets of spaces such as $\mathbb{R}^2$. The framework is that of metric spaces; this is adopted because metric spaces are both rigorously and intuitively accessible, yet full of surprises. They provide a suitable setting for fractal geometry. The concepts introduced include openness, closedness, compactness, convergence, completeness, connectedness, and equivalence of metric spaces. An important theme concerns properties which are preserved under equivalent metrics. Chapter Two concludes by presenting the most exciting idea: a metric space, denoted by $\mathscr{H}$, whose elements are the nonempty compact subsets of a metric space. Under the right conditions this space is complete, Cauchy sequences converge, and fractals can be found!

Chapter Three deals with transformations on metric spaces. First the goal is to develop intuition and practical experience with the actions of elementary transformations on subsets of spaces. Particular attention is devoted to affine transformations and Möbius transformations in $\mathbb{R}^2$. Then the contraction mapping principle is revealed, followed by the construction of contraction mappings on $\mathscr{H}$. Fractals are discovered as the fixed points of certain set maps. We learn how fractals are generated by the application of "simple" transformations on "simple" spaces, and yet they are geometrically complicated. It is explained what an iterated function system (IFS) is, and how it can define a fractal. Iterated function systems provide a convenient framework for the description, classification, and communication of fractals. Two algorithms, the "Chaos Game" and the Deterministic Algorithm, for computing pictures of fractals, are presented. Attention is then turned to the inverse problem: given a compact subset of $\mathbb{R}^2$, fractal, how do you go about finding a fractal approximation to it? Part of the answer is provided by the Collage Theorem. Finally, the thought of the wind blowing through a fractal tree leads to discovery of conditions under which fractals depend continuously on the parameters which define them.

Chapter Four is devoted to dynamics on fractals. The idea of addresses of points on certain fractals is developed. In particular, the reader learns about the metric space to which addresses belong. Nearby addresses correspond to nearby points on the fractal. This observation is made precise by the construction of a continuous transformation from the space of addresses to the fractal. Then dynamical systems on metric spaces are introduced. The ideas of orbits, repulsive cycles, and equivalent dynamical systems are described. The concept of the shift dynamical system associated with an IFS is introduced and explored. This is a visual and simple idea in which the author and the reader are led to wonder about the complexity and beauty of the available orbits. The equivalence of this dynamical system with a corresponding system on the space of addresses is established. This equivalence takes no account of the geometrical complexity of the dance of the orbit on the fractal. The chapter then moves towards its conclusion, the definition of a chaotic dynamical system and the realization that "most" orbits of the shift dynamical system on a fractal are chaotic. To this end, two simple and delightful ideas are shown to the reader. The Shadow Theorem illustrates how apparently random orbits may actually be the "shadows" of deterministic motions in higher dimensional spaces. The Shadow*ing* Theorem demonstrates how a rottenly inaccurate orbit may be trailed by a precise orbit, which clings like a secret agent. These ideas are used to make an explanation of why the "Chaos Game" computes fractals.

Chapter Five introduces the concept of fractal dimension. The fractal dimension of a set is a number which tells how densely the set occupies the metric space in which it lies. It is invariant under various stretchings and squeezings of the underlying space. This makes the fractal dimension meaningful as an experimental observable; it possesses a certain robustness, and is independent of the measurement units. Various theoretical properties of the fractal dimension, including some explicit formulas, are developed. Then the reader is shown how to calculate the fractal dimension of real-world data; and an application to a turbulent jet exhaust is described. Lastly, the Hausdorff-Besicovitch dimension is introduced. This is another number which can be associated with a set. It is more robust and less practical than the fractal dimension. Some mathematicians love it; most experimentalists hate it; and we are intrigued.

Chapter Six is devoted to fractal interpolation. The aim of the chapter is to teach the student practical skill in using a new technology for making complicated curves and fitting experimental data. It is shown how geometrically complex graphs of continuous functions can be constructed to pass through specified data points. The functions are represented by succinct formulas. The main existence theorems and computational algorithms are provided. The functions are known as fractal interpolation functions. It is explained how they can be readily computed, stored, manipulated, and communicated. "Hidden variable" fractal interpolation functions are introduced and illustrated; they are defined by the shadows of the graphs of three-dimensional fractal paths. These geometrical ideas are extended to introduce space-filling curves.

Chapter Seven gives an introduction to Julia sets. Julia sets are deterministic fractals which arise from the iteration of analytic functions. The objective is to show the reader how to understand these fractals, using the ideas of Chapters Three and Four. In so doing, we have the pleasure of explaining and illustrating the Escape Time Algorithm. This algorithm is a means for computergraphical experimentation on dynamical systems which act on two-dimensional spaces. It provides illumination and coloration, a searchlight to probe dynamical systems for fractal structures and regions of chaos. The algorithm relies on the existence of "repelling sets" for continuous transformations which map open sets to open sets. The applications of Julia sets to biological modelling and to understanding Newton's method are considered.

Chapter Eight is concerned with how to make maps of certain spaces, known as parameter spaces, where every point in the space corresponds to a fractal. The fractals depend "smoothly" on the location in the parameter space. How can one make a picture which provides useful information about what kinds of fractals are located where? If both the space in which the fractals lie, and the parameter space, are two-dimensional, the parameter space can sometimes be "painted" to reveal an associated Mandelbrot set. Mandelbrot sets are defined, and three different examples are explored, including the one which was discovered by Mandelbrot. A computergraphical technique for producing images of these sets is described. Some basic theorems are proved.

Chapter Nine is an introduction to measures on fractals, and to measures in general. The chapter is an outline which can be used by a professor as the basis of a course in fractal measure theory. It can also be used in a standard measure theory course as a source of applications and examples. One goal is to demonstrate that measure theory is a workaday tool in science and engineering. Models for real world images can be made using measures. The variations in color and brightness, and the complex textures in a color picture can be successfully modelled by measures which can be written down explicitly in terms of succinct "formulas." These measures are desirable for image engineering applications, and have a number of advantages over non-negative "density" functions. Section 9.1 provides an intuitive description of what measures are, and motivates the rest of the chapter. The context is that of Borel measures on compact metric spaces. Fields, sigma-fields, and measures are defined. Carathéodory's extension theorem is introduced and used to explain what a Borel measure is. Then the integral of a continuous real-valued function, with respect to a measure, is defined. The reader learns to evaluate some integrals. Next the space $\mathscr{P}$ of normalized Borel measures on a compact metric space is defined. With an appropriate metric, $\mathscr{P}$ becomes a compact metric space. Succinctly defined contraction mappings on this space lead to measures which live on fractals. Integrals with respect to these measures can be evaluated with the aid of Elton's ergodic theorem. The book ends with a description of the application of these measures to computer graphics.

This book teaches the tools, methods, and theory of deterministic geometry. It is useful for describing *specific* objects and structures. Models are represented by succint "formulas." Once the formula is known, the model can be reproduced. We do not consider statistical geometry. The latter aims at discovering general statistical laws which govern families of similar-looking structures, such as *all* cumulus clouds, *all* maple leaves, or *all* mountains.

In deterministic geometry, structures are defined, communicated, and analysed, with the aid of elementary transformations such as affine transformations, scalings, rotations, and congruences. A fractal set generally contains infinitely many points whose organization is so complicated that it is not possible to describe the set by specifying directly where each point in it lies. Instead, the set may be defined by "the relations between the pieces." It is rather like describing the solar system by quoting the law of gravitation and stating the initial conditions. Everything follows from that. It appears always to be better to describe in terms of relationships.

2 Metric Spaces; Equivalent Spaces; Classification of Subsets; and the Space of Fractals

2.1 SPACES

In fractal geometry we are concerned with the structure of subsets of various very simple "geometrical" spaces. Such a space is denoted by $\mathbf{X}$. It is the space on which we think of drawing our fractals; it is the place where fractals live. What is a fractal? For us, for now, it is just a subset of a space. Whereas the space is simple, the fractal subset may be geometrically complicated.

Definition 1. A *space* $\mathbf{X}$ is a set. The *points* of the space are the elements of the set.

Although this definition does not say it, the nomenclature "space" implies that there is some structure to the set, some sense of which points are close to which. We give some examples to show the sort of thing this may mean. Throughout this text $\mathbb{R}$ denotes the set of real numbers, and " $\in$ " means "belongs to."

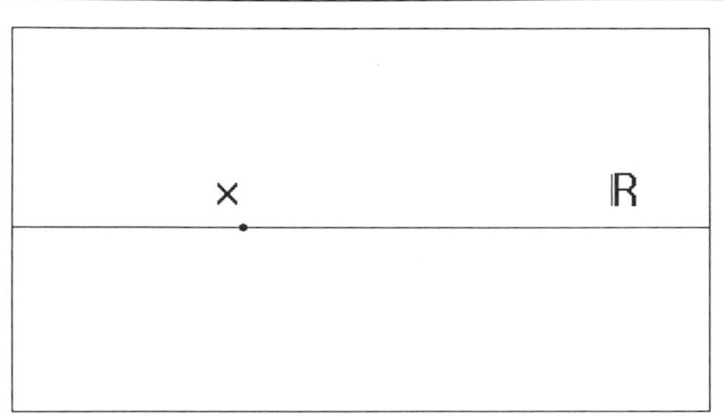

Figure 2.1.1
A point x in the space $\mathbb{R}$.

Examples

1.1. $X = \mathbb{R}$. Each "point" $x \in X$ is a real number. Equally well it is a dot on a line.

1.2. $X = C[0,1]$, the set of continuous functions which take the real closed interval $[0,1] = \{x \in \mathbb{R}: 0 \leq x \leq 1\}$ into the real line $\mathbb{R}$. A "point" $f \in X$ is a function f: $[0,1] \xrightarrow{\text{cts.}} \mathbb{R}$ where f may be represented by its graph.

Notice that here $f \in X$ is not a point on the x-axis, it is the whole function. A continuous function on an interval is characterized by the fact that its graph is unbroken; as a picture it contains no rips or tears and it can be drawn without removing the pencil from the paper.

1.3. $X = \mathbb{R}^2$, the Euclidean plane, the coordinate plane of calculus. Any pair of real numbers $x_1, x_2 \in \mathbb{R}$ determine a single point in $\mathbb{R}^2$. A point

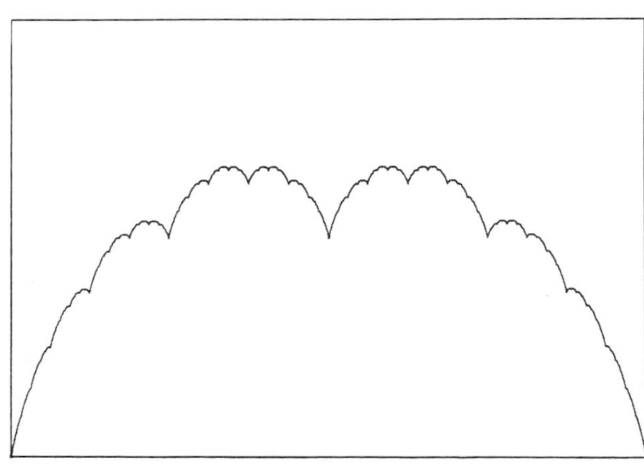

Figure 2.1.2
A point f in the space of continuous functions on $[0,1]$.

Figure 2.1.3
A point x in the space $\mathbb{R}^2$.

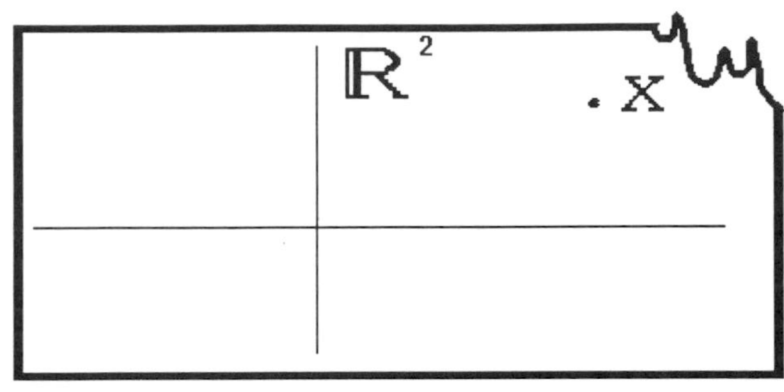

$x \in X$ is represented in several equivalent ways: $x = (x_1, x_2) = \binom{x_1}{x_2} =$ a point in a figure such as Figure 2.1.3.

The spaces in examples 1.1, 1.2, and 1.3 are each *linear spaces*. There is an obviously defined way, in each case, of adding two points in the space to obtain a new one in the same space. In 1.1 if x, and $y \in \mathbb{R}$ then $x + y$ is also in $\mathbb{R}$; in 1.2 we define $(f + g)(x) = f(x) + g(x)$; and in 1.3 we define $x + y = \binom{x_1}{x_2} + \binom{y_1}{y_2} = \binom{x_1 + y_1}{x_2 + y_2}$. Similarly, in each of the above examples, we can multiply members of X by a scalar, that is, by a real number $\alpha \in \mathbb{R}$. For example, in 1.2 $(\alpha f)(x) = \alpha f(x)$ for any $\alpha \in \mathbb{R}$, and $\alpha f \in C[0,1]$ whenever $f \in C[0,1]$. Example 1.1 is a one-dimensional linear space; 1.2 is an ∞-dimensional linear space (can you think why the dimension is infinite?); and 1.3 is a two-dimensional linear space. A linear space is also called a vector space. The scalars may be complex numbers instead of real numbers.

1.4. The complex plane, $X = \mathbb{C}$, where any point $x \in X$ is represented

$$x = x_1 + ix_2 \quad \text{where } i = \sqrt{-1},$$

for some pair of real numbers $x_1, x_2 \in \mathbb{R}$. Any pair of numbers $x_1, x_2 \in \mathbb{R}$ determine a point of $\mathbb{C}$. It is obvious that $\mathbb{C}$ is essentially the same as $\mathbb{R}^2$; but there is an implied distinction. In $\mathbb{C}$ we can multiply two points x, y and obtain a new point in $\mathbb{C}$. Specifically, we define

$$x \cdot y = (x_1 + ix_2)(y_1 + iy_2) = (x_1 y_1 - x_2 y_2) + i(x_2 y_1 + x_1 y_2).$$

1.5. $X = \hat{\mathbb{C}}$, the Riemann sphere. Formally $\hat{\mathbb{C}} = \mathbb{C} \cup \{\infty\}$; that is, all the points of $\mathbb{C}$ together with "The Point at Infinity." Here is a way of constructing and thinking about $\hat{\mathbb{C}}$. Place a sphere on the plane $\mathbb{C}$, with the South Pole on the origin, and the North Pole N vertically above it. To a given point $x \in \mathbb{C}$ we associate a point x' on the sphere by constructing the straight line from N to x and marking where this line intersects the sphere. This associates a unique point $x' = h(x)$ with each

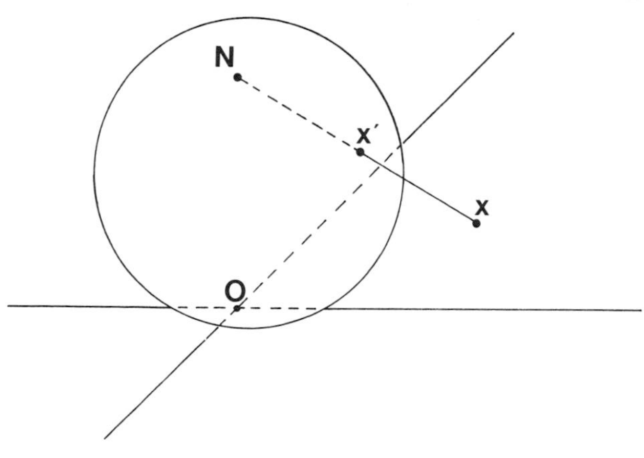

Figure 2.1.4 *Construction of a geometrical representation for the Riemann sphere. N is the North Pole, and corresponds to the "Point at Infinity."*

point $x \in \mathbb{C}$. The transformation h: $\mathbb{C} \to$ sphere is clearly continuous in the sense that nearby points go to nearby points. Points further and further away from 0 in the plane $\mathbb{C}$ end up closer and closer to N. $\hat{\mathbb{C}}$ consists of the completion of the range of h by including N on the sphere: "The Point at Infinity (∞)" can be thought of as a giant circle, infinitely far out in $\mathbb{C}$, whose image under h is N. It is easier to think of $\hat{\mathbb{C}}$ being the whole of the sphere, rather than as the plane together with ∞. It is of interest that h: $\mathbb{C} \to$ sphere is *conformal*: it preserves angles. The image under h of a triangle in the plane is a curvaceous triangle on the sphere. Although the sides of the triangle on the sphere are curvaceous they meet in well-defined angles, as one can visualize by imagining the globe to be magnified enormously. The angles of the curvaceous triangle are the same as the corresponding angles of the triangle in the plane.

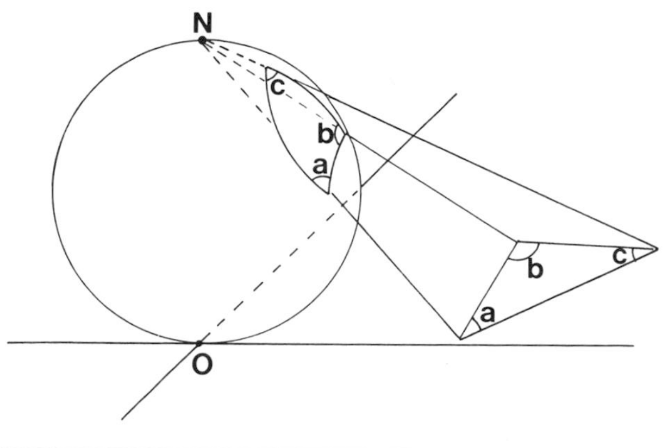

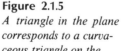

Figure 2.1.5 *A triangle in the plane corresponds to a curvaceous triangle on the sphere.*

Examples & Exercises

1.6. $\mathbf{X} = \Sigma$, the *code space* on N symbols. N is a positive integer. The symbols are the integers $\{0, 1, 2, \ldots, N-1\}$. A typical point in $\mathbf{X}$ is a semi-infinite word such as

$$x = 2\;17\;0\;0\;1\;21\;15\;(N-1)\;3\;0\ldots .$$

There are infinitely many symbols in this sequence. In general, for a given element $x \in \mathbf{X}$, we can write

$$x = x_1x_2x_3x_4x_5x_6x_7x_8\ldots \quad \text{where each } x_i \in \{0, 1, 2, \ldots, N\}.$$

1.7. A few other favorite spaces are defined as follows:

(a) A disk in the plane with center at the origin and with finite radius $R > 0$:

$$● = \{x \in \mathbb{R}^2 : x_1^2 + x_2^2 \le R^2\}.$$

(b) A "filled" square:

$$■ = \{x \in \mathbb{R}^2 : 0 \le x_1 \le 1, 0 \le x_2 \le 1\}.$$

(c) An interval:

$[a, b] = \{x \in \mathbb{R} : a \le x \le b\}$, where a and b are real numbers with $a < b$.

(d) Body space:

$= \{x \in \mathbb{R}^3$: coordinate points implied by a cadaver frozen in $\mathbb{R}^3\}$.

(e) Sierpinski space:

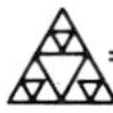

$= \{x \in \mathbb{R}^2$: x is a point on a certain fixed Sierpinski triangle$\}$.

Sierpinski triangles occur often in this text as displayed above. There is a Sierpinski triangle in Figure 3.4.3.

1.8. Show that the examples in 1.5, 1.6, and 1.7 are not vector spaces, using addition and multiplication by reals as defined in the usual way.

1.9. The notation $A \subset \mathbf{X}$ means A is a *subset* of $\mathbf{X}$; that is, if $x \in A$ then $x \in \mathbf{X}$, or $x \in A \Rightarrow x \in \mathbf{X}$. Here, and elsewhere, "$\Rightarrow$" means "implies." The symbol $\varnothing$ means the empty set. It is defined to be the set such that the statement "$x \in \varnothing$" is always false. We use the notation $\{x\}$ to denote the set consisting of a single point $x \in \mathbf{X}$. Show that if $x \in \mathbf{X}$, then $\{x\}$ is a subset of $\mathbf{X}$.

1.10. Any set of points makes a space, if we care to define it as such. The points are what we choose them to be. Why, do you think, have the spaces defined above been picked out as important? Describe other spaces which are equally important.

1.11. Let $\mathbf{X}_1$ and $\mathbf{X}_2$ be spaces. These can be used to make a new space denoted $\mathbf{X}_1 \times \mathbf{X}_2$, called the Cartesian product of $\mathbf{X}_1$ and $\mathbf{X}_2$. A point in $\mathbf{X}_1 \times \mathbf{X}_2$ is represented by the ordered pair (x_1, x_2) where $x_1 \in \mathbf{X}_1$ and $x_2 \in \mathbf{X}_2$. For example, $\mathbb{R}^2$ is the Cartesian product of $\mathbb{R}$ and $\mathbb{R}$.

2.2 METRIC SPACES

We use the notation "$\forall$" to mean "for all." We also introduce the notation $A \setminus B$ to mean the set A "take away" the set B. That is, $A \setminus B = \{x \in A: x \notin B\}$.

Definition 1. A metric space $(\mathbf{X}, d)$ is a space $\mathbf{X}$ together with a real-valued function $d: \mathbf{X} \times \mathbf{X} \to \mathbb{R}$, which measures the *distance* between pairs of points x and y in $\mathbf{X}$. We require that d obeys the following axioms:

(i) $d(x, y) = d(y, x) \forall x, y \in \mathbf{X}$
(ii) $0 < d(x, y) < \infty \forall x, y \in \mathbf{X}, x \neq y$
(iii) $d(x, x) = 0 \forall x \in \mathbf{X}$
(iv) $d(x, y) \leq d(x, z) + d(z, y) \forall x, y, z \in \mathbf{X}$

Such a function d is called a *metric*.

The concept of shortest paths between points in a space, *geodesics*, is dependent on the metric. The metric may determine a *geodesic structure* of the space. Geodesics on a sphere are great circles; in the plane with the Euclidean metric they are straight lines.

Examples & Exercises

2.1. Show that the following are all metrics in the space $\mathbf{X} = \mathbb{R}$:

a) $d(x, y) = |x - y|$ (Euclidean metric)
b) $d(x, y) = 2 \cdot |x - y|$
c) $d(x, y) = |x^3 - y^3|$

2.2. Show that the following are metrics in the space $\mathbf{X} = \mathbb{R}^2$:

a) $d(x, y) = \sqrt{(x_1 - y_1)^2 + (x_2 - y_2)^2}$ (Euclidean metric)
b) $d(x, y) = |x_1 - y_1| + |x_2 - y_2|$ (Manhattan metric)

Why is the name Manhattan used in connection with (b)?

2.3. Show that $d(x, y) = |xy|$ does not define a metric in $\mathbb{R}$.

2.4. Let $\mathbb{R}^2 \setminus \{O\}$ denote the punctured plane. Define $d(x, y)$ as follows:

$$d(x, y) = |r_1 - r_2| + |\theta|$$

where r_1 = Euclidean distance from x to O, r_2 = Euclidean distance from y to O, where O is the origin, and θ is the smallest angle subtended by the two straight lines connecting x and y to the origin. Show that d is a metric.

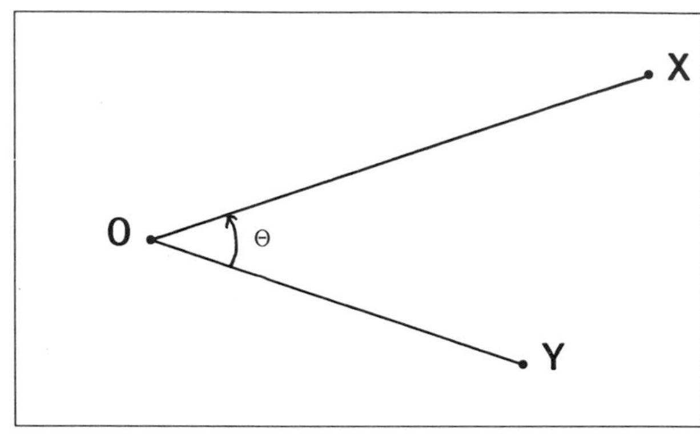

Figure 2.2.1 (*The angle θ, and the distances r_1, r_2 used to construct a metric on the punctured plane.*) *Acute angle subtended by two straight lines.*

2.5. On the code space Σ define

$$d(x, y) = d(x_1x_2x_3\ldots, y_1y_2y_3\ldots) = \sum_{i=1}^{\infty} \frac{|x_i - y_i|}{(N+1)^i}.$$

Show that every pair of points in Σ is a *finite* distance apart. That is, d is indeed a function which takes $\Sigma \times \Sigma$ into $\mathbb{R}$. Verify that (Σ, d) is a metric space. Try to envisage a possible geometry for Σ. (Do not confuse the symbol Σ for the space, with the symbol for summation $\Sigma_{i=1}^{\infty}$.)

2.6. In $X =$, define $d(x, y)$ to be the Euclidean length of the shortest path lying entirely within X which connects x and y. Show that this is a metric. Discuss the utility of this metric in anatomy. The distance from a toenail to a fingertip does not much depend on the position of the body, whereas the usual spatial distance does.

2.7. Invent a function d: ■ × ■ → $\mathbb{R}$ which is not a metric. Define a metric for the space , namely an annulus, which makes it seem like the curved wall of a cylinder: .

2.8. Show that a metric on $X = \hat{\mathbb{C}}$ is defined by the shortest great circle distances on the sphere. Compare the distances from 0, and from $1 + i$, to ∞.

Definition 2. Two metrics d_1 and d_2 on a space X are equivalent if there exist constants $0 < c_1 < c_2 < \infty$ such that

$$c_1 d_1(x, y) \le d_2(x, y) \le c_2 d_1(x, y)\ \forall (x, y) \in X \times X.$$

Exercises & Examples

2.9. Definition 2 looks unsymmetrical; it does not appear to make the same requirements on d_1 as it does on d_2. Show that this is an illusion by establishing that if the definition holds then there are constants $0 < e_1$

$< e_2 < \infty$ so that

$$e_1 d_2(x, y) < d_1(x, y) < e_2 d_2(x, y) \; \forall (x, y) \in X \times X.$$

2.10. Are the Manhattan and Euclidean metrics equivalent on ■ $\subset \mathbb{R}^2$? What about on $\mathbb{R}^2$?

2.11. Show that the metric in 2.4 is *not* equivalent to the Euclidean metric on ● $\setminus \{0\}$.

One notion underlying the concept of equivalent metrics is that any pair of equivalent metrics gives the same notion of which points are close together, and which are far apart. It is as though there were a standard way for boundedly deforming the space, whereby distances are determined both before and after deformation.

For example, consider a pair of points x, and y, in ■ $\subset \mathbb{R}^2$. Let the Euclidean distance between these points be $d_1(x, y)$. Think of a thin rubber sheet lying over ■. This sheet is stretched in some repeatable fashion, carrying copies of the points x and y to new locations, as illustrated in Figure 2.2.2. The Euclidean distance between these moved points is called $d_2(x, y)$. The condition of equivalence is the requirement that there is no extreme (infinite) stretching or compression of the space.

This leads us to the idea of equivalent metric spaces.

Definition 3. Two metric spaces (X_1, d_1) and (X_2, d_2) are *equivalent* if there is a function h: $X_1 \to X_2$ which is one-to-one and onto (i.e., it is invertible), such that the metric $\tilde{d}_1$ on X_1 defined by

$$\tilde{d}_1(x, y) = d_2(h(x), h(y)) \; \forall \, x, y \in X_1$$

is equivalent to d_1.

One can think of Definition 3 as requiring that X_1 and X_2 are related to one another by a bounded deformation, and nowhere is there an arbitrarily

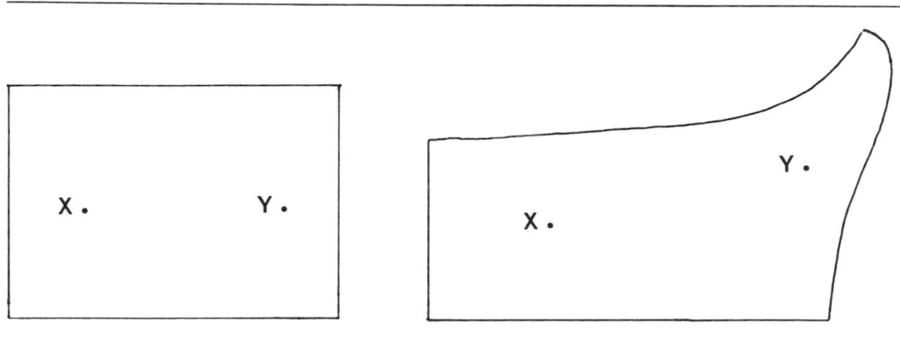

Figure 2.2.2 *A thin rubber sheet lies over the ■ in the plane and is stretched. The Euclidean distances between points are determined before and after deformation, yielding two metrics. These metrics may be equivalent if the deformation leads to no rips, tears, or infinite stretching.*

large compression or stretching. There is also no overlapping, folding or ripping.

Definition 4. A function f: $X_1 \to X_2$ from a metric space (X_1, d_1) into a metric space (X_2, d_2) is *continuous* if, for each $\epsilon > 0$ and $x \in X_1$, there is a $\delta > 0$ so that

$$d_1(x, y) < \delta \Rightarrow d_2(f(x), f(y)) < \epsilon.$$

If f is also *one-to-one* and *onto*, and thus *invertible*, and if also the inverse f^{-1} of f is continuous, then we say that f is a *homeomorphism* between X_1 and X_2. In such a case we say that X_1 and X_2 are homeomorphic.

The assertion that two spaces are equivalent is much stronger than the statement that they are homeomorphic: to be equivalent there must be a bounded relationship between ϵ and δ independent of x.

Examples & Exercises

2.12. Let $X_1 = [1,2]$ and $X_2 = [0,1]$. Let d_1 denote the Euclidean metric and let $d_2(x, y) = 2 \cdot |x - y|$ in X_2. Show that (X_1, d_1) and (X_2, d_2) are equivalent metric spaces.

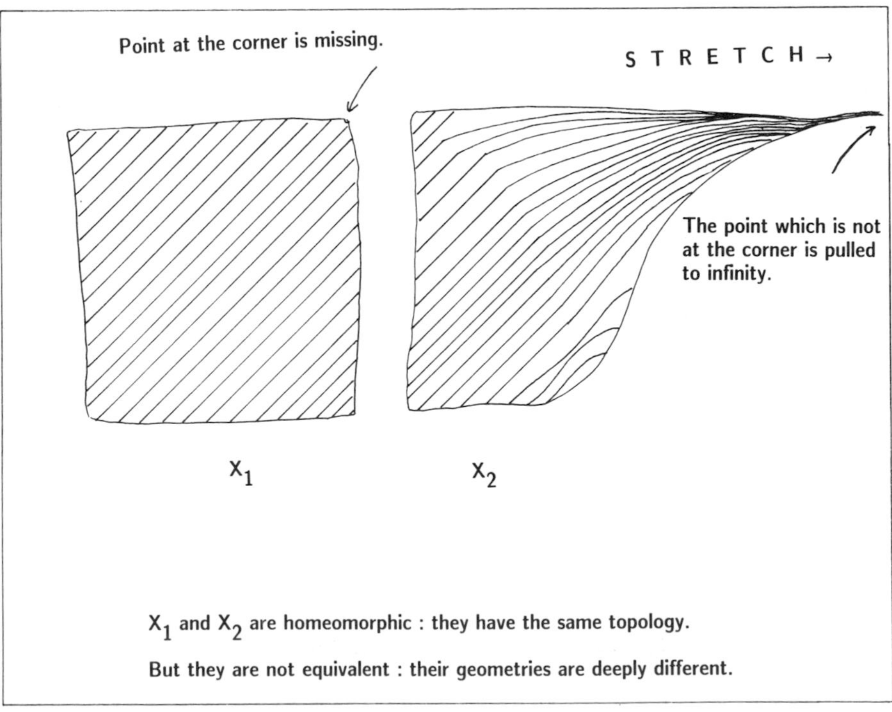
Point at the corner is missing.
S T R E T C H →
The point which is not at the corner is pulled to infinity.
X1
X2
X1 and X2 are homeomorphic : they have the same topology.
But they are not equivalent : their geometries are deeply different.

Figure 2.2.3 *This picture suggests two metric spaces X_1 and X_2 which have the same topology, but which are not metrically equivalent: their "geometries" are deeply different.*

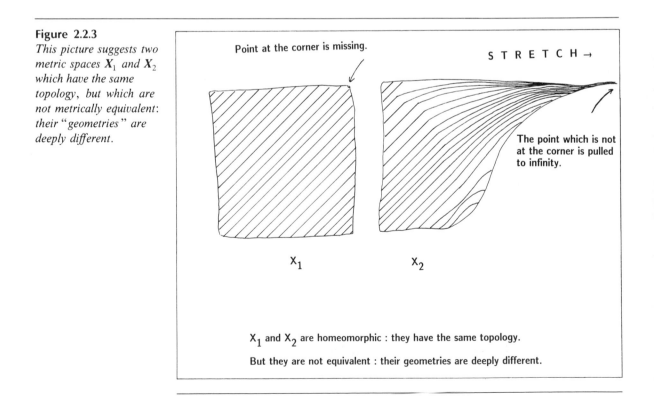

2.13. Show that (■, Euclidean) and (■, Manhattan) are equivalent metric spaces.

2.14. Show that ($\mathbb{C}$, Euclidean) and ($\mathbb{R}^2$, Manhattan) are equivalent metric spaces.

2.15. Define two different metrics on the space $X = (0,1] = \{x \in \mathbb{R}: 0 < x \leq 1\}$ by

$$d_1(x, y) = |x - y| \quad \text{and} \quad d_2(x, y) = \left|\frac{1}{x} - \frac{1}{y}\right|.$$

Show that (X, d_1) and (X, d_2) are not equivalent metric spaces.

2.16. Figure 2.2.4 suggests a subset (black) of (■, Euclidean). It also shows the space and set deformed by a metric equivalence. Discuss the properties of the image which would be invariant under (a) any metric equivalence, and (b) any homeomorphism. To what extent might one be able to "see" these invariances? Think about how much deformation an image can withstand while remaining recognizably the same image. Look at reflections of sets and images in the back of a shiny spoon.

Figure 2.2.4(a) *What features of the set (black) are invariant under a metric equivalence transformation? Two sets which are metrically equivalent to (a) are shown in (b) and (c).*

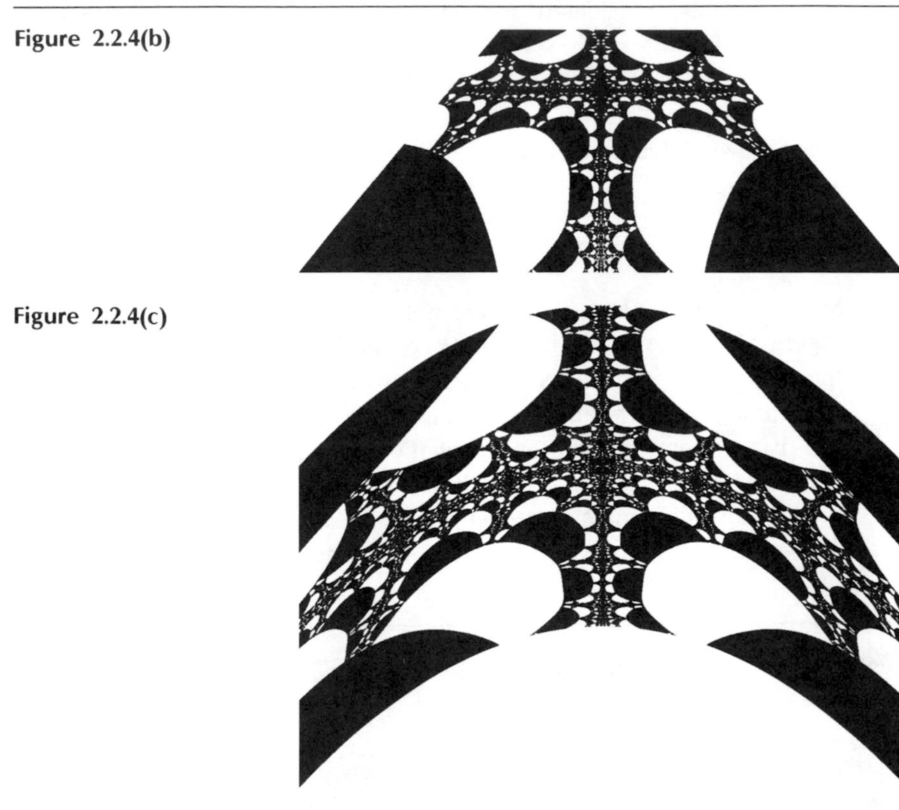

Figure 2.2.4(b)

Figure 2.2.4(c)

2.17. Show that if two metric spaces are metrically equivalent then there is a homeomorphism between them.

2.3 CAUCHY SEQUENCES, LIMIT POINTS, CLOSED SETS, PERFECT SETS, AND COMPLETE METRIC SPACES

Fractal geometry is concerned with the description, classification, analysis, and observation of subsets of metric spaces $(\mathbf{X}, d)$. The metric spaces are usually, but not always, of an inherently "simple" geometrical character; the subsets are typically geometrically "complicated." There are a number of general properties of subsets of metric spaces, which occur over and over again, which are very basic, and which form part of the vocabulary for describing fractal sets and other subsets of metric spaces. Some of these properties, such as openness and closedness, which we are going to introduce, are of a *topological* character. That is to say, they are invariant under homeomorphism.

For us what is important, however, is that there is another class of properties which are invariant under metric space equivalence. These include openness, closedness, boundedness, completeness, compactness, and perfection; these properties are introduced in this and the next section. Later we will discover another such property: the fractal dimension of a set. If a subset of a metric space has one of these properties, and the space is deformed with bounded distortion, then the corresponding subset in the deformed space still has that same property.

We are also about another business in this section. In our search for fractals we are always going to look in a certain type of metric space known as "complete." We need to understand this concept.

Definition 1. A sequence $\{x_n\}_{n=1}^{\infty}$ of points in a metric space $(\mathbf{X}, d)$ is called a *Cauchy sequence* if, for any given number $\epsilon > 0$, there is an integer $N > 0$ so that

$$d(x_n, x_m) < \epsilon \qquad \text{for all } n, m > N.$$

In other words, the further along the sequence one goes, the closer together become the points in the sequence. Mentally one pictures something like the image in Figure 2.3.1.

However, just because a sequence of points moves closer together as one goes along the sequence, we must not infer that they are approaching a point. Perhaps they are trying to approach a point that is not there?

Definition 2. A sequence $\{x_n\}_{n=1}^{\infty}$ of points in a metric space $(\mathbf{X}, d)$ is said to *converge* to a point $x \in \mathbf{X}$ if, for any given number $\epsilon > 0$, there is an integer $N > 0$ so that

$$d(x_n, x) < \epsilon \qquad \text{for all } n > N.$$

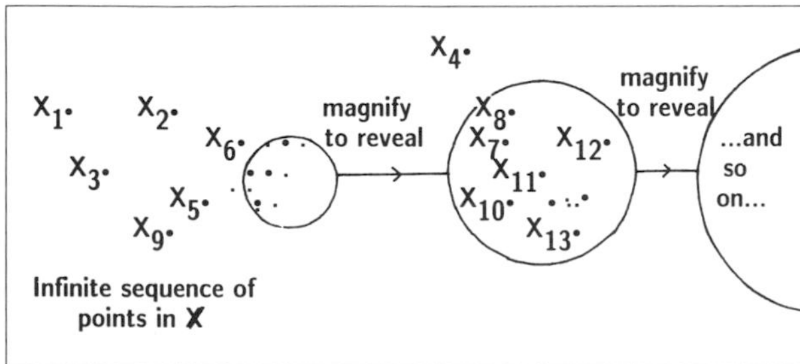

Figure 2.3.1 *Image representing successive magnifications on a Cauchy sequence, an infinite sequence of points in* $\mathbf{X}$. *Just because the points are getting closer and closer together as one looks in at higher magnification does not mean that there is a point* x *to which the sequence is converging!*

In this case the point $x \in \mathbf{X}$, to which the sequence converges, is called the *limit* of the sequence, and we use the notation

$$x = \operatorname*{Lim}_{n \to \infty} x_n.$$

The limit x of a convergent sequence $\{x_n\}_{n=1}^{\infty}$ has this property: Let

$$B(x, \epsilon) = \{y \in \mathbf{X}: d(x, y) \leq \epsilon\}$$

denote a closed ball of radius $\epsilon > 0$ centered at x, as illustrated in Figure 2.3.2.
Any such ball centered at x contains all of the points x_n after some index N, where N typically becomes larger and larger as ϵ becomes smaller and smaller. See Figure 2.3.3.

Theorem 1. *If a sequence of points $\{x_n\}_{n=1}^{\infty}$ in a metric space $(\mathbf{X}, d)$ converges to a point $x \in \mathbf{X}$, then $\{x_n\}_{n=1}^{\infty}$ is a Cauchy sequence.*

Definition 3. A metric space $(\mathbf{X}, d)$ is *complete* if every Cauchy sequence $\{x_n\}_{n=1}^{\infty}$ in $\mathbf{X}$ has a limit $x \in \mathbf{X}$.

In other words, there actually exists, in the space, a point x to which the Cauchy sequence is converging. This point x is of course the limit of the sequence. If $\{x_n\}_{n=1}^{\infty}$ is a Cauchy sequence of points in $\mathbf{X}$ and if $\mathbf{X}$ is complete, then there is a point $x \in \mathbf{X}$ such that, for each $\epsilon > 0$, $B(x, \epsilon)$ contains x_n for infinitely many integers n.

We will sometimes use the notation $\{x_n\}$ in place of $\{x_n\}_{n=1}^{\infty}$ and *lim* in place of $lim_{n \to \infty}$ when it is clear from the context what the domain of the index is.

Exercises & Examples

3.1. Prove that if $\{x_n\}_{n=1}^{\infty}$ is a Cauchy sequence of points in $\mathbf{X}$ and if $\mathbf{X}$ is complete, then there is a point $x \in \mathbf{X}$ such that, for each $\epsilon > 0$, $B(x, \epsilon)$ contains x_n for infinitely many integers n.

3.2. Show that ($\mathbb{R}$, Euclidean metric) is a complete metric space.

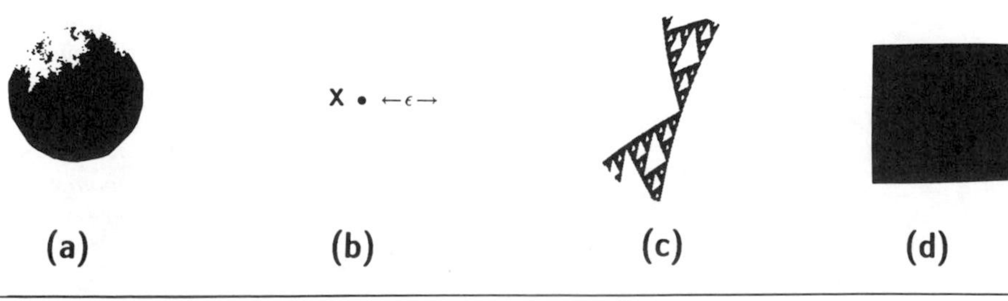

Figure 2.3.2
Uncelebrated small ball $B(x, \epsilon)$ with its center at x and radius ϵ. Beware! Balls do not usually look like balls. It depends on the metric and on the space. Balls (a)–(c) represent balls (marked in black) in spaces $\mathbf{X}$ which are subsets of $\mathbb{R}^2$, with the Euclidean metric. In (a) $\mathbf{X}$ has a ragged boundary, viewed as a subset of $\mathbb{R}^2$. In (b) the point $x \in \mathbf{X}$ is isolated. In (c) $\mathbf{X}$ is a curvaceous Sierpinski triangle. The ball depicted in (d) is in $\mathbb{R}^2$, but the metric is $d(x, y) =$ $\text{Max}\{|x_1 - y_1|, |x_2 - y_2|\}$.

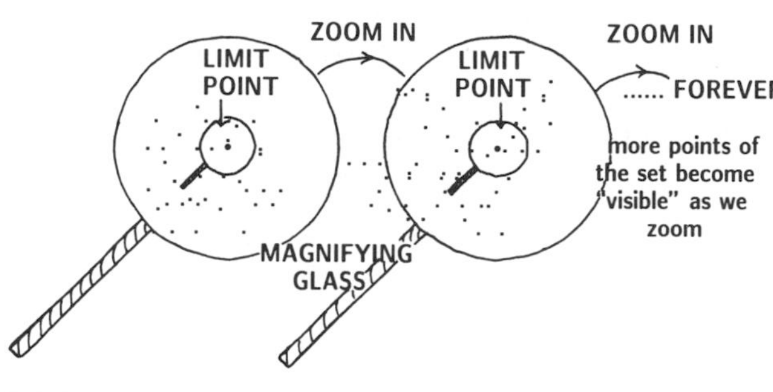

Figure 2.3.3 *Magnifying glass looking at a magnifying glass near a limit point.*

3.3. Show that ($\mathbb{R}^2$, Euclidean metric) is a complete metric space.
3.4. Show that (■, Euclidean metric) is a complete metric space.
3.5. Show that ($\hat{\mathbb{C}}$, metric on sphere) is a complete metric space.
3.6. Show that (Σ, code space) is a complete metric space.
3.7. Show that $(C[0,1], D)$ is a complete metric space, where the metric D is defined by

$$D(f, g) = \text{Max}\{|f(s) - g(s)|: s \in [0,1]\}.$$

3.8. Let (X_1, d_1) and (X_2, d_2) be equivalent metric spaces. Suppose (X_1, d_1) is complete. Show that (X_2, d_2) is complete.
3.9. Show that there are many different "shortest paths" between most pairs of points in (■, Manhattan).
3.10. Prove Theorem 2.3.1.
3.11. Prove that any sequence in a metric space can have at most one limit.

Definition 4. Let $S \subset X$ be a subset of a metric space (X, d). A point $x \in X$ is called a *limit point* of S if there is a sequence $\{x_n\}_{n=1}^{\infty}$ of points $x_n \in S \setminus \{x\}$ such that $\text{Lim}_{n \to \infty} x_n = x$.

Definition 5. Let $S \subset X$ be a subset of a metric space (X, d). The *closure* of S, denoted $\bar{S}$, is defined to be $\bar{S} = S \cup \{\text{Limit points of } S\}$. S is *closed* if it contains all of its limit points, that is, $S = \bar{S}$. S is *perfect* if it is equal to the set of all its limit points.

Exercises & Examples

3.12. Show that 0 is a limit of the sequence $\{x_n = 1/n\}_{n=1}^{\infty}$ in the metric space ([0, 1], Euclidean) but not in the metric space ((0, 1], Euclidean).
3.13. A metric space (X, d) consists of a single point $X = \{a\}$, together with a metric defined by $d(a, a) = 0$. Show that X contains a Cauchy sequence and the limit of the Cauchy sequence, but that it possesses no limit points. Hence show that X is closed and complete but not perfect.

3.14. Show that the sequence $\{x_n = n\}_{n=1}^{\infty}$ has no limit in ($\mathbb{R}$, Euclidean) but that it does when the points are treated as belonging to ($\hat{\mathbb{C}}$, spherical).

3.15. Show that if h: $\mathbf{X}_1 \to \mathbf{X}_2$ makes the metric spaces $(\mathbf{X}_1, d_1)$ and $(\mathbf{X}_2, d_2)$ equivalent, then the statements "$x \in \mathbf{X}_1$ is a limit point of $S \subset \mathbf{X}_1$" and "$h(x) \in \mathbf{X}_2$ is a limit point of $h(S) \subset \mathbf{X}_2$" are equivalent. Here we use the notation

$$h(S) = \{h(s): s \in S\}.$$

3.16. Find all of the limit points of the set $\{x_n = (1/n + (-1)^n, 1/n + (-1)^{2n}): n = 1, 2, 3, \dots\}$ in the metric space (■, Euclidean).

3.17. Show that the subset $S = \{x = 1/n: n = 1, 2, 3, \dots\}$ is closed in ((0, 1], Euclidean).

3.18. Show that $S = [0, 1]$ is a perfect subset of ($\mathbb{R}$, Euclidean).

3.19. Show that $S = \{1/n: n = 1, 2, 3, \dots\} \cup \{0\}$ is not a *perfect* subset of ($\mathbb{R}$, Euclidean), but that $S = \bar{S}$.

3.20. Show that $S = \Sigma$ is a perfect subset of (Σ, code space metric).

3.21. Let S be a subset of a complete metric space $(\mathbf{X}, d)$. Then (S, d) is a metric space. Show that (S, d) is complete if, and only if, S is closed in $\mathbf{X}$.

2.4 COMPACT SETS, BOUNDED SETS, OPEN SETS, INTERIORS, AND BOUNDARIES

We continue the description of the basic properties to be used to describe sets and subsets of metric spaces. Where are the fractals? What are they? They are everywhere, and soon you will be able to see them: not just the pictures, which are shadows, but in your mind's eye you will see what they really *are*.

Definition 1. Let $S \subset \mathbf{X}$ be a subset of a metric space $(\mathbf{X}, d)$. S is *compact* if every infinite sequence $\{x_n\}_{n=1}^{\infty}$ in S contains a subsequence having a limit in S.

Definition 2. Let $S \subset \mathbf{X}$ be a subset of a metric space $(\mathbf{X}, d)$. S is *bounded* if there is a point $a \in \mathbf{X}$ and a number $R > 0$ so that

$$d(a, x) < R \;\forall\, x \in \mathbf{X}.$$

Definition 3. Let $S \subset \mathbf{X}$ be a subset of a metric space $(\mathbf{X}, d)$. S is *totally bounded* if, for each $\epsilon > 0$, there is a finite set of points $\{y_1, y_2, \dots, y_n\} \subset S$ such that whenever $x \in \mathbf{X}$, $d(x, y_i) < \epsilon$ for some $y_i \in \{y_1, y_2, \dots, y_n\}$. This set of points $\{y_1, y_2, \dots, y_n\}$ is called an *ϵ-net*.

Theorem 1. *Let (X, d) be a complete metric space. Let $S \subset X$. Then S is compact if and only if it is closed and totally bounded.*

Proof. Suppose that S is closed and totally bounded. Let $\{x_i \in S\}$ be an infinite sequence of points in S. Since S is totally bounded we can find a finite collection of closed balls of radius 1 such that S is contained in the union of these balls. By the Pigeon-Hole Principle (a huge number of pigeons laying eggs in two letter boxes $\Rightarrow$ at least one letter box contains a huge number of angry pigeons), one of the balls, say B_1, contains infinitely many of the points x_n. Choose N_1 so that $x_{N_1} \in B_1$. It is easy to see that B_1 is totally bounded. So we can cover B_1 by a finite set of balls of radius $1/2$. By the Pigeon-Hole Principle, one of the balls, say B_2, contains infinitely many of the points x_n. Choose N_2 so that $x_{N_2} \in B_2$ and $N_2 > N_1$. We continue in this fashion to construct a *nested* sequence of balls,

$$B_1 \supset B_2 \supset B_3 \supset B_4 \supset B_5 \supset B_6 \supset B_7 \supset B_8 \supset B_9 \supset \cdots \supset B_n \supset \cdots$$

where B_n has radius $1/2^{n-1}$, and a sequence of integers $\{N_n\}_{n=1}^{\infty}$ such that $x_{N_n} \in B_n$. It is easy to see that $\{x_{N_n}\}_{n=1}^{\infty}$, which is a subsequence of the original sequence $\{x_n\}$, is a Cauchy sequence in S. Since S is closed, $\{x_{N_n}\}$ converges to a point x in S. (Notice that $\{x\}$ is exactly $\bigcap_{n=1}^{\infty} B_n$.) Thus, S is compact.

Conversely, suppose that S is compact. Let $\epsilon > 0$. Suppose that there does not exist an ϵ-net for S. Then there is an infinite sequence of points $\{x_n \in S\}$ with $d(x_i, x_j) \geq \epsilon$ for all $i \neq j$. But this sequence must possess a convergent subsequence $\{x_{N_i}\}$. By Theorem 2.3.1 this sequence is a Cauchy sequence, and so we can find a pair of integers N_i and N_j with $N_i \neq N_j$ so that $d(x_{N_i}, x_{N_j}) < \epsilon$. But $d(x_{N_i}, x_{N_j}) \geq \epsilon$, so we have a contradiction. Thus there *does* exist an ϵ-net. This completes the proof.

Definition 4. Let $S \subset \mathbf{X}$ be a subset of a metric space $(\mathbf{X}, d)$. S is *open* if for each $x \in S$ there is an $\epsilon > 0$ such that $B(x, \epsilon) = \{y \in \mathbf{X}: d(x, y) \leq \epsilon\} \subset S$.

Exercises & Examples

4.1. Show that if $(\mathbf{X}, d)$ is a metric space then $\mathbf{X}$ is closed. Give an example of a metric space which is closed but not complete.

4.2. Let S be a closed subset of a complete metric space $(\mathbf{X}, d)$. Show that (S, d) is a complete metric space.

4.3. Let $(\mathbf{X}_1, d_1)$ and $(\mathbf{X}_2, d_2)$ be equivalent metric spaces, and let a transformation θ: $\mathbf{X}_1 \to \mathbf{X}_2$ provide this equivalence. Let $S \subset \mathbf{X}_1$ be closed. Show that $\theta(S) = \{\theta(s): s \in S\}$ is closed. This idea is illustrated in Figure 2.4.1.

4.4. If $(\mathbf{X}, d)$ is a metric space then $\mathbf{X}$ is open.

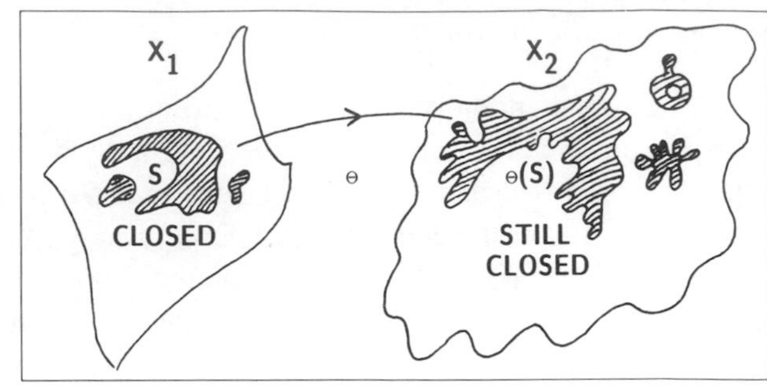

Figure 2.4.1 *A transformation θ between two metric spaces, establishing the equivalence of the spaces and carrying the closed set S onto a closed set $\theta(S)$.*

Proof. Let $x \in \mathbf{X}$. Clearly $B(x,1) \subset \mathbf{X}$.

4.5. If $(\mathbf{X}, d)$ is a metric space, then "$S \subset \mathbf{X}$ is open" is the same as "$\mathbf{X} \setminus S$ is closed."

Proof. Suppose "$S \subset \mathbf{X}$ is open." Suppose $\{x_n\}$ is a sequence in $\mathbf{X} \setminus S$ with a limit $x \in \mathbf{X}$. We must show that $x \in \mathbf{X} \setminus S$. Assume that $x \in S$. Then every ball $B(x, \epsilon)$ with $\epsilon > 0$ contains a point $x_n \in \mathbf{X} \setminus S$ which means that S is not open. This is a contradiction. The assumption is false. Therefore $x \in \mathbf{X} \setminus S$. Therefore "$\mathbf{X} \setminus S$ is closed."

Suppose "$\mathbf{X} \setminus S$ is closed." Let $x \in S$. We want to show there is a ball $B(x, \epsilon) \subset S$. Assume there is no ball $B(x, \epsilon) \subset S$. Then for every integer $n = 1, 2, 3, \ldots$ we can find a point $x_n \in B(x, 1/n) \cap (\mathbf{X} \setminus S)$. Clearly $\{x_n\}$ is a sequence in $\mathbf{X} \setminus S$, with limit $x \in \mathbf{X}$. Since $\mathbf{X} \setminus S$ is closed we conclude that $x \in \mathbf{X} \setminus S$. This contradicts $x \in S$. The assumption that there is no ball $B(x, \epsilon) \subset S$ is false. Therefore there is a ball $B(x, \epsilon) \subset S$. Therefore "S is open."

4.6. Every bounded subset S of ($\mathbb{R}^2$, Euclidean) has the Bolzano-Weierstrass property: "Every infinite sequence $\{x_n\}_{n=1}^{\infty}$ of points of S contains a subsequence which is a Cauchy sequence." The proof is suggested by the picture in Figure 2.4.2.

We deduce that every closed bounded subset of ($\mathbb{R}^2$, Euclidean) is compact. In particular, every metric space of the form (closed bounded subset of $\mathbb{R}^2$, Euclidean) is a complete metric space. Show that we can make a rigorous proof by using Theorem 1. Begin by proving that any bounded subset of $\mathbb{R}^n$ is totally bounded.

4.7. Let $(\mathbf{X}, d)$ be a metric space. Let $f\colon \mathbf{X} \to \mathbf{X}$ be continuous. Let A be a compact nonempty subset of $\mathbf{X}$. Show that $f(A)$ is a compact nonempty subset of $\mathbf{X}$. (This result is proved later as Lemma 3.7.2.)

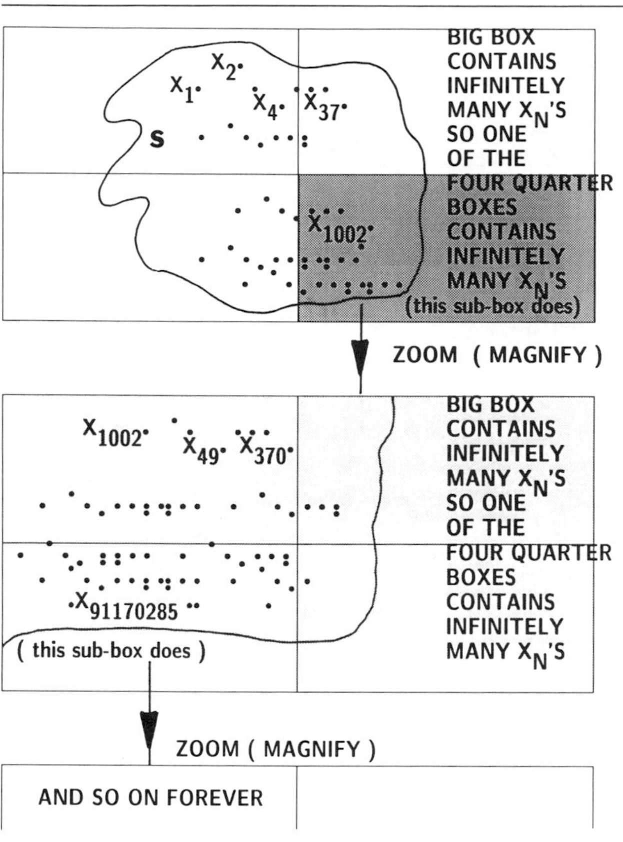

Figure 2.4.2 *Demonstration of the Bolzano-Weierstrass Theorem. (Government warning: This is not a proof.)*

4.8. Let $S \subset (X_1, d_1)$ be open, and let (X_2, d_2) be a metric space equivalent to (X_1, d_1), the equivalence being provided by a function $h: X_1 \to X_2$. Show that $h(S)$ is an open subset of X_2.

4.9. Let (X, d) be a metric space. Let $C \subset X$ be a compact subset of X. Let $\{C_n: n = 1, 2, 3, \ldots\}$ be a set of open subsets of X such that "$x \in C$" implies "$x \in C_n$ for some n." $\{C_n\}$ is called a countable open cover of C. Show that there is a finite integer N so that "$x \in C$" implies "$x \in C_n$ for some integer $n < N$."

Proof. Assume that an integer N does not exist such that "$x \in C$" implies "$x \in C_n$ for some $n < N$." Then for each N we can find

$$x_N \in C \setminus \bigcup_{n=1}^{N} C_n .$$

Since $\{x_N\}_{N=1}^{\infty}$ is in C it possesses a subsequence with a limit $y \in C$. Clearly y does not belong to any of the subsets C_n. Hence "$y \in C$" does not imply

"$y \in C_n$ for some integer n." We have a contradiction. This completes the proof.

The following even stronger statement is true. Let (X, d) be a metric space. Let $C \subset X$ be compact. Let $\{C_i : i \in I\}$ denote *any* collection of open sets such that whenever $x \in C$, it is true that $x \in C_i$ for some index $i \in I$. Then there is a *finite* subcollection, say $\{C_1, C_2, \ldots, C_n\}$ such that $C \subset \bigcup_{i=1}^{N} C_i$. The point is that the original collection of open sets need not even be countably infinite. A good discussion of compactness in metric spaces can be found in Mendelson [1963, Chapter V].

4.10. Let $X = (0,1) \cup \{2\}$. That is, X consists of an open interval in $\mathbb{R}$, together with an 'isolated' point. Show that the subsets $(0,1)$ and $\{2\}$ of $(X$, Euclidean) are open. Show that $(0,1)$ is closed in X. Show that $\{2\}$ is closed in X. Show that $\{2\}$ is compact in X but $(0,1)$ is not compact in X.

Definition 5. Let $S \subset X$ be a subset of a metric space (X, d). A point $x \in X$ is a *boundary point* of S if for every number $\epsilon > 0$, $B(x, \epsilon)$ contains a point in $X \setminus S$ and a point in S. The set of all boundary points of S is called the *boundary* of S, and is denoted ∂S.

Definition 6. Let $S \subset X$ be a subset of a metric space (X, d). A point $x \in S$ is called an *interior point* of S if there is a number $\epsilon > 0$ such that $B(x, \epsilon) \subset S$. The set of interior points of S is called the *interior* of S, and is denoted S^0.

Exercises & Examples

4.11. Let S be a subset of a metric space (X, d). Show that $\partial S = \partial(X \setminus S)$. Deduce that $\partial X = \varnothing$.

4.12. Show that the property of being a boundary of a set is invariant under metric equivalence.

4.13. Let (X, d) be the real line with the Euclidean metric. Let S denote the set of all rational points in X (i.e., real numbers which can be written p/q where p and q are integers with $q \neq 0$). Show that $\partial S = X$.

4.14. Find the boundary of $\mathbb{C}$ viewed as a subset of $(\hat{\mathbb{C}}$, spherical metric).

4.15. Let S be a closed subset of a metric space. Show that $\partial S \subset S$.

4.16. Let S be an open subset of a metric space. Show that $\partial S \cap S = \varnothing$.

4.17. Let S be an open subset of a metric space. Show that $S^0 = S$. Conversely, show that if $S^0 = S$, then S is open.

4.18. Let S be a closed subset of a metric space. Show that $S = S^0 \cup \partial S$.

4.19. Show that the property of being the interior of a set is invariant under metric equivalence.

4.20. Show that the boundary of a set S in a metric space always divides the space into two disjoint open sets whose union, with the boundary ∂S, is the whole space. Illustrate this result in the following cases, in the metric space ($\mathbb{R}^2$, Euclidean): (a) $S = \{(x, y) \in \mathbb{R}^2 : x^2 + y^2 < 1\}$; (b) $S = \mathbb{R}^2$.

4.21. Show that the boundary of a set is closed.

4.22. Let S be a subset of a compact metric space. Show that ∂S is compact.

4.23. Figure 2.4.3 shows how we think of boundaries and interiors. What features of the picture are misleading?

4.24. To what extent does Mercator's projection provide a metric equivalence to a Cartesian map of the world?

4.25. Locate the boundary of the set of points marked in black in Figure 2.4.4.

4.26. Prove the assertion made in the caption to Figure 2.4.5.

2.5 CONNECTED SETS, DISCONNECTED SETS, AND PATHWISE CONNECTED SETS

Definition 1. A metric space $(\mathbf{X}, d)$ is *connected* if the only two subsets of $\mathbf{X}$ that are simultaneously open and closed are $\mathbf{X}$ and $\varnothing$. A subset $S \subset \mathbf{X}$ is connected if the metric space (S, d) is connected. S is *disconnected* if it is not connected. S is *totally disconnected* provided that the only nonempty connected subsets of S are subsets consisting of single points.

Definition 2. Let $S \subset \mathbf{X}$ be a subset of a metric space $(\mathbf{X}, d)$. Then S is *pathwise connected* if, for each pair of points x and y in S, there is a continuous function $f\colon [0, 1] \to S$, from the metric space ([0, 1], Euclidean) into the metric space (S, d), such that $f(0) = x$ and $f(1) = y$. Such a function f is called a *path* from x to y in S. S is *pathwise disconnected* if it is not pathwise connected.

One can also define *simply connected* and *multiply connected*. Let S be pathwise connected. A pair of points $x, y \in S$ is *simply connected* in S if,

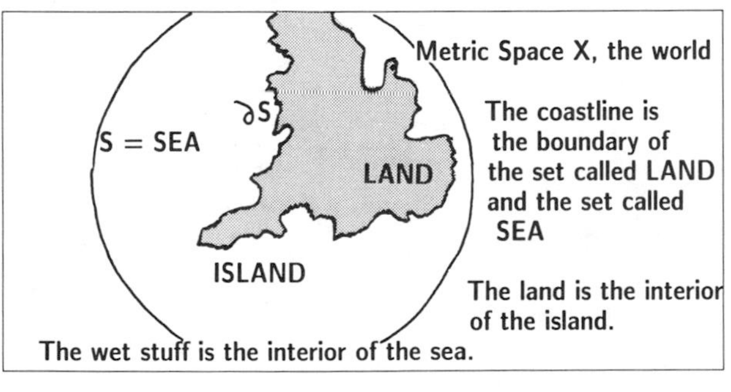
Metric Space X, the world
∂S
S = SEA
LAND
The coastline is the boundary of the set called LAND and the set called SEA
ISLAND
The land is the interior of the island.
The wet stuff is the interior of the sea.

Figure 2.4.3 *How well can topological concepts such as open, boundary, etc., be used to model land, sea, and coastlines?*

Figure 2.4.4
Should the black part be called open and the white part closed? Locate the boundary of the set of points marked in black.

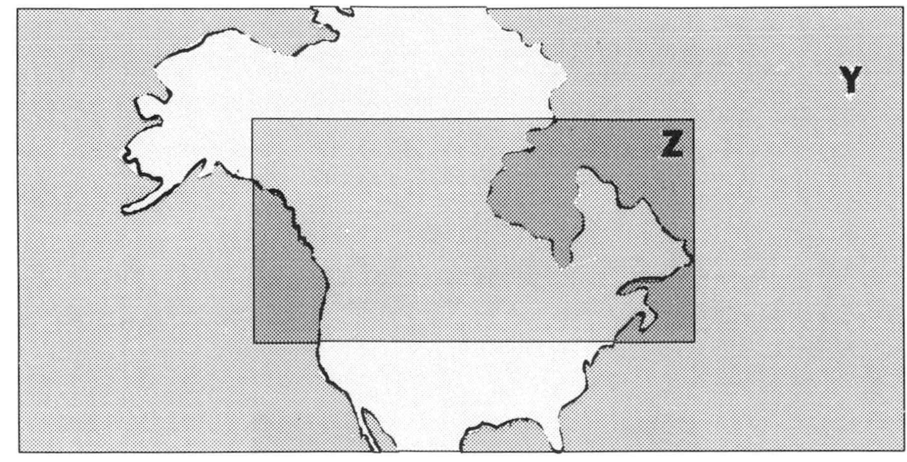

Figure 2.4.5
The interior of the "land" set is an open set in the metric space (Y = ▭, *Euclidean*). *The smaller filled rectangle denotes a subset* Z = ■ *of* Y. *The intersection of the interior of the land with* Z *is an open set in the metric space* (Z, *Euclidean*), *despite the fact that it includes some points of the "border" of* ■.

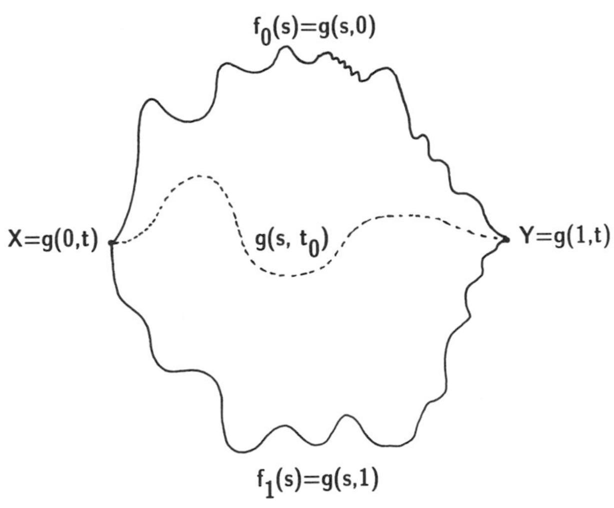

Figure 2.5.1
A path f_0 which connects the points x and y is continuously deformed, while remaining "attached" to x and y, to become a second path f_1.

given any two paths f_0 and f_1 connecting x, y in S, we can continuously deform f_0 to f_1 without leaving the subset S. What does this mean?

Let there be given the two points $x, y \in S$ and the two paths f_0, f_1 connecting x, y in S. In other words, f_0, f_1 are two continuous functions mapping the unit interval $[0, 1]$ into S so that $f_0(0) = f_1(0) = x$ and $f_0(1) = f_1(1) = y$. By a *continuous deformation* of f_0 and f_1 within S we mean a function g continuously mapping the Cartesian product $[0, 1] \times [0, 1]$ into S, so that

(a) $g(s, 0) = f_0(s)$ $\quad (0 \le s \le 1)$
(b) $g(s, 1) = f_1(s)$ $\quad (0 \le s \le 1)$
(c) $g(0, t) = x$ $\quad (0 \le t \le 1)$
(d) $g(1, t) = y$ $\quad (0 \le t \le 1)$.

Thus, we say that two points x, y in S are *simply connected* in S if, given any two paths f_0, f_1 going from x to y in S, there exists a function g as just described. This idea is illustrated in Figure 2.5.1.

If x, y are not simply connected in S, then we say that x, y are *multiply connected* in S.

S itself is called simply connected if every pair of points x, y in S is simply connected in S. Otherwise, S is called multiply connected. In the latter case we can imagine that S contains a "hole", as illustrated in Figure 2.5.2.

Exercises & Examples

5.1. Show that the properties of being (pathwise) connected, disconnected, simply connected and multiply connected are invariant under metric equivalence.

Figure 2.5.2
In a multiply connected space there exist paths which cannot be continuously deformed from one to another. There is some kind of "hole" in the space.

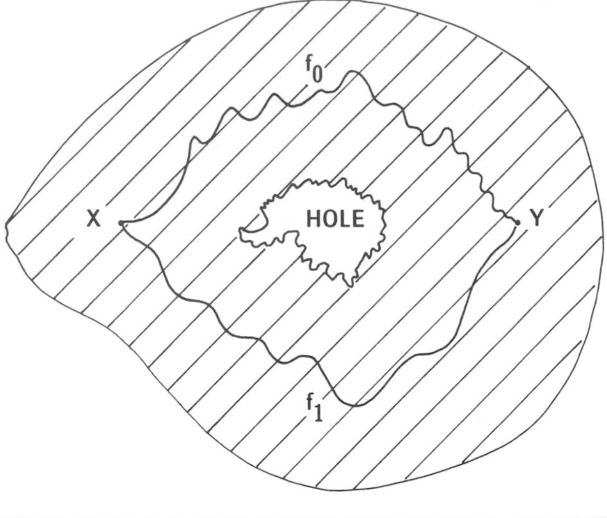

Figure 2.5.3
Locate the largest connected subsets of this subset of $\mathbb{R}^2$.

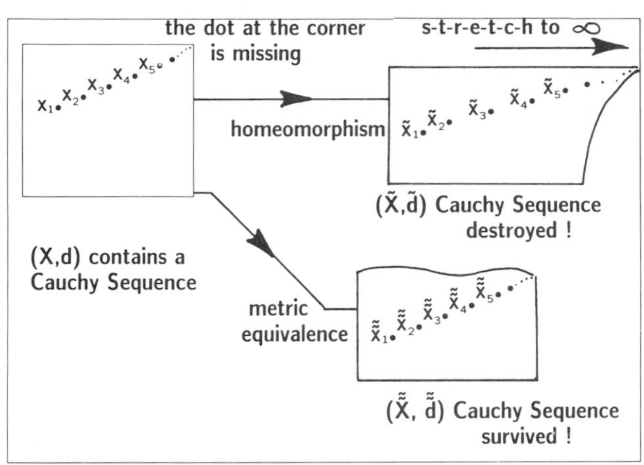

Figure 2.5.4
A Cauchy sequence being preserved by a metric equivalence, and destroyed by a certain homeomorphism.

5.2. Show that the metric space (■, Euclidean) is simply connected.

5.3. Show that the metric space ($X = (0, 1) \cup \{2\}$, Euclidean) is disconnected.

5.4. Show that the metric space (Σ, code space metric) is totally disconnected.

5.5. Show that the metric space (, Manhattan) is multiply connected.

5.6. Suppose $S_1 \supset S_2 \supset \cdots \supset S_n \supset \cdots$ is a nested sequence of nonempty connected subsets. Is $\bigcap_{n=1}^{\infty} S_n$ necessarily connected?

5.7. Identify pathwise connected subsets of the metric space suggested in Figure 2.5.3.

5.8. Is (, Euclidean) simply or multiply connected?

5.9. Discuss which set-theoretic properties (open, closed, connected, compact, bounded, ...) would be best suited for a model of a cloud, treated as a subset of $\mathbb{R}^3$.

5.10. The property that $\{x_n\}_{n=1}^{\infty}$ is a Cauchy sequence in the metric space (X, d) is not invariant under homeomorphism but is invariant under metric equivalence, as illustrated in Figure 2.5.4.

2.6 THE METRIC SPACE ($\mathscr{H}(X), h$): THE SPACE WHERE FRACTALS LIVE

We come to the ideal space in which to study fractal geometry. To start with, and always at the deepest level, we work in some complete metric space, such as ($\mathbb{R}^2$, Euclidean) or ($\hat{\mathbb{C}}$, spherical), which we denote by (X, d). But then, when we wish to discuss pictures, drawings, "black-on-white" subsets of the space, it becomes natural to introduce the space $\mathscr{H}$.

Definition 1. Let $(\boldsymbol{X}, d)$ be a complete metric space. Then $\mathscr{H}(\boldsymbol{X})$ denotes the space whose points are the compact subsets of $\boldsymbol{X}$, other than the empty set.

Exercises & Examples

6.1. Show that if x and $y \in \mathscr{H}(\boldsymbol{X})$, then $x \cup y$ is in $\mathscr{H}(\boldsymbol{X})$. Show that $x \cap y$ need not be in $\mathscr{H}(\boldsymbol{X})$. A picture of this situation is given in Figure 2.6.1.

6.2. What is the difference between a subset of $\mathscr{H}(\boldsymbol{X})$ and a compact nonempty subset of $\boldsymbol{X}$?

Definition 2. Let $(\boldsymbol{X}, d)$ be a complete metric space, $x \in \boldsymbol{X}$, and $B \in \mathscr{H}(\boldsymbol{X})$. Define

$$d(x, B) = \text{Min}\{d(x, y): y \in B\}.$$

Then $d(x, B)$ is the *distance from* the point x *to* the set B.

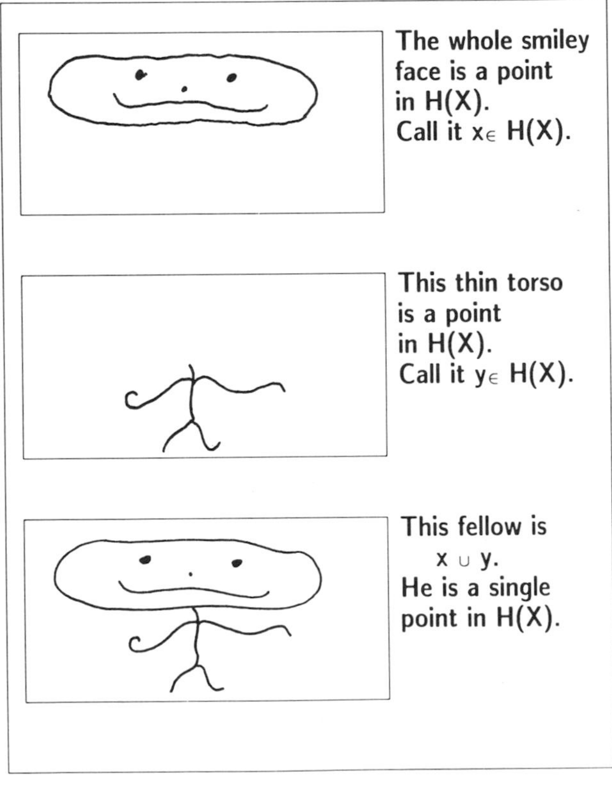

The whole smiley face is a point in H(X). Call it x∈ H(X).
This thin torso is a point in H(X). Call it y∈ H(X).
This fellow is x ∪ y. He is a single point in H(X).

Figure 2.6.1 *Points in the space $\mathscr{H}(\mathbb{R}^2)$ may be interpreted as black-and-white images. Unions of points yield new points. Be careful with intersections, however.*

How do we know that the set of real numbers $\{d(x, y): y \in B\}$ contains a minimum value, as claimed in the definition? This follows from the compactness and nonemptyness of the set $B \in \mathscr{H}(X)$. Consider the function $f: B \to \mathbb{R}$ defined by

$$f(y) = d(x, y) \qquad \text{for all } y \in B.$$

From the definition of the metric it follows that f is continuous, viewed as a transformation from the metric space (B, d) to the metric space ($\mathbb{R}$, Euclidean). Let $P = \text{Inf}\{f(y): y \in B\}$, where "Inf" is defined in Example 2.6.20, and also in Definition 3.6.2. Since $f(y) \geq 0$ for all $y \in B$, it follows that P is finite. We claim there is a point $\hat{y} \in B$ such that $d(x, \hat{y}) = P$. We can find an infinite sequence of points $\{y_n: n = 1, 2, 3, \ldots\} \subset B$ such that $|f(y_n) - P| < 1/n$. Using the compactness of B, we find that $\{y_n: n = 1, 2, 3, \ldots\}$ has a subsequence with limit $\hat{y} \in B$. Using the continuity of f we discover that $f(\hat{y}) = P$, which is what we needed to show.

Color Plate 2.6.1 shows a picture of the metric space (■, Manhattan). It has been colored as follows. Let $\mathscr{F}$ denote a certain subset of ■ whose 'geometry' is that of a piece of a fern. Then the color of each point $a \in$ ■ is fixed by the value of $d(a, \mathscr{F})$.

Definition 3. Let (X, d) be a complete metric space. Let $A, B \in \mathscr{H}(X)$. Define

$$d(A, B) = \text{Max}\{d(x, B): x \in A\}.$$

$d(A, B)$ is the *distance from* the set $A \in \mathscr{H}(X)$ *to* the set $B \in \mathscr{H}(X)$.

Just as above, using the compactness of A and B, we can prove that this definition is meaningful. In particular, there are points $\hat{x} \in A$ and $\hat{y} \in B$ such that $d(A, B) = d(\hat{x}, \hat{y})$.

Exercises & Examples

6.3. Show that $B, C \in \mathscr{H}(X)$, with $B \subset C$ implies $d(x, C) \leq d(x, B)$.

6.4. Calculate $d(x, B)$ if (X, d) is the space ($\mathbb{R}^2$, Euclidean), $x \in \mathbb{R}^2$ is the point $(1, 1)$, and B is a closed disk of radius $\frac{1}{2}$ centered at the point $(\frac{1}{2}, 0)$.

6.5. Same as 6.4, but use the Manhattan metric.

6.6. Calculate $d(x, B)$ if (X, d) is ($\mathbb{R}$, Euclidean), $x = \frac{1}{2}$, and

$$B = \left\{x_n = 3 + (-1)^n \frac{n}{n^2 + 1}: n = 1, 2, 3, \ldots\right\} \cup \{3\}.$$

6.7. Let $A, B \in \mathscr{H}(X)$ where (X, d) is a metric space. Show that, in general, $d(A, B) \neq d(B, A)$. Conclude that d does not provide a metric on

Figure 2.6.2 *This fractal image contains a pair of disjoint subsets of $\blacksquare \subset \mathbb{R}^2$, "black" and "white". Let A denote the closure of the set in black and let B denote the closure of its complement. Find a pair of points $x \in A$ and $y \in B$, such that $d(x, y) = d(A, B)$. Find a pair of points $\tilde{x} \in A$ and $\tilde{y} \in B$ such that $d(\tilde{y}, \tilde{x}) = d(B, A)$. Why do we "close" the sets before we begin?*

$\mathscr{H}(X)$. It is not symmetrical: the distance from A to B need not equal the distance from B to A.

6.8. Figure 2.6.2 shows two subsets A, and B, of ($\blacksquare \subset \mathbb{R}^2$, Euclidean). A is the white part and B is the black part. (a) Estimate the location of a pair of points, $x \in A$ and $y \in B$, such that $d(x, y) = d(A, B)$. (b) Estimate the location of a pair of points, $\tilde{x} \in A$ and $\tilde{y} \in B$, such that $d(\tilde{x}, \tilde{y}) = d(B, A)$.

Figure 2.6.3 *Find a pair of points $\hat{x}$ and $\hat{y}$, one in the dark fern and one in the pale fern, such that the Hausdorff distance between the two fern images is the same as the distance between the points.*

6.9. Figure 2.6.3, shows two fern-like subsets, A and B, of $(\mathbb{R}^2$, Manhattan). Locate points $x \in A$ and $y \in B$ such that: (a) $d(x, y) = d(A, B)$; (b) $d(x, y) = d(B, A)$.

6.10. Find d(France, USA) and d(USA, France) on ($\hat{\mathbb{C}}$, spherical metric). Which is larger? Also compare d(Georgia, USA) to d(USA, Georgia).

6.11. Let (X, d) be a complete metric space. Let A and B be points in $\mathscr{H}(X)$ such that $A \neq B$. Show that either $d(A, B) \neq 0$ or $d(B, A) \neq 0$. Show that if $A \subset B$ then $d(A, B) = 0$.

6.12. Let (X, d) be a complete metric space. Show that if A, B, and $C \in \mathscr{H}(X)$ then $B \subset C \Rightarrow d(A, C) \leq d(A, B)$. (Hint: Use Example 6.3.)

6.13. Let (X, d) be a complete metric space. Show that if A, B, and $C \in \mathscr{H}(X)$ then

$$d(A \cup B, C) = d(A, C) \vee d(B, C).$$

We use the notation $x \vee y$ to mean the maximum of the two real numbers x and y.

Proof. $d(A \cup B, C) = \text{Max}\{d(x, C)\colon x \in A \cup B\} = \text{Max}\{d(x, C)\colon x \in A\} \vee \text{Max}\{d(x, C)\colon x \in B\}$.

6.14. Let A, B and C belong to $\mathscr{H}(X)$ where (X, d) is a metric space. Show that

$$d(A, B) \leq d(A, C) + d(C, B).$$

Also determine whether or not the inequality

$$d(A, B) \leq d(C, A) + d(C, B)$$

is true in general.

Definition 4. Let (X, d) be a complete metric space. Then the *Hausdorff distance* between points A and B in $\mathscr{H}(X)$ is defined by

$$h(A, B) = d(A, B) \vee d(B, A).$$

Exercises & Examples

6.15. Show that h is a metric on the space $\mathscr{H}(X)$.

Proof. Let $A, B, C \in \mathscr{H}(X)$. Clearly $h(A, A) = d(A, A) \vee d(A, A) = d(A, A) = \text{Max}\{d(x, A)\text{: } x \in A\} = 0$. $h(A, B) = d(a, b)$ for some $a \in A$ and $b \in B$, using the compactness of A and B. Hence $0 \leq h(A, B) < \infty$. If $A \neq B$ we can assume there is an $a \in A$ so that $a \notin B$. Then $h(A, B) \geq d(a, B) > 0$. To show that $h(A, B) \leq h(A, C) + h(C, B)$ we first show that $d(A, B) \leq d(A, C) + d(C, B)$. We have for any $a \in A$

$$\begin{aligned} d(a, B) &= \text{Min}\{d(a, b)\text{: } b \in B\} \\ &\leq \text{Min}\{d(a, c) + d(c, b)\text{: } b \in B\}\ \forall c \in C \\ &= d(a, c) + \text{Min}\{d(c, b)\text{: } b \in B\}\ \forall c \in C, \text{ so} \\ d(a, b) &\leq \text{Min}\{d(a, c)\text{: } c \in C\} + \text{Max}\{\text{Min}\{d(c, b)\text{: } b \in B\}\text{: } c \in C\} \\ &= d(a, C) + d(C, B), \text{ so} \\ d(A, B) &\leq d(A, C) + d(C, B). \end{aligned}$$

Similarly

$$\begin{aligned} d(B, A) &\leq d(B, C) + d(C, A), \text{ whence} \\ h(A, B) &= d(A, B) \vee d(B, A) \leq d(B, C) \vee d(C, B) + d(A, C) \vee d(C, A) \\ &= h(B, C) + h(A, C), \text{ as desired.} \end{aligned}$$

6.16. Show that $h(A \cup B, C \cup D) \leq h(A, C) \vee h(B, D)$, for all A, B, C, and $D \in \mathscr{H}(X)$.

6.17. Let (X, d) be a compact metric space. Show that $(\mathscr{H}(X), h)$ is a compact metric space, where h is the Hausdorff metric on the space $\mathscr{H}(X)$.

6.18. Show that $h(A, B) = d(a, b)$ for some $a \in A$ and $b \in B$.

6.19. The same situation as in 6.10, but this time locate a pair of points $\hat{x} \in A$ and $\hat{y} \in B$ such that $d(\hat{x}, \hat{y}) = h(A, B)$, the Hausdorff distance from A to B.

6.20. Let $S \subset \mathbb{R}$, with $S \neq \varnothing$. The *supremum* of S is denoted by $\operatorname{Sup} S$. The *infimum* of S is denoted by $\operatorname{Inf} S$. If there is no real number which is greater than all the numbers in S then $\operatorname{Sup} S = +\infty$; otherwise $\operatorname{Sup} S = \operatorname{Min}\{x \in \mathbb{R}\colon x \geq s \;\forall s \in S\}$. If there is no real number which is less than all of the numbers in S then $\operatorname{Inf} S = -\infty$; otherwise $\operatorname{Inf} S = \operatorname{Max}\{x \in \mathbb{R}\colon x \leq s \;\forall s \in S\}$. Show that $\operatorname{Sup} S$ and $\operatorname{Inf} S$ are well-defined. Show that if S is compact then $\operatorname{Sup} S = \operatorname{Max} S$ and $\operatorname{Inf} S = \operatorname{Min} S$. Further exercises on Sup and Inf are given following Definition 3.6.2.

By replacing Max by Sup and Min by Inf, respectively, throughout the definition of the Hausdorff metric, define a "distance" between arbitrary pairs of subsets of a metric space. Give several reasons why this "distance" is not usually a metric.

2.7 THE COMPLETENESS OF THE SPACE OF FRACTALS

We refer to $(\mathscr{H}(\mathbf{X}), h)$ as "the space of fractals". It is too soon to be formal about the exact meaning of "a fractal". At the present stage of development of science and mathematics, the idea of a fractal is most useful as a broad concept. Fractals are not defined by a short legalistic statement, but by the many pictures and contexts which refer to them. For us, for the first eight chapters of this book, any subset of $(\mathscr{H}(\mathbf{X}), h)$ is a fractal. However, as with the concept of "a space", more meaning is suggested than is formalized.

In this section our principal goal is to establish that the space of fractals $(\mathscr{H}(\mathbf{X}), h)$ is a complete metric space. We also want to characterize convergent sequences in $\mathscr{H}(\mathbf{X})$. To achieve these goals using only the tools introduced so far is quite difficult. Indeed, at this juncture, we want to introduce another notion; namely, the idea of *extending* certain Cauchy subsequences.

Definition 1. Let $S \subset \mathbf{X}$ and let $\Gamma \geq 0$. Then $S + \Gamma = \{y \in \mathbf{X}\colon d(x, y) \leq \Gamma$ for some $x \in S\}$. $S + \Gamma$ is sometimes called, for example, in the theory of set morphology, the *dilation* of S *by a ball of radius* Γ.

Lemma 1. *Let A and B in $\mathscr{H}(\mathbf{X})$ where $(\mathbf{X}, d)$ is a metric space. Let $\epsilon > 0$. Then*

$$h(A, B) \leq \epsilon \Leftrightarrow A \subset B + \epsilon \text{ and } B \subset A + \epsilon.$$

Proof. Begin by showing that $d(A, B) \leq \epsilon \Leftrightarrow A \subset B + \epsilon$. Suppose "$d(A, B) \leq \epsilon$." Then $\text{Max}\{d(a, B): a \in A\} \leq \epsilon$ implies $d(a, B) \leq \epsilon$ for all $a \in A$. Hence for each $a \in A$ we have $a \in B + \epsilon$. Hence "$A \subset B + \epsilon$." Suppose "$A \subset B + \epsilon$." Consider $d(A, B) = \text{Max}\{d(a, B): a \in A\}$. Let $a \in A$. Since $A \subset B + \epsilon$, there is a $b \in B$ so that $d(a, b) \leq \epsilon$. Hence $d(a, B) \leq \epsilon$. This is true for each $a \in A$. So "$d(A, B) \leq \epsilon$." This completes the proof.

Let $\{A_n: n = 1, 2, \ldots, \infty\}$ be a Cauchy sequence of sets in $(\mathscr{H}(\boldsymbol{X}), h)$. That is, given $\epsilon > 0$, there is N so that $n, m \geq N$ implies

$$A_n + \epsilon \supset A_m \quad \text{and} \quad A_m + \epsilon \supset A_n,$$

i.e., $h(A_n, A_m) \leq \epsilon$. We are concerned with Cauchy sequences $\{x_n\}_{n=1}^{\infty}$ in $\boldsymbol{X}$ with the property that $x_n \in A_n$ for each n. In particular, we need the following property which allows the *extension* of a Cauchy *subsequence* $\{x_{n_j} \in A_{n_j}\}_{j=1}^{\infty}$, with the property that $x_{n_j} \in A_{n_j}$ for each j, to a Cauchy sequence $\{x_n \in A_n\}_{n=1}^{\infty}$.

Lemma 2. (The Extension Lemma) *Let $(\boldsymbol{X}, d)$ be a metric space. Let $\{A_n: n = 1, 2, \ldots, \infty\}$ be a Cauchy sequence of points in $(\mathscr{H}(\boldsymbol{X}), h)$. Let $\{n_j\}_{j=1}^{\infty}$ be an infinite sequence of integers*

$$0 < n_1 < n_2 < n_3 < \cdots.$$

Suppose that we have a Cauchy sequence $\{x_{n_j} \in A_{n_j}: j = 1, 2, 3, \ldots\}$ in $(\boldsymbol{X}, d)$. Then there is a Cauchy sequence $\{\tilde{x}_n \in A_n: n = 1, 2, \ldots\}$ such that $\tilde{x}_{n_j} = x_{n_j}$, for all $j = 1, 2, 3, \ldots$.

Proof. We give the construction of the sequence $\{\tilde{x}_n \in A_n: n = 1, 2, \ldots\}$. For each $n \in \{1, 2, \ldots, n_1\}$ choose $\tilde{x}_n \in \{x \in A_n: d(x, x_{n_1}) = d(x_{n_1}, A_n)\}$. That is, $\tilde{x}_n$ is the closest point (or one of the closest points) in A_n to x_{n_1}. The existence of such a closest point is ensured by the compactness of A_n. Similarly, for each $j \in \{2, 3, \cdots\}$ and each $n \in \{n_j + 1, \ldots, n_{j+1}\}$ choose $\tilde{x}_n \in \{x \in A_n: d(x, x_{n_j}) = d(x_{n_j}, A_n)\}$.

Now we show that $\{\tilde{x}_n\}$ has the desired properties, that it is indeed an extension of $\{x_{n_j}\}$ to $\{A_n\}$. Clearly $\tilde{x}_{n_j} = x_{n_j}$ and $x_n \in A_n$, by construction. To show that it is a Cauchy sequence let $\epsilon > 0$ be given. There is N_1 so that $n_k, n_j \geq N_1$ implies $d(x_{n_k}, x_{n_j}) \leq \epsilon/3$. There is N_2 so that $m, n \geq N_2$ implies

$$d(A_m, A_n) \leq \epsilon/3.$$

Let $N = \text{Max}\{N_1, N_2\}$ and note that, for $m, n \geq N$,

$$d(\tilde{x}_m, \tilde{x}_n) \leq d(\tilde{x}_m, x_{n_j}) + d(x_{n_j}, x_{n_k}) + d(x_{n_k}, \tilde{x}_n)$$

where $m \in \{n_{j-1} + 1, n_{j-1} + 2, \ldots, n_j\}$ and $n \in \{n_{k-1} + 1, n_{k-1} +$

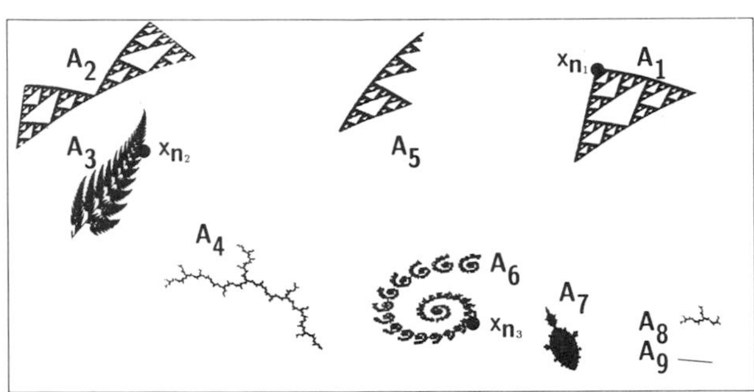

Figure 2.7.1 *The beginning of a Cauchy sequence $\{A_n\}$ of sets in $\mathscr{H}(\mathbb{R}^2)$ is shown. A Cauchy subsequence of points $\{x_{n_i}\}$ belonging to a subsequence of the sets is also indicated. Make a photocopy of the figure, and mark on it the extension of the subsequence of points to the visible sets in $\{A_n\}$.*

$2, \ldots, n_k\}$. Since $h(A_m, A_{n_j}) < \epsilon/3$ there exists $y \in A_m \cap (\{x_{n_j}\} + \epsilon/3)$ so that $d(\tilde{x}_m, x_{n_j}) \le \epsilon/3$. Similarly $d(x_{n_k}, \tilde{x}_n) \le \epsilon/3$. Hence $d(\tilde{x}_n, \tilde{x}_n) \le \epsilon$ for all $m, n > N$. This completes the proof.

Exercises & Examples

7.1. A Cauchy sequence $\{A_n\}$ of sets in $(\mathscr{H}(\mathbb{R}^2), h)$ is sketched in the Figure 2.7.1. The underlying metric space is ($\mathbb{R}^2$, Euclidean). A Cauchy subsequence $\{x_{n_j} \in A_{n_j}\}$ is also shown. Sketch, in the same Figure, an extension $\{\tilde{x}_n\}$, of this subsequence, to $\{A_n\}$.

7.2. Repeat 7.1 but this time with reference to Figure 2.7.2.

The central result we have been driving for is this:

Theorem 1. (The Completeness of the Space of Fractals) *Let $(\mathbf{X}, d)$ be a complete metric space. Then $(\mathscr{H}(\mathbf{X}), h)$ is a complete metric space. Moreover, if $\{A_n \in \mathscr{H}(\mathbf{X})\}_{n=1}^{\infty}$ is a Cauchy sequence then*

$$A = \lim_{n \to \infty} A_n \in \mathscr{H}(\mathbf{X})$$

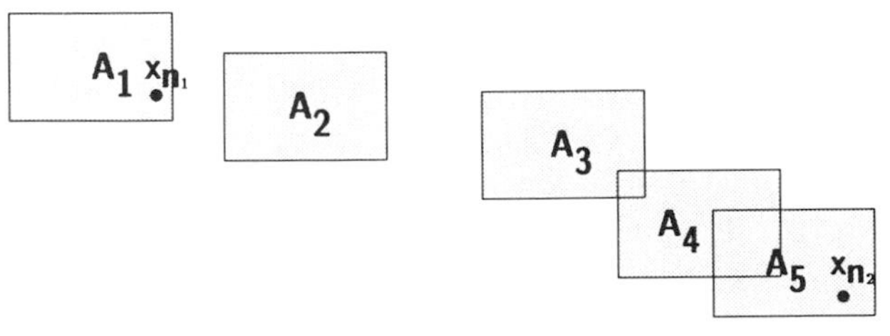

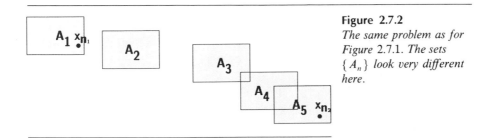

Figure 2.7.2 *The same problem as for Figure 2.7.1. The sets $\{A_n\}$ look very different here.*

can be characterized as follows:

$$A = \{ x \in \mathbf{X}: \textit{there is a Cauchy sequence } \{ x_n \in A_n \} \textit{ that converges to } x \}.$$

Proof. Let $\{A_n\}$ be a Cauchy sequence in $\mathscr{H}(\mathbf{X})$ and let A be defined as in the statement of the theorem. We break the proof up into the following parts:

(a) $A \neq \varnothing$;
(b) A is closed and hence complete since $\mathbf{X}$ is complete;
(c) for $\epsilon > 0$ there is N such that for $n \geq N$, $A \subset A_n + \epsilon$;
(d) A is totally bounded and thus by (b) is compact;
(e) $\operatorname{Lim} A_n = A$.

Proof of (a). We shall prove this part by proving the existence of a Cauchy sequence $\{a_i \in A_i\}$ in $\mathbf{X}$. Towards this end find a sequence of positive integers $N_1 < N_2 < N_3 < \cdots < N_n < \cdots$ such that

$$h(A_m, A_n) < \frac{1}{2^i} \qquad \text{for } m, n > N_i.$$

Choose $x_{N_1} \in A_{N_1}$. Then, since $h(A_{N_1}, A_{N_2}) \leq \frac{1}{2}$, we can find an $x_{N_2} \in A_{N_2}$ such that $d(x_{N_1}, x_{N_2}) \leq \frac{1}{2}$. Assume that we have selected a finite sequence $x_{N_i} \in A_{N_i}$; $i = 1, 2, \ldots, k$ for which $d(x_{N_{i-1}}, x_{N_i}) \leq 1/2^{i-1}$. Then since $h(A_{N_k}, A_{N_{k+1}}) \leq 1/2^k$, and $x_{N_k} \in A_{N_k}$, we can find $x_{N_{k+1}} \in A_{N_{k+1}}$ such that $d(x_{N_k}, x_{N_{k+1}}) \leq 1/2^k$. For example let $x_{N_{k+1}}$ be the point in $A_{N_{k+1}}$ which is closest to x_{N_k}. By induction we can find an infinite sequence $\{x_{N_i} \in A_{N_i}\}$ such that $d(x_{N_i} x_{N_{i+1}}) \leq 1/2^i$. To see that $\{x_{N_i}\}$ is a Cauchy sequence in $\mathbf{X}$, let $\epsilon > 0$ and choose N_ϵ such that $\sum_{i=N_\epsilon}^{\infty} 1/2^i < \epsilon$. Then for $m > n \geq N_\epsilon$ we have

$$\begin{aligned} d(x_{N_m}, x_{N_n}) &\leq d(x_{N_m}, x_{N_{m+1}}) + d(x_{N_{m+1}}, x_{N_{m+2}}) + \cdots + d(x_{N_{n-1}} x_{N_n}) \\ &< \sum_{i=N_\epsilon}^{\infty} \frac{1}{2^i} < \epsilon. \end{aligned}$$

By the Extension Lemma, there exists a convergent subsequence $\{a_i \in A_i\}$ for which $a_{N_i} = x_{N_i}$. Then $\operatorname{Lim} a_i$ exists and by definition is in A. Thus $A \neq \varnothing$.

Proof of (b). To show that A is closed, suppose $\{a_i \in A\}$ is a sequence that converges to a point a. We will show that $a \in A$, hence making A closed. For each positive integer i, there exists a sequence $\{x_{i,n} \in A_i\}$ such that $\operatorname{Lim}_{n \to \infty} x_{i,n} = a_i$. There exists an *increasing* sequence of positive numbers $\{N_i\}_{i=1}^{\infty}$ such that $d(a_{N_i}, a) < 1/i$. Furthermore, for each N_i there is an integer m_i such that $d(x_{N_i, m_i}, a_{N_i}) \leq 1/i$. Thus $d(x_{N_i, m_i}, a) \leq 2/i$. If we let $y_{N_i} = x_{N_i, m_i}$ we see that $y_{N_i} \in A_{N_i}$ and $\operatorname{Lim}_{i \to \infty} y_{N_i} = a$. By the Extension Lemma $\{y_{N_i}\}$ can be extended to a convergent sequence $\{z_i \in A_i\}$, and so $a \in A$. Thus we have shown A is closed.

Proof of (c). Let $\epsilon > 0$. There exists an N such that for $m, n \geq N$, $h(A_m, A_n) \leq \epsilon$. Now let $n \geq N$. Then for $m \geq n$, $A_m \subset A_n + \epsilon$. We need to show that $A \subset A_n + \epsilon$. To do this, let $a \in A$. There is a sequence $\{a_i \in A_i\}$ that converges to a. We may assume N is also large enough so that for $m \geq N$, $d(a_m, a) < \epsilon$. Then $a_m \in A_n + \epsilon$ since $A_m \subset A_n + \epsilon$. Since A_n is compact, it can be shown that $A_n + \epsilon$ is closed. Then since $a_m \in A_n + \epsilon$ for all $m \geq N$, a must also be in $A_n + \epsilon$. This completes the proof that $A \subset A_n + \epsilon$ for n large enough.

Proof of (d). Suppose A were not totally bounded. Then for some $\epsilon > 0$ there would not exist a finite ϵ-net. We could then find a sequence $\{x_i\}_{i=1}^{\infty}$ in A such that $d(x_i, x_j) \geq \epsilon$ for $i \neq j$. We shall show that this gives a contradiction. By (c) there exists an n large enough so that $A \subset A_n + \epsilon/3$. For each x_i, there is a corresponding $y_i \in A_n$ for which $d(x_i, y_i) \leq \epsilon/3$. Since A_n is compact some subsequence $\{y_{n_i}\}$ of $\{y_i\}$ converges. Then we can find points in the sequence $\{y_{n_i}\}$ as close together as we wish. In particular we can find two points y_{n_i} and y_{n_j} such that $d(y_{n_i}, y_{n_j}) < \epsilon/3$. But then

$$d(x_{n_i}, x_{n_j}) \leq d(x_{n_i}, y_{n_i}) + d(y_{n_i}, y_{n_j}) + d(y_{n_j}, x_{n_j}) < \frac{\epsilon}{3} + \frac{\epsilon}{3} + \frac{\epsilon}{3}$$

and we have a contradiction to the way $\{x_{n_i}\}$ was chosen. Thus A is totally bounded and by part (b) compact.

Proof of (e). From part (d), $A \in \mathscr{H}(X)$. Hence by part (c) and Lemma 1 the proof that $\text{Lim}\, A_i = A$ will be complete if we show that for $\epsilon > 0$, there exists an N such that for $n \geq N$, $A_n \subset A + \epsilon$. To show this let $\epsilon > 0$ and find N such that for $m, n \geq N$, $h(A_m, A_n) \leq \epsilon/2$. Then for $m, n \geq N$, $A_m \subset A_n + \epsilon/2$. Let $n \geq N$. We will show that $A_n \subset A + \epsilon$. Let $y \in A_n$. There exists an increasing sequence $\{N_i\}$ of integers such that $n < N_1 < N_2 < N_3 < \cdots < N_k < \cdots$ and for $m, k \geq N_j$, $A_m \subset A_k + \epsilon/2^{j+1}$. Note that $A_n \subset A_{N_1} + \epsilon/2$. Since $y \in A_n$, there is an $x_{N_1} \in A_{N_1}$ such that $d(y, x_{N_1}) \leq \epsilon/2$. Since $x_{N_1} \in A_{N_1}$, there is a point $x_{N_2} \in A_{N_2}$ such that $d(x_{N_1}, x_{N_2}) \leq \epsilon/2^2$. In a similar manner we can use induction to find a sequence $x_{N_1}, x_{N_2}, x_{N_3}, \ldots$ such that $x_{N_j} \in A_{N_j}$ and $d(x_{N_j}, x_{N_{j+1}}) < \epsilon/2^{j+1}$. Using the triangle inequality a number of times we can show that

$$d(y, x_{N_j}) \leq \epsilon \qquad \text{for all } j,$$

and also that $\{x_{N_j}\}$ is a Cauchy sequence. $\{x_{N_j}\}$ converges to a point x which is in A. Moreover $d(y, x_{N_j}) \leq \epsilon$ implies that $d(y, x) \leq \epsilon$. We have thus shown that $A_n \subset A + \epsilon$ for $n \geq N$. This completes the proof that $\text{Lim}\, A_n = A$ and consequently that $(\mathscr{H}(X), h)$ is a complete metric space.

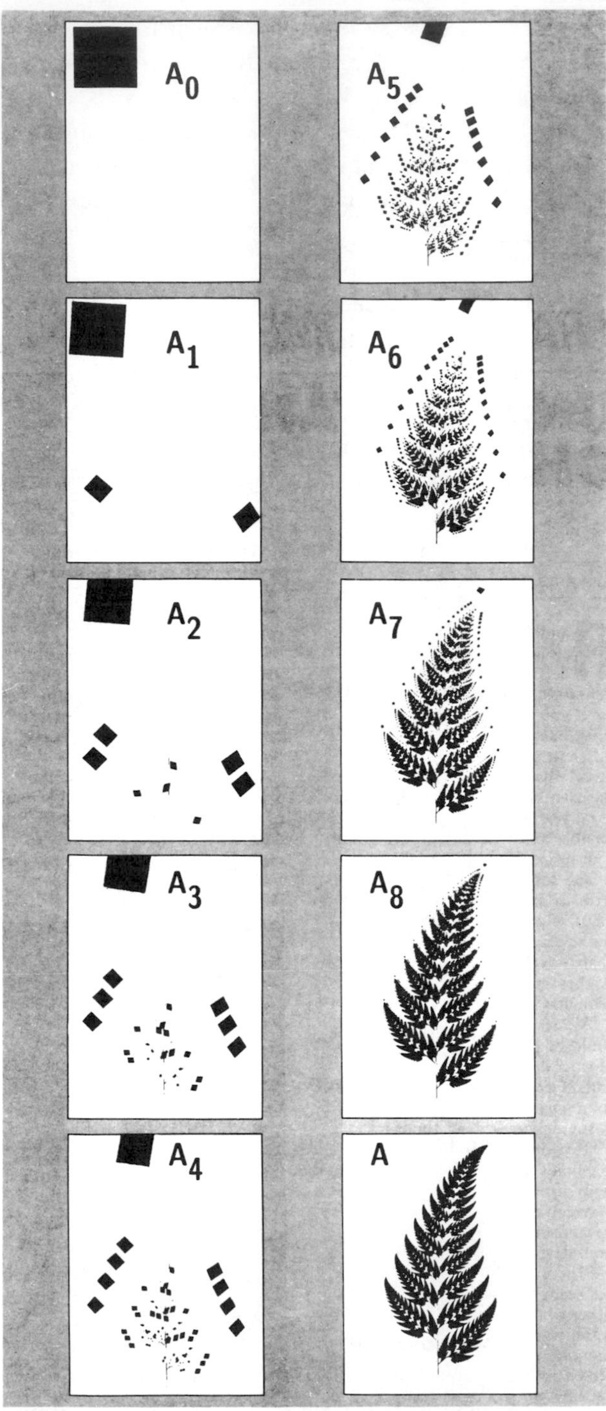

Figure 2.7.3
A Cauchy sequence of sets $\{A_n\}$ in the space $\mathscr{H}(\mathbb{R}^2)$ converging to a fern-like set.

Exercises & Examples

7.3. A tree waves in the wind. A special camera photographs the tree at times $t_n = (1 - 1/n)$ secs, $n = 1, 2, 3, \ldots$. Show that with reasonable assumptions the sequence of pictures thus obtained form a Cauchy sequence $\{A_n\}_{n=1}^{\infty}$ in $\mathscr{H}(\mathbb{R}^2)$. What does $A = \text{Lim}_{n\to\infty} A_n$ look like?

7.4. The Sierpinski Triangle △ is a compact subset of ($\mathbb{R}^2$, Euclidean). Hence (△ , Euclidean) is a compact metric space. Give an example of an infinite set in $(\mathscr{H}($ △ $), h)$. Demonstrate a Cauchy sequence $\{A_n \in \mathscr{H}($ △ $)\}$ which is contained in your set, and describe its limit.

7.5. Figure 2.7.3 shows a convergent sequence of sets in $\mathscr{H}(\blacksquare)$ converging to a fern. Pick a point in A. Find a Cauchy sequence $\{x_n \in A_n\}$ which converges to it.

2.8 ADDITIONAL THEOREMS ABOUT METRIC SPACES

We state here a number of theorems which we shall use later on. Full proofs are not provided. They can be found in most introductory topology texts. We particularly recommend Kasriel [1971] and Mendelson [1963]. These theorems may be treated as exercises in metric space theory.

Theorem 1. *Let $(\mathbf{X}, d)$ be a metric space. Let $\{x_n\}$ be a Cauchy sequence convergent to $x \in \mathbf{X}$ (or equivalently let $\{x_n\}$ be a sequence and x be a point, such that $\text{Lim}_{n\to\infty} d(x, x_n) = 0$). Let $f\colon \mathbf{X} \to \mathbf{X}$ be continuous. Then*

$$\lim_{n\to\infty} f(x_n) = f(x).$$

Proof. See your first calculus book.

Theorem 2. *Let $(\mathbf{X}_1, d_1)$ and $(\mathbf{X}_2, d_2)$ be metric spaces. Let $f\colon \mathbf{X}_1 \to \mathbf{X}_2$ be continuous. Let $E \subset \mathbf{X}_1$ be compact. Then $f\colon E \to \mathbf{X}$ is **uniformly** continuous: that is, given $\epsilon > 0$ there is a number $\delta > 0$ so that*

$$d_2(f(x), f(y)) < \epsilon \qquad \text{whenever } d_1(x, y) < \delta \text{ for all } x, y \in E.$$

Proof. Use the fact that any countable open cover of E contains a finite subcover.

Theorem 3. *Let $(\mathbf{X}_i, d_i)$ be metric spaces for $i = 1, 2, 3$. Let $f\colon \mathbf{X}_1 \times \mathbf{X}_2 \to \mathbf{X}_3$ be continuous in $x_1 \in \mathbf{X}_1$ for each fixed $x_2 \in \mathbf{X}_2$ and continuous in $x_2 \in \mathbf{X}_2$ for* *each fixed* $x_1 \in \mathbf{X}_1$. *Then f is continuous on the metric space* $(\mathbf{X} = \mathbf{X}_1 \times \mathbf{X}_2, d)$ *where* $d((x_1, x_2), (y_1, y_2)) = \text{Max}\{d_1(x_1, y_1), d_2(x_2, y_2)\}$.

Proof. Use $d(f(x_1, x_2), f(y_1, y_2)) \leq d(f(x_1, x_2), f(y_1, x_2)) + d(f(y_1, x_2), f(y_1, y_2))$, but check first that d is a metric.

Theorem 4. *Let* $(\mathbf{X}_i, d_i)$ *be metric spaces for* $i = 1, 2$ *and let the metric space* $(\mathbf{X}, d)$ *be defined as in Theorem* 3. *If* $K_1 \subset \mathbf{X}_1$ *and* $K_2 \subset \mathbf{X}_2$ *are compact then* $K_1 \times K_2 \subset \mathbf{X}$ *is compact.*

Proof. Deal with the component in K_1 first.

Theorem 5. *Let* $(\mathbf{X}_i, d_i)$ *be compact metric spaces for* $i = 1, 2$. *Let* $f\colon \mathbf{X}_1 \to \mathbf{X}_2$ *be continuous, one-to-one and onto. Then f is a homeomorphism.*

Transformations on Metric Spaces; Contraction Mappings; and the Construction of Fractals

3

3.1 TRANSFORMATIONS ON THE REAL LINE

Fractal geometry studies "complicated" subsets of geometrically "simple" spaces such as $\mathbb{R}^2, \mathbb{C}, \mathbb{R}, \hat{\mathbb{C}}$. In deterministic fractal geometry the focus is on those subsets of a space which are generated by, or possess invariance properties under, simple geometrical transformations of the space into itself. A simple geometrical transformation is one which is easily conveyed or explained to someone else. Usually they can be completely specified by a small set of parameters. Examples include affine transformations in $\mathbb{R}^2$, which are expressed using 2×2 matrices and 2-vectors, and rational transformations on the Riemann Sphere, which require the specification of the coefficients in a pair of polynomials.

Definition 1. Let $(\boldsymbol{X}, d)$ be a metric space. A *transformation* on $\boldsymbol{X}$ is a function $f: \boldsymbol{X} \to \boldsymbol{X}$, which assigns exactly one point $f(x) \in \boldsymbol{X}$ to each point $x \in \boldsymbol{X}$. If $S \subset \boldsymbol{X}$ then $f(S) = \{f(x): x \in S\}$. f is *one-to-one* if $x, y \in \boldsymbol{X}$ with $f(x) = f(y)$ implies $x = y$. Function f is *onto* if $f(\boldsymbol{X}) = \boldsymbol{X}$. f is called *invertible* if it is one-to-one and onto: in this case it is possible to define a transformation $f^{-1}: \boldsymbol{X} \to \boldsymbol{X}$, called the inverse of f, by $f^{-1}(y) = x$ where $x \in \boldsymbol{X}$ is the unique point such that $y = f(x)$.

Definition 2. Let f: $\boldsymbol{X} \to \boldsymbol{X}$ be a transformation on a metric space. The *forward iterates* of f are transformations $f^{\circ n}$: $\boldsymbol{X} \to \boldsymbol{X}$ defined by $f^{\circ 0}(x) = x$, $f^{\circ 1}(x) = f(x)$, $f^{\circ(n+1)}(x) = f \circ f^{\circ n}(x) = f(f^{\circ n}(x))$ for $n = 0, 1, 2, \ldots$. If f is invertible then the *backward iterates* of f are transformations $f^{\circ(-m)}(x)$: $\boldsymbol{X} \to \boldsymbol{X}$ defined by $f^{\circ(-1)}(x) = f^{-1}(x)$, $f^{\circ(-m)}(x) = (f^{\circ m})^{-1}(x)$ for $m = 1, 2, 3, \ldots$.

In order to work in fractal geometry one needs to be familiar with the basic families of transformations in $\mathbb{R}$, $\mathbb{R}^2$, $\mathbb{C}$, and $\hat{\mathbb{C}}$. One needs to know well the relationship between "formulas" for transformations and the geometric changes, stretchings, twistings, foldings, and skewings of the underlying fabric, the metric space upon which they act. It is more important to understand what the transformations do to sets than how they act on individual points. So, for example, it is more useful to know how an affine transformation in $\mathbb{R}^2$ acts on a straight line, a circle, or a triangle, than to know to where it takes the origin.

Exercises & Examples

1.1. Let f: $\boldsymbol{X} \to \boldsymbol{X}$ be an invertible transformation. Show that

$$f^{\circ m} \circ f^{\circ n} = f^{\circ(m+n)} \qquad \text{for all integers } m \text{ and } n.$$

1.2. A transformation f: $\mathbb{R} \to \mathbb{R}$ is defined by $f(x) = 2x$ for all $x \in \mathbb{R}$. Is f invertible? Find a formula for $f^{\circ n}(x)$ which applies for all integers n.

1.3. A transformation f: $[0, 1] \to [0, 1]$ is defined by $f(x) = \frac{1}{2}x$. Is this transformation one-to-one? Onto? Invertible?

1.4. The mapping f: $[0, 1] \to [0, 1]$ is defined by $f(x) = 4x \cdot (1 - x)$. Is this transformation one-to-one? Onto? Is it invertible?

1.5. Let $\mathscr{C}$ denote the *Classical Cantor Set*. This subset of the metric space $[0, 1]$ is obtained by successive deletion of middle third open subintervals as follows. We construct a nested sequence of closed sets

$$I_0 \supset I_1 \supset I_2 \supset I_3 \supset I_4 \supset I_5 \supset I_6 \supset I_7 \cdots \supset I_N \supset \cdots$$

where

$I_0 = [0, 1]$,
$I_1 = [0, \frac{1}{3}] \cup [\frac{2}{3}, \frac{3}{3}]$,
$I_2 = [0, \frac{1}{9}] \cup [\frac{2}{9}, \frac{3}{9}] \cup [\frac{6}{9}, \frac{7}{9}] \cup [\frac{8}{9}, \frac{9}{9}]$,
$I_3 = [0, \frac{1}{27}] \cup [\frac{2}{27}, \frac{3}{27}] \cup [\frac{6}{27}, \frac{7}{27}] \cup [\frac{8}{27}, \frac{9}{27}] \cup [\frac{18}{27}, \frac{19}{27}] \cup [\frac{20}{27}, \frac{21}{27}] \cup [\frac{24}{27}, \frac{25}{27}] \cup [\frac{26}{27}, \frac{27}{27}]$,
$I_4 = I_3$ take away the middle open third of each interval in I_3,
⋮
$I_N = I_{N-1}$ take away the middle open third of each interval in I_{N-1}.
⋮

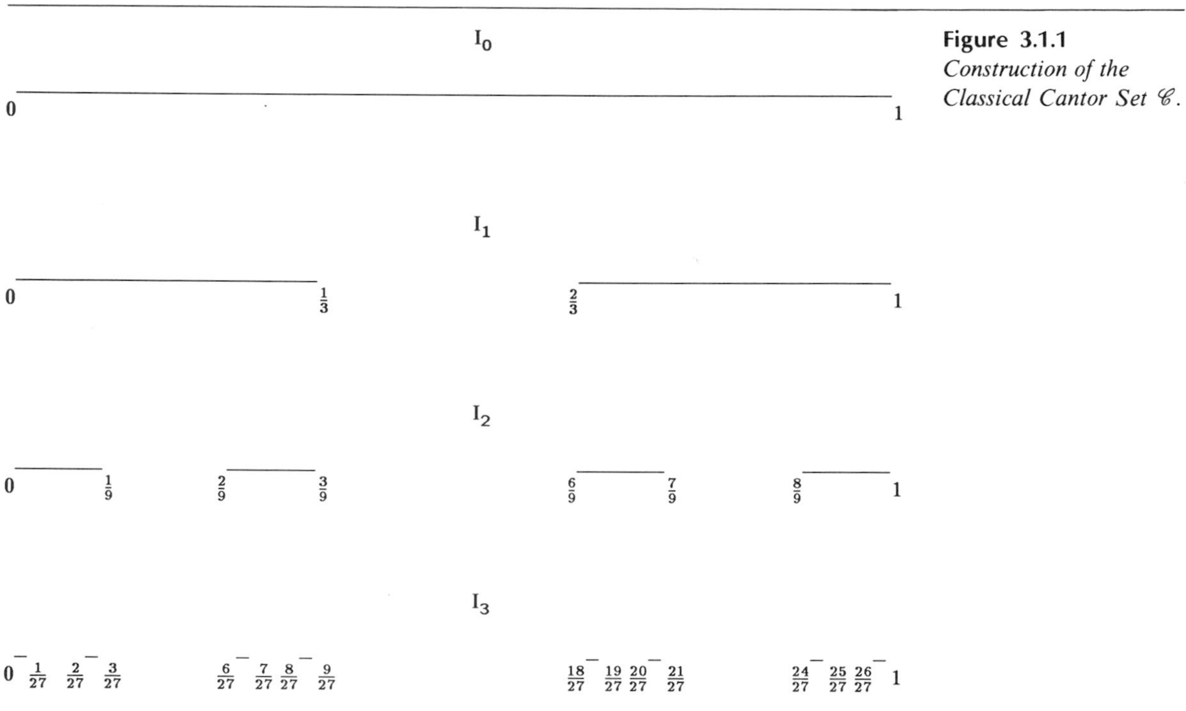

Figure 3.1.1 *Construction of the Classical Cantor Set* $\mathscr{C}$.

This construction is illustrated in Figure 3.1.1. We define

$$\mathscr{C} = \bigcap_{n=0}^{\infty} I_n.$$

$\mathscr{C}$ contains the point $x = 0$, so it is nonempty. In fact $\mathscr{C}$ is a perfect set which contains uncountably many points, as discussed in Chapter 4. $\mathscr{C}$ is an official fractal and we will often refer to it.

We are now able to work in the metric space ($\mathscr{C}$, Euclidean). A transformation $f\colon \mathscr{C} \to \mathscr{C}$ is defined by $f(x) = \frac{1}{3}x$. Show that this transformation is one-to-one but not onto. Also, find another affine transformation (see example 1.7) which maps $\mathscr{C}$ one-to-one into $\mathscr{C}$.

1.6. $f\colon \mathbb{R}^2 \to \mathbb{R}^2$ is defined by $f(x_1, x_2) = (2x_1, x_2^2 + x_1)$ for all $(x_1, x_2) \in \mathbb{R}^2$. Show that f is not invertible. Give a formula for $f^{\circ 2}(x)$.

1.7. *Affine transformations* in $\mathbb{R}^1$ are transformations of the form $f(x) = a \cdot x + b$ where a and b are real constants. Given the interval $I = [0, 1]$, $f(I)$ is a new interval of length $|a|$, and f rescales by a. The left endpoint 0 of the interval is moved to b, and $f(I)$ lies to the left or right of b according to whether a is positive or negative respectively (see Figure 3.1.2).

We think of the action of an affine transformation on all of $\mathbb{R}$ as follows. The whole line is stretched away from the origin if $|a| > 1$, or

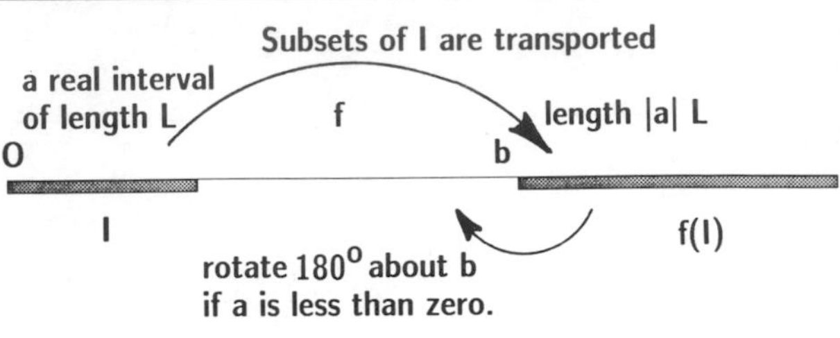

Figure 3.1.2 *The action of the affine transformation $f\colon \mathbb{R} \to \mathbb{R}$ defined by $f(x) = ax + b$.*

contracted towards it if $|a| < 1$; flipped through 180° about O if $a < 0$; and then *translated* (shifted as a whole) by an amount b (shift to the left if $b < 0$, and to the right if $b > 0$).

1.8. Describe the set of affine transformations which take the real interval $X = [1, 2]$ into itself. Show that if f and g are two such transformations then $f \circ g$ and $g \circ f$ are also affine transformations on $[1, 2]$. Under what conditions does $f \circ g(X) \cup g \circ f(X) = X$?

1.9. A sequence of intervals $\{I_n\}_{n=0}^{\infty}$ is indicated in Figure 3.1.3. Find an affine transformation $f\colon \mathbb{R} \to \mathbb{R}$ so that $f^{\circ n}(I_0) = I_n$ for $n = 0, 1, 2, 3, \ldots$. Use a straight-edge and dividers to help you. Also show that $\{I_n\}_{n=1}^{\infty}$ is a Cauchy sequence in $(\mathscr{H}(\mathbb{R}), h)$, where h is the Hausdorff distance on $\mathscr{H}(\mathbb{R})$ induced by the Euclidean metric on $\mathbb{R}$. Evaluate $I = \text{Lim}_{n \to \infty} I_n$.

1.10. Consider the geometric series $\sum_{n=0}^{\infty} b \cdot a^n = b + a \cdot b + a^2 b + a^3 b + a^4 b + \cdots > 0, 0 < a < 1$. This is associated with a sequence of intervals $I_0 = [0, b]$, $I_n = f^{\circ n}(I_0)$ where $f(x) = ax + b$, $n = 1, 2, 3, \ldots$; as illustrated in Figure 3.1.4.

Let $I = \cup_{n=0}^{\infty} I_n$ and let l denote the total length of I. Show that $f(I) = I \setminus I_0$, and hence deduce that $al = l - b$ so that $l = b/(1 - a)$. Deduce at once that

$$\sum_{n=0}^{\infty} b \cdot a^n = b/(1 - a).$$

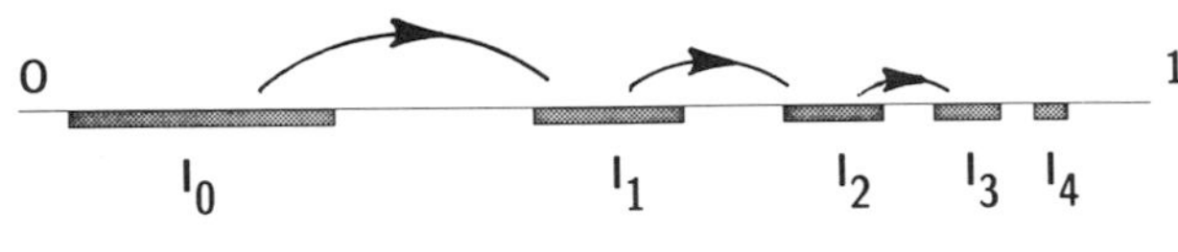

Figure 3.1.3 *This Figure suggests a sequence of intervals $\{I_n\}_{n=0}^{\infty}$. Find an affine transformation $f\colon \mathbb{R} \to \mathbb{R}$ so that $f^{\circ n}(I_0) = I_n$ for $n = 0, 1, 2, 3, \ldots$. Use a straight-edge and dividers to help you.*

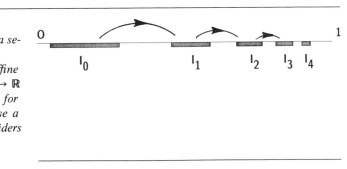

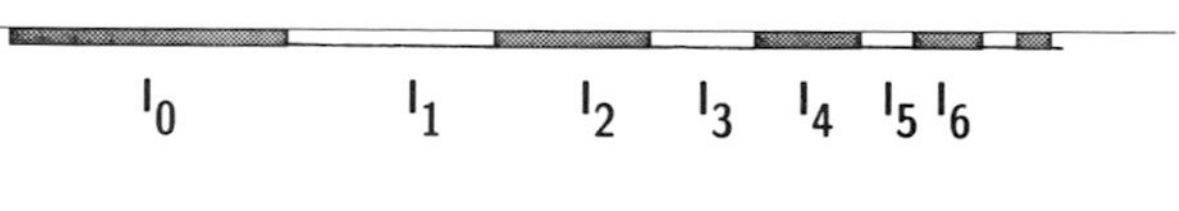

Figure 3.1.4
Picture of a convergent geometric series in $\mathbb{R}^1$ *(see Ex.* 1.9).

Thus we see from a geometrical point of view a well-known result about *geometric series*. Make a similar geometrical argument to cover the case $-1 < a < 0$.

Definition 3. A transformation $f\colon \mathbb{R} \to \mathbb{R}$ of the form

$$f(x) = a_0 + a_1 x + a_2 x^2 + a_3 x^3 + \cdots + a_N x^N,$$

where the coefficients a_i ($i = 0, 1, 2, \ldots, N$) are real numbers, $a_N \neq 0$, and N is a nonnegative integer, is called a *polynomial transformation*. N is called the *degree* of the transformation.

Exercises & Examples

1.11. Show that if $f\colon \mathbb{R} \to \mathbb{R}$ and $g\colon \mathbb{R} \to \mathbb{R}$ are polynomial transformations, then so is $f \circ g$. If f is of degree N, calculate the degree of $f^{\circ m}(x)$ for $m = 1, 2, 3, \ldots$.

1.12. Show that for $n > 1$ a polynomial transformation $f\colon \mathbb{R} \to \mathbb{R}$ of degree n is not generally invertible.

1.13. Show that far enough out (i.e., for large enough $|x|$), a polynomial transformation $f\colon \mathbb{R} \to \mathbb{R}$ always stretches intervals. That is, view f as a transformation from ($\mathbb{R}$, Euclidean) into itself. Show that if I is an interval of the form $I = \{x\colon |x - a| \leq b\}$ for fixed $a, b \in \mathbb{R}$, then for any number $M > 0$ there is number $\beta > 0$ such that if $b > \beta$, then the ratio (length of $f(I)$)/(length of I) is larger than M. This idea is illustrated in Figure 3.1.5.

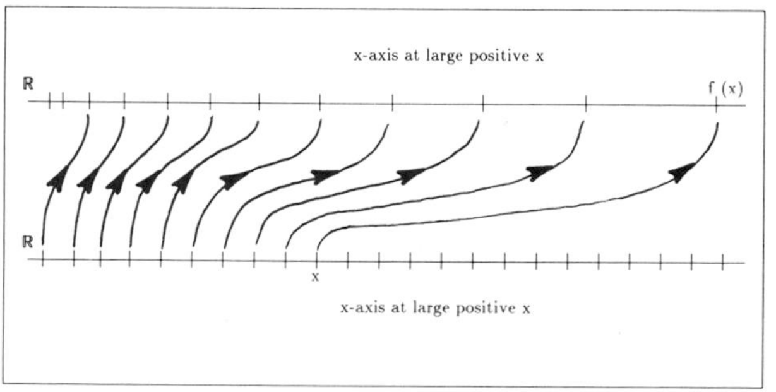

Figure 3.1.5
A polynomial transformation $f\colon \mathbb{R} \to \mathbb{R}$ *of degree* > 1 *stretches* $\mathbb{R}$ *more and more the further out one goes.*

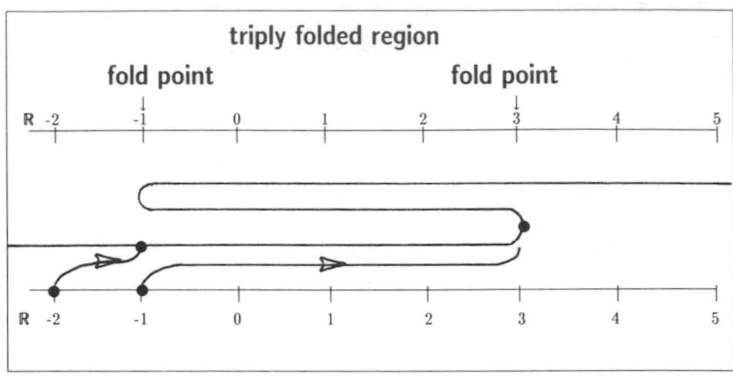

Figure 3.1.6 *The polynomial transformation* $f(x) = x^3 - 3x + 1$.

1.14. A polynomial transformation f: $\mathbb{R} \to \mathbb{R}$ of degree n can produce at most $(n - 1)$ *folds*. For example $f(x) = x^3 - 3x + 1$ behaves as shown in Figure 3.1.6.

1.15. Find a family of polynomial transformations of degree 2 which map the interval $[0, 2]$ into itself, such that, with one exception, if $y \in f([0, 2])$ then there exist two distinct points x_1 and x_2 in $[0, 2]$ with $f(x_1) = f(x_2) = y$.

1.16. Show that the one-parameter family of polynomial transformations f_λ: $[0, 2] \to [0, 2]$, where

$$f_\lambda(x) = \lambda \cdot x \cdot (2 - x),$$

and the parameter λ belongs to $[0, 2]$, indeed takes the interval $[0, 2]$ into itself. Locate the value of x at which the fold occurs. Sketch the behavior of the family, in the spirit of Figure 3.1.6.

1.17. Let f: $\mathbb{R} \to \mathbb{R}$ be a polynomial transformation of degree n. Show that values of x which are transformed into fold points are solutions of

$$\frac{df}{dx}(x) = 0, \qquad x \in \mathbb{R}.$$

Solutions of this equation are called (real) *critical points* of the function f. If c is a critical point then $f(c)$ is a *critical value*. Show that a critical value need not be a fold point.

1.18. Find a polynomial transformation such that Figure 3.1.7 is true.

1.19. Recall that a polynomial transformation of an interval f: $I \subset \mathbb{R} \to I$ is normally represented as in Figure 3.1.8. This will be useful when we study iterates $\{f^{\circ n}(x)\}_{n=1}^{\infty}$. However the folding point of view helps us to understand the idea of the deformation of space.

1.20. Polynomial transformations can be lifted to act on subsets of $\mathbb{R}^2$ in a simple way: we can define for example $F(x) = (f_1(x_1), f_2(x_2))$ where f_1 and f_2 are polynomial transformations in $\mathbb{R}$, so that F: $\mathbb{R}^2 \to \mathbb{R}^2$. Desired foldings in two orthogonal directions can be produced; or

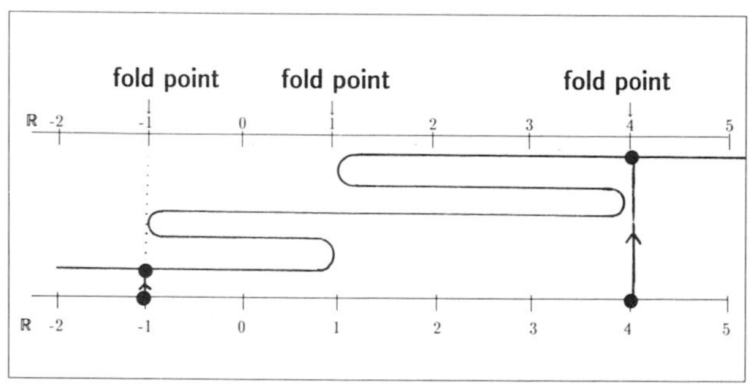

Figure 3.1.7
Find a polynomial transformation $f: \mathbb{R} \to \mathbb{R}$, so that this figure correctly represents the way it folds the real line.

shrinking in one direction and folding in another. Show that the transformation $F(x_1, x_2) = (\frac{8}{5}x_1^3 - \frac{36}{5}x_1^2 + \frac{48}{5}x_1)$ acts on the triangular set S in Figure 3.1.9 as shown.

The real line can be extended to a space which is topologically a circle by including "The Point at Infinity." One way to do this is to think of $\mathbb{R}$ as a subset of $\hat{\mathbb{C}}$, and then include the North Pole on $\hat{\mathbb{C}}$. We define this space to be $\hat{\mathbb{R}} = \mathbb{R} \cup \{\infty\}$, and will usually give it the spherical metric.

Definition 4. A transformation $f: \hat{\mathbb{R}} \to \hat{\mathbb{R}}$ defined in the form

$$f(x) = \frac{ax+b}{cx+d}, \qquad a, b, c, d \in \mathbb{R}, \qquad ad \neq bc,$$

is called a *linear fractional transformation* or a *Möbius transformation*. If $c \neq 0$ then $f(-d/c) = \infty$, and $f(\infty) = a/c$. If $c = 0$ then $f(\infty) = \infty$.

Exercises & Examples

1.21. Show that a Möbius transformation is invertible.

1.22. Show that if f_1 and f_2 are both Möbius transformations then so is $f_1 \circ f_2$.

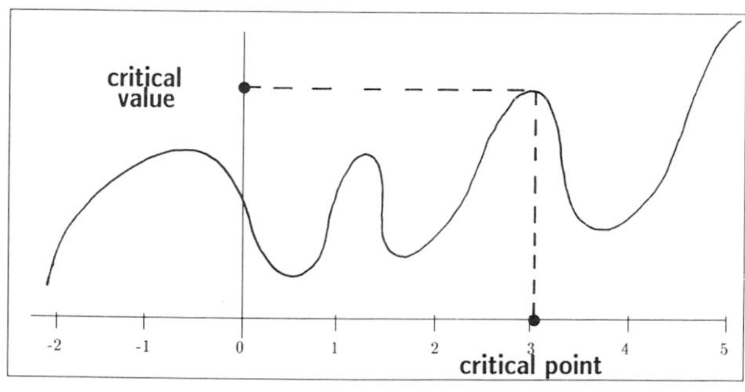

Figure 3.1.8
The usual way of picturing a polynomial transformation.

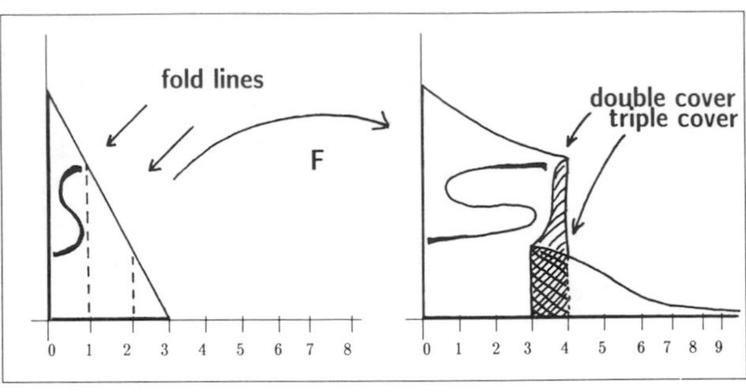

Figure 3.1.9 *A polynomial transformation acting on a set S in the plane*

1.23. What does $f(x) = 1/x$ do to $\hat{\mathbb{R}}$ on the sphere?

1.24. Show that the set of Möbius transformations f such that $f(\infty) = \infty$ is the set of affine transformations.

1.25. Find a Möbius transformation $f\colon \hat{\mathbb{R}} \to \hat{\mathbb{R}}$ so that $f(1) = 2$, $f(2) = 0$, $f(0) = \infty$. Evaluate $f(\infty)$.

1.26. Figure 3.1.11 shows a Sierpinski triangle before and after the polynomial transformation $x \to ax(x - b)$ has been applied to the x-axis. Evaluate the real constants a and b. Notice how well fractals can be used to illustrate how a transformation acts.

3.2 AFFINE TRANSFORMATIONS IN THE EUCLIDEAN PLANE

Definition 1. A transformation $w\colon \mathbb{R}^2 \to \mathbb{R}^2$ of the form

$$w(x_1, x_2) = (ax_1 + bx_2 + e, cx_1 + dx_2 + f)$$

where a, b, c, d, e, and f are real numbers, is called a (two-dimensional) *affine* transformation.

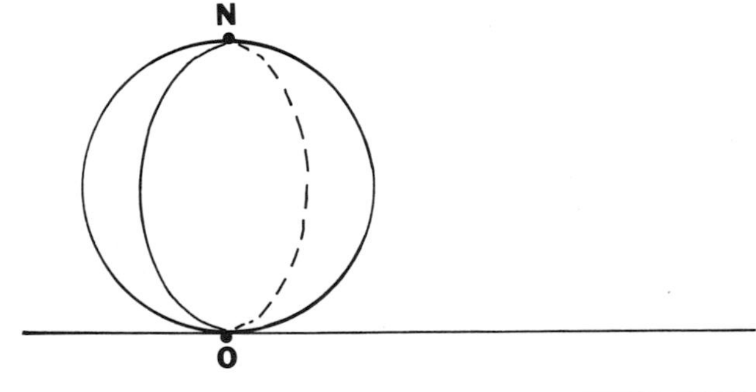

Figure 3.1.10 $\mathbb{R} \cup \{\infty\}$ *becomes a circle on the sphere.*

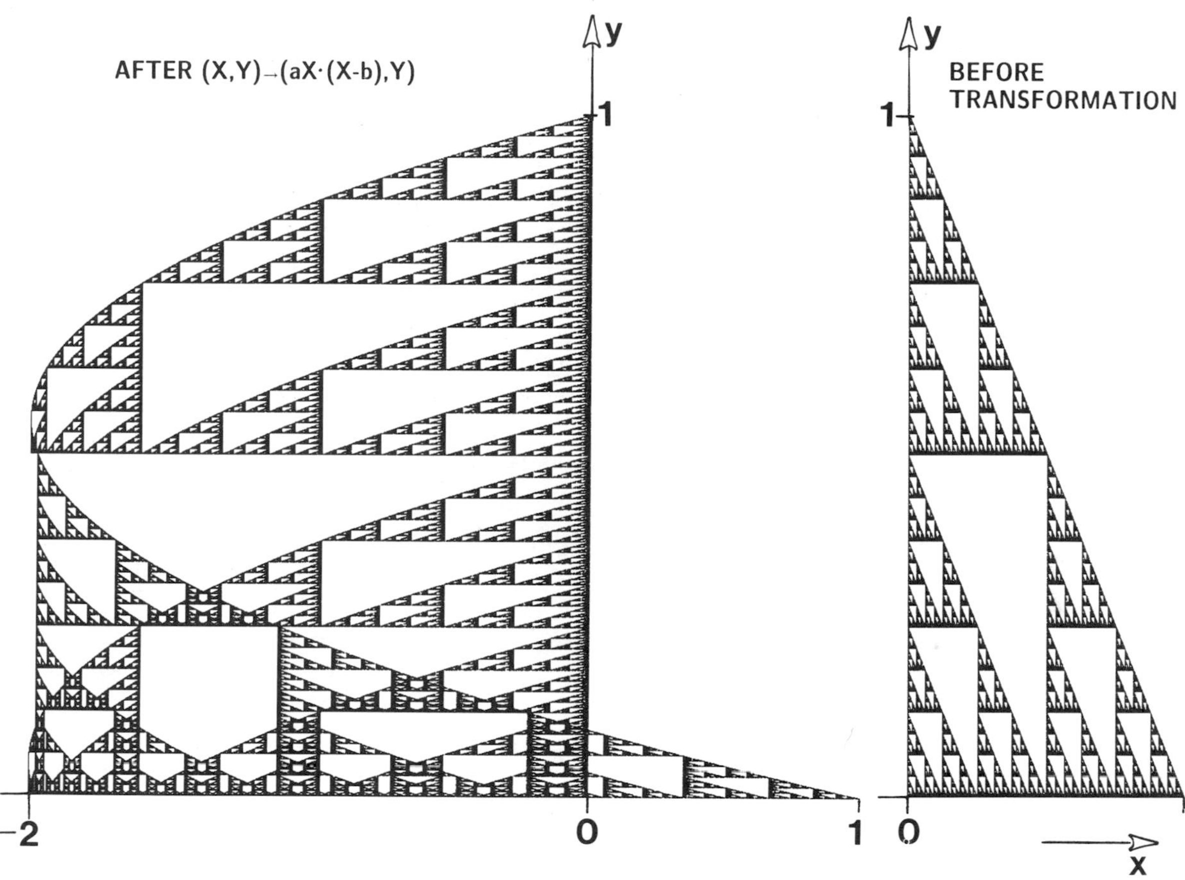

Figure 3.1.11
A Sierpinski triangle before and after the polynomial transformation $x \mapsto ax(x-b)$ is applied to the x-axis. Evaluate the real constants a and b.

We will often use the following equivalent notations

$$w(x) = w\begin{pmatrix} x_1 \\ x_2 \end{pmatrix} = \begin{pmatrix} a & b \\ c & d \end{pmatrix}\begin{pmatrix} x_1 \\ x_2 \end{pmatrix} + \begin{pmatrix} e \\ f \end{pmatrix} = Ax + t.$$

Here $A = \begin{pmatrix} a & b \\ c & d \end{pmatrix}$ is a two-dimensional 2×2 real matrix and t is the column vector $\begin{pmatrix} e \\ f \end{pmatrix}$ which we do not distinguish from the coordinate pair $(e, f) \in \mathbb{R}^2$. Such transformations have important geometrical and algebraic properties. Here and in all that follows we shall assume that the reader is familiar with matrix multiplication.

The matrix A can always be written in the form

$$\begin{pmatrix} a & b \\ c & d \end{pmatrix} = \begin{pmatrix} r_1 \cos\theta_1 & -r_2 \sin\theta_2 \\ r_1 \sin\theta_1 & r_2 \cos\theta_2 \end{pmatrix}$$

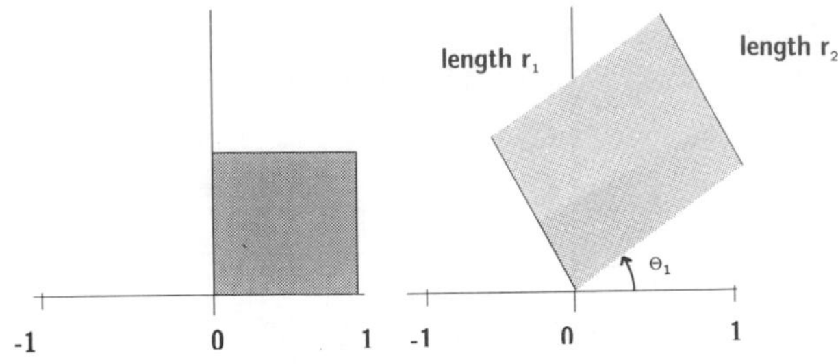

Figure 3.2.1
An affine transformation takes parallelograms into parallelograms.

where (r_1, θ_1) are the polar coordinates of the point (a, c) and $(r_2, (\theta_2 + \pi/2))$ are the polar coordinates of the point (b, d). The *linear* transformation

$$\begin{pmatrix} x_1 \\ x_2 \end{pmatrix} \to A\begin{pmatrix} x_1 \\ x_2 \end{pmatrix}$$

in $\mathbb{R}^2$ maps any parallelogram with a vertex at the origin to another parallelogram with a vertex at the origin, as illustrated in Figure 3.2.1. Notice that the parallelogram may be "turned over" by the transformation, as illustrated in Figure 3.2.2.

The general affine transformation $w(x) = Ax + t$ in $\mathbb{R}^2$ consists of a linear transformation, A, which deforms space relative to the origin, as

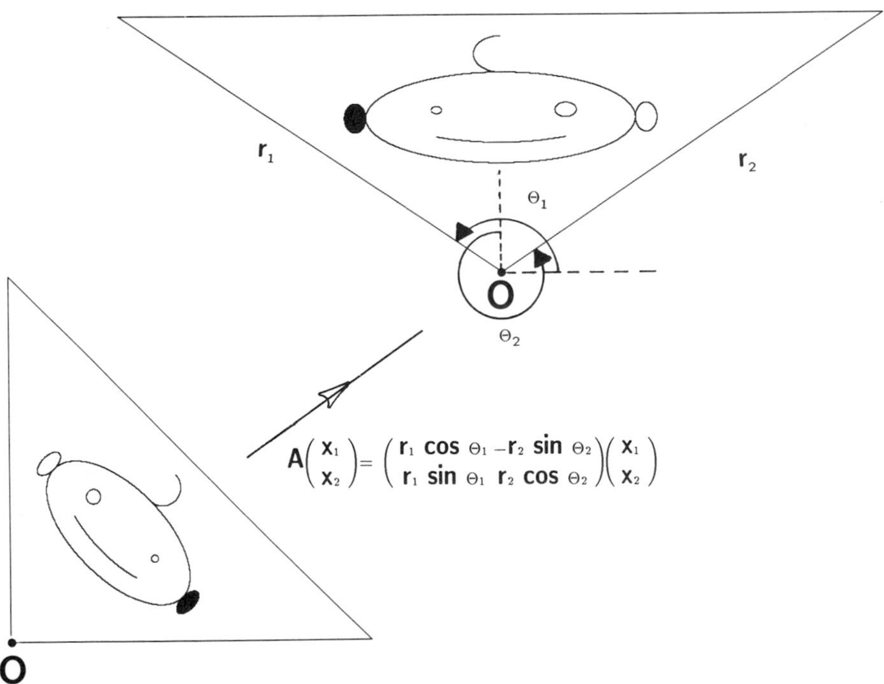

Figure 3.2.2
A linear transformation can turn pictures over.

described above, followed by a *translation* or *shift* specified by the vector t (see Figure 3.2.3).

How can one find an affine transformation which approximately transforms one given set into another given set in $\mathbb{R}^2$? Let's show how to find the affine transformation which almost takes the big leaf to the little leaf in Figure 3.2.4. This figure actually shows a photocopy of two real Ivy leaves. We wish to find the numbers a, b, c, d, e, and f defined above, so that

$$w\,(\text{BIG LEAF}) \quad \text{approximately equals} \quad \text{LITTLE LEAF}.$$

Begin by introducing x and y coordinate axes, as already shown in Figure 3.2.4. Mark three points on the big leaf (we've chosen the leaf tip, a side spike, and the point where the stem joins the leaf) and determine their coordinates (x_1, x_2), (y_1, y_2), and (z_1, z_2). Mark the corresponding points on the little leaf, assuming that a caterpillar hasn't eaten them, and determine their coordinates; say $(\tilde{x}_1, \tilde{x}_2)$, $(\tilde{y}_1, \tilde{y}_2)$, and $(\tilde{z}_1, \tilde{z}_2)$ respectively. Then a, b, and e are obtained by solving the three linear equations

$$\begin{aligned} x_1 a + x_2 b + e &= \tilde{x}_1, \\ y_1 a + y_2 b + e &= \tilde{y}_1, \\ z_1 a + z_2 b + e &= \tilde{z}_1; \end{aligned}$$

while c, d, and f satisfy

$$\begin{aligned} x_1 c + x_2 d + f &= \tilde{x}_2, \\ y_1 c + y_2 d + f &= \tilde{y}_2, \\ z_1 c + z_2 d + f &= \tilde{z}_2. \end{aligned}$$

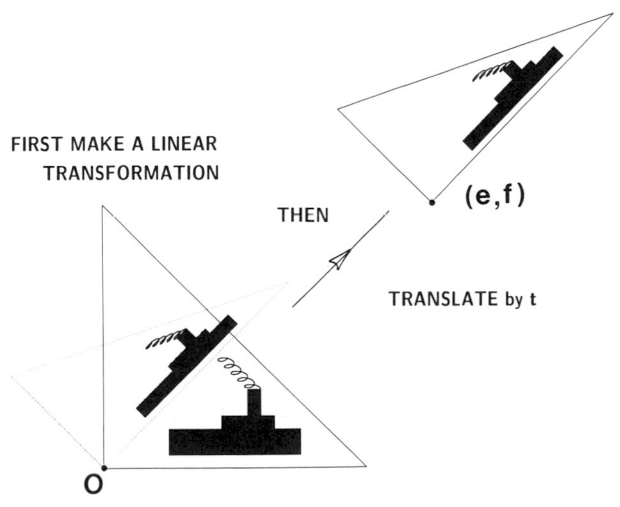

Figure 3.2.3 *An affine transformation consists of a linear transformation followed by a translation.*

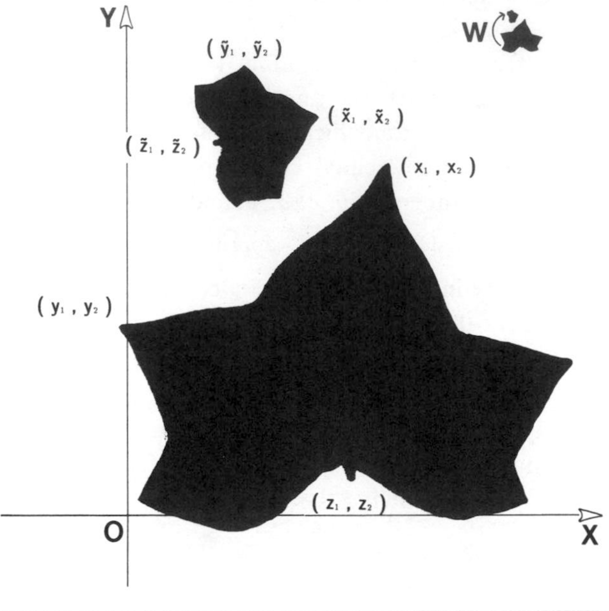

Figure 3.2.4 *Two Ivy leaves lying on the Euclidean Plane determine an affine transformation.*

Exercises & Examples

2.1. Find an affine transformation in $\mathbb{R}^2$ which takes the triangle with vertices at $(0,0)$, $(0,1)$, $(1,0)$ to the triangle with vertices at $(4,5)$, $(-1,2)$ and $(3,0)$. Show what this transformation does to a circle inscribed in the first triangle.

2.2. Show that a necessary and sufficient condition for the affine transformation

$$\begin{pmatrix} a & b \\ c & d \end{pmatrix}\begin{pmatrix} x_1 \\ x_2 \end{pmatrix} + \begin{pmatrix} e \\ f \end{pmatrix} = Ax + t$$

to be invertible is $\det A \neq 0$, where $\det A = (ad - bc)$ is the determinant of the 2×2 matrix A.

2.3. Show that if $f_1: \mathbb{R}^2 \to \mathbb{R}^2$ and $f_2: \mathbb{R}^2 \to \mathbb{R}^2$ are both affine transformations, then so is

$$f_3 = f_1 \circ f_2 .$$

If $f_i(x) = A_i x + t_i$, $i = 1, 2, 3$ where A_i is a 2×2 real matrix, express A_3 in terms of A_1 and A_2.

Definition 2. A transformation $w: \mathbb{R}^2 \to \mathbb{R}^2$ is called a *similitude* if it is an affine transformation having one of the special forms

$$w\begin{pmatrix} x_1 \\ x_2 \end{pmatrix} = \begin{pmatrix} r\cos\theta & -r\sin\theta \\ r\sin\theta & r\cos\theta \end{pmatrix}\begin{pmatrix} x_1 \\ x_2 \end{pmatrix} + \begin{pmatrix} e \\ f \end{pmatrix}$$

$$w\begin{pmatrix} x_1 \\ x_2 \end{pmatrix} = \begin{pmatrix} r\cos\theta & r\sin\theta \\ r\sin\theta & -r\cos\theta \end{pmatrix}\begin{pmatrix} x_1 \\ x_2 \end{pmatrix} + \begin{pmatrix} e \\ f \end{pmatrix}$$

for some translation $(e, f) \in \mathbb{R}^2$, some real number $r \neq 0$, and some angle θ, $0 \leq \theta < 2\pi$. θ is called the rotation angle while r is called the *scale factor* or *scaling*. The linear transformation

$$R_\theta\begin{pmatrix} x_1 \\ x_2 \end{pmatrix} = \begin{pmatrix} \cos\theta & -\sin\theta \\ \sin\theta & \cos\theta \end{pmatrix}\begin{pmatrix} x_1 \\ x_2 \end{pmatrix}$$

is a *rotation*. The linear transformation

$$R\begin{pmatrix} x_1 \\ x_2 \end{pmatrix} = \begin{pmatrix} 1 & 0 \\ 0 & -1 \end{pmatrix}\begin{pmatrix} x_1 \\ x_2 \end{pmatrix}$$

is a *reflection*.

Figure 3.2.5 shows some of the things that a similitude can do. Notice that a similitude preserves angles.

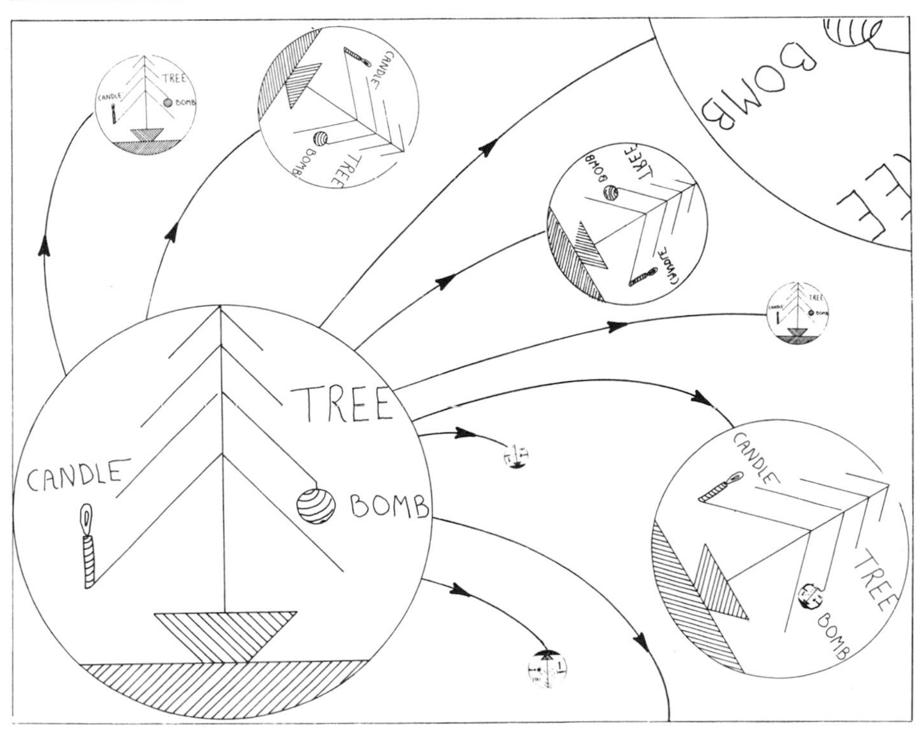

Figure 3.2.5 *Some of the things that a similitude can do.*

Exercises & Examples

2.4. Find the scaling ratios r_1, r_2 and the rotation angles θ_1, θ_2 for the affine transformation which takes the triangle $(0,0)$, $(0,1)$, $(1,0)$ onto the straight line segment from $(1,1)$ to $(2,2)$ in $\mathbb{R}^2$ in such a way that both $(0,1)$ and $(1,0)$ go to $(1,1)$.

2.5. Let S be a region in $\mathbb{R}^2$ bounded by a polygon or other 'nice' boundary. Let $w: \mathbb{R}^2 \to \mathbb{R}^2$ be an affine transformation, $w(x) = Ax + t$. Show that

$$(\text{area of } w(S)) = |det\, A| \cdot (\text{area of } S)$$

(see Figure 3.2.6). Show that det $A < 0$ has the interpretation that S is "flipped over" by the transformation. (Hint: Suppose first that S is a triangle.)

2.6. Show that if $w: \mathbb{R}^2 \to \mathbb{R}^2$ is a similitude, $w(x) = Ax + t$, where t is the translation and A is a 2×2 matrix, then A can always be written either $A = rR_\theta$ or $A = rRR_\theta$.

2.7. View the railway tracks image in Figure 3.2.7 as a subset S of $\mathbb{R}^2$. Find a similitude $w: \mathbb{R}^2 \to \mathbb{R}^2$ such that $w(S) \subset S$, $w(S) \neq S$.

2.8. We use the notation introduced in Definition 2.2.2. Find a nonzero real number r, an angle θ, and a translation vector t such that the similitude $wx = rR_\theta x + t$ on $\mathbb{R}^2$ obeys

$$w(\triangle) \subset \triangle, \text{ with } w(\triangle) \neq \triangle$$

where $\triangle$ denotes a Sierpinski Triangle with vertices at $(0,0)$, $(1,0)$, and $(\frac{1}{2}, 1)$.

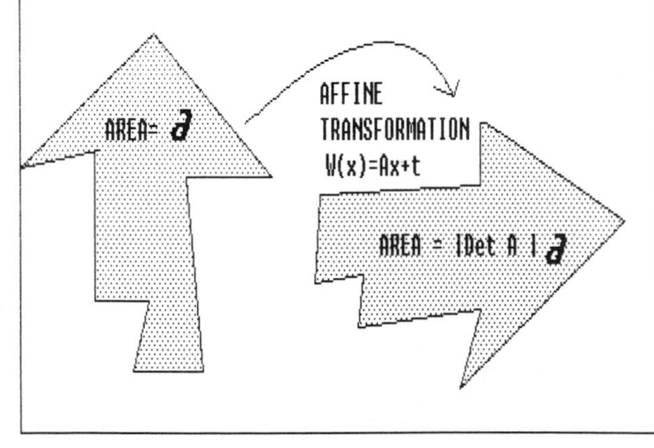

Figure 3.2.6 *The scaling factor by which an affine transformation changes area is determined by the determinant of its linear part.*

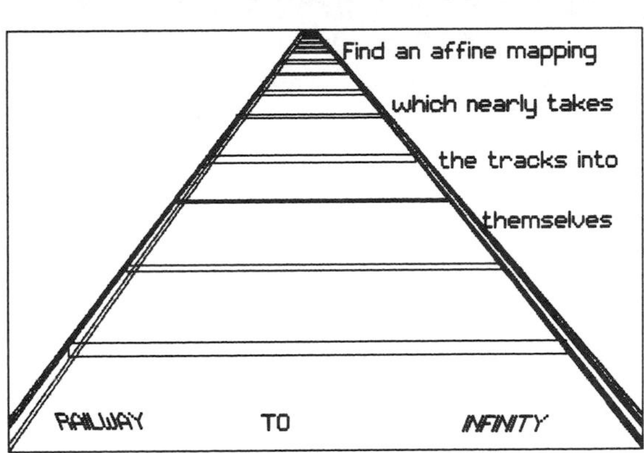

Figure 3.2.7 *Railway to Infinity. Can you find an affine transformation which nearly maps the tracks into themselves?*

2.9. Show that if $w\colon \mathbb{R}^2 \to \mathbb{R}^2$ is affine, $w(x) = Ax + t$, then it can be reexpressed

$$w(x) = \begin{pmatrix} r_1 & 0 \\ 0 & r_2 \end{pmatrix} R_\theta \begin{pmatrix} r_3 & 0 \\ 0 & r_4 \end{pmatrix} \begin{pmatrix} x_1 \\ x_2 \end{pmatrix} + t$$

where $r_i \in \mathbb{R}$ and $0 \le \theta < 2\pi$. We call a transformation of the form

$$w\begin{pmatrix} x_1 \\ x_2 \end{pmatrix} = \begin{pmatrix} r_1 & 0 \\ 0 & r_2 \end{pmatrix} \begin{pmatrix} x_1 \\ x_2 \end{pmatrix}$$

a coordinate rescaling.

2.10. Let S denote the two-dimensional orchard subset of $\mathbb{R}^2$ shown in Figure 3.2.8. Find two fundamentally different affine transformations which map S into S but not onto S. Define the transformations by specifying how they act on three points.

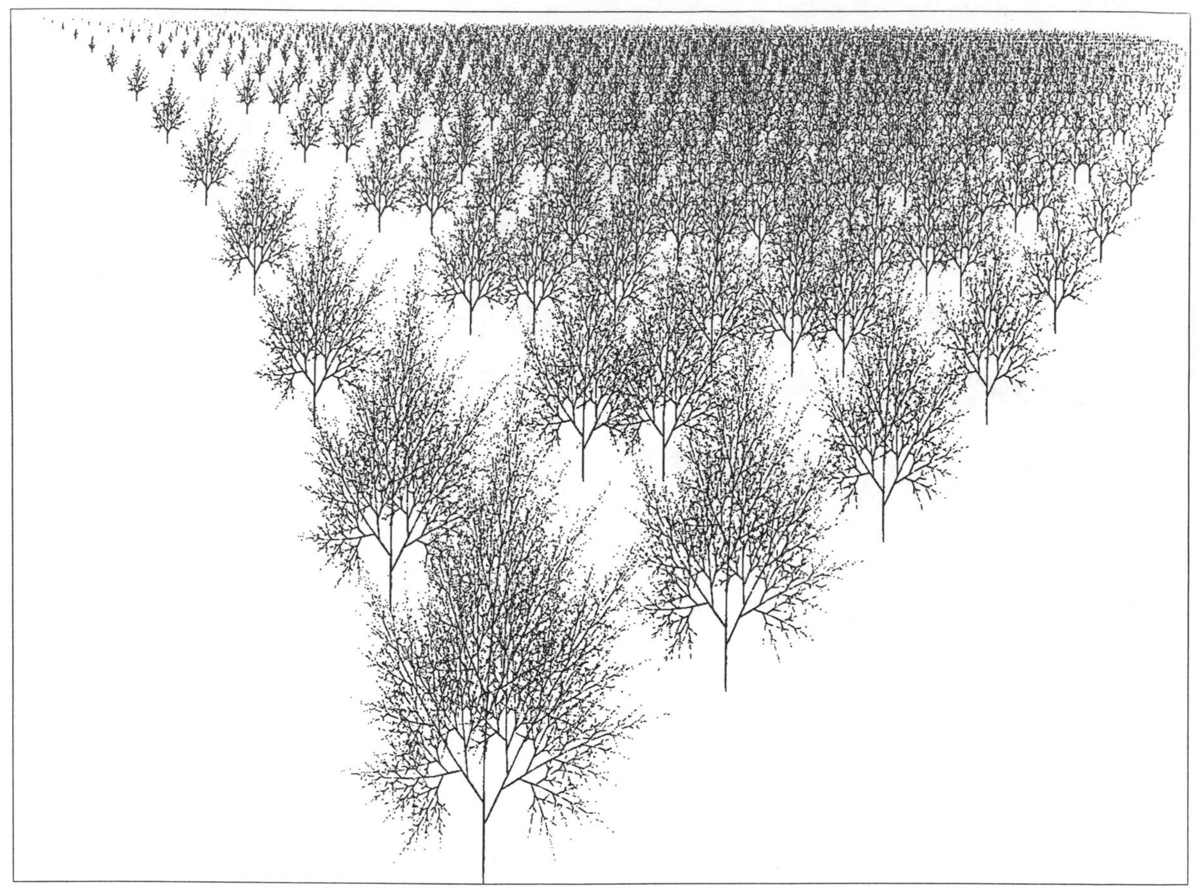

Figure 3.2.8
Orchard Subset of $\mathbb{R}^2$. *Can you find some interesting affine transformations which map this set into itself?*

2.11. Let $w(x) = Ax$ denote a linear transformation in the metric space $(\mathbb{R}^2, D)$ where

$$A = \begin{pmatrix} a & b \\ c & d \end{pmatrix}.$$

Define the *norm* of a point $x \in \mathbb{R}^2$ to be $|x| = D(x, O)$ where O denotes the origin. Define the norm of the linear transformation A by

$$|A| = \text{Max}\left\{\frac{|Ax|}{|x|} : x \in \mathbb{R}^2, x \neq 0\right\}$$

when this maximum exists. Show that $|A|$ is defined when D is the Euclidean metric and when it is the Manhattan metric. Find an expression for $|A|$ in terms of a, b, c, and d in each case. Make a geometrical interpretation of the $|A|$. Show that when $|A|$ exists we have

$$|Ax| \leq |A| \cdot |x| \qquad \text{for all } x \in \mathbb{R}^2.$$

3.3 MÖBIUS TRANSFORMATIONS ON THE RIEMANN SPHERE

Definition 6. A transformation $f\colon \hat{\mathbb{C}} \to \hat{\mathbb{C}}$ defined by

$$f(z) = \frac{(az + b)}{(cz + d)}$$

where a, b, c, and $d \in \mathbb{C}$, $ad - bc \neq 0$, is called a *Möbius transformation* on $\hat{\mathbb{C}}$. If $c \neq 0$ then $f(-d/c) = \infty$, and $f(\infty) = a/c$. If $c = 0$ then $f(\infty) = \infty$.

As shown by the following exercises and examples, one can think of a Möbius transformation as follows. Map the whole plane $\mathbb{C}$ together with the point at infinity, onto the sphere $\hat{\mathbb{C}}$, as described in Chapter 2. A sequence of operations is then applied to the sphere. Each operation is elementary and has the property that it takes circles to circles. The possible operations are: rotation about an axis, rescaling (uniformly expand or contract the sphere), and translation (the whole sphere is picked up and moved to a new place on the plane, without rotation). Finally, the sphere is mapped back onto the plane in the usual way. Since the mappings back and forth from the plane to the sphere take straight lines and circles in the plane to circles on the sphere, we see that a Möbius transformation transforms the set of straight lines and circles in the plane onto itself. We also see that a Möbius transformation is invertible. It is wonderful how the quite complicated geometry of Möbius transformations is handled by straightforward complex algebra, where we simply manipulate expressions of the form $(az + b)/(cz + d)$.

Exercises & Examples

3.1. Show that the most general Möbius transformation which maps ∞ to ∞ is of the form $f(z) = az + b$, $a, b \in \mathbb{C}$, $a \neq 0$, and that this is a similitude. Show that any two-dimensional similitude which does not involve a reflection can be written in this form. That is, disregarding changes in notation,

$$\begin{aligned} f(z) &= f(x_1 + ix_2) = (a_1 + ia_2)(x_1 + ix_2) + (x_1 + ib_2) \\ &= re^{i\theta}(x_1 + ix_2) + (b_1 + ib_2), \qquad (i = \sqrt{-1}) \\ &= \begin{pmatrix} r\cos\theta & -r\sin\theta \\ r\sin\theta & r\cos\theta \end{pmatrix}\begin{pmatrix} x_1 \\ x_2 \end{pmatrix} + \begin{pmatrix} b_1 \\ b_2 \end{pmatrix}. \end{aligned}$$

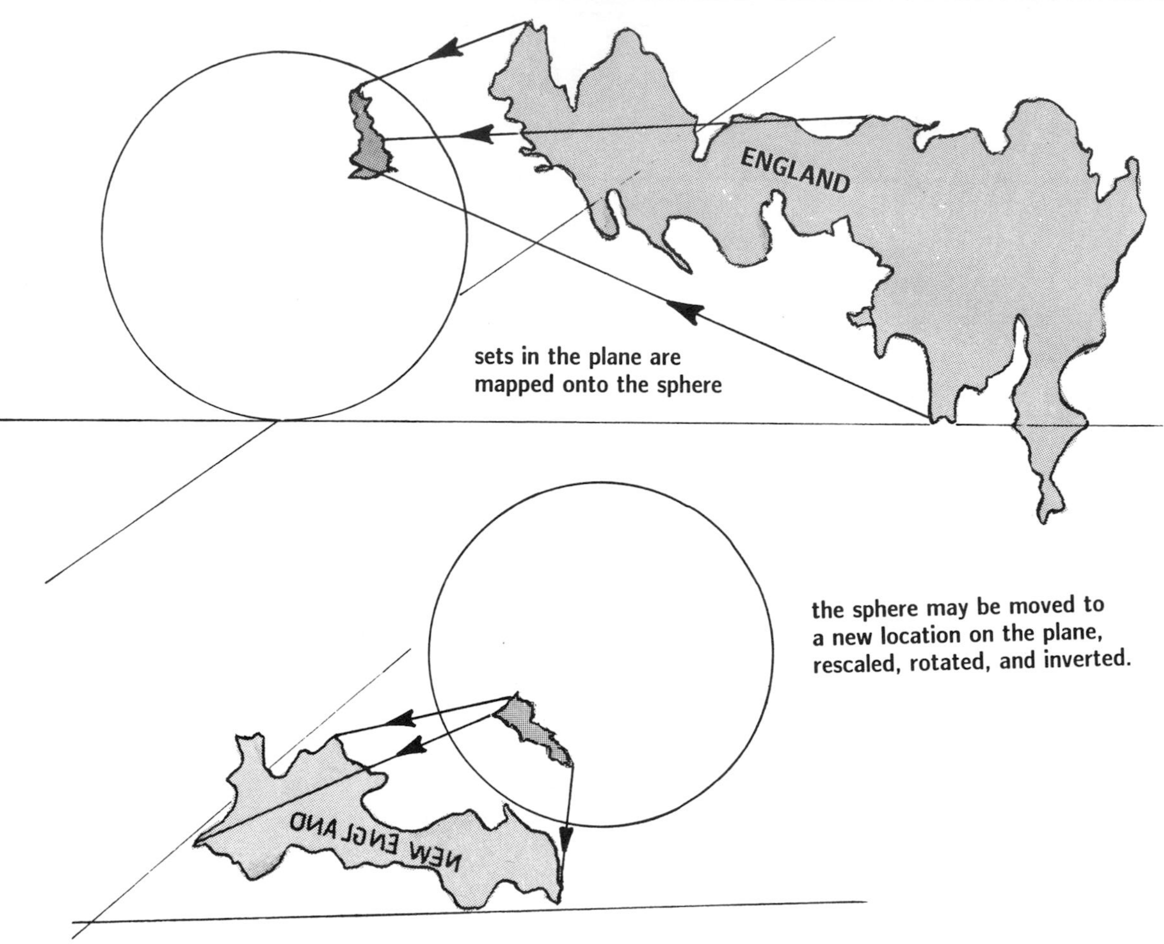

ENGLAND
sets in the plane are
mapped onto the sphere
the sphere may be moved to
a new location on the plane,
rescaled, rotated, and inverted.

Figure 3.3.1
A Möbius transformation acting on England to produce a new country.

Find r and θ in terms of a_1 and a_2. Show that the transformation can be achieved as illustrated in Figure 3.3.2.

3.2. Show that the Möbius transformation $f(z) = 1/z$ corresponds to first mapping the plane to the sphere, in such a way that the unit circle $\{z \in \mathbb{C}\colon |z| = 1\}$ goes to the equator, followed by an inversion of the sphere (turn it upside down by rotating about an axis through $+1$ and -1 on the equator), and finally mapping back to the plane.

3.3. Show that any Möbius transformation which is not a similitude may be written

$$f(z) = e + \frac{f}{z+g} \qquad \text{for some } e, f, g \in \mathbb{C},\ f \neq 0.$$

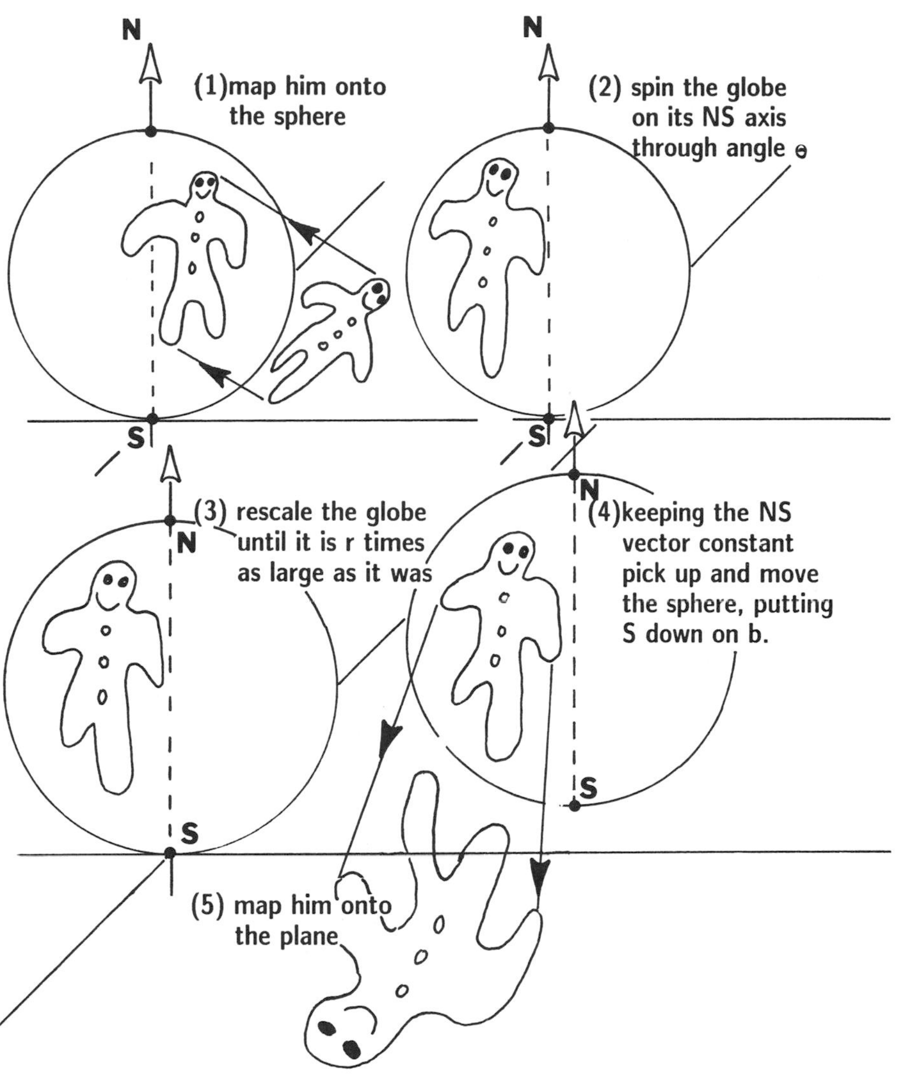

Figure 3.3.2 *The mechanism of the similitude $f(z) = re^{i\theta}z + b$ in terms of the sphere.*

3.4. Sketch what happens to the picture in Figure 3.3.3 under the Möbius transformation $f(z) = 1/z$.

3.5. What happens to Figure 3.3.3 under the Möbius transformation $f(z) = 1 + iz$.

3.6. Show algebraically that a Möbius transformation $f\colon \hat{\mathbb{C}} \to \hat{\mathbb{C}}$ is always invertible.

3.7. Show that if f_1 and f_2 are Möbius transformations then $f_1 \circ f_2$ is a Möbius transformation.

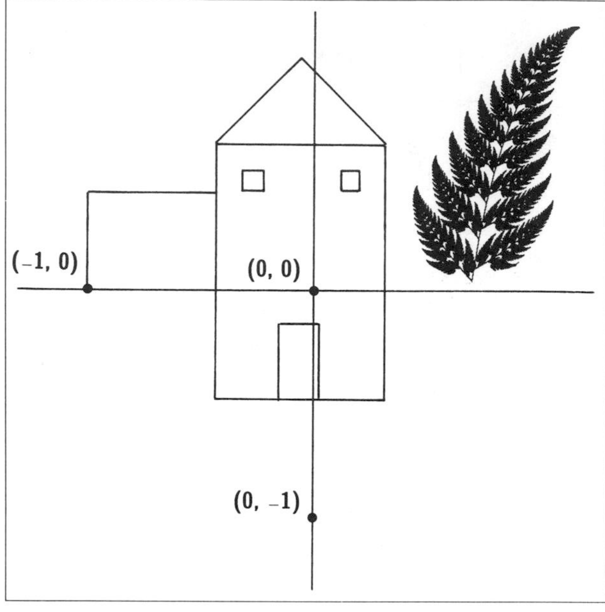

Figure 3.3.3 *Up the Garden Path. What does the Möbius transformation $z \mapsto 1 + iz$ do to this picture?*

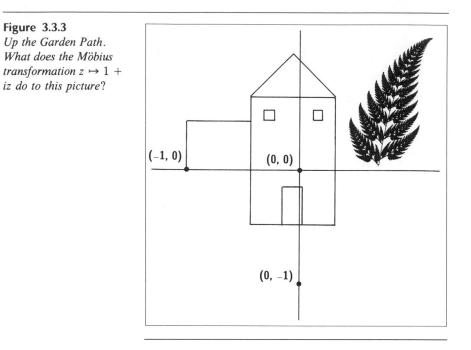

3.8. Find a Möbius transformation which takes the real line to the unit circle centered at the origin.

3.9. Evaluate $f^{\circ n}(z)$ if $f(z) = 1/(1+z)$, $n \in \{-2, -1, 0, 1, 2, 3, \ldots\}$.

3.10. Interpret the Möbius transformation $f(z) = i + 1/(z - i)$ in terms of operations on the sphere.

3.4 ANALYTIC TRANSFORMATIONS

In this section we continue the discussion of transformations on the metric spaces ($\mathbb{C}$, Euclidean) and ($\hat{\mathbb{C}}$, Spherical). We introduce a generalization of the Möbius transformations, called analytic transformations. We concentrate on the behaviour of quadratic transformations. It is recommended that, during a first reading or first course, the reader obtains a good mental picture of how the quadratic transformation acts on the sphere. The reader may then want to study this section more closely after reading about Julia Sets in Chapter 7.

The similitude $f\colon \hat{\mathbb{C}} \to \hat{\mathbb{C}}$, defined by the formula $f(z) = 3z + 1$ is an example of an analytic transformation. It maps circles to circles magnified by a factor three. A disk with center at z_0 is taken to a disk with center at $f(z_0) = 3z_0 + 1$. The transformation is continuous, and it maps open sets to open sets. Nowhere does it "fold back along the dotted line."

The similitude $f\colon \hat{\mathbb{C}} \to \hat{\mathbb{C}}$ defined by $f(z) = (3 + 3i)z + (1 - 2i)$ is similarly described. The circles and disks are now rotated by 45° in addition to being magnified and translated.

Loosely a transformation on $\hat{\mathbb{C}}$ is analytic if it is continuous and locally it "behaves like" a similitude. If you take a very small region indeed (How small? Small enough! There is a smallness such that what is about to be said is true!) and you watch what the transformation does to that tiny region, you will typically find that it is magnified or shrunk, rotated, and translated, in almost exactly the same manner that some similitude would do the job. The similitude will always be of the special type discussed in example 3.1 above.

We make this description more precise. Let us decide to look at what our transformation does in the vicinity of a point $z_0 \in \hat{\mathbb{C}}$. Assume that z_0 is not a critical point, defined below. Let T denote a tiny region, a disk for example, which contains the point z_0. Let $f(T)$ be its image under the transformation. Then one can rescale T by a factor which makes it roughly the size of the unit square, and one can rescale $f(T)$ by the same factor. The assertion of the previous paragraph is that the action of the transformation, viewed as taking T, rescaled, onto $f(T)$, rescaled, can be described more and more accurately by a similitude. If you like, one could consider a picture P drawn in T and examine the transformed image $f(P)$. If P and $f(P)$ are rescaled by the same factor so that P is the size of the unit square, then $f(P)$ looks more and more like a similitude applied to P. This description becomes more and more precise the tinier the region under discussion.

Consider the quadratic transformation $f\colon \hat{\mathbb{C}} \to \hat{\mathbb{C}}$ defined by

$$f(z) = z^2 = (x_1 + ix_2)^2 = (x_1^2 - x_2^2) + 2x_1x_2i = f_1(x_1, x_2) + f_2(x_1, x_2)i,$$

where $f_1(x_1, x_2) = (x_1^2 - x_2^2)$, is called the real part of $f(z)$, and $f_2(x_1, x_2) = 2x_1x_2$ is called the imaginary real part of f. Pictures of what this transformation does to some Sierpinski triangles in $\mathbb{C}$ appear in Figure 3.4.1.

Two features are to be noticed. (I) Provided that we stay away from the Origin, the transformation behaves locally like a similitude: for points z close to z_0, $f(z)$ is approximated by the similitudc

$$w(z) = az + b \qquad \text{where } a = 2z_0 \text{ and } b = -z_0^2.$$

This fact shows up in Figure 3.4.1: Upon close examination (we suggest the use of a magnifying glass) of the transformed Sierpinski triangles, one sees that they are built up out of small triangles whose shapes are only slightly different from that of their preimages. The only place where this is not true is at the forward image of the origin, which is a critical point. (II) The transformation maps the space twice around the origin.

One can track analytically what happens to the point

$$z = R\cos t + iR\sin t$$

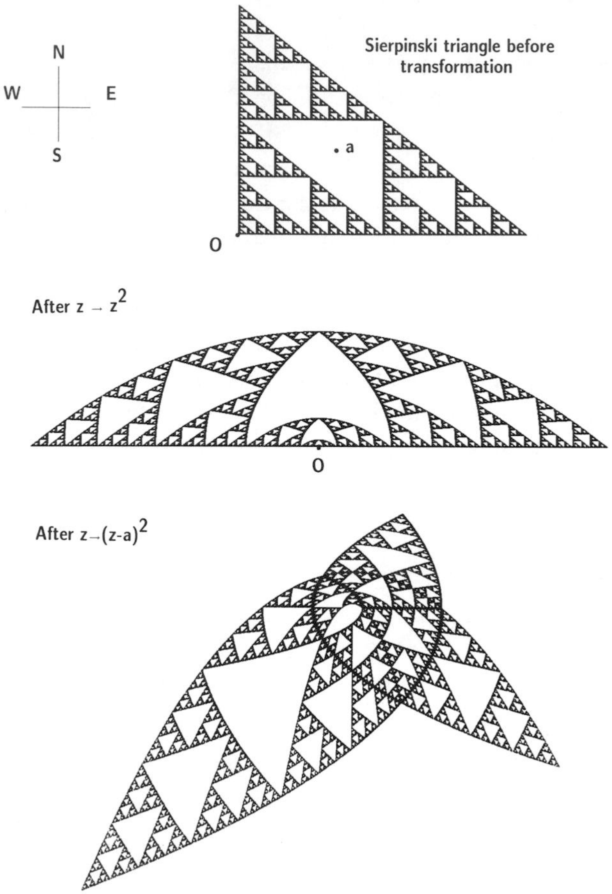

Figure 3.4.1 *Quadratic transformations are described by showing how they act on a Sierpinski triangle. Use a magnifying glass to check that the transformations behave locally like similitudes.*

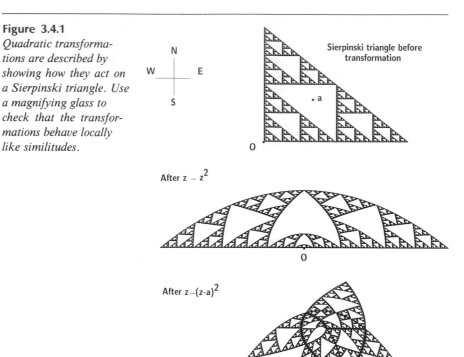

where $R > 0$. As the time parameter t goes from zero to 2π, z moves anticlockwise once around the circle of radius R. The transformed point $f(z)$ is given by

$$f(z) = R^2 \cos 2t + iR^2 \sin 2t.$$

As the time parameter t goes from 0 to 2π, $f(z)$ goes twice around the circle of radius R^2.

On the Riemann sphere the transformation $z \mapsto z^2$ can be described as follows. Let us say that the Equator corresponds to the circle of unit radius in the plane, that the South Pole corresponds to the Origin, and that the North Pole corresponds to the Point at Infinity. Then the transformation leaves both Poles fixed. The Line of Longitude L connecting the Poles, which corresponds to the positive real axis, is mapped into itself, and the Equator is mapped into itself. Here is what we must picture. First, points which lie above the Equator are moved closer to the North Pole, points which lie below the Equator are moved closer to the South Pole, and the Equator is not shifted. Second, the skin of the sphere is cut along the Line of Longitude L. One side of the cut is held fixed while the other side is pulled around the sphere (following the terminator when the Sun is high above the Equator), uniformly stretching the space, until the edge of the cut is back over L. The two lips of the cut are rejoined. The sphere has been mapped twice over itself. The Poles are the critical points of the transformation; they are the points about which wrapping occurs. This description is illustrated in Figure 3.4.2.

The most general quadratic transformation on the sphere is expressible by a formula of the form $f(z) = Az^2 + Bz + C$ where A, B, and C are complex numbers. One can show there is a change of coordinates, $z \mapsto \theta(z)$, where θ is a similitude, such that $f(z)$ becomes expressible in the special form $f(z) =$

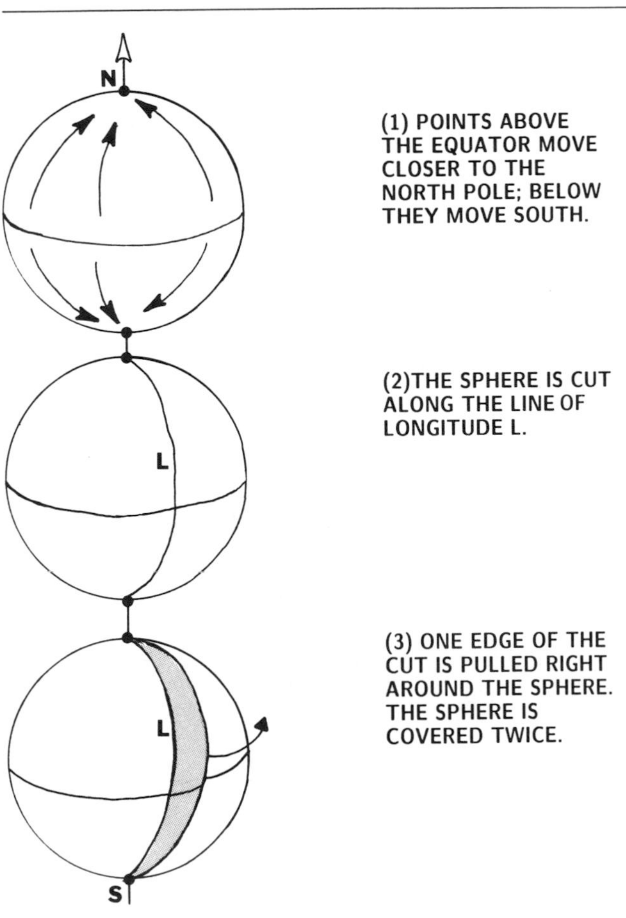

N
(1) POINTS ABOVE THE EQUATOR MOVE CLOSER TO THE NORTH POLE; BELOW THEY MOVE SOUTH.
L
(2)THE SPHERE IS CUT ALONG THE LINE OF LONGITUDE L.
L
(3) ONE EDGE OF THE CUT IS PULLED RIGHT AROUND THE SPHERE. THE SPHERE IS COVERED TWICE.
S

Figure 3.4.2 *The action of the quadratic transformation* $z \mapsto z^2$ *in terms of the sphere.*

$z^2 + \tilde{C}$ for some complex number $\tilde{C}$; see Exercise 3.5 (5.18). Hence the description of the most general quadratic transformation on the sphere can be made in the same terms as above, except that at the end there is a translation by some constant amount $\tilde{C}$. This translation leaves the Point at Infinity fixed.

The quadratic transformation $f(z) = z^2$ maps the twice punctured plane $\mathbb{C}$ onto itself twice. Each point $z \in \mathbb{C}$ has two preimages. Hence $f\colon \hat{\mathbb{C}} \to \hat{\mathbb{C}}$ is not an invertible transformation. In such situations we can define a set-valued inverse function.

Definition 1. Let $f\colon \hat{\mathbb{C}} \to \hat{\mathbb{C}}$ be an analytic transformation such that $f(\hat{\mathbb{C}}) = \hat{\mathbb{C}}$. Then the *set-valued inverse* of f is the mapping $f^{-1}\colon \mathscr{H}(\hat{\mathbb{C}}) \to \mathscr{H}(\hat{\mathbb{C}})$ defined by

$$f^{-1}(A) = \{w \in \hat{\mathbb{C}}: f(w) \in A\} \qquad \text{for all } A \in \mathscr{H}(\hat{\mathbb{C}}).$$

In Figure 3.4.3 we illustrate the transformation f^{-1} acting on the Space of Fractals, in the case of the quadratic transformation $f(z) = z^2$.

One can obtain explicit formulas for $f^{-1}(z)$ when f is a quadratic transformation. For example, for $f(z) = z^2$, $f^{-1}(0) = O$, $f^{-1}(\infty) = \infty$, and $f^{-1}(z) = \{w_1(z), w_2(z)\}$ for $z \in \hat{\mathbb{C}} \setminus \{0, \infty\}$. Here $w_1(x_1 + ix_2) = a(x_1, x_2) + ib(x_1, x_2)$, and $w_2(x_1, x_2) = -a(x_1, x_2) - ib(x_1, x_2)$, where

$$a(x_1, x_2) = \sqrt{\frac{\sqrt{x_1^2 + x_2^2} + x_1}{2}} \qquad \text{when } x_2 \geq 0,$$

$$a(x_1, x_2) = -\sqrt{\frac{\sqrt{x_1^2 + x_2^2} + x_1}{2}} \qquad \text{when } x_2 < 0,$$

$$b(x_1, x_2) = \sqrt{\frac{\sqrt{x_1^2 + x_2^2} - x_1}{2}}.$$

Each of the two functions $w_1(z)$ and $w_2(z)$ is itself analytic on $\mathbb{C} \setminus [0, \infty]$.

The following definition formalizes what is meant by an analytic transformation on the complex plane. We recommend further reading, for example [Rudi 1966].

Definition 2. Let $(\mathbb{C}, d)$ denote the complex plane with the Euclidean metric. A transformation $f\colon \mathbb{C} \to \mathbb{C}$ is called *analytic* if for each $z_0 \in \mathbb{C}$ there is a similitude of the form

$$w(z) = az + b, \qquad \text{for some pair of numbers } a, b \in C$$

such that $d(f(z), w(z))/d(z, z_0) \to 0$ as $z \to z_0$. The numbers a and b depend on z_0. If, corresponding to a certain point $z_0 = c$, we have $a = 0$, then c is called a *critical point* of the transformation; and $f(c)$ is called a *critical value*.

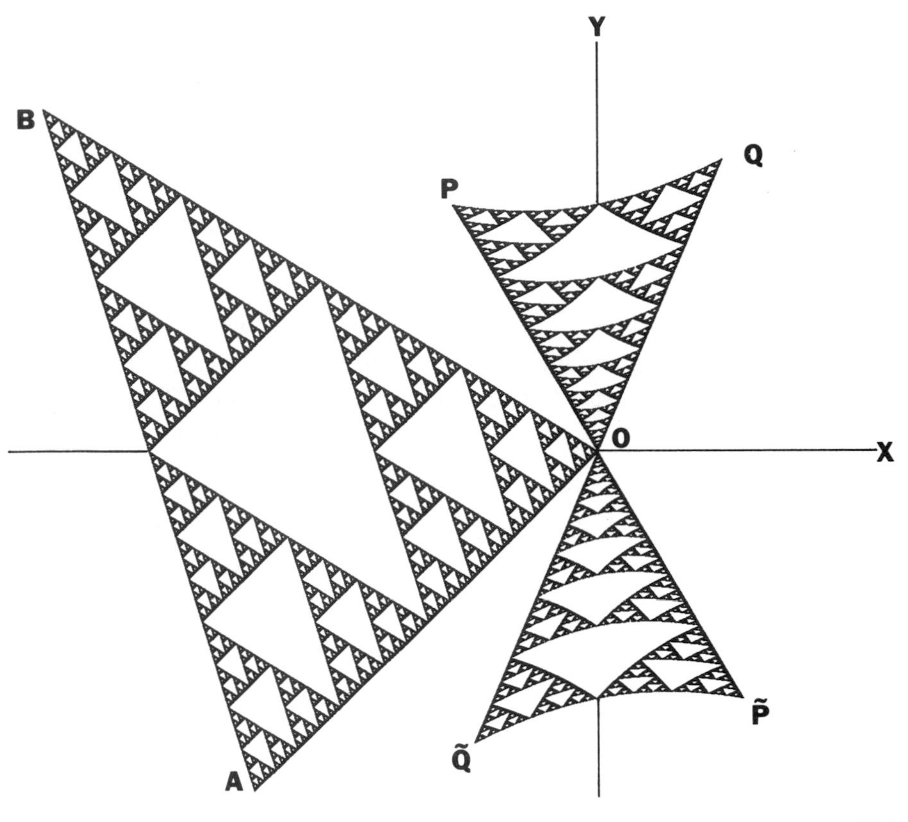

Figure 3.4.3
The set valued inverse, f^{-1}, of the quadratic transformation $f(z) = z^2$, maps the Sierpinski triangle AOB into the $POQ \cup \tilde{P}O\tilde{Q}$. More generally f^{-1} maps the Space of Fractals into itself. Look carefully at this image! There are several important features of analytic transformations illustrated here.

If the analytic transformation $f(z)$ is a rational transformation, which means that it is expressible as a ratio of two polynomials in z, such as

$$\text{(i)} \quad f(z) = 1 + 2i + 27z^2 - 9z^3,$$

$$\text{(ii)} \quad f(z) = \frac{1+z}{1-z},$$

$$\text{(iii)} \quad f(z) = \frac{1+z+z^2}{1-z+z^2};$$

then the numbers a and b in the similitude $w(z)$ of Definition 1 are given by the formulas

$$a = f'(z_0) \text{ and } b = f(z_0) - az_0.$$

The derivative $f'(z)$ of the rational function $f(z)$ can be calculated by treating z as though it were the real variable x and applying the standard differentiation rules of calculus. The critical points $c \in \mathbb{C}$ are the solutions of the equation $f'(c) = 0$.

For example, close enough to any point $z_0 \in \mathbb{C}$ such that $f'(z_0) \neq 0$, the cubic transformation (i) is well described by the similitude

$$w(z) = \left(54z_0 - 27z_0^2\right)z + \left(1 + 2i - 27z_0^2 + 18z_0^3\right)$$

The finite critical points associated with (i) may be obtained by solving

$$54c - 27c^2 = 0$$

and are accordingly $c = 0 + i0$ and $c = 2 + i0$. By making the change of coordinates $z' = 1/z$ (see section 3.5), one can also analyse the behaviour near the Point at Infinity. It turns out that $c = \infty$ is always a critical point for a polynomial transformation $f(z)$ on $\hat{\mathbb{C}}$. The space is "wrapped" an integral number of times about the image of a critical point. For example the cubic transformation (i) wraps space twice about each of the points $f(0 + i0) = 1 + 2i$, and $f(2 + i0) = 37 + 2i$, and it wraps it three times about $f(\infty) = \infty$.

Exercises & Examples

4.1. Sketch a globe representing $\hat{\mathbb{C}}$, including a subset which looks like Africa, and show what happens to the subset under the quadratic transformation $f(z) = z^2$.

4.2. Verify the following explicit formulas for $f^{-1}(z)$, corresponding to $f(z) = z^2 - 1$: $f^{-1}(-1) = O$; $f^{-1}(\infty) = \infty$; and $f^{-1}(z) = \{w_1(z), w_2(z)\}$ for $z \in \hat{\mathbb{C}} \setminus \{-1, \infty\}$, where $w_1(x_1 + ix_2) = a(x_1, x_2) + ib(x_1, x_2)$, and $w_2(x_1, x_2) = -a(x_1, x_2) - ib(x_1, x_2)$. Here

$$a(x_1, x_2) = \sqrt{\frac{\sqrt{(1 + x_1)^2 + x_2^2} + 1 + x_1}{2}} \qquad \text{when } x_2 \geq 0,$$

$$a(x_1, x_2) = -\sqrt{\frac{\sqrt{(1 + x_1)^2 + x_2^2} + 1 + x_1}{2}} \qquad \text{when } x_2 < 0,$$

and

$$b(x_1, x_2) = \sqrt{\frac{\sqrt{(1 + x_1)^2 + x_2^2} - 1 - x_1}{2}}.$$

We remark that both $w_1(z)$ and $w_2(z)$ are analytic on $\mathbb{C} \setminus [-1, \infty)$.

4.3. Locate the critical points and critical values of the quadratic transformation $f(z) = z^2 + 1$.

4.4. Draw a side view of a man with an arm stretched out in front of him, holding a knife. The blade should point down. Choose the origin of coordinates to be his naval. Draw another picture to explain how *hara-kiri* may be achieved by applying the inverse of the quadratic transformation $f(z) = z^2$ to your image.

4.5. Find a similitude which approximates the behaviour of the given analytic transformation in the vicinity of the given point: (a) $f(z) = z^2$ near $z_0 = 1$; (b) $f(z) = 1/z$ near $z_0 = 1 + i$; (c) $f(z) = (z - 1)^3$ near $z_0 = 1 - i$.

3.5 HOW TO CHANGE COORDINATES

In describing transformations on spaces we usually make use of an underlying coordinate system. Most spaces have a coordinate system by means of which the points in the space are located. This underlying coordinate system is implied by the specification of the space. For example, $X = [1, 2]$ provides a collection of points together with the natural coordinate x restricted by $1 \leq x \leq 2$. We can think of the space, made of points $x \in X$, or equivalently we can think of the system of coordinates. If the space X is $\mathbb{R}^2$ or $\mathbb{C}$ then the underlying coordinate system may be Cartesian coordinates. If $X = \hat{\mathbb{C}}$ then the coordinate system may be angular coordinates on the sphere.

In each case the underlying coordinate system is itself a subset of a metric space. We denote this metric space by X_C. Usually we do not consciously distinguish between a point $x \in X$ and its coordinate $x \in X_C$. Notice however that the space X_C may contain points (coordinates) which do not correspond to any point in the space X. For example, in the case of the space $X = \blacksquare$, it is natural to take $X_C = \mathbb{R}^2$; then points $x \in X$ in the space correspond to coordinates $x = (x_1, x_2) \in X_C$ restricted by $0 \leq x_1 \leq 1$ and $0 \leq x_2 \leq 1$. However the coordinates $(3, 5) \in X_C$ do not correspond to a point in X. We would like the reader to think of the space itself as "lying above" its coordinate system, as suggested in Figure 3.5.1.

A change of coordinate system may be described by a transformation θ: $X_C \to X_C$. We can think of a change of coordinates being effected by physically moving each point $x \in X$ so that it no longer lies above $x \in X_C$ but instead above the coordinate $x' = \theta(x) \in X_C$. Thus we must now distinguish between a point x lying in the space, X, and its coordinate $x \in X_C$. Then we want to think of the change of coordinates θ: $X_C \to X_C$ as moving X relative to the underlying coordinate space X_C, as illustrated in Figure 3.5.2.

Example

5.1. Let $X = [1, 2]$ and take X_C to be $\mathbb{R}$. Let θ: $\mathbb{R} \to \mathbb{R}$ be defined by $\theta(x) = 2x + 1$. Then the coordinate of the point $x = 1.5$ becomes changed to 4. We want to think of the space X as being moved relative to the coordinate space X_C, which is held fixed, as illustrated in Figure 3.5.3.

Let θ: $X_C \to X_C$ denote a change of coordinates. In order that the new coordinate system be useful it is usually necessary that θ, treated as a transformation from X to $\theta(X)$, be one-to-one and onto, and hence invertible. Let f: $X \to X$ be a transformation on a metric space X. We want to consider how the transformation f should be expressed after the change of coordinates. Let x denote simultaneously a point in X and the coordinates of that point. Let $f(x)$ denote simultaneously the point to

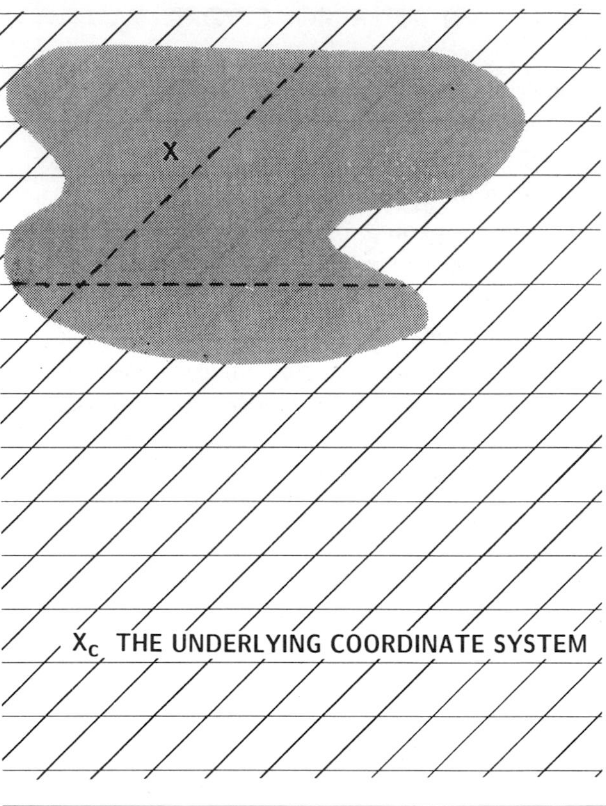

Figure 3.5.1 *The underlying coordinate system $\mathbf{X}_C$ for the space $\mathbf{X}$.*

which x is transformed by f, and the coordinates of that point. Let x' denote the point $x \in \mathbf{X}$ in the new coordinate system. That is, $x' = \theta(x) \in \mathbf{X}_C$ denotes the new coordinates of the point x. Let $f'(x')$ denote the same transformation $f\colon \mathbf{X} \to \mathbf{X}$, but expressed in the new coordinate system. Then the relation between the two coordinate systems is expressed by the commutative diagram in Figure 3.5.4, and is illustrated in Figure 3.5.5.

Theorem 1. *Let $\mathbf{X}$ be a space and let $\mathbf{X}_C \supset \mathbf{X}$ be a coordinate space for $\mathbf{X}$. Let a change of coordinates be provided by a transformation $\theta\colon \mathbf{X}_C \to \mathbf{X}_C$. Let θ be invertible when treated as a transformation from $\mathbf{X}$ to $\theta(\mathbf{X})$. Let the coordinates of a point $x \in \mathbf{X}$ be denoted by x before the change of coordinates, and by x' after the change of coordinates, so that*

$$x' = \theta(x).$$

Let $f\colon \mathbf{X} \to \mathbf{X}$ be a transformation on the space $\mathbf{X}$. Let $x \mapsto f(x)$ be the formula

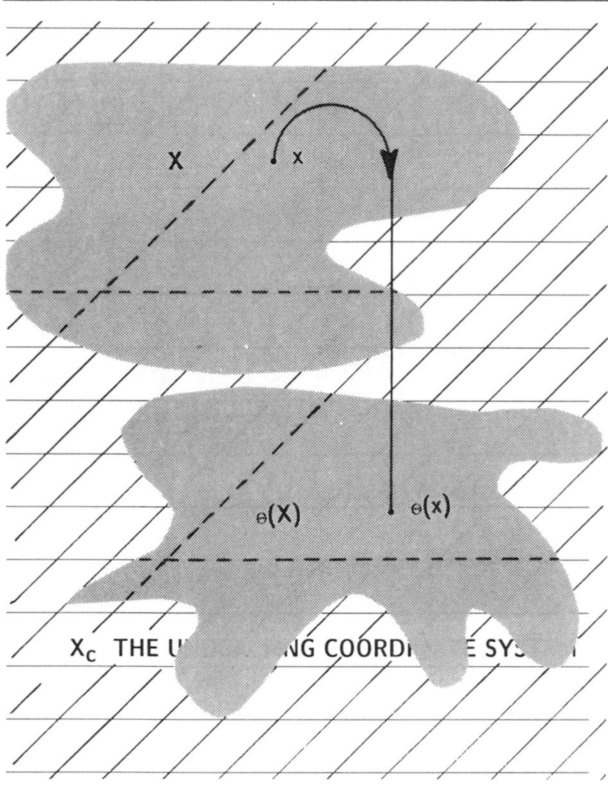

Figure 3.5.2
A change of coordinates in terms of $\boldsymbol{X}$ *and* $\boldsymbol{X_C}$*. We think of* $\boldsymbol{X}$ *as being moved relative to the underlying coordinate space* $\boldsymbol{X_C}$*.*

for f expressed in the original coordinates. Let $x' \mapsto f'(x')$ *be the formula for f expressed in the new coordinates. Then*

$$f(x) = (\theta^{-1} \circ f' \circ \theta)(x)$$
$$f'(x') = (\theta \circ f \circ \theta^{-1})(x').$$

Exercises & Examples

5.2. Consider an affine transformation $f(x) = ax + b$, $a \neq 0$, $a \neq 1$, $a, b \in \mathbb{R}$. This has a fixed point $x_f \in \mathbb{R}$ defined by $f(x_f) = x_f$. We find $x_f = b/(1 - a)$. x_f is clearly the interesting point in the action of an

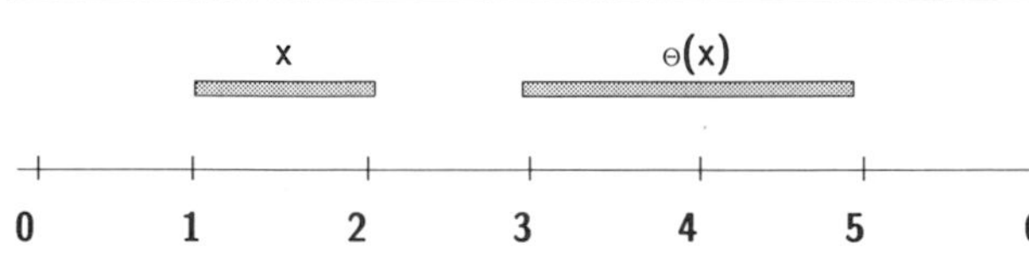

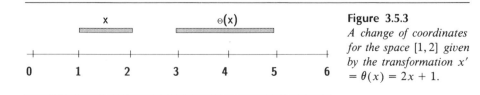

Figure 3.5.3
A change of coordinates for the space $[1, 2]$ *given by the transformation* $x' = \theta(x) = 2x + 1$.

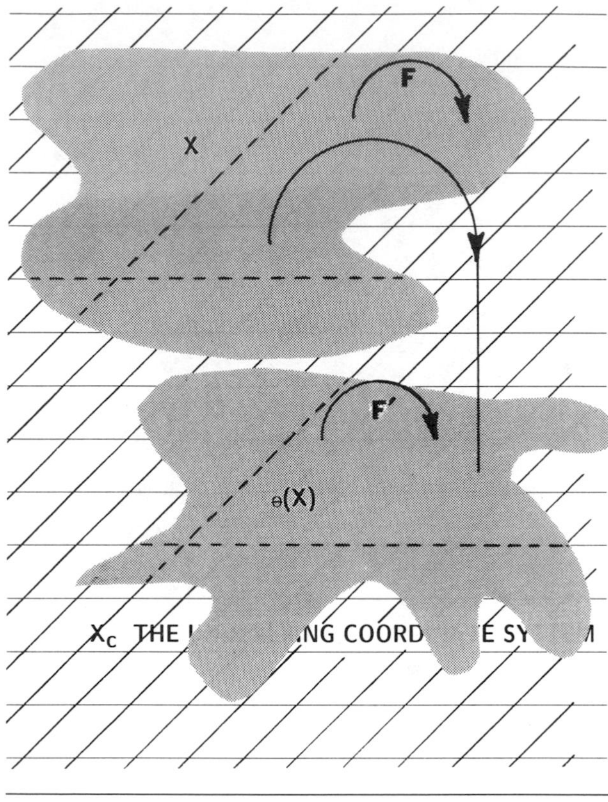

Figure 3.5.4
*The transformation F acting on **X** is equivalent to F′ acting on θ(**X**).*

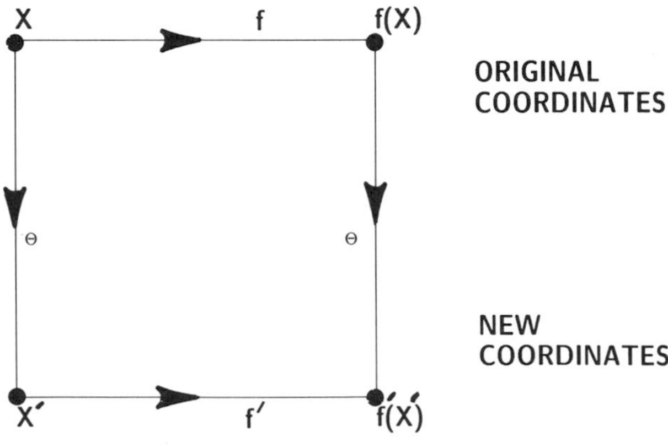

Figure 3.5.5
Commutative diagram for the coordinate change θ: $\boldsymbol{X}_C \to \boldsymbol{X}_C$.

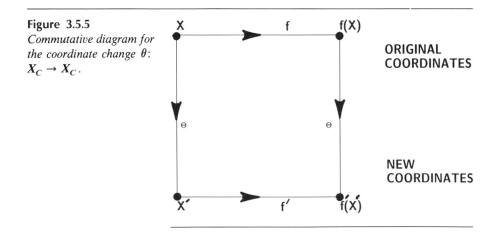

affine transformation on $\mathbb{R}$. Accordingly let us change coordinates to move x_f to the origin: that is $x' = \theta(x) = x - x_f$. What does f look like in this new coordinate system?

$$f'(x') = (\theta \circ f \circ \theta^{-1})(x') = \theta \circ f(x' + x_f) = a(x' + x_f) + b - x_f;$$

$f'(x') = ax'$ which is simply a rescaling! Now using the first formula we get

$$f(x) = a(x - x_f) + x_f$$

and

$$f^{\circ n}(x) = a^n(x - x_f) + x_f \qquad \text{for all } n \in \{0, \pm 1, \pm 2, \pm 3, \dots\}.$$

We now see a new way of vizualizing an affine transformation on $\mathbb{R}$: for example, if $a > 1$ we see the image in Figure 3.5.6.

5.3. Let $X = [1, 2]$ and let a change of coordinates be defined by $x' = 2x - 1$. Let a transformation f: $X \to X$ be defined by $f(x) = (x - 1)^2 + 1$. Express f in the new coordinate system.

Definition 1. Let f: $X \to X$ be a transformation on a metric space. A point $x_f \in X$ such that $f(x_f) = x_f$ is called a *fixed point* of the transformation.

The fixed points of a transformation are very important. They tell us which parts of the space are pinned in place, not moved, by the transformation. The fixed points of a transformation restrict the motion of the space under nonviolent, non-ripping transformations of bounded deformation.

Exercises & Examples

5.4. Find the fixed points x_1 and x_2 of the Möbius transformation

$$f(z) = \frac{(z + 2)}{(4 - z)}$$

on $\hat{\mathbb{C}}$. Make a change of coordinates so that x_1 becomes the origin and x_2 becomes the Point at Infinity. Hence interpret the action of $f(z)$ on the sphere in geometrical terms.

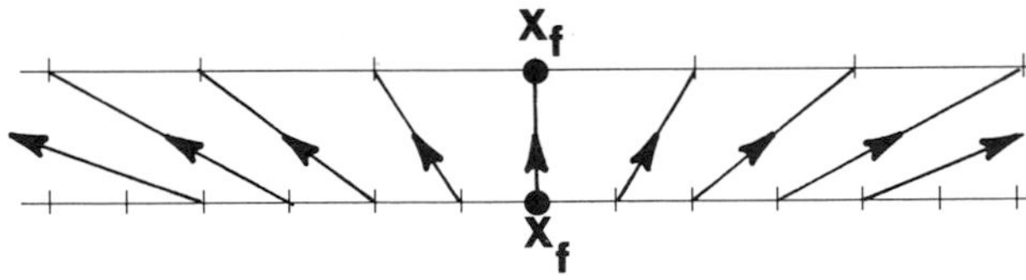

Figure 3.5.6
An affine transformation on $\mathbb{R}$. We see rescaling (magnification or diminution) centered at the fixed point, together with a flip of 180° if $a < 0$.

5.5. Let $W(x) = Ax + t$ be a two-dimensional affine transformation acting on the space $X = \mathbb{R}^2$ where det(A-I) $\neq 0$. Find the fixed point x_f. Change coordinates so that x_f becomes the origin of coordinates. Hence describe the action geometrically of a two-dimensional nondegenerate affine transformation. What can happen if det(A-I) $= 0$?

5.6. Analyze the behavior of the affine transformation $w(z) = 7z + 1$ on $\hat{\mathbb{C}}$ near the Point at Infinity by making the change of coordinates $h(z) = 1/z$.

5.7. Two one-parameter families of transformations on $\mathbb{R}$ are $f_\mu(x) = x^2 - \mu$ and $g_\lambda(x) = \lambda x(1 - x)$, where μ and λ are real parameters. Find a change of coordinates and a function $\mu = \mu(\lambda)$ so that $f'_{\mu(\lambda)}(x') = g_\lambda(x')$ is valid for an appropriate interval on the λ-axis.

5.8. Find the real fixed points of $g(x) = x^2 - \frac{1}{2}$. Analyze the behavior of g near each of its fixed points by changing coordinates so as to move first one then the other to the origin. Another method for looking at the behaviour of g near a fixed point is to approximate $g(x)$ by the first two terms of its Taylor series expansion about the fixed point. Compare these methods.

5.9. Let w: $\mathbb{R}^2 \to \mathbb{R}^2$ denote the affine transformation

$$w\begin{pmatrix} x_1 \\ x_2 \end{pmatrix} = \begin{pmatrix} 1 & 2 \\ 2 & 3 \end{pmatrix}\begin{pmatrix} x_1 \\ x_2 \end{pmatrix} + \begin{pmatrix} 1 \\ 1 \end{pmatrix}.$$

Make a change of coordinates so that the transformation is simply a coordinate rescaling. What are the rescaling factors?

***Definition* 2.** Let F denote a set of transformations on a metric space $\boldsymbol{X}$. F is called a *semigroup* if $f, g \in F$ implies $f \circ g \in F$. F is called a *group* if it is a semigroup of invertible transformations, and $f \in F$ implies $f^{-1} \in F$.

We introduce this definition because we will use semigroups (and groups) of transformations both to characterize and to compute fractal subsets of $\boldsymbol{X}$. However we do not use any deep theorems from group theory.

Exercises & Examples

5.10. Let f: $\boldsymbol{X} \to \boldsymbol{X}$ be a transformation on a metric space. Show that the set of transformations $\{f^{\circ n}: n = 0, 1, 2, 3, \dots\}$ forms a semigroup.

5.11. A transformation T: $\Sigma \to \Sigma$ on code space is defined by

$$T(x_1x_2x_3x_4x_5 \;\cdots) = x_2x_3x_4x_5x_6 \;\cdots$$

and is called a *shift operator*. Describe the semigroup of transformations $\{T^{\circ n}: n = 0, 1, 2, 3, \dots\}$. What are the *fixed points* of $T^{\circ 3}$ if the code space is built up from the two symbols $\{0, 1\}$?

5.12. Show that the set of Möbius transformations on $\hat{\mathbb{R}}$ form a group.

5.13. Show that the set of Möbius transformations on $\hat{\mathbb{C}}$ form a group.

5.14. Show that the set of invertible affine transformations on $\mathbb{R}^2$ form a group.

5.15. Show that the set of transformations $f\colon \mathbb{R}^2 \to \mathbb{R}^2$ such that $f(\quad) \subset$ form a semigroup.

5.16. Show that a group of transformations is provided by the set of affine transformations of the form $w(x) = Ax + t$ where $A = \begin{pmatrix} a & 0 \\ b & c \end{pmatrix}$ for $a, b, c \in \mathbb{R}$, with $ac \neq 0$, and the translation vector t is arbitrary.

5.17. The most general analytic quadratic transformation $f\colon \hat{\mathbb{C}} \to \hat{\mathbb{C}}$ can be expressed by a formula of the form $f(z) = Az^2 + Bz + C$ where A, B, and C are complex numbers, and $A \neq 0$. Show that by means of a suitable change of coordinates, $z' = \theta(z)$, where θ is a similitude, show that $f(z)$ can be reexpressed as a quadratic transformation of the special form $f'(z) = (z')^2 + \tilde{C}$ for some complex number $\tilde{C}$.

3.6 THE CONTRACTION MAPPING THEOREM

Definition 1. A transformation $f\colon \mathbf{X} \to \mathbf{X}$ on a metric space $(\mathbf{X}, d)$ is called *contractive* or *a contraction mapping* if there is a constant $0 \le s < 1$ such that

$$d(f(x), f(y)) \le s \cdot d(x, y) \ \forall\, x, y \in \mathbf{X}.$$

Any such number s is called a *contractivity factor* for f.

It would be convenient to be able to talk about the largest number and the smallest number in a set of real numbers. However a set such as $S = (-\infty, 3)$ does not possess either. This difficulty is overcome by the following definition.

Definition 2. Let S denote a set of real numbers. Then the *infimum* of S is equal to $-\infty$ if S contains negative numbers of arbitrarily large magnitude. Otherwise the infimum of $S = \text{Max}\{x \in \mathbb{R}\colon x \le s \text{ for all } s \in S\}$. The infimum of S always exists, because of the nature of the real number system, and it is denoted by *Inf S*. The *supremum* of S is similarly defined. It is equal to $+\infty$ if S contains arbitrarily large numbers; otherwise it is the minimum of the set of numbers which are greater than or equal to all of the numbers in S. The supremum of S always exists and it is denoted by *Sup S*.

Exercises & Examples

6.1. Find the supremum and the infimum of the following sets of real numbers: (a) $(-\infty, 3)$; (b) $\mathscr{C}$, the Classical Cantor Set; (c) $\{1, 2, 3, 4, \ldots\}$; (d) the positive real numbers.

6.2. Let $f\colon \mathbf{X} \to \mathbf{X}$ be a contraction mapping on a metric space $(\mathbf{X}, d)$. Show that $\operatorname{Inf}\{s \in \mathbb{R}: s \text{ is a contractivity factor for } f\}$ is a contractivity factor for f.

6.3. Show that if $f\colon \mathbf{X} \to \mathbf{X}$ and $g\colon \mathbf{X} \to \mathbf{X}$ are contraction mappings on a space $(\mathbf{X}, d)$, with contractivity factors s and t respectively, then $f \circ g$ is a contraction mapping with contractivity factor st.

Theorem 1. [The Contraction Mapping Theorem] *Let $f\colon \mathbf{X} \to \mathbf{X}$ be a contraction mapping on a complete metric space $(\mathbf{X}, d)$. Then f possesses exactly one fixed point $x_f \in \mathbf{X}$ and moreover for any point $x \in \mathbf{X}$, the sequence $\{f^{\circ n}(x)\colon n = 0, 1, 2, \ldots\}$ converges to x_f. That is,*

$$\lim_{n \to \infty} f^{\circ n}(x) = x_f, \quad \text{for each } x \in \mathbf{X}.$$

Figure 3.6.1 illustrates the idea of a contractive transformation on a compact metric space.

Proof. Let $x \in \mathbf{X}$. Let $0 \le s < 1$ be a contractivity factor for f. Then

$$d(f^{\circ n}(x), f^{\circ m}(x)) \le s^{m \wedge n} d(x, f^{\circ |n-m|})(x) \tag{3.6.1}$$

for all $m, n = 0, 1, 2, \ldots,$ where we have fixed $x \in \mathbf{X}$. The notation $u \wedge v$ denotes the minimum of the pair of real numbers u and v. In particular, for $k = 0, 1, 2, \ldots,$ we have

$$\begin{aligned} d(x, f^{\circ k}(x)) &\le d(x, f(x)) + (f(x), f^{\circ 2}(x)) + \cdots + d(f^{\circ (k-1)}(x), f^{\circ k}(x)) \\ &\le (1 + s + s^2 + \cdots + s^{k-1}) d(x, f(x)) \\ &\le (1 - s)^{-1} d(x, f(x)) \end{aligned}$$

so substituting into equation (3.6.1) we now obtain

$$d(f^{\circ n}(x), f^{\circ m}(x)) \le s^{m \wedge n} \cdot (1 - s)^{-1} \cdot d(x, f(x))$$

from which it immediately follows that $\{f^{\circ n}(x)\}_{n=0}^{\infty}$ is a Cauchy sequence. Since $\mathbf{X}$ is complete, this Cauchy sequence possesses a limit $x_f \in \mathbf{X}$, and we have

$$\lim_{n \to \infty} f^{\circ n}(x) = x_f.$$

Now we shall show that x_f is a fixed point of f. Since f is contractive it is continuous and hence

$$f(x_f) = f\left(\lim_{n \to \infty} f^{\circ n}(x)\right) = \lim_{n \to \infty} f^{\circ (n+1)}(x) = x_f.$$

Finally, can there be more than one fixed point? Suppose there are. Let x_f and y_f be two fixed points of f. Then $x_f = f(x_f)$, $y_f = f(y_f)$, and

$$d(x_f, y_f) = d(f(x_f), f(x_f)) \le s d(x_f, y_f)$$

whence $(1 - s) d(x_f, y_f) \le 0$, which implies $d(x_f, y_f) = 0$ and hence $x_f = y_f$. This completes the proof.

Figure 3.6.1(a) *Illustrates the idea of a contractive transformation on a compact metric space.*

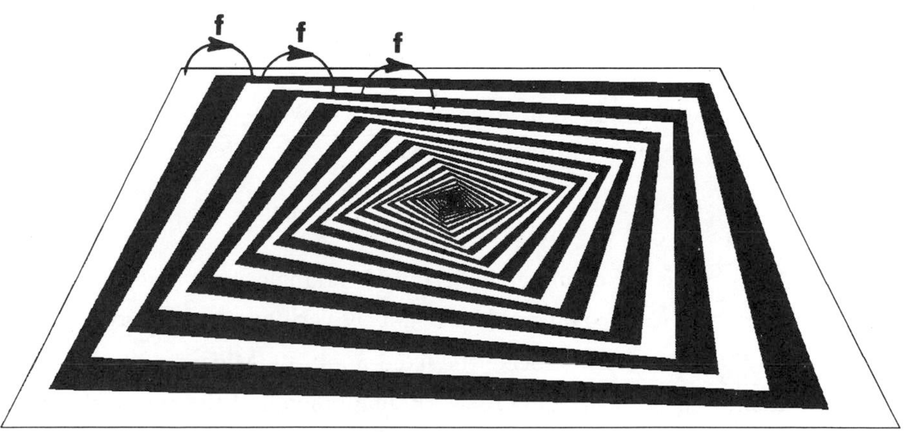

Figure 3.6.1(b) *A contraction mapping doing its work, drawing all of a compact metric space **X** towards the fixed point.*

Exercises & Examples

6.4. Let $w(x) = Ax + t$ be an affine transformation in two dimensions. Make the change of coordinates $h(x) = x' = x - x_f$, under the assumption that $\det(I - A) \neq 0$, and show that $w'(x') = h \circ w \circ h^{-1}(x') = Ax'$, that $w(x) = (h^{-1} \circ w' \circ h)(x) = A(x - x_f) + x_f$, and hence that

$$w^{\circ n}(x) = A^n(x - x_f) + x_f \qquad \text{for } n = 0, 1, 2, 3, \ldots \tag{3.6.2}$$

Give conditions on A such that it is contractive (a) in the Euclidean metric, (b) in the Manhattan metric. Show that if $|A| < 1$, where $|A|$ denotes any appropriate norm of A viewed as a linear operator on a two-dimensional vector space, then $\{w^{\circ n}(x)\}$ is a Cauchy sequence which converges to x_f, for each $x \in \mathbb{R}^2$.

6.5. Let f: ■ → ■ be a contraction mapping on (■, Euclidean). Show that Figure 3.6.1 gives the right idea.

6.6. Let $f: \mathbb{R} \to \mathbb{R}$ be the affine transformation $f(x) = \frac{1}{2}x + \frac{1}{2}$. Verify f is a contraction mapping, and deduce

$$\lim_{n \to \infty} f^{\circ n}(x) = x_f \qquad \text{for each } x \in \mathbb{R},$$

Use this formula with $x = 0$ to obtain a geometrical series for the fixed point $x_f \in \mathbb{R}$. Observe however that $f(\mathbb{R}) = \mathbb{R}$; indeed, that f is invertible.

6.7. Let $(\mathbf{X}, d)$ be a compact metric space which contains more than one point. Show that the situation in Example 6.6 cannot occur for any contraction mapping $f: \mathbf{X} \to \mathbf{X}$. That is, show that $f(\mathbf{X}) \subset \mathbf{X}$ but $f(\mathbf{X}) \neq \mathbf{X}$. That is, show that a contraction mapping on a nontrivial compact metric space is not invertible. Hint: Use the compactness of the space to show that there is a point in the space which is furthest away from the fixed point. Then show that there is a point that is not in $f(\mathbf{X})$.

6.8. Show that the set of contraction mappings on a metric space form a semigroup.

6.9. Show that the affine transformation w: → defined by $w(x) = Ax + t$ is a contraction, where

$$A = \begin{pmatrix} \frac{1}{2}\cos 120° & -\frac{1}{2}\sin 120° \\ \frac{1}{2}\sin 120° & \frac{1}{2}\cos 120° \end{pmatrix} \text{ and } t = \begin{pmatrix} \frac{1}{2} \\ 0 \end{pmatrix}$$

Here is an equilateral Sierpinski Triangle with a vertice at the origin and at $(1, 0)$. You need to begin by verifying that w does indeed map into itself! Locate the fixed point x_f. Make a picture of this contraction mapping "doing its work, mapping all of the compact metric space towards the fixed point". Use different colors to denote the successive regions $f^{\circ(n)}(\ ) \setminus f^{\circ(n+1)}(\ )$ for $n = 0, 1, 2, 3, \ldots$.

6.10. Define a mapping on the code space of two symbols $\{0, 1\}$ by $f(x_1x_2x_3x_4\ldots) = 1x_1x_2x_3x_4\ldots$. (Recall that the metric is $d(x, y) = \sum_{i=1}^{\infty}|x_i - y_i|/3^i$ or equivalent.) Show that f is a contraction mapping. Locate the fixed point of f.

6.11. Let $(\mathbf{X}, d)$ be a compact metric space, and let $f: \mathbf{X} \to \mathbf{X}$ be a contraction mapping. Show that $\{f^{\circ n}(X)\}_{n=0}^{\infty}$ is a Cauchy sequence of points in $(\mathscr{H}(X), h)$ and $\mathrm{Lim}_{n\to\infty} f^{\circ n}(X) = \{x_f\}$, where x_f is the fixed point of f.

6.12. Let (X, d) be a compact metric space. Let $f\colon X \to X$ have the property $\mathrm{Lim}_{n\to\infty} f^{\circ n}(X) = \{x_f\}$. Find a metric $\tilde{d}$ on X such that f is a contraction mapping, and the identity is a homeomorphism from $(\overline{X}, d) \to (\overline{X}, \tilde{d})$

6.13. Let $Ax = \begin{pmatrix} a & b \\ c & d \end{pmatrix}\begin{pmatrix} x_1 \\ x_2 \end{pmatrix}$ with $a, b, c, d \in \mathbb{R}$, all strictly positive, be a linear transformation on $\mathbb{R}^2$. Show that A maps the positive quadrant $\{(x_1, x_2)\colon x_1 \geq 0,\ x_2 \geq 0\}$ into itself. Let a mapping $f\colon [0, 90°] \to [0, 90°]$ be defined by

$$A\begin{pmatrix} \cos\theta \\ \sin\theta \end{pmatrix} = (\text{some positive number})\begin{pmatrix} \cos f(\theta) \\ \sin f(\theta) \end{pmatrix}.$$

Show that f is a contraction mapping on the metric space ([0, 90°], Euclidean). Deduce that there exists a unique positive number λ, and an angle $0 < \theta < 90°$ such that $A\begin{pmatrix} \cos\theta \\ \sin\theta \end{pmatrix} = \lambda\begin{pmatrix} \cos\theta \\ \sin\theta \end{pmatrix}$. See Figure 3.6.2.

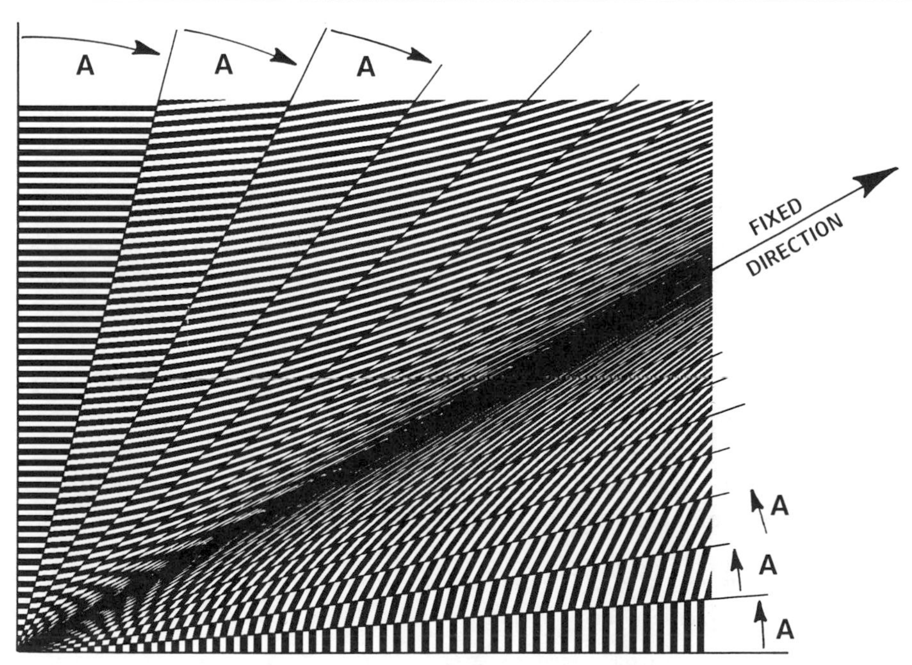

Figure 3.6.2
The existence of a positive eigenvalue of an "angle-squeezing" linear transformation.

3.7 CONTRACTION MAPPINGS ON THE SPACE OF FRACTALS

Let $(\mathbf{X}, d)$ be a metric space and let $(\mathscr{H}(\mathbf{X}), h(d))$ denote the corresponding space of nonempty compact subsets, with the Hausdorff metric $h(d)$. We introduce the notation $h(d)$ to show that d is the underlying metric for the Hausdorff metric h. For example, we may discuss $(\mathscr{H}(\hat{\mathbb{C}}), h(\text{spherical}))$ or $(\mathscr{H}(\mathbb{R}^2), h(\text{Manhattan}))$. We will drop this additional notation when we evaluate Hausdorff distances.

We have repeatedly refused to define fractals: we have agreed that they are subsets of simple geometrical spaces, such as $(\mathbb{R}^2, \text{Euclidean})$ and $(\hat{\mathbb{C}}, \text{Spherical})$. If we were to define a *deterministic fractal*, we might say that it is a fixed point of a contractive transformation on $(\mathscr{H}(\mathbf{X}), h(d))$. We would require that the underlying metric space $(\mathbf{X}, d)$ be "geometrically simple". We would require also that the contraction mapping be constructed from simple, easily specified, contraction mappings on $(\mathbf{X}, d)$, as described below.

Lemma 1. *Let w: $\mathbf{X} \to \mathbf{X}$ be a contraction mapping on the metric space $(\mathbf{X}, d)$. Then w is continuous.*

Proof. Let $\epsilon > 0$ be given. Let $s > 0$ be a contractivity factor for w. Then

$$d(w(x), w(y)) \leq sd(x, y) < \epsilon$$

whenever $d(x, y) < \delta$ where $\delta = \epsilon/s$. This completes the proof.

Lemma 2. *Let w: $\mathbf{X} \to \mathbf{X}$ be a continuous mapping on the metric space $(\mathbf{X}, d)$. Then w maps $\mathscr{H}(\mathbf{X})$ into itself.*

Proof. Let S be a nonempty compact subset of $\mathbf{X}$. Then clearly $w(S) = \{w(x)\colon x \in S\}$ is nonempty. We want to show that $w(S)$ is compact. Let $\{y_n = w(x_n)\}$ be an infinite sequence of points in S. Then $\{x_n\}$ is an infinite sequence of points in S. Since S is compact there is a subsequence $\{x_{N_n}\}$ which converges to a point $\hat{x} \in S$. But then the continuity of w implies that $\{y_{N_n} = f(x_{N_n})\}$ is a subsequence of $\{y_n\}$ which converges to $\hat{y} \in w(S)$. This completes the proof.

The following lemma tells us how to make a contraction mapping on $(\mathscr{H}(\mathbf{X}), h)$ out of a contraction mapping on $(\mathbf{X}, d)$.

Lemma 3. *Let w: $\mathbf{X} \to \mathbf{X}$ be a contraction mapping on the metric space $(\mathbf{X}, d)$ with contractivity factor s. Then w: $\mathscr{H}(\mathbf{X}) \to \mathscr{H}(\mathbf{X})$ defined by*

$$w(B) = \{w(x)\colon x \in B\} \ \forall B \in \mathscr{H}(\mathbf{X})$$

is a contraction mapping on $(\mathscr{H}(\mathbf{X}), h(d))$ with contractivity factor s.

Proof. From Lemma 1 it follows that w: $\mathbf{X} \to \mathbf{X}$ is continuous. Hence by Lemma 2 w maps $\mathscr{H}(\mathbf{X})$ into itself.

Now let $B, C \in \mathscr{H}(\mathbf{X})$. Then

$$d(w(B), w(C)) = \text{Max}\{\text{Min}\{d(w(x), w(y)): y \in C\}: x \in B\}$$
$$\leq \text{Max}\{\text{Min}\{s \cdot d(x, y): y \in C\}: x \in B\} = s \cdot d(B, C).$$

Similarly $d(w(C), w(B)) \leq s \cdot d(C, B)$. Hence

$$h(w(B), w(C)) = d(w(B), w(C)) \vee d(w(C), w(B)) \leq s \cdot d(B, C) \vee d(C, B)$$
$$\leq s \cdot h(B, C).$$

This completes the proof.

The following lemma gives a characteristic property of the Hausdorff metric which we will shortly need. The proof follows at once from Exercise 2.6.13.

Lemma 4. *For all B, C, D, and E, in $\mathscr{H}(\mathbf{X})$*

$$h(B \cup C, D \cup E) \leq h(B, D) \vee h(C, E)$$

where as usual h is the Hausdorff metric.

The next lemma provides an important method for combining contraction mappings on $(\mathscr{H}(\mathbf{X}), h)$ to produce new contraction mappings on $(\mathscr{H}(\mathbf{X}), h)$. This method is distinct from the obvious one of composition.

Lemma 5. *Let $(\mathbf{X}, d)$ be a metric space. Let $\{w_n$: $n = 1, 2, \ldots, N\}$ be contraction mappings on $(\mathscr{H}(\mathbf{X}), h)$. Let the contractivity factor for w_n be denoted by s_n for each n. Define W: $\mathscr{H}(\mathbf{X}) \to \mathscr{H}(\mathbf{X})$ by*

$$W(B) = w_1(B) \cup w_2(B) \cup \cdots \cup w_N(B)$$
$$= \bigcup_{n=1}^{N} w_n(B), \qquad \textit{for each } B \in \mathscr{H}(\mathbf{X}).$$

Then W is a contraction mapping with contractivity factor $s = \text{Max}\{s_n$: $n = 1, 2, \ldots, N\}$.

Proof. We demonstrate the claim for $N = 2$. An inductive argument then completes the proof. Let $B, C \in \mathscr{H}(X)$. We have

$$h(W(B), W(C)) = h(w_1(B) \cup w_2(B), w_1(C) \cup w_2(C))$$
$$\leq h(w_1(B), w_1(C)) \vee h(w_2(B), w_2(C)) \quad \text{(by Lemma 4)}$$
$$\leq s_1 h(B, C) \vee s_2 h(B, C) \leq s h(B, C).$$

This completes the proof.

Definition 1. A (hyperbolic) *iterated function system* consists of a complete metric space (X, d) together with a finite set of contraction mappings w_n: $\mathbf{X} \to \mathbf{X}$, with respective contractivity factors s_n, for $n = 1, 2, \ldots, N$. The abbreviation "IFS" is used for "iterated function system." The notation for the IFS just announced is $\{\mathbf{X};\ w_n,\ n = 1, 2, \ldots, N\}$ and its contractivity factor is $s = \text{Max}\{s_n\text{: } n = 1, 2, \ldots, N\}$.

We put the word "hyperbolic" in parentheses in this definition because it is sometimes dropped in practice. Moreover, we will sometimes use the nomenclature "IFS" to mean simply a finite set of maps acting on a metric space, with no particular conditions imposed upon the maps.

The following theorem summarizes the main facts so far about a hyperbolic IFS.

Theorem 1. *Let $\{\mathbf{X};\ w_n,\ n = 1, 2, \ldots, N\}$ be a hyperbolic iterated function system with contractivity factor s. Then the transformation W: $\mathscr{H}(\mathbf{X}) \to \mathscr{H}(\mathbf{X})$ defined by*

$$W(B) = \bigcup_{n=1}^{N} w_n(B)$$

for all $B \in \mathscr{H}(X)$, is a contraction mapping on the complete metric space $(\mathscr{H}(\mathbf{X}), h(d))$ with contractivity factor s. That is

$$h(W(B), W(C)) \leq s \cdot h(B, C)$$

for all $B, C \in \mathscr{H}(\mathbf{X})$. Its unique fixed point, $A \in \mathscr{H}(\mathbf{X})$, obeys

$$A = W(A) = \bigcup_{n=1}^{N} w_n(A),$$

and is given by $A = \text{Lim}_{n\to\infty} W^{\circ n}(B)$ for any $B \in \mathscr{H}(\mathbf{X})$.

Definition 2. The fixed point $A \in \mathscr{H}(\mathbf{X})$ described in the theorem is called the *attractor* of the IFS.

Sometimes we will use the name "attractor" in connection with an IFS which is simply a finite set of maps acting on a complete metric space $\mathbf{X}$. By this we mean that one can make an assertion which is analagous to the last sentence of Theorem 3.7.1.

We wanted to use the words "deterministic fractal" in place of "attractor" in Definition 2. We were tempted, but resisted. The nomenclature "iterated function system" is meant to remind one of the name "dynamical system." We will introduce dynamical systems in Chapter 4. Dynamical systems often possess attractors, and when these are interesting to look at, they are called *strange attractors*.

Exercises & Examples

7.1. This exercise takes place in the metric spaces ($\mathbb{R}$, Euclidean) and $(\mathscr{H}(\mathbb{R}), h(\text{Euclidean}))$. Consider the IFS $\{\mathbb{R}^2; w_1, w_2\}$ where $w_1(x) = \frac{1}{3}x$ and $w_2(x) = \frac{1}{3}x + \frac{2}{3}$. Show that this is indeed an IFS with contractivity factor $s = \frac{1}{3}$. Let $B_0 = [0, 1]$. Calculate $B_n = W^{\circ n}(B_0)$, $n = 1, 2, 3, \ldots$. Deduce that $A = \text{Lim}_{n\to\infty} B_n$ is the classical Cantor set. Verify directly that $A = \frac{1}{3}A \cup \{\frac{1}{3}A + \frac{2}{3}\}$. Here we use the following notation: for a subset A of $\mathbb{R}$, $xA = \{xy: y \in A\}$ and $A + x = \{y + x: y \in A\}$.

7.2. With reference to example 7.1, show that if $w_1(x) = s_1 x$ and $w_2(x) = (1 - s_1)x + s_1$ where s_1 is a number such that $0 < s_1 < 1$, then $B_1 = B_2 = B_3 = \cdots$. Find the attractor.

7.3. Repeat example 7.1 with $w_1(x) = \frac{1}{3}x$ and $w_2(x) = \frac{1}{2}x + \frac{1}{2}$. In this case $A = \text{Lim}_{n\to\infty} B_n$ will not be the classical Cantor set, but it will be something like it. Describe A. Show that A contains no intervals. How many points does A contain?

7.4. Consider the IFS $\{\mathbb{R}, \frac{1}{4}x + \frac{3}{4}, \frac{1}{2}x, \frac{1}{4}x + \frac{1}{4}\}$. Verify that the attractor looks like the image in Figure 3.7.1.
Show precisely how the set in Figure 3.7.1 is a union of three "shrunken copies of itself." This attractor is interesting: it contains countably many holes and countably many intervals.

7.5. Show that the attractor of an IFS having the form $\{\mathbb{R}; w_1(x) = ax + b, w_2(x) = (cx + d)\}$ where a, b, c, and $d \in \mathbb{R}$, is either connected or totally disconnected.

7.6. Does there exist an IFS of three affine maps in $\mathbb{R}^2$ whose attractor is the union of two disjoint closed intervals?

7.7. Consider the IFS

$$\left\{\mathbb{R}^2; \begin{pmatrix} \frac{1}{2} & 0 \\ 0 & \frac{1}{2} \end{pmatrix}\begin{pmatrix} x \\ y \end{pmatrix} + \begin{pmatrix} \frac{1}{2} \\ \frac{1}{2} \end{pmatrix}, \begin{pmatrix} \frac{1}{2} & 0 \\ 0 & \frac{1}{2} \end{pmatrix}\begin{pmatrix} x \\ y \end{pmatrix}\right\}.$$

Let $A_0 = \{(\frac{1}{2}, y): 0 \leq y \leq 1\}$, and let $W^{\circ n}(A_0) = A_n$, where W is defined on $\mathscr{H}(\mathbb{R}^2)$ in the usual way. Show that the attractor is $A =$

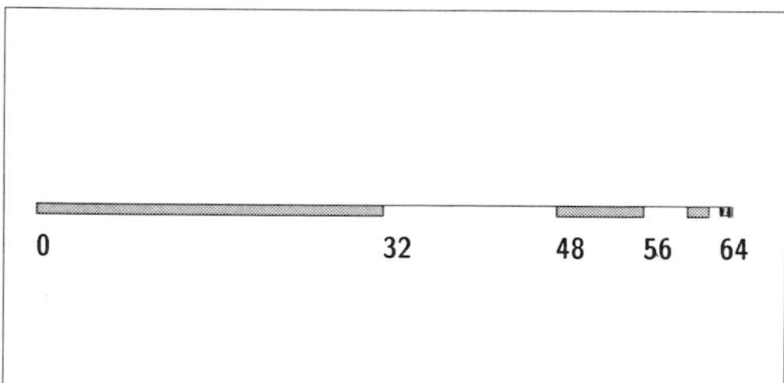

Figure 3.7.1 *Attractor for three affine maps on the real line. Can you find the maps?*

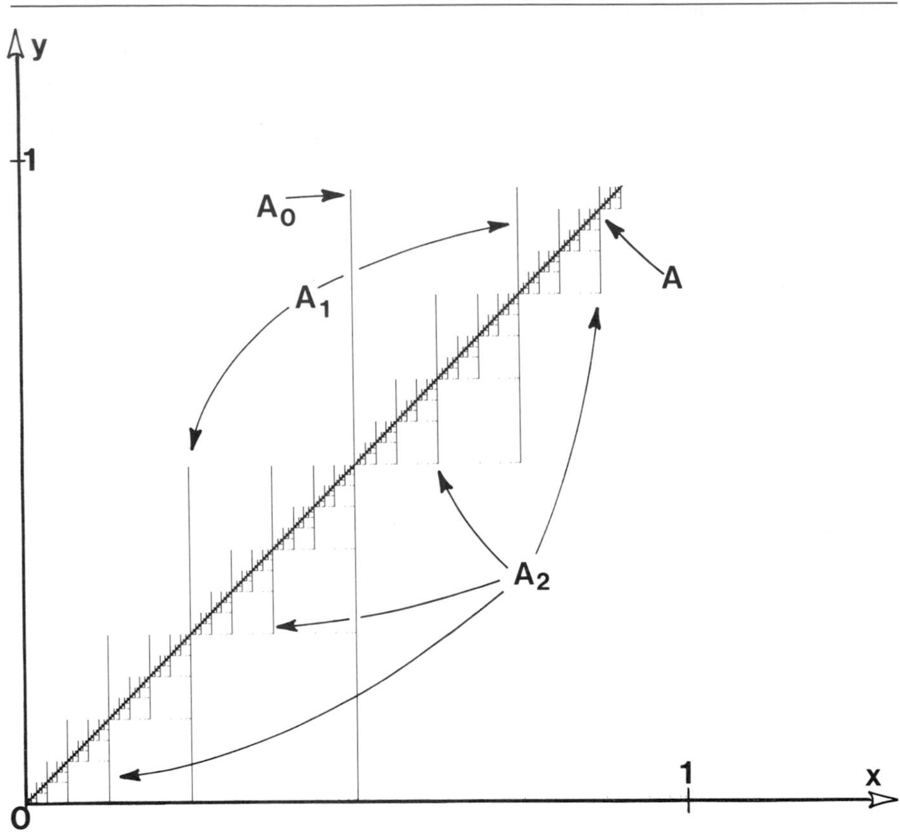

Figure 3.7.2
A sequence of sets converging to a line segment.

$\{(x, y): x = y,\ 0 \le x \le 1\}$ and that Figure 3.7.2 is correct. Draw a sequence of pictures to show what happens if $A_0 = \{(x, y) \in \mathbb{R}^2: 0 \le x \le 1, 0 \le y \le 1\}$.

7.8. Consider the attractor for the IFS $\{\mathbb{R};\ w_1(x) = 0,\ w_2(x) = \frac{2}{3}x + \frac{1}{3}\}$. Show that it consists of a countable increasing sequence of real points $\{x_n: n = 0, 1, 2, \ldots\}$ together with $\{1\}$. Show that x_n can be expressed as the nth partial sum of an infinite geometric series. Give a succinct formula for x_n.

7.9. Describe the attractor A for the IFS $\{[0, 2];\ w_1(x) = \frac{1}{9}x^2,\ w_2(x) = \frac{3}{4}x + \frac{1}{2}\}$ by describing a sequence of sets which converges to it. Show that A is totally disconnected. Show that A is perfect. Find the contractivity factor for the IFS.

7.10. Let (r, θ), $0 \le r < \infty$, $0 \le \theta < 2\pi$ denote the polar coordinates of a point in the plane, $\mathbb{R}^2$. Define $w_1(r, \theta) = (\frac{1}{2}r + \frac{1}{2}, \frac{1}{2}\theta)$, and $w_2(r, \theta) =$

$(\frac{2}{3}r + \frac{1}{3}, \frac{2}{3}\theta + 2\pi/3)$. Show that $\{\mathbb{R}^2; w_1, w_2\}$ is not a hyperbolic IFS because both maps w_1 and w_2 are not continuous on the whole plane. Show that $\{\mathbb{R}^2; w_1, w_2\}$ nevertheless has an attractor; find it (just consider r and θ separately).

7.11. Show that the sequence of sets illustrated in Figure 3.7.3 can be written in the form $A_n = W^{\circ n}(A_0)$ for $n = 1, 2, \ldots$ and find W: $\mathscr{H}(\mathbb{R}^2) \to \mathscr{H}(\mathbb{R}^2)$.

7.12. Describe the collection of functions which constitute the attractor A for the IFS

$$\{C[0,1];\ w_1(f(x)) = \tfrac{1}{2}f(x),\ w_2(f(x)) = \tfrac{1}{2}f(x) + 2x(1-x)\}.$$

Find the contractivity factor for the IFS.

7.13. Let $C^0[0,1] = \{f \in C[0,1]\colon f(0) = f(1) = 0\}$, and define $d(f, g) = \text{Max}\{|f(x) - g(x)|\colon x \in [0,1]\}$. Define w_1: $C^0[0,1] \to C^0[0,1]$ by $w_1(f(x)) = \frac{1}{2}f(2x \bmod 1) + 2x(1-x)$ and $w_2(f(x)) = \frac{1}{2}f(x)$. Show that $\{C^0[0,1]; w_1, w_2\}$ is an IFS, find its contractivity factor, and find its attractor. Draw a picture of the attractor.

7.14. Find conditions such that the Möbius transformation $w(x) = (ax + b)/(cz + d)$, $a, b, c, d \in \mathbb{C}$, $ad - bc \neq 0$, provides a contraction mapping on the unit disk $\boldsymbol{X} = \{z \in \mathbb{C}\colon |z| \leq 1\}$. Find an upper bound for the contractivity factor. Construct an IFS using two Möbius transformations on $\boldsymbol{X}$, and describe its attractor.

7.15. Show that a Möbius transformation on $\hat{\mathbb{C}}$ is never a contraction in the spherical metric.

7.16. Let (Σ, d) be the code space on three symbols $\{0, 1, 2\}$, with metric

$$d(x, y) = \frac{\sum_{n=1}^{\infty} |x_n - y_n|}{4^n}.$$

Define w_1: $\Sigma \to \Sigma$ by $w_1(x) = 0x_1x_2x_3\ldots$ and $w_2(x) = 2x_1x_2x_3\ldots$. Show that w_1 and w_2 are both contraction mappings and find their contractivity factors. Describe the attractor of the IFS $\{\Sigma; w_1, w_2\}$. What happens if we include in the IFS a third transformation defined by $w_3x = 1x_1x_2x_3, \ldots$?

7.17. Let [Sierpinski triangle] $\subset \mathbb{R}^2$ denote the compact metric space consisting of an equilateral Sierpinski triangle with vertices at $(0,0)$, $(1,0)$ and $(\frac{1}{2}, \sqrt{3}/2)$, and consider the IFS $\left\{ \text{[Sierpinski triangle]}, \frac{1}{2}z + \frac{1}{2}, \frac{1}{2}e^{2\pi i/3}z + \frac{1}{2} \right\}$ where we use complex number notation. Let $A_0 =$ [Sierpinski triangle], and $A_n = W^{\circ n}(A_0)$ for $n = 1, 2, 3, \ldots$. Describe A_1, A_2, and the attractor A. What happens if the third transformation $w_3(z) = \frac{1}{2}z + \frac{1}{4} + (\sqrt{3}/4)i$ is included in the IFS?

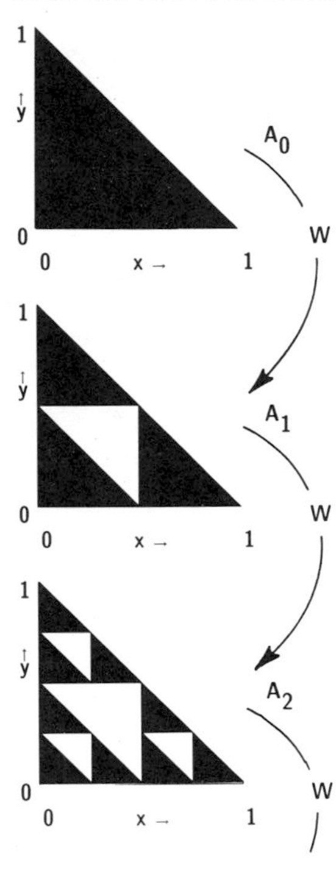

Figure 3.7.3
The first three sets A_0, A_1, and A_2 in a convergent sequence of sets in $\mathscr{H}(\mathbb{R}^2)$. Can you find a transformation W: $\mathscr{H}(\mathbb{R}^2) \to \mathscr{H}(\mathbb{R}^2)$ such that $A_{n+1} = W(A_n)$?

3.8 TWO ALGORITHMS FOR COMPUTING FRACTALS FROM ITERATED FUNCTION SYSTEMS

In this section we take time out from the mathematical development to provide two algorithms for rendering pictures of attractors of an IFS on the graphics display device of a microcomputer or workstation. The reader should establish a computergraphical environment which includes one or both of the software tools which are suggested in this section.

The algorithms presented are (1) the Deterministic Algorithm, and (2) the Random Iteration Algorithm. The deterministic algorithm is based on the idea of computing directly a sequence of sets $\{A_n = W^{\circ n}(A)\}$ starting from an initial set A_0. The Random Iteration Algorithm is founded in ergodic theory; its mathematical basis will be presented in Chapter 9. An intuitive explanation of why it works will be presented in Chapter 4. We defer important questions concerning discretization and accuracy. Such questions are considered to some extent in later chapters.

For simplicity we restrict attention to hyperbolic IFS of the form $\{\mathbb{R}^2; w_n: n = 1, 2, \ldots, N\}$, where each mapping is an affine transformation. We illustrate the algorithms for an IFS whose attractor is a Sierpinski triangle. Here is an example of such an IFS.

$$w_1\begin{bmatrix} x_1 \\ x_2 \end{bmatrix} = \begin{bmatrix} 0.5 & 0 \\ 0 & 0.5 \end{bmatrix}\begin{bmatrix} x_1 \\ x_2 \end{bmatrix} + \begin{bmatrix} 1 \\ 1 \end{bmatrix},$$

$$w_2\begin{bmatrix} x_1 \\ x_2 \end{bmatrix} = \begin{bmatrix} 0.5 & 0 \\ 0 & 0.5 \end{bmatrix}\begin{bmatrix} x_1 \\ x_2 \end{bmatrix} + \begin{bmatrix} 1 \\ 50 \end{bmatrix},$$

$$w_3\begin{bmatrix} x_1 \\ x_2 \end{bmatrix} = \begin{bmatrix} 0.5 & 0 \\ 0 & 0.5 \end{bmatrix}\begin{bmatrix} x_1 \\ x_2 \end{bmatrix} + \begin{bmatrix} 50 \\ 50 \end{bmatrix}.$$

This notation for an IFS of affine maps is cumbersome. Let us agree to write

$$w_i(x) = w_i\begin{bmatrix} x_1 \\ x_2 \end{bmatrix} = \begin{bmatrix} a_i & b_i \\ c_i & d_i \end{bmatrix}\begin{bmatrix} x_1 \\ x_2 \end{bmatrix} + \begin{bmatrix} e_i \\ f_i \end{bmatrix} = A_i x + t_i.$$

Then Table 3.8.1 is a tidier way of conveying the same iterated function system.

Table 3.8.1 also provides a number p_i associated with w_i for $i = 1, 2, 3$. These numbers are in fact probabilities. In the more general case of the IFS

Table 3.8.1
IFS code for a Sierpinski triangle.

w	a	b	c	d	e	f	p
1	0.5	0	0	0.5	1	1	0.33
2	0.5	0	0	0.5	1	50	0.33
3	0.5	0	0	0.5	50	50	0.34

$\{X;\ w_n\colon\ n = 1, 2, \ldots, N\}$ there would be N such numbers $\{p_i\colon\ i = 1, 2, \ldots, N\}$ which obey

$$p_1 + p_2 + p_3 + \cdots + p_N = 1 \quad \text{and} \quad p_i > 0 \qquad \text{for } i = 1, 2, \ldots, N.$$

These probabilities play an important role in the computation of images of the attractor of an IFS using the Random Iteration Algorithm. They play no role in the Deterministic Algorithm. Their mathematical significance is discussed in later chapters. For the moment we will use them only as a computational aid, in connection with the Random Iteration Algorithm. To this end we take their values to be given approximately by

$$p_i \approx \frac{|\det A_i|}{\sum_{i=1}^{N} |\det A_i|} = \frac{|a_i d_i - b_i c_i|}{\sum_{i=1}^{N} |a_i d_i - b_i c_i|} \qquad \text{for } i = 1, 2, \ldots, N.$$

Here the symbol $\approx$ means "approximately equal to." If, for some i, $\det A_i = 0$, then p_i should be assigned a small positive number, such as 0.001. Other situations should be treated empirically. We refer to the data in Table 3.8.1 as an IFS *code*. Other IFS codes are given in Tables 3.8.2, 3.8.3 and 3.8.4.

Table 3.8.2
IFS code for a Square.

w	a	b	c	d	e	f	p
1	0.5	0	0	0.5	1	1	0.25
2	0.5	0	0	0.5	50	1	0.25
3	0.5	0	0	0.5	1	50	0.25
4	0.5	0	0	0.5	50	50	0.25

Table 3.8.3
IFS code for a Fern.

w	a	b	c	d	e	f	p
1	0	0	0	0.16	0	0	0.01
2	0.85	0.04	−0.04	0.85	0	1.6	0.85
3	0.2	−0.26	0.23	0.22	0	1.6	0.07
4	−0.15	0.28	0.26	0.24	0	0.44	0.07

Table 3.8.4
IFS code for a Fractal Tree.

w	a	b	c	d	e	f	p
1	0	0	0	0.5	0	0	0.05
2	0.42	−0.42	0.42	0.42	0	0.2	0.4
3	0.42	0.42	−0.42	0.42	0	0.2	0.4
4	0.1	0	0	0.1	0	0.2	0.15

(1) The Deterministic Algorithm

Let $\{X;\ w_1, w_2, \ldots, w_N\}$ be a hyperbolic IFS. Choose a compact set $A_0 \subset \mathbb{R}^2$. Then compute successively $A_n = W^{\circ n}(A)$ according to

$$A_{n+1} = \bigcup_{j=1}^{N} w_j(A_n) \qquad \text{for } n = 1, 2, \ldots .$$

Thus construct a sequence $\{A_n\text{: } n = 0, 1, 2, 3, \ldots\} \subset \mathscr{H}(X)$. Then by Theorem 3.7.1 the sequence $\{A_n\}$ converges to the attractor of the IFS in the Hausdorff metric.

We illustrate the implementation of the algorithm. The following program computes and plots successive sets A_{n+1} starting from an initial set A_0, in this case a square, using the IFS code in Table 3.8.1. The program is written in BASIC. It should run without modification on an IBM PC with a Color Graphics Adaptor or Enhanced Graphics Adaptor, and Turbobasic. It also can be modified to run on any personal computer with graphics display capability. On any line, the words which are preceded by a ' are comments, they are not part of the program.

PROGRAM 3.8.1 (Example of the Deterministic Algorithm)

```
screen 1: cls   'initialize graphics
dim s(100, 100): dim t(100, 100)   'allocate two arrays of pixels
a(1) = 0.5:b(1) = 0:c(1) = 0:d(1) = 0.5:e(1) = 1:f(1) = 1   'input the
    IFS code
a(2) = 0.5:b(2) = 0:c(2) = 0:d(2) = 0.5:e(2) = 50:f(2) = 1
a(3) = 0.5:b(3) = 0:c(3) = 0:d(3) = 0.5:e(3) = 50:f(3) = 50
for i = 1 to 100   'input the initial set A(0), in this case a square, into
    the array t(i, j)
t(i, 1) = 1: pset(i, 1)   'A(0) can be used as a condensation set
t(1, i) = 1:pset(1, i)   'A(0) is plotted on the screen
t(100, i) = 1:pset(100, i)
t(i, 100) = 1:pset(i, 100)
next: do
for i = 1 to 100   'apply W to set A(n) to make A(n + 1) in the array s(i, j)
for j = 1 to 100: if t(i, j) = 1 then
s(a(1)*i + b(1)*j + e(1), c(1)*i + d(1)*j + f(1)) = 1   'and apply W
    to A(n)
s(a(2)*i + b(2)*j + e(2), c(2)*i + d(2)*j + f(2)) = 1
s(a(3)*i + b(3)*j + e(3), c(3)*i + d(3)*j + f(3)) = 1
end if: next j: next i
cls   'clears the screen—omit to obtain sequence with a A(0) as a condensa-
    tion set (see section 3.9)
```

```
for i = 1 to 100: for j = 1 to 100
t(i,j) = s(i,j)  'put A(n + 1) into the array t(i,j)
s(i,j) = 0  'reset the array s(i,j) to zero
if t(i,j) = 1 then
pset(i,j)  'plot A(n + 1)
end if: next: next
loop until instat  'if a key has been pressed then stop, otherwise compute
     A(n + 1) = W(A(n + 1))
```

The result of running a higher resolution version of this program on a Masscomp 5600 workstation, and then printing the contents of the graphics screen is presented in Figure 3.8.1. In this case we have kept each successive image produced by the program.

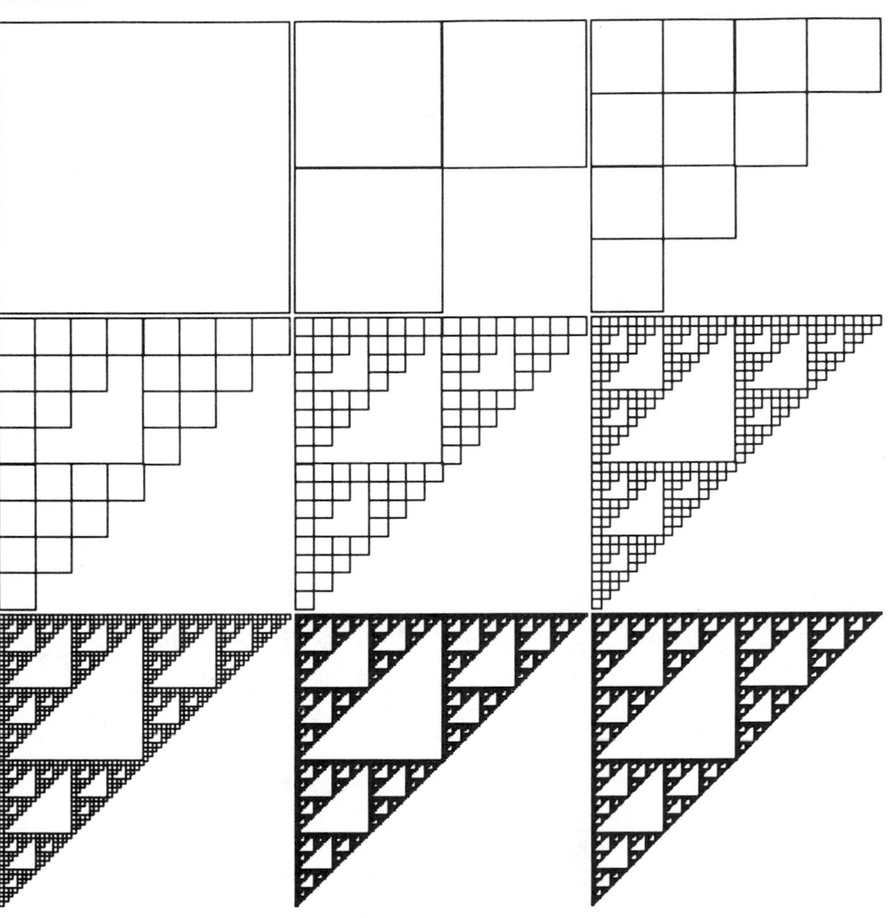

Figure 3.8.1
The result of running the Deterministic Algorithm (Program 1) *with various values of N, for the IFS code in Table* 1.

Notice that the program begins by drawing a box in the array $s(i, j)$. This box has no influence on the finally computed image of a Sierpinski triangle. One could just as well have started from any other (nonempty) set of points in the array $s(i, j)$, as illustrated in Figure 3.8.2.

To adapt Program 3.8.1 so that it runs with other IFS codes it usually will be necessary to change coordinates to ensure that each of the transformations of the IFS map the pixel array $s(i, j)$ into itself. Change of coordinates in an IFS is discussed in 3.10, example 10.5. As it stands in Program 3.8.1, the array $s(i, j)$ is a discretized representation of the square in $\mathbb{R}^2$ with lower left corner at $(1, 1)$ and upper right corner at $(100, 100)$. Failure to correctly adjust coordinates will lead to unpredictable and exciting results!

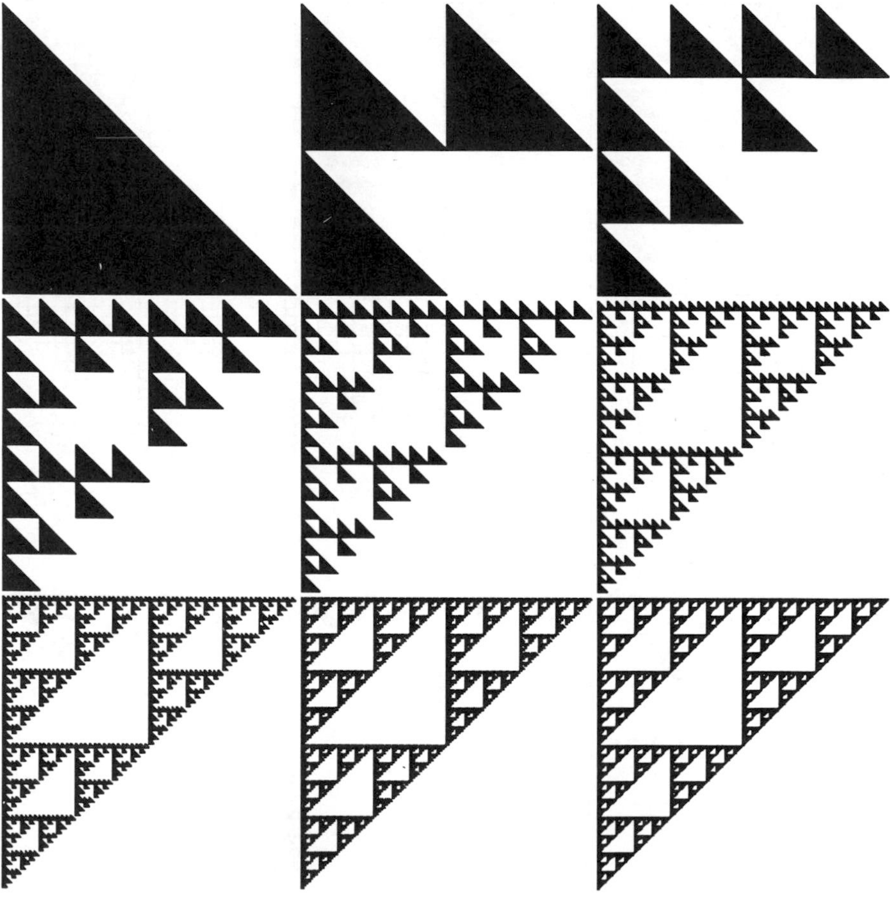

Figure 3.8.2 *The result of running the Deterministic Algorithm (Program* 1*), again for the IFS code in Table* 1*, but starting from a different initial array. The final result is always the same*!

(2) The Random Iteration Algorithm

Let $\{X;\ w_1, w_2, \ldots, w_N\}$ be a hyperbolic IFS, where probability $p_i > 0$ has been assigned to w_i for $i = 1, 2, \ldots, N$, where $\Sigma_{i=1}^{N} p_i = 1$. Choose $x_0 \in X$ and then choose recursively, independently,

$$x_n \in \{w_1(x_{n-1}), w_2(x_{n-1}), \ldots, w_N(x_{n-1})\} \qquad \text{for } n = 1, 2, 3, \ldots,$$

where the probability of the event $x_n = w_i(x_{n-1})$ is p_i. Thus construct a sequence $\{x_n\colon n = 0, 1, 2, 3, \ldots\} \subset X$.

↣ The reader should skip the rest of this paragraph and come back to it after reading section 3.9. If $\{X, w_0, w_1, w_2, \ldots, w_N\}$ is an IFS with condensation map w_0, and associated condensation set $C \in \mathscr{H}(X)$ then the algorithm is modified by (a) attaching a probability $p_0 > 0$ to w_0, so now $\Sigma_{i=0}^{N} p_i = 1$; (b) whenever $w_0(x_{n-1})$ is selected for some n, choose x_n "at random" from C. Thus, in this case too, we construct a sequence $\{x_n\colon n = 0, 1, 2, \ldots\}$ of points in X.

The sequence $\{x_n\}_{n=0}^{\infty}$ "converges to" the attractor of the IFS, under various conditions, in a manner which will be made precise in Chapter 9.

We illustrate the implementation of the algorithm. The following program computes and plots a thousand points on the attractor corresponding to the IFS code in Table 3.8.1. The program is also written in BASIC and runs without modification on an IBM PC with Enhanced Graphics Adaptor and Turbobasic. On any line the words which are preceded by a ' are comments: they are not part of the program.

PROGRAM 3.8.2

```
a[1] = 0.5: b[1] = 0: c[1] = 0: d[1] = .5: e[1] = 1: f[1] = 1   'Iterated
     Function System Data
a[2] = 0.5: b[2] = 0: c[2] = 0: d[2] = .5: e[2] = 50: f[2] = 1
a[3] = 0.5: b[3] = 0: c[3] = 0: d[3] = .5: e[3] = 50: f[3] = 50
screen 1: cls   'initialize computer graphics
window (0, 0)-(100, 100)   'set plotting window to 0 < x < 1, 0 < y < 1
x = 0: y = 0: numits = 1000   'initialize (x, y) and define the number of
     iterations, numits
for n = 1 to numits   'Random Iteration begins!
k = int(3*rnd-0.00001) + 1   'choose one of the numbers 1, 2, and 3
     with equal probability
'apply affine transformation number k to (x, y)
newx = a[k]*x + b[k]*y + e[k]: newy = c[k]*x + d[k]*y + f[k]
x = newx: y = newy   'set (x, y) to the point thus obtained
if n > 10 then pset (x, y)   'plot (x, y) after the first 10 iterations
next: end
```

The result of running an adaptation of this program to data in Table 3.8.3 on a Masscomp workstation, and then printing the contents of the graphics screen is presented in Figure 3.8.3. Notice that if the size of the plotting window is decreased, for example by replacing the window call by WINDOW (0, 0)-(50, 50), then only a portion of the image is plotted, but at a higher resolution. Thus we have a simple means for "zooming in" on images of IFS attractors. The number of iterations may be increased to improve the quality of the computed image.

Exercises & Examples

8.1. Rewrite Programs 3.8.1 and 3.8.2 in a form suitable for your own computer environment, then run it and obtain hard copy of the output. Compare their performance.

8.2. Modify Programs 3.8.1 and 3.8.2 so that they will compute images associated with the IFS code given in Table 3.8.2.

8.3. Modify Program 3.8.2 so that it will compute images associated with the IFS codes given in Tables 3.8.3 and 3.8.4.

8.4. By changing the window size in Program 3.8.2, obtain images of "zooms" on the Sierpinski triangle. For example, use the following windows: (1, 1)-(50, 50); (1, 1)-(25, 25); (1, 1)-(12, 12); ... (1, 1)-(N, N). How must the total number of iterations be adjusted as a function of N in order that (approximately) the number of points which land within the window remains constant? Make a graph of the total number of iterations against the window size.

Figure 3.8.3
The result of running the Chaos Algorithm for increasing numbers of iterations. The randomly dancing point starts to suggest the structure of the attractor of the IFS given in Table 3.8.3.

8.5. What should happen, theoretically, to the sequences of images computed by Program 1 if the set A_0 is changed? What happens in practice? Make a computational experiment to see if there is any difference in say A_{10} corresponding to two different choices for A_0.

8.6. Rewrite Program 3.8.2 so that it applies the transformation w_i with probability p_i, where the probabilities are input by the user. Compare the number of iterations needed to produce a "good" rendering of the Sierpinski triangle, for the cases (a) $p_1 = 0.33,\ p_2 = 0.33,\ p_3 = 0.34$; (b) $p_1 = 0.2,\ p_2 = 0.46,\ p_3 = 0.34$; (c) $p_1 = 0.1,\ p_2 = 0.56,\ p_3 = 0.34$.

3.9 CONDENSATION SETS

There is another important way of making contraction mappings on $\mathscr{H}(X)$.

Definition 1. Let (X, d) be a metric space and let $C \in \mathscr{H}(X)$. Define a transformation w_0: $\mathscr{H}(X) \to \mathscr{H}(X)$ by $w_0(B) = C$ for all $B \in \mathscr{H}(X)$. Then w_0 is called a *condensation transformation* and C is called the associated *condensation set*.

Observe that a condensation transformation w_0: $\mathscr{H}(X) \to \mathscr{H}(X)$ is a contraction mapping on the metric space $(\mathscr{H}(X), h(d))$, with contractivity factor equal to zero, and that it possesses a unique fixed point, namely the condensation set.

Definition 2. Let $\{X, w_1, w_2, \ldots, w_N\}$ be a hyperbolic IFS with contractivity factor $0 \le s < 1$. Let w_0: $\mathscr{H}(X) \to \mathscr{H}(X)$ be a condensation transformation. Then $\{X, w_0, w_1, \ldots, w_N\}$ is called a *hyperbolic IFS with condensation*, with contractivity factor s.

Theorem 3.7.1 can be modified to cover the case of an IFS with condensation.

Theorem 3.7.1′. *Let $\{X;\ w_n\colon\ n = 0, 1, 2, \ldots, N\}$ be a hyperbolic iterated function system with condensation, with contractivity factor s. Then the transformation W: $\mathscr{H}(X) \to \mathscr{H}(X)$ defined by*

$$W(B) = \bigcup_{n=0}^{N} w_n(B)\ \forall\ B \in \mathscr{H}(X)$$

is a contraction mapping on the complete metric space $(\mathscr{H}(X), h(d))$ with contractivity factor s. That is

$$h(W(B), W(C)) \le s \cdot h(B, C)\ \forall\ B, C \in \mathscr{H}(X).$$

Its unique fixed point $A \in \mathscr{H}(\mathbf{X})$ obeys

$$A = W(A) = \bigcup_{n=0}^{N} w_n(A),$$

and is given by $A = \text{Lim}_{n \to \infty} W^{\circ n}(B)$ for any $B \in \mathscr{H}(\mathbf{X})$.

Exercises & Examples

9.1. A sequence of sets $\{A_n \subset \mathbf{X}\}_{n=0}^{\infty}$, where $(\mathbf{X}, d)$ is a metric space, is said to be *increasing* if $A_0 \subset A_1 \subset A_2 \subset \cdots$, and *decreasing* if $A_0 \supset A_1 \supset A_2 \supset \cdots$. The inclusions are not necessarily strict. A decreasing sequence of sets $\{A_n \subset \mathscr{H}(\mathbf{X})\}_{n=0}^{\infty}$ is a Cauchy sequence (prove it!). If $\mathbf{X}$ is compact, then an increasing sequence of sets $\{A_n \subset \mathscr{H}(\mathbf{X})\}_{n=0}^{\infty}$ is a Cauchy sequence (prove it!). Let $\{\mathbf{X};\ w_0, w_1, \ldots, w_N\}$ be a hyperbolic IFS with condensation set C, and let $\mathbf{X}$ be compact. Let $W_0(B) = \cup_{n=0}^{N} w_n(B)$ $\forall B \in \mathscr{H}(\mathbf{X})$ and let $W(B) = \cup_{n=1}^{N} w_n(B)$. Define $\{C_n = W_0^{\circ n}(C)\}_{n=0}^{\infty}$. Then Theorem 3.7.1′ tells us $\{C_n\}$ is a Cauchy sequence in $\mathscr{H}(\mathbf{X})$ which converges to the attractor of the IFS. Independently of the theorem observe that

$$C_n = C \cup W(C) \cup W^{\circ 2}(C) \cup \cdots \cup W^{\circ n}(C)$$

provides an increasing sequence of compact sets. It follows immediately that the limit set A obeys $W_0(A) = A$.

9.2. This example takes place in ($\mathbb{R}^2$, Euclidean). Let $C =$ $=$ $A_0 \subset \mathbb{R}^2$ denote a set which looks like a scorched pine tree standing at the origin, with its trunk perpendicular to the x-axis. Let

$$w_1\begin{pmatrix} x \\ y \end{pmatrix} = \begin{pmatrix} 0.75 & 0 \\ 0 & 0.75 \end{pmatrix}\begin{pmatrix} x \\ y \end{pmatrix} + \begin{pmatrix} 0.25 \\ 0 \end{pmatrix}.$$

Show that $\{\mathbb{R}^2;\ w_0, w_1\}$ is an IFS with condensation and find its contractivity factor. Let $A_n = W^{\circ n}(A_0)$ for $n = 1, 2, 3, \ldots$ where $W(B) = \cup_{n=0}^{N} w_n(B)$ for $B \in \mathscr{H}(\mathbb{R}^2)$. Show that A_n consists of the first $(n+1)$ pine trees reading from left to right in Figure 3.9.1.

If the first tree required 0.1% of the ink in the artist's pen to draw, and if the artist had been very meticulous in drawing the whole attractor correctly, find the total amount of ink used to draw the whole attractor.

9.3. What happens to the trees in Figure 3.9.1 if $w_1\binom{x}{y}$ is replaced by

$$w_1\begin{pmatrix} x \\ y \end{pmatrix} = \begin{pmatrix} 0.5 & 0 \\ 0 & 0.75 \end{pmatrix}\begin{pmatrix} x \\ y \end{pmatrix} + \begin{pmatrix} 0.5 \\ 0 \end{pmatrix}$$

in Exercise 9.2.

9.4. Find the attractor for the IFS with condensation $\{\mathbb{R};\ w_0, w_1\}$, where the

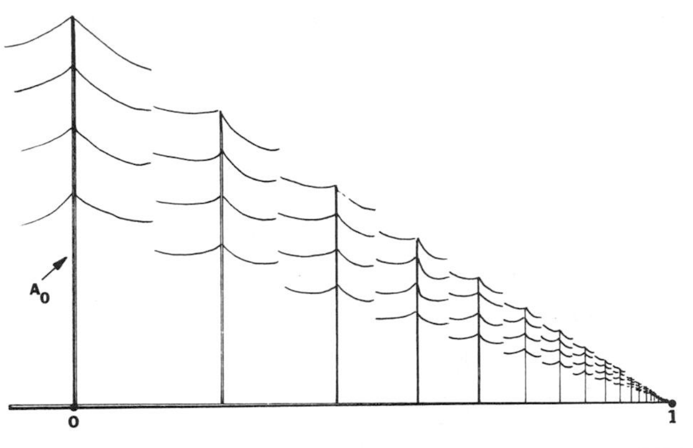

Figure 3.9.1
A geometric series of pine trees, the attractor of an IFS with condensation.

condensation set is the interval [0, 1] and $w_1(x) = \frac{1}{2}x + 2$. What happens if $w_1(x) = \frac{1}{2}x$?

9.5. Find an IFS with condensation which generates the treelike set in Figure 3.9.2. Give conditions on r and θ such that the tree is simply connected. Show that the tree is either simply connected or infinitely connected.

9.6. Find an IFS with condensation which generates Figure 3.9.3.

9.7. You are given a condensation map $w_0(x)$ in $\mathbb{R}^2$ which provides the largest tree in Figure 3.2.8. Find a hyperbolic IFS with condensation, of the form

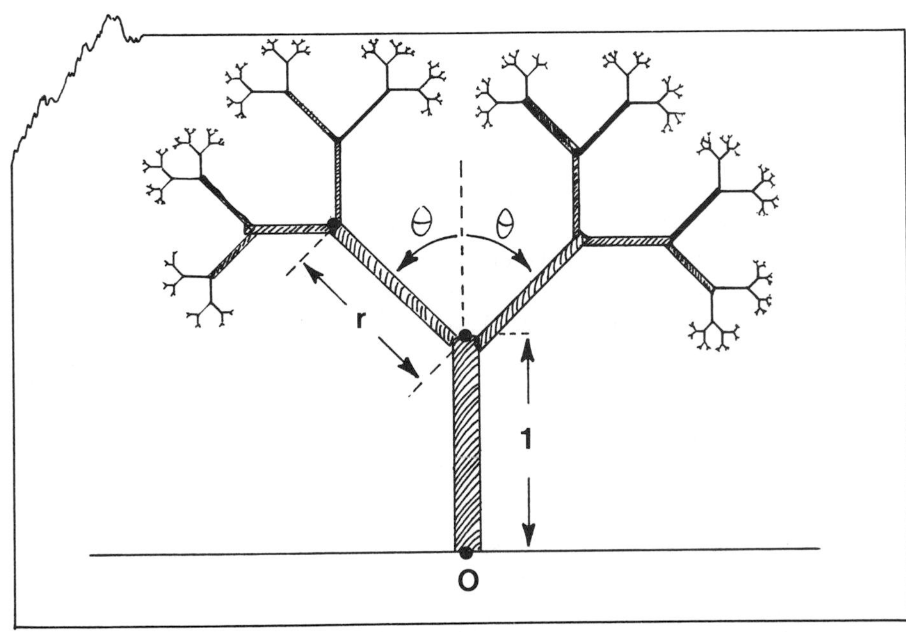

Figure 3.9.2
Sketch of a fractal tree, the attractor of an IFS with condensation.

Figure 3.9.3
An endless spiral of little men.

$\{\mathbb{R}^2; w_0, w_1, w_2\}$, which produces the whole orchard. What is the contractivity factor for this IFS? Find the attractor of the IFS $\{\mathbb{R}^2; w_1, w_2\}$.

9.8. Explain why removing the command which clears the screen ("cls") from Program 3.8.1 will result in the computation of an image associated with an IFS with condensation. Identify the condensation set. Run your version of Program 3.8.1 with the "cls" command removed.

3.10 HOW TO MAKE FRACTAL MODELS WITH THE HELP OF THE COLLAGE THEOREM

The following theorem is central to the design of IFS's whose attractors are close to given sets.

Theorem 1. [The Collage Theorem, [Barnsley, 1985b]] *Let* $(\mathbf{X}, d)$ *be a* *complete metric space. Let $L \in \mathscr{H}(\mathbf{X})$ be given, and let $\epsilon \geq 0$ be given. Choose an IFS (or IFS with condensation) $\{\mathbf{X}; (w_0), w_1, w_2, \ldots, w_N\}$ with contractivity factor $0 \leq s < 1$, so that*

$$h\left(L, \bigcup_{\substack{n=1\\(n=0)}}^{N} w_n(L)\right) \leq \epsilon,$$

where $h(d)$ is the Hausdorff metric. Then

$$h(L, A) \leq \epsilon/(1-s)$$

where A is the attractor of the IFS. Equivalently,

$$h(L, A) \leq (1-s)^{-1} h\left(L, \bigcup_{\substack{n=1\\(n=0)}}^{N} w_n(L)\right) \qquad \textit{for all } L \in \mathscr{H}(\mathbf{X}).$$

The proof of the Collage Theorem is given in the next section. The theorem tells us that to find an IFS whose attractor is "close to" or "looks like" a given set, one must endeavor to find a set of transformations—contraction mappings on a suitable space within which the given set lies—such that the union, or collage, of the images of the given set under the transformations is near to the given set. Nearness is measured using the Hausdorff metric.

Exercises & Examples

10.1. This example takes place in ($\mathbb{R}$, Euclidean). Observe that $[0,1] = [0, \frac{1}{2}] \cup [\frac{1}{2}, 1]$. Hence $[0,1]$ is the attractor for any pair of contraction mappings $w_1: \mathbb{R} \to \mathbb{R}$ and $w_2: \mathbb{R} \to \mathbb{R}$ such that $w_1([0,1]) = [0, \frac{1}{2}]$ and $w_2([0,1]) = [\frac{1}{2}, 1]$. For example $w_1(x) = \frac{1}{2}x$ and $w_2(x) = \frac{1}{2}x + \frac{1}{2}$ does the trick. The unit interval is a collage of two smaller "copies" of itself.

10.2. Suppose we are using a trial-and-error procedure to adjust the coefficients in two affine transformations $w_1(x) = ax + b$, $w_2(x) = cx + d$, where $a, b, c, d \in \mathbb{R}$, to look for an IFS $\{\mathbb{R}; w_1, w_2\}$ whose attractor is $[0,1]$. We might come up with $w_1(x) = 0.51x - 0.01$ and $w_2(x) = 0.47x + 0.53$. How far from $[0,1]$ will the attractor for the IFS be? To find out compute

$$h\left([0,1], \bigcup_{i=1}^{2} w_i([0,1])\right) = h([0,1], [-0.01, 0.5] \cup [0.53, 1])$$
$$= 0.015,$$

and observe that the contractivity factor of the IFS is $s = 0.51$. So by the Collage Theorem, if A is the attractor,

$$h([0,1], A) \leq 0.015/0.49 < 0.04.$$

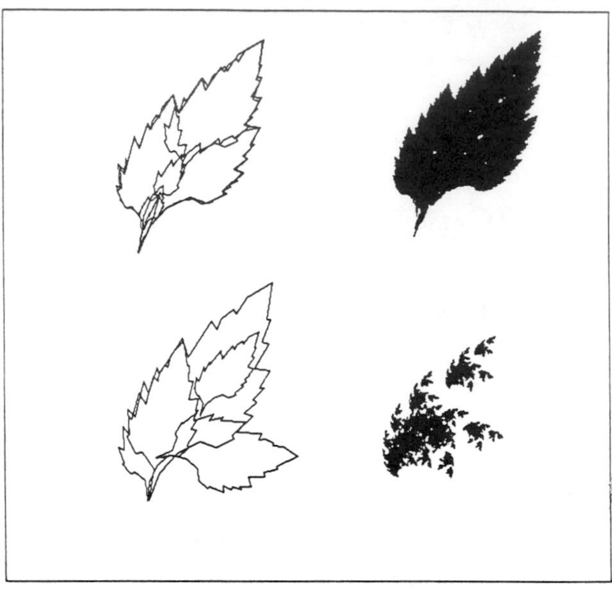

Figure 3.10.1
The Collage Theorem applied to a region bounded by a polygonalized leaf boundary.

10.3. Figure 3.10.1 shows a target set $L \subset \mathbb{R}^2$, a leaf, represented by the polygonalized boundary of the leaf. Four affine transformations, contractive, have been applied to the boundary at lower left, producing the four smaller deformed leaf boundaries. The Hausdorff distance between the union of the four copies and the original is approximately 1.0 units, where the width of the whole frame is taken to be 10 units. The contractivity of the associated IFS $\{\mathbb{R}^2, w_1, w_2, w_3, w_4\}$ is approximately 0.6. Hence the Hausdorff distance h (Euclidean) between the original target leaf L and the attractor A of the IFS will be less than 2.5 units. (This is not promising much!) The actual attractor, translated to the right, is shown at lower right. Not surprisingly, it does not look much like the original leaf! An improved collage is shown at the upper left. The distance $h(L, \cup_{n=1}^{4} w_n(L))$ is now less than 0.02 units whilst the contractivity of the IFS is still approximately 0.6. Hence $h(L, A)$ should now be less than 0.05 units and we expect that the attractor should look quite like L at the resolution of the figure. A, translated to the right, is shown at the upper right.

10.4. To find an IFS whose attractor is a region bounded by a right angle triangle, observe the collage in Figure 3.10.2.

10.5. A nice proof of Pythagoras' Theorem is obtained from the collage in Figure 3.10.2. Clearly both transformations involved are similitudes.

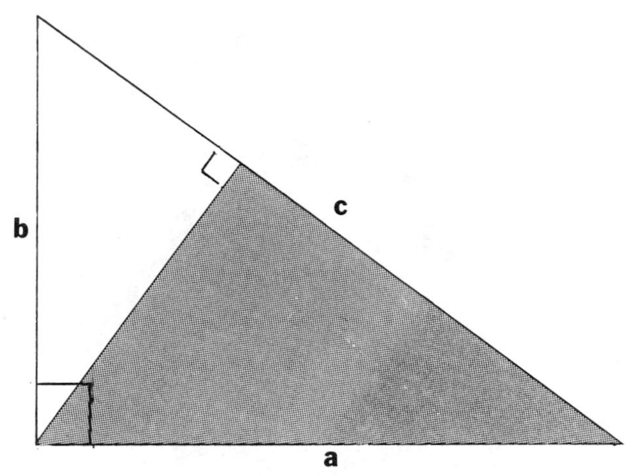

Figure 3.10.2
The region bounded by a right-angle triangle is the union of the results of two similitudes applied to it.

The contractivity factors of these similitudes involved are (b/c) and (a/c). Hence the area $\mathscr{A}$ obeys $\mathscr{A} = (b/c)^2\mathscr{A} + (a/c)^2\mathscr{A}$. This implies $c^2 = a^2 + b^2$ since $\mathscr{A} > 0$.

10.6. Figures 3.10.3 through 3.10.7 provide exercises in the application of the Collage Theorem. Condensation sets are not allowed when working these examples!

10.7. It is straightforward to see how the Collage Theorem gives us sets of

Figure 3.10.3
Use the Collage Theorem to help you find an IFS consisting of two affine maps in $\mathbb{R}^2$ whose attractor is close to this set.

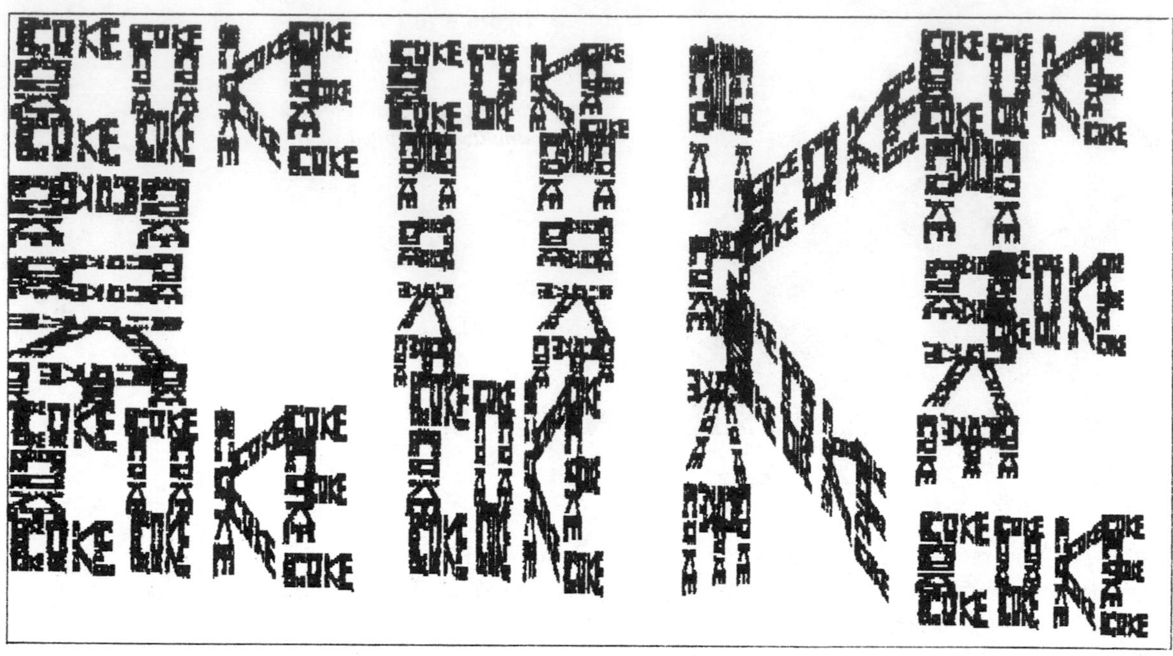

Figure 3.10.4
This image represents the attractor of fourteen affine transformations in $\mathbb{R}^2$. Use the Collage Theorem to help you find them.

Figure 3.10.5
Use the Collage Theorem to help find a hyperbolic IFS of the form $\{\mathbb{R}^2; w_1, w_2, w_3\}$, where w_1, w_2, and w_3 are similitudes in $\mathbb{R}^2$, whose attractor is represented here. You choose the coordinate system.

Figure 3.10.6
Find an IFS of the form $\{\mathbb{R}^2; w_1, w_2, w_3, w_4\}$, *where the* w_i*'s are affine transformations on* $\mathbb{R}^2$, *whose attractor when rendered contains this image. Check your conclusion using Program* 3.8.2.

maps for IFS's which generate . Menger Sponges, look like this: . Find an IFS for which a sponge is the attractor.

10.8. The IFS which generates the *Black Spleenwort* fern, shown in Figures

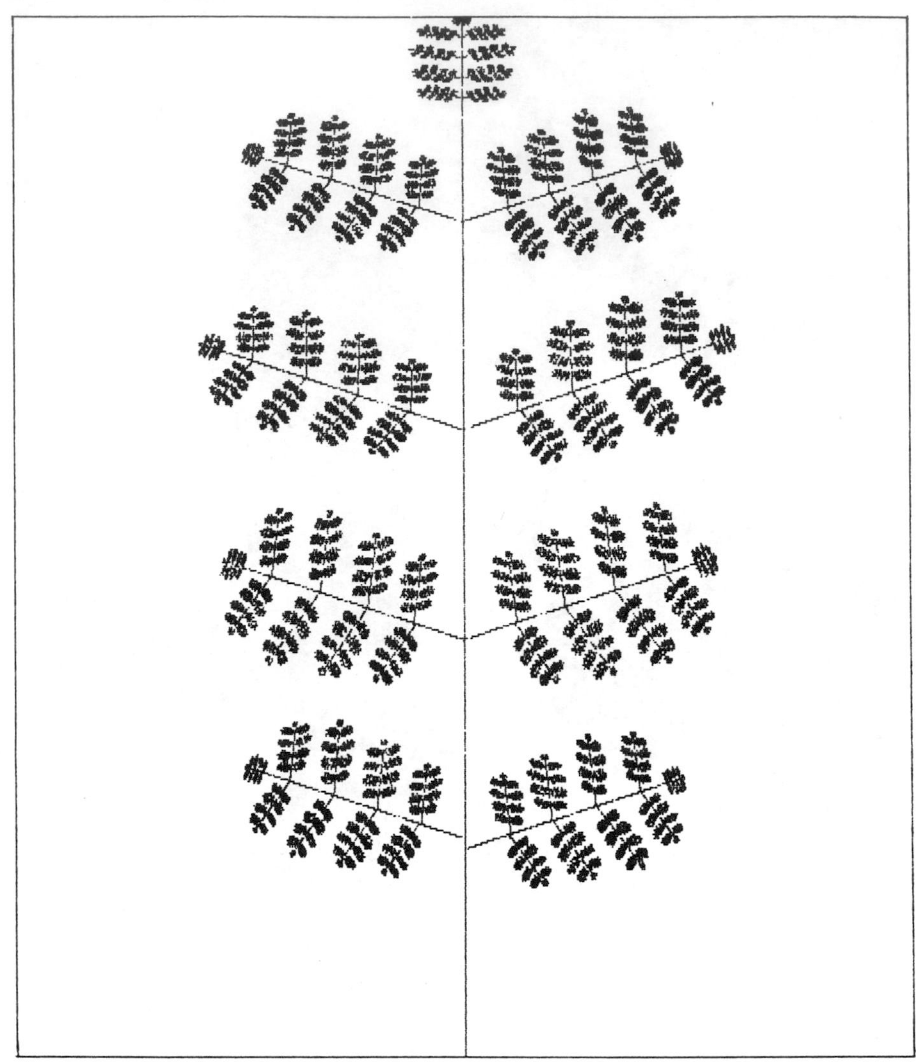

Figure 3.10.7 *How many affine transformations in $\mathbb{R}^2$ are needed to generate this attractor? You do not need to use a condensation set.*

3.10.8(a) and (b), consists of four affine maps in the form

$$w_i\begin{pmatrix} x \\ y \end{pmatrix} = \begin{pmatrix} r\cos\theta & -s\sin\phi \\ r\sin\theta & s\cos\phi \end{pmatrix}\begin{pmatrix} x \\ y \end{pmatrix} + \begin{pmatrix} h \\ k \end{pmatrix} \qquad (i = 1, 2, 3, 4);\ \text{see Table 3.10.1.}$$

10.9. Find a collage of affine transformations in $\mathbb{R}^2$, corresponding to Figure 3.10.9.

10.10. A collage of a leaf is shown in Figure 3.10.10(a). This collage implies

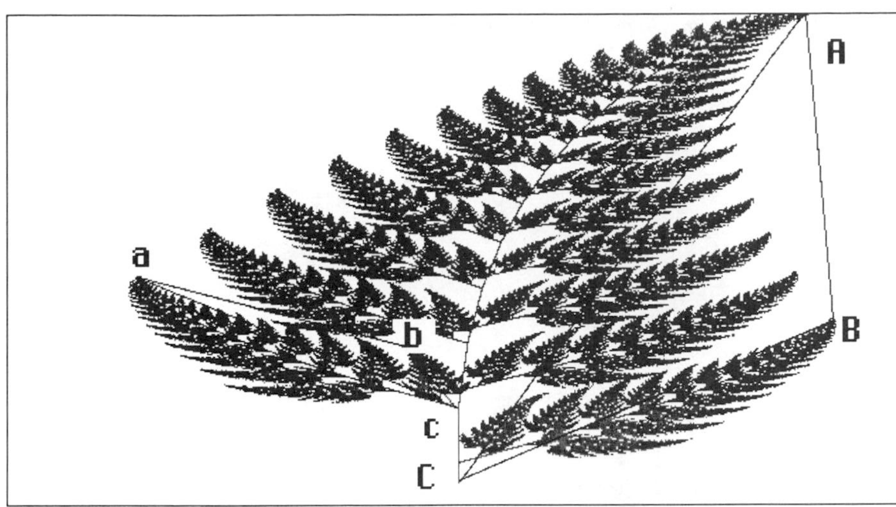

Figure 3.10.8(a) *The Black Spleenwort fern. This image illustrates one of the four affine transformations in the IFS whose attractor was used to render the fern. The transformation takes the triangle ABC to the triangle abc. The Collage Theorem provides the other three transformations. The IFS code for this image is given in Table* 3.8.3. *Observe that the stem is the image of the whole set under one of the transformations. Determine to which map number in Table* 3.8.3 *the stem corresponds.*

Figure 3.10.8(b) *The Black Spleenwort fern and a close-up.*

the IFS $\{\mathbb{C}; w_1, w_2, w_3, w_4\}$ where, in complex notation,

$$w_i(z) = s_i z + (1 - s_i) a_i \qquad \text{for } i = 1, 2, 3, 4.$$

Verify that in this formula a_i is the fixed point of the transformation. The values found for s_i and a_i are listed in Table 3.10.2. Check that these make sense in relation to the collage. The attractor for the IFS is shown in Figure 3.10.10(b).

Table 3.10.1
The IFS code for the Black Spleenwort, expressed in scale and angle format.

	Translations		*Rotations*		*Scalings*	
Map	h	k	θ	ϕ	r	s
1	0.0	0.0	0	0	0.0	0.16
2	0.0	1.6	−2.5	−2.5	0.85	0.85
3	0.0	1.6	49	49	0.3	0.34
4	0.0	0.44	120	−50	0.3	0.37

10.11. The attractor in Figure 3.10.11 is determined by two affine maps. Locate the fixed points of two such affine transformations on $\mathbb{R}^2$.

10.12. Figure 3.10.12 shows the attractor for an IFS $\{\mathbb{R}^3;\ w_i,\ i = 1, 2, 3, 4\}$ where each w_i is a three-dimensional affine transformation. See also Color Plate 3.10.1. The attractor is contained in the region $\{(x_1, x_2, x_3) \in \mathbb{R}^3\colon -10 \le x_1 \le 10, 0 \le x_2 \le 10, -10 \le x_3 \le 10\}$.

Figure 3.10.9
Use the Collage Theorem to find the four affine transformations corresponding to this image. Can you find a transformation which will put in the "missing corner?"

Table 3.10.2
Scaling factors and fixed points for the collage in Figure 3.10.10.

s	a
0.6	0.45 + 0.9i
0.6	0.45 + 0.3i
0.4 − 0.3i	0.60 + 0.3i
0.4 + 0.3i	0.30 + 0.3i

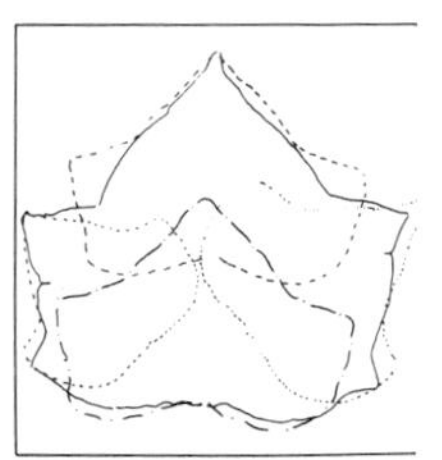

(a) Collage

(b) Attractor

Figure 3.10.10
A collage of a leaf is obtained using four similitudes, as illustrated in (a). The corresponding IFS is presented in complex notation in Table 3.10.2. The attractor of the IFS is rendered in (b).

$$w_1\begin{bmatrix}x_1\\x_2\\x_3\end{bmatrix}=\begin{bmatrix}0&0&0\\0&0.18&0\\0&0&0\end{bmatrix}\begin{bmatrix}x_1\\x_2\\x_3\end{bmatrix}+\begin{bmatrix}0\\0\\0\end{bmatrix}$$

$$w_2\begin{bmatrix}x_1\\x_2\\x_3\end{bmatrix}=\begin{bmatrix}0.85&0&0\\0&0.85&0.1\\0&-0.1&0.85\end{bmatrix}\begin{bmatrix}x_1\\x_2\\x_3\end{bmatrix}+\begin{bmatrix}0\\1.6\\0\end{bmatrix}$$

$$w_3\begin{bmatrix}x_1\\x_2\\x_3\end{bmatrix}=\begin{bmatrix}0.2&0.2&0\\0.2&0.2&0\\0&0&0.3\end{bmatrix}\begin{bmatrix}x_1\\x_2\\x_3\end{bmatrix}+\begin{bmatrix}0\\0.8\\0\end{bmatrix}$$

$$w_4\begin{bmatrix}x_1\\x_2\\x_3\end{bmatrix}=\begin{bmatrix}-0.2&0.2&0\\0.2&0.2&0\\0&0&0.3\end{bmatrix}\begin{bmatrix}x_1\\x_2\\x_3\end{bmatrix}+\begin{bmatrix}0\\0.8\\0\end{bmatrix}$$

10.13. Find an IFS of similitudes in $\mathbb{R}^2$ such that the attractor is represented by the shaded region in Figure 3.10.13. The collage should be "just-

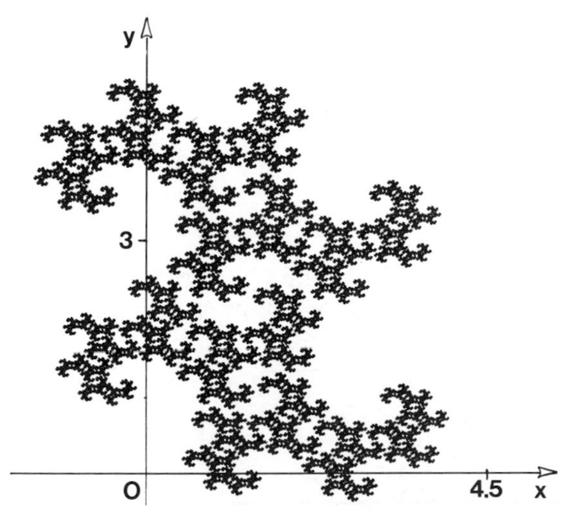

Figure 3.10.11
Locate the fixed points of a pair of affine transformations in $\mathbb{R}^2$ whose attractor is rendered here.

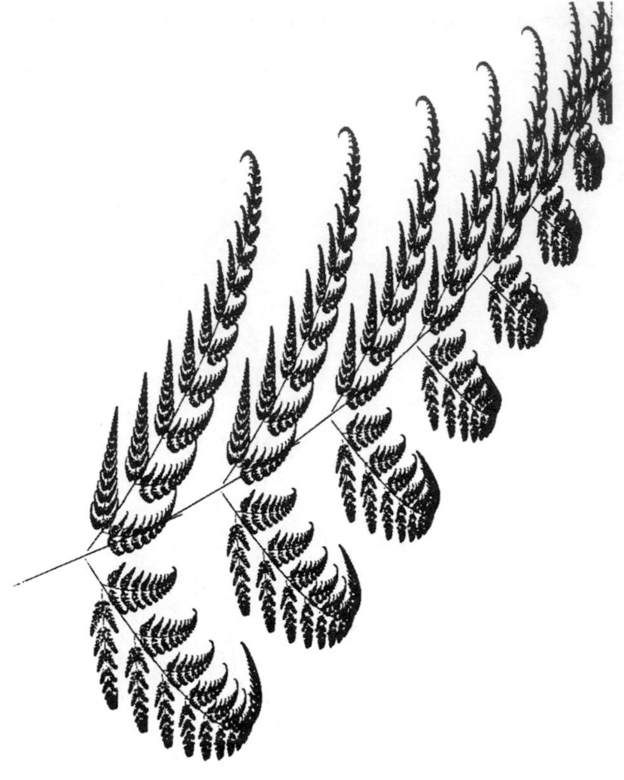

Figure 3.10.12
Single three-dimensional fern. The attractor of an IFS of affine maps in $\mathbb{R}^3$.

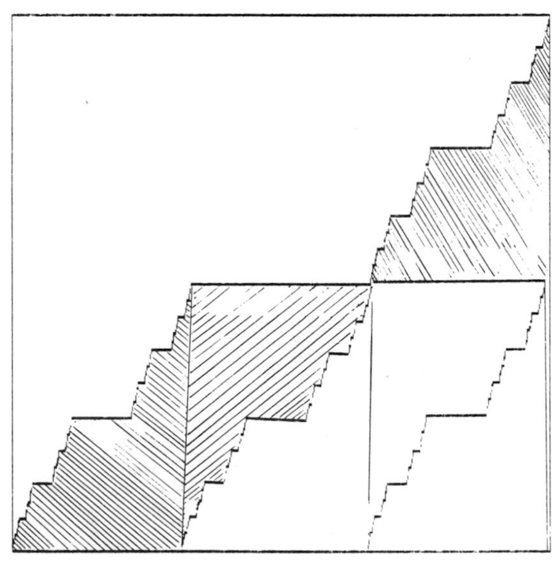

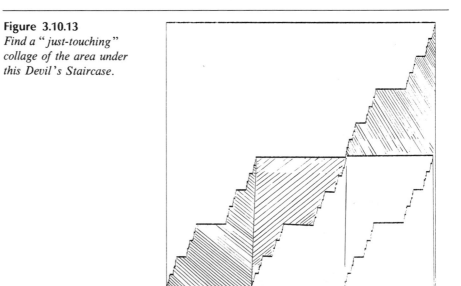

Figure 3.10.13
Find a "just-touching" collage of the area under this Devil's Staircase.

touching," by which we mean that the transforms of the region provide a tiling of the region: they should fit together like the pieces of a jigsaw puzzle.

10.14. This exercise suggests how to change the coordinates of an IFS. Let $\{X_1, d_1\}$ and $\{X_2, d_2\}$ be metric spaces. Let $\{X_1; w_1, w_2, \ldots, w_N\}$ be

Figure 3.10.14
Determine some of the affine transformations used in the design of this fractal scene. For example, where do the dark sides of the largest mountain come from?

Figure 3.10.15
"Typical" fractals are not pretty: use the Collage Theorem to find an IFS whose attractor approximates this set.

Figure 3.10.16
Determine the affine transformations for an IFS corresponding to this fractal. Can you see, just by looking at the picture, if the linear part of any of the transformations has a negative determinant?

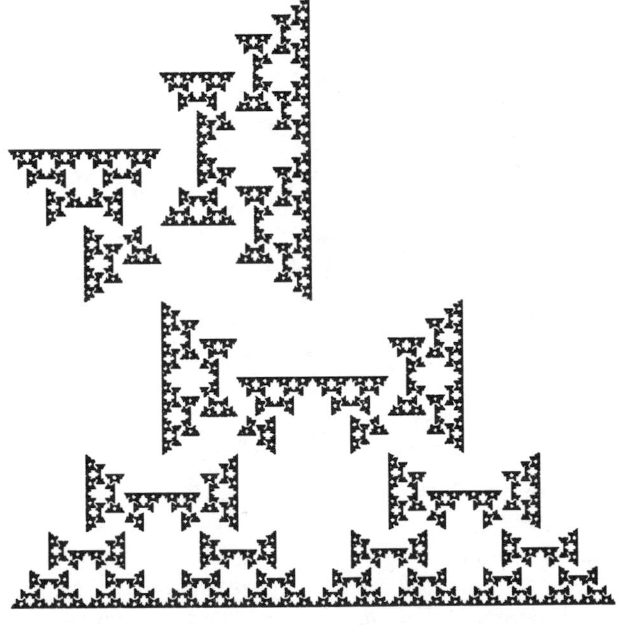

Figure 3.10.17
Use the Collage Theorem to analyse this fractal. On how many different scales is the whole image apparently repeated here? How many times is the smallest clearly discernible copy repeated?

a hyperbolic IFS with attractor A_1. Let θ: $X_1 \to X_2$ be an invertible continuous transformation. Consider the IFS $\{X_2;\ \theta \circ w_1 \circ \theta^{-1}, \theta \circ w_2 \circ \theta^{-1}, \ldots, \theta \circ w_N \circ \theta^{-1}\}$. Use θ to define a metric on X_2 such that the new IFS is indeed a hyperbolic IFS. Prove that if $A_2 \in \mathscr{H}(X_2)$ is the attractor of the new IFS, then $A_2 = \theta(A_1)$. Thus we can readily construct an IFS whose attractor is a transform of the attractor of another IFS.

Figure 3.10.18 *Consider the white areas in this figure to represent a set S in* $\mathbb{R}^2$. *Locate the boundary of the largest pathwise connected subset of S. It is recommended that you work with a photocopy of the image, a magnifying glass, and a fine red felt-tip pen.*

10.15. Find some of the affine transformations used in the design of the fractal scene in Figure 3.10.14.

10.16. Use the Collage Theorem to find an IFS whose attractor approximates the set in Figure 3.10.15.

10.17. Solve the problems proposed in the captions of (a) Figure 3.10.16, (b) Figure 3.10.17, (c) Figure 3.10.18.

3.11 BLOWING IN THE WIND: THE CONTINUOUS DEPENDENCE OF FRACTALS ON PARAMETERS

The Collage Theorem provides a way of approaching the inverse problem: Given a set L, find an IFS for which L is the attractor. The underlying mathematical principle is very easy: the proof of the Collage Theorem is just the proof of the following lemma.

Lemma 1. *Let $(\mathbf{X}, d)$ be a complete metric space. Let $f\colon \mathbf{X} \to \mathbf{X}$ be a contraction mapping with contractivity factor $0 \le s < 1$, and let the fixed point of f be $x_f \in \mathbf{X}$. Then*

$$d(x, x_f) \le (1-s)^{-1} \cdot d(x, f(x)) \qquad \text{for all } x \in \mathbf{X}.$$

Proof. The distance function $d(a, b)$ for fixed $a \in \mathbf{X}$ is continuous in $b \in \mathbf{X}$. Hence

$$\begin{aligned} d(x, x_f) &= d\Big(x, \lim_{n\to\infty} f^{\circ n}(x)\Big) = \lim_{n\to\infty} d(x, f^{\circ n}(x)) \\ &\le \lim_{n\to\infty} \sum_{m=1}^{n} d\big(f^{\circ(m-1)}(x), f^{\circ(m)}(x)\big) \\ &\le \lim_{n\to\infty} d(x, f(x))(1 + s + \cdots + s^{n-1}) \le (1-s)^{-1} d(x, f(x)) \end{aligned}$$

This completes the proof.

The following results are important and closely related to the above material. They establish the continuous dependence of the attractor of a hyperbolic IFS on parameters in the maps which constitute the IFS.

Lemma 2. *Let (P, d_p) and $(\mathbf{X}, d)$ be metric spaces, the latter being complete. Let $w\colon P \times \mathbf{X} \to \mathbf{X}$ be a family of contraction mappings on $\mathbf{X}$ with contractivity factor $0 \le s < 1$. That is, for each $p \in P$, $w(p, \cdot)$ is a contraction mapping on $\mathbf{X}$. For each fixed $x \in \mathbf{X}$ let w be continuous on P. Then the fixed point of w depends continuously on p. That is, $x_f\colon P \to \mathbf{X}$ is continuous.*

Proof. Let $x_f(p)$ denote the fixed point of w for fixed $p \in P$. Let $p \in P$ and $\epsilon > 0$ be given. Then for all $q \in P$,

$$\begin{aligned} d(x_f(p), x_f(q)) &= d\big(w(p, x_f(p)), w(q, x_f(q))\big) \\ &\le d\big(w(p, x_f(p)), w(q, x_f(p))\big) \\ &\quad + d\big(w(q, x_f(p)), w(q, x_f(q))\big) \\ &\le d\big(w(p, x_f(p)), w(q, x_f(p))\big) + s d(x_f(p), x_f(q)) \end{aligned}$$

which implies

$$d(x_f(p), x_f(q)) \leq (1-s)^{-1} d(w(p, x_f(p)), w(q, x_f(p))).$$

The right-hand side here can be made arbitrarily small by restricting q to be sufficiently close to p. (Notice that if there is a real constant C such that

$$d(w(p, x), w(q, x)) \leq C d(p, q) \qquad \text{for all } p, q \in P, \text{ for all } x \in \mathbf{X}$$

then $d(x_f(p), x_f(q)) \leq (1-s)^{-1} \cdot C \cdot d(p, q)$, which is a useful estimate.) This completes the proof.

Exercises & Examples

11.1. The fixed point of the contraction mapping w: $\mathbb{R} \to \mathbb{R}$ defined by $w(x) = \frac{1}{2}x + p$ depends continuously on the real parameter p. Indeed, $x_f = 2p$.

11.2. Show that the fixed function for the transformation w: $C^0[0,1] \to C^0[0,1]$ defined by $w(f(x)) = pf(2x \bmod 1) + x(1-x)$ is continuous in p for $p \in (-1, 1)$. Here, $C^0[0,1] = \{f \in C[0,1]\text{: } f(0) = f(1) = 0\}$ and the distance is $d(f, g) = \text{Max}\{|f(x) - g(x)|\text{: } x \in [0,1]\}$.

Lemma 3. *Let $(\mathbf{X}, d)$ be a metric space and suppose we have continuous transformations w_n: $\mathbf{X} \to \mathbf{X}$ $(n = 1, 2, \ldots, N)$ depending continuously on a parameter $p \in P$, where (P, d_p) is a compact metric space. That is $w_n(p, x)$ depends continuously on p for fixed $x \in \mathbf{X}$. Then the transformation W: $\mathscr{H}(\mathbf{X}) \to \mathscr{H}(\mathbf{X})$ defined by*

$$W(p, B) = \bigcup_{n=1}^{N} w_n(p, B) \ \forall B \in \mathscr{H}(\mathbf{X})$$

is also continuous in p. That is, $W(p, B)$ is continuous in p for each $B \in \mathscr{H}(\mathbf{X})$, in the metric space $(\mathscr{H}(\mathbf{X}), h(d))$.

Proof. It suffices to consider the case $N = 1$, and then extend the result using Lemma 3.7.4. For $B \in \mathscr{H}(\mathbf{X})$ we have for $p, q \in P$, and given $\epsilon > 0$,

$$\begin{aligned} d(w_1(p, B), w_1(q, B)) &= \operatorname*{Max}_{x \in B} \operatorname*{Min}_{y \in B} d(w_1(p, x), w_1(q, y)) \\ &\leq \operatorname*{Max}_{x \in B} \operatorname*{Min}_{y \in B} \{ d(w_1(p, x), w_1(p, y)) \\ &\qquad + d(w_1(p, y), w_1(q, y)) \} \end{aligned}$$

Now $P \times B$ is compact and w_1: $P \times B \to \mathbf{X}$ is continuous. Hence w_1 is uniformly continuous: There is a number $\delta > 0$ so that $d(w_1(p, y), w_1(q, y))$

$< \epsilon$ for all $y \in B$, whenever $d_p(p, q) < \delta$. So assuming $d_p(p, q) < \delta$ we have

$$d(w_1(p, B), w_1(q, B)) < \underset{x \in B}{\text{Max}} \underset{y \in B}{\text{Min}} \{ d(w_1(p, x), w_1(p, y)) + \epsilon \}$$

$$\leq d(w_1(p, B), w_1(p, B)) + \epsilon = \epsilon.$$

Similarly

$$d(w_1(q, B), w_1(p, B)) < \epsilon \text{ for } d_p(p, q) < \delta, \qquad \text{and we deduce}$$

$$h(w_1(p, B), w_1(q, B)) < \epsilon \text{ for } d_p(p, q) < \delta.$$

This completes the proof.

We now combine Lemmas 2 and 3 to obtain the result we want.

Theorem 1. *Let $(\mathbf{X}, d)$ be a metric space. Let $\{\mathbf{X}; (w_0), w_1, w_2, \ldots, w_N\}$ be a hyperbolic IFS (with condensation), of contractivity s. For $n = 1, 2, \ldots, N$, let w_n depend continuously on a parameter $p \in P$, where P is a compact metric space. Then the attractor $A(p) \in \mathscr{H}(\mathbf{X})$ depends continuously on $p \in P$, with respect to the Hausdorff metric $h(d)$.*

In other words, small changes in the parameters will lead to small changes in the attractor, provided that the system remains hyperbolic. This is very important because it tells us that we can continuously control the attractor of an IFS, by adjusting parameters in the transformations, as is done in image compression applications. It also means we can smoothly interpolate between attractors: this is useful for image animation, for example. The frames from the video "A Cloud Study" [Barnsley, 1987a] shown in Color Slide 3.11.1 provides an illustration of the application of this technique.

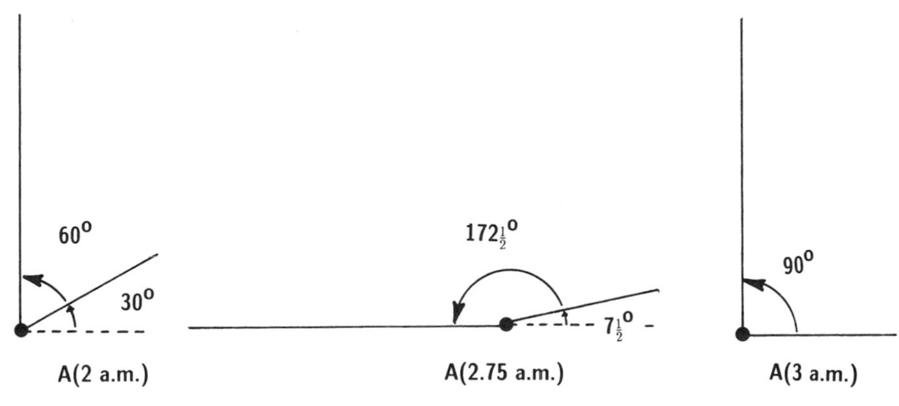

Figure 3.11.1 *A one-parameter family of IFS which tells the time!*

Exercises & Examples

11.3. Construct a one-parameter family of IFS, of the form $\{\mathbb{R}^2; w_1, w_2, w_3\}$, where each w_i is affine and the parameter p lies in the interval $[0, 24]$. The attractor should tell the time, as illustrated in Figure 3.11.1. $A(p)$ denotes the attractor at time p.

11.4. Imagine a slightly more complicated clockface, generated using a one-parameter family of IFS of the form $\{\mathbb{R}^2; w_0, w_1, w_2, w_3\}$, $p \in [0, 24]$. w_0 creates the clockface, w_1 and w_2 are as in Exercise 11.3, and w_3 is a similitude which places a copy of the clockface on the hour hand, as illustrated in Figure 3.11.2. Then as p goes from 0 to 24, the hour hand sweeps through 360°, and the hour hand on the smaller clockface sweeps through 720°, and the hour hand on the yet smaller clockface sweeps

Figure 3.11.2 *This fractal clockface depends continuously on time in the Hausdorff metric.*

through 1080°, and so on. Thus as p advances, there exist lines on the attractor which are rotating at arbitrarily great speeds. Nonetheless we have continuous dependence of the image on p in the Hausdorff metric! At what times do all of the hour hands point in the same direction?

11.5. Find a one-parameter family of IFS in $\mathbb{R}^2$, whose attractors include the three trees in Figure 3.11.3.

11.6. Run your version of Program 3.8.1 or Program 3.8.2, making small changes in the IFS code. Convince yourself that resulting rendered images "vary continuously" with respect to these changes.

11.7. Solve the following problems with regard to the images (a)-(f) in Figure 3.11.4. Recall that a "just-touching" collage in $\mathbb{R}^2$ is one where the

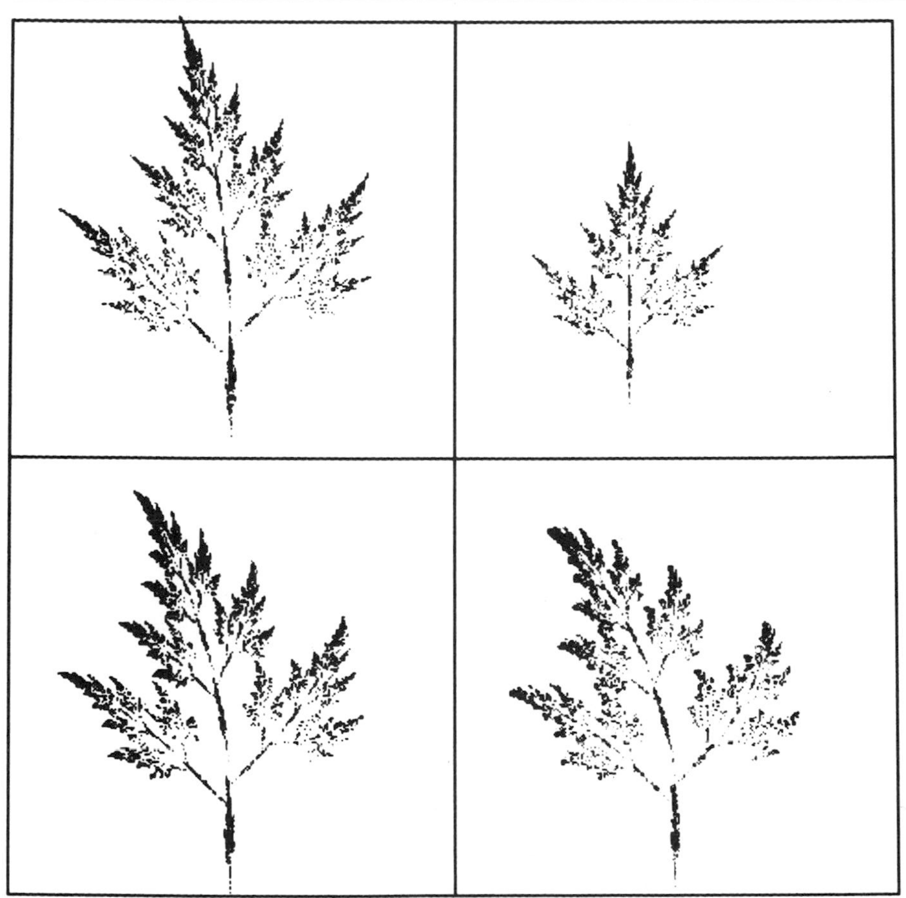

Figure 3.11.3 *Blowing in the wind. Find a one-parameter family of IFS whose attractors include the trees shown here. The Random Iteration Algorithm was used to compute these images.*

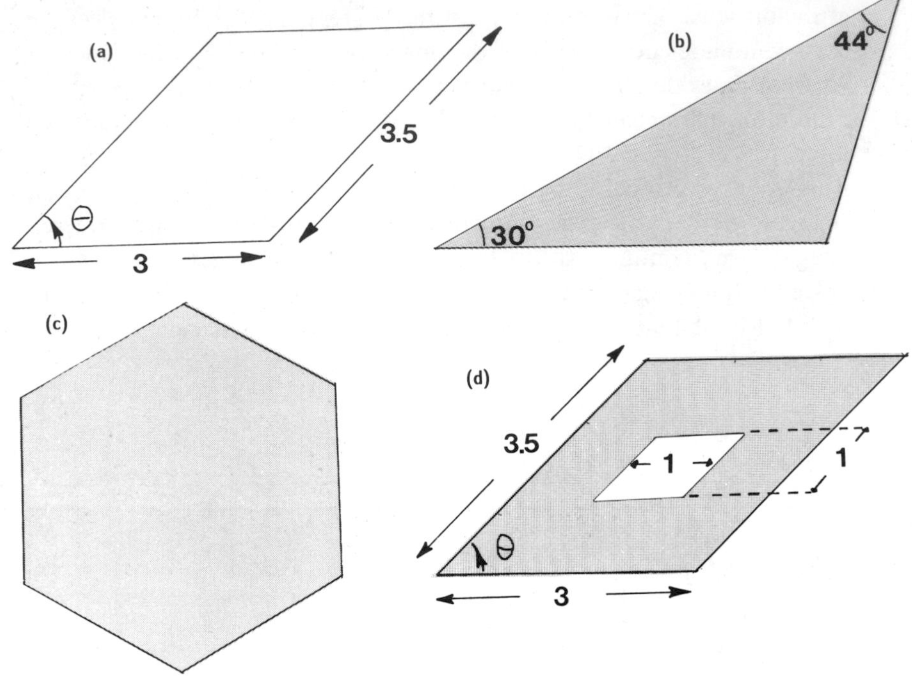

Figure 3.11.4(a)–(d) *Classical Collages. Can you find an IFS corresponding to each of these classical geometrical objects?*

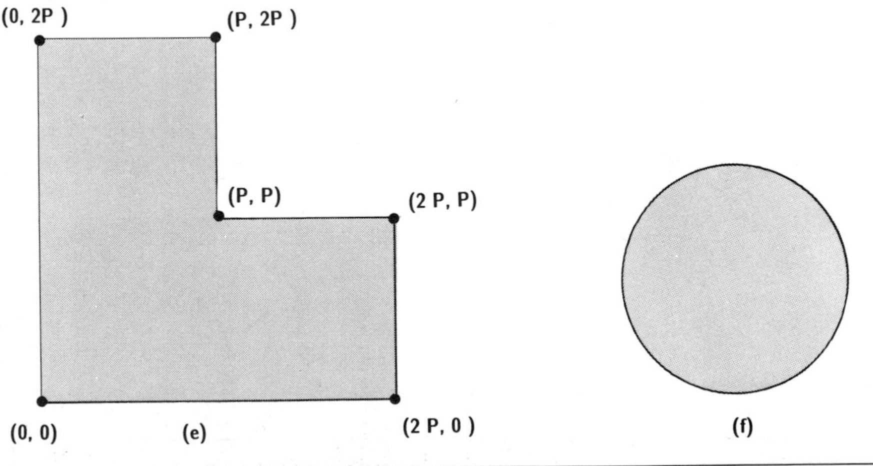

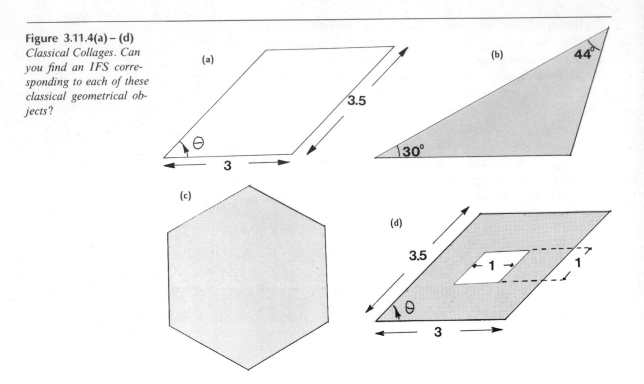

Figure 3.11.4(e), (f) *Classical Collages. Can you find an IFS corresponding to each of these Euclidean objects?*

Plate 2.6.1
Color photograph of the metric space (■, Manhattan) painted by $d(x, \mathscr{F})$ where $\mathscr{F}$ is a fern-like subset of ■.

Plate 3.10.1
Three dimensional ferns.

Plate 7.1.1
The result of running a version of Program 7.1.1 on a Masscomp 5600 workstation with Aurora graphics.

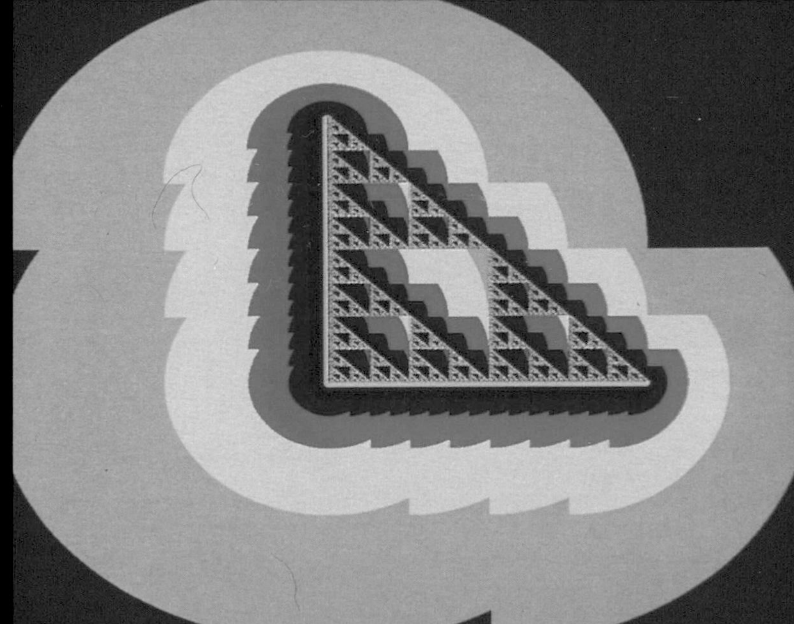

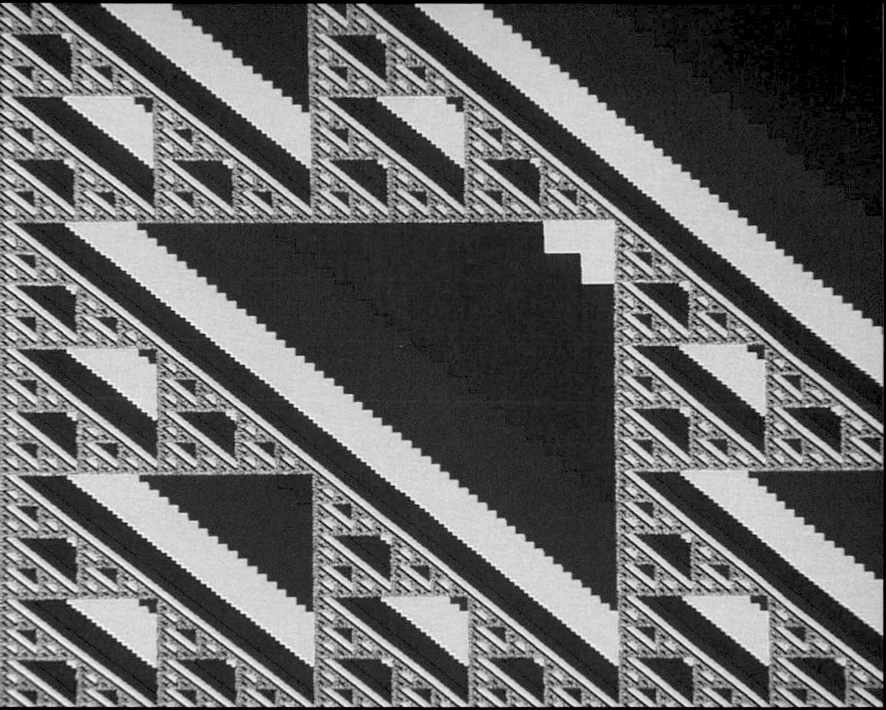

Plate 7.1.2
The result of running a version of Program 7.1.1, with $(a, b) = (0, 0), (c, d) = (5 \times 10^{-18}, 5 \times 10^{-18})$ *and numits* $= 65$. *This viewing window is minute, the colors arranged in a wonderful pattern: we are stunned by the geometrical complexity of this image at such high magnification.*

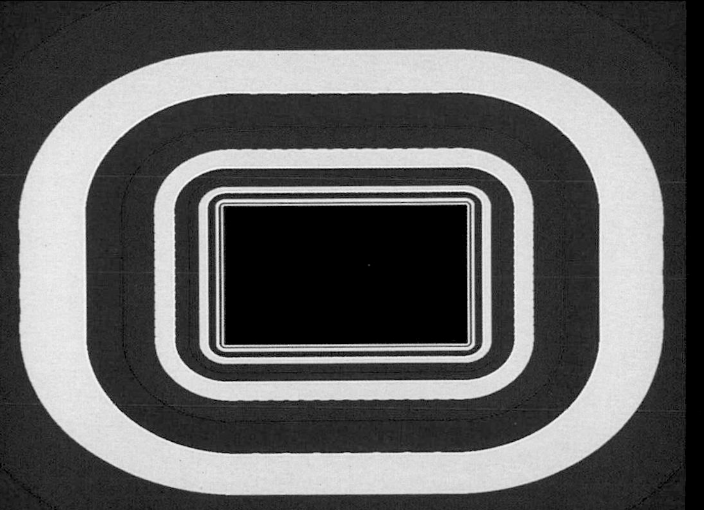

Plate 7.1.3
A rectangular "Julia Set" computed using the Escape Time Algorithm applied to an IFS.

Plate 7.1.4
The Escape Time Algorithm applied to an IFS. The attractor is embedded in the space like a jewel.

Plate 7.1.5
The Escape Time Algorithm applied to an IFS of affine transformations. Such images can be magnified enormously, without loss of mystery.

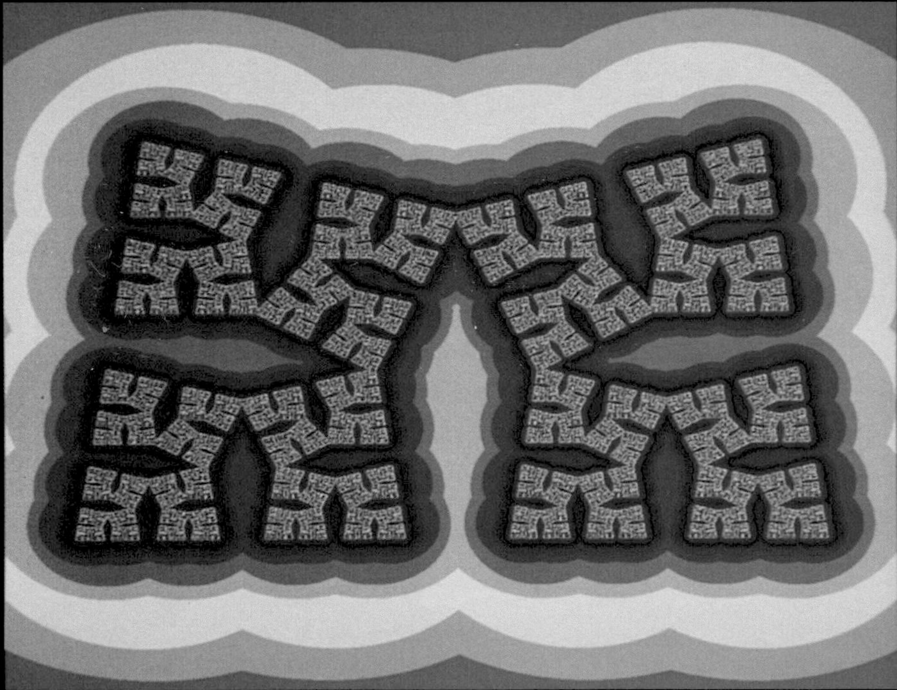

Plate 7.1.6
The Escape Time Algorithm applied to an IFS of affine transformations results in a good textile design.

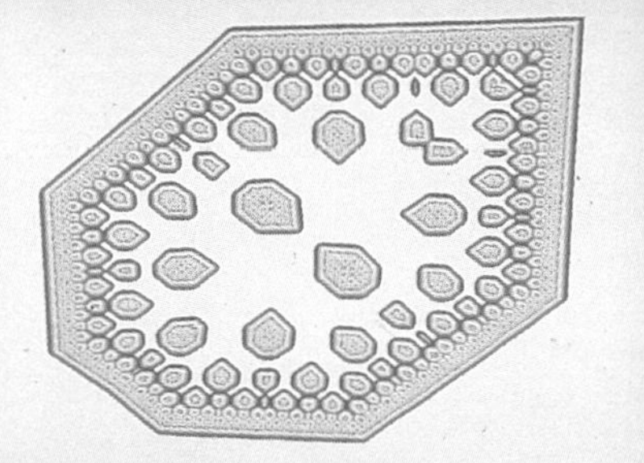

Plate 7.1.7
The Escape Time Algorithm is applied to an overlapping IFS of two affine transformations. Can such sets be used to model sections of plants, as can be seen under a microscope?

Plate 7.1.8
The Escape Time Algorithm is applied to an IFS of two maps. Here the magnification is extreme. Do not race by. Look closely and think.

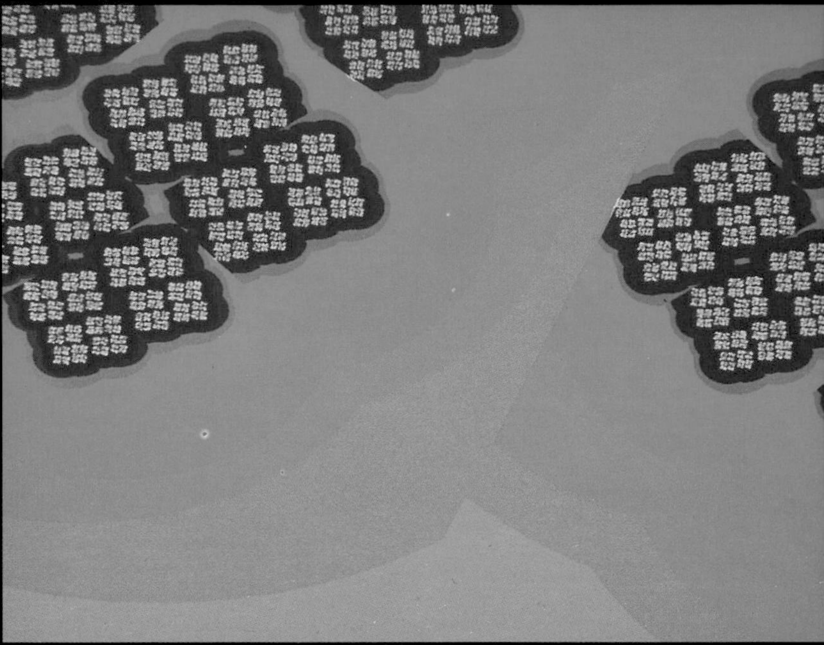

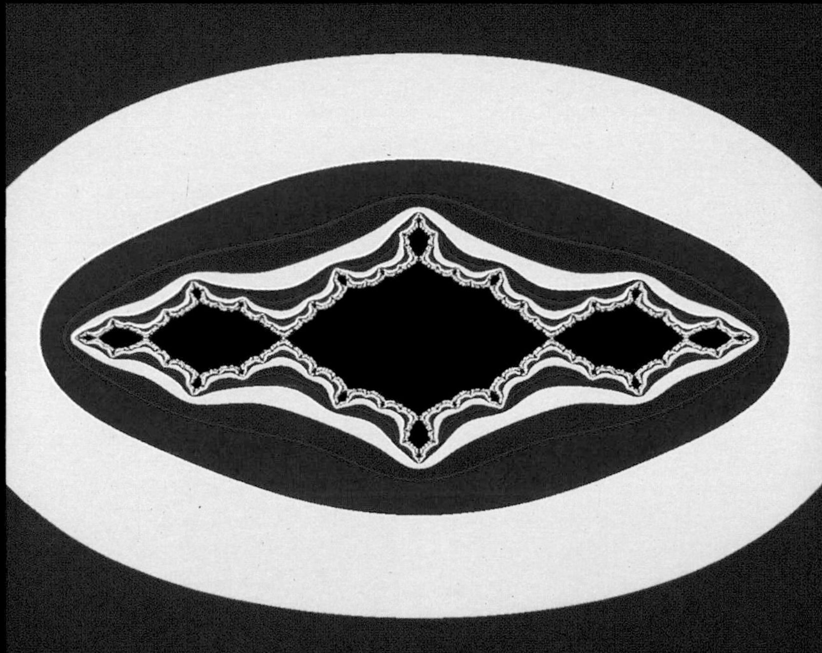

Plate 7.1.9
The Escape Time Algorithm is applied to produce an image of the Julia set for $z^2 - 1$.

Plate 7.3.1
The Escape Time Algorithm is used to analyse the Newton transformation associated with the polynomial $z^4 - 1$. *The Riemann sphere is mapped onto a rectangle using Mercator's projection. The North Pole, at the top corresponds to the Point at Infinity, w*[illegible] *South Pole, at the bottom, corresponds to the Origin. The roots are* $a_1 = 1$, $a_2 = -1$, $a_3 = i$, $a_4 = -$[illegible] *Points whose orbits converge to* a[illegible] *are plotted in white; points whose orbits converge to* a_2 *are plotted in dark blue; points whose orbits converge to* a_3 *are plotted in light blue; and points whose orbits converge* a_4 *are plotted in turquoise.*

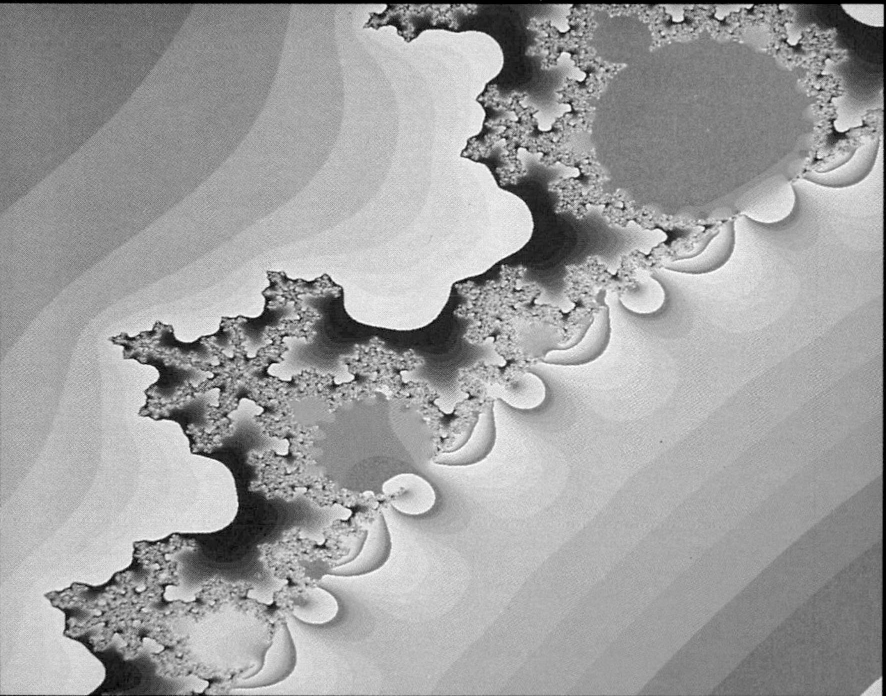

Plate 8.3.1
Part of the Mandelbrot set for $z^2 - \lambda$, *painted in colors according to "escape times*[illegible] *The green land represents values of the complex parameter* [illegible]*hich the associated Julia set is connected. The blue sea corresponds to totally disconnected Julia sets. The exciting place is the coastline, where the white foam of the breaking waves symbolizes the transition between connection and disconnection.*

Plate 8.4.1
Part of the parameter space $P = \{\lambda \in C: |\lambda| < 1\}$ *for the family of dynamical systems*

$$f_\lambda(z) = \begin{cases} (z-1)/\lambda & \text{if } \operatorname{Re} z \geq 0; \\ (z+1)/\lambda^* & \text{if } \operatorname{Re} z < 0. \end{cases}$$

colored according to the "escape time" of a point $O \in \boldsymbol{R}^2$.

Plate 8.4.2
An extreme close-up on a piece of the parameter space in Color Plate 8.4.1.

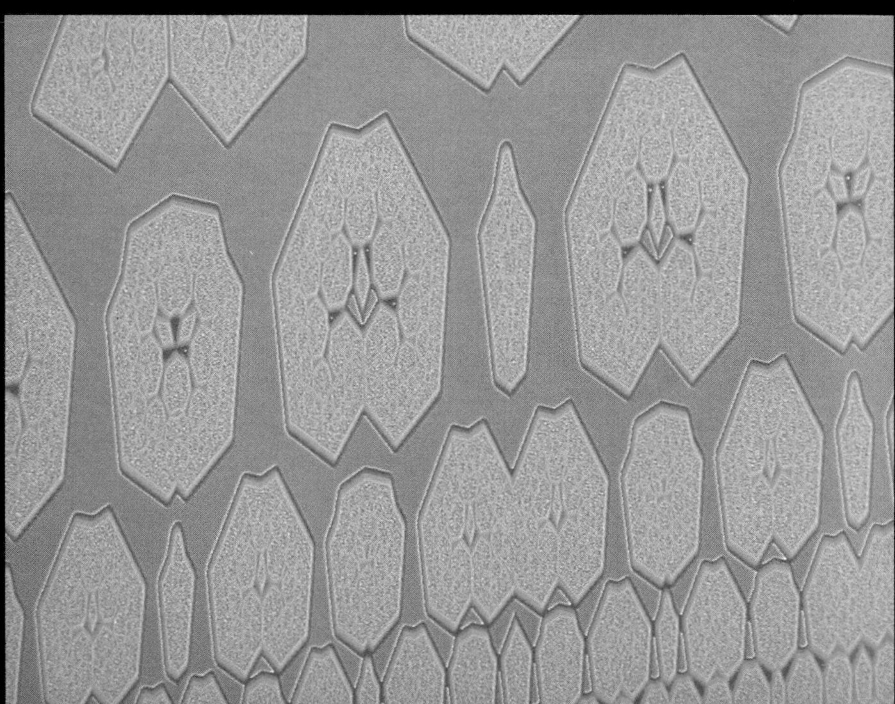

transforms of the target set do not overlap. They fit together like the pieces of a jigsaw puzzle.

(a) Find a one-parameter family collage of affine transformations.
(b) Find a "just-touching" collage of affine transformations.
(c) Find a collage using similitudes only. What is the smallest number of affine transformations in $\mathbb{R}^2$, such that the boundary is the attractor?
(d) Find a one-parameter family collage of affine transformations.
(e) Find a "just-touching" collage, using similitudes only, parameterized by the real number p.
(f) Find a collage for circles and disks.

4 Chaotic Dynamics on Fractals

4.1 THE ADDRESSES OF POINTS ON FRACTALS

We begin by considering informally the concept of the *addresses* of points on the attractor of a hyperbolic IFS. Figure 4.1.1 shows the attractor of the IFS

$$\{C; w_1(z) = (0.13 + 0.64i)z, w_2(z) = (0.13 + 0.64i)z + 1\}.$$

This attractor, A, is the union of two disjoint sets, $w_1(A)$and $w_2(A)$, lying to the left and right, respectively, of the line ab. In turn, each of these two sets is made of two disjoint sets:

$$w_1(A) = w_1(w_1(A)) \cup w_1(w_2(A)), \qquad w_2(A) = w_2(w_1(A)) \cup w_2(w_2(A)).$$

This leads to the idea of addressing points in terms of the sequences of transformations, applied to A, which lead to them. All points belonging to A, in the subset $w_1(w_1(A))$, are situated on the piece of the attractor which lies below dc and to the left of ab, and their addresses all begin with $11\ldots$. Clearly, the more precisely we specify geometrically where a point in A lies, the more bits to the address we can provide. For example, every point to the right of ab, below ef, to the left of gh, has an address which begins $212\ldots$. In Theorem 4.2.1 we prove that, in examples such as this one, it is possible to

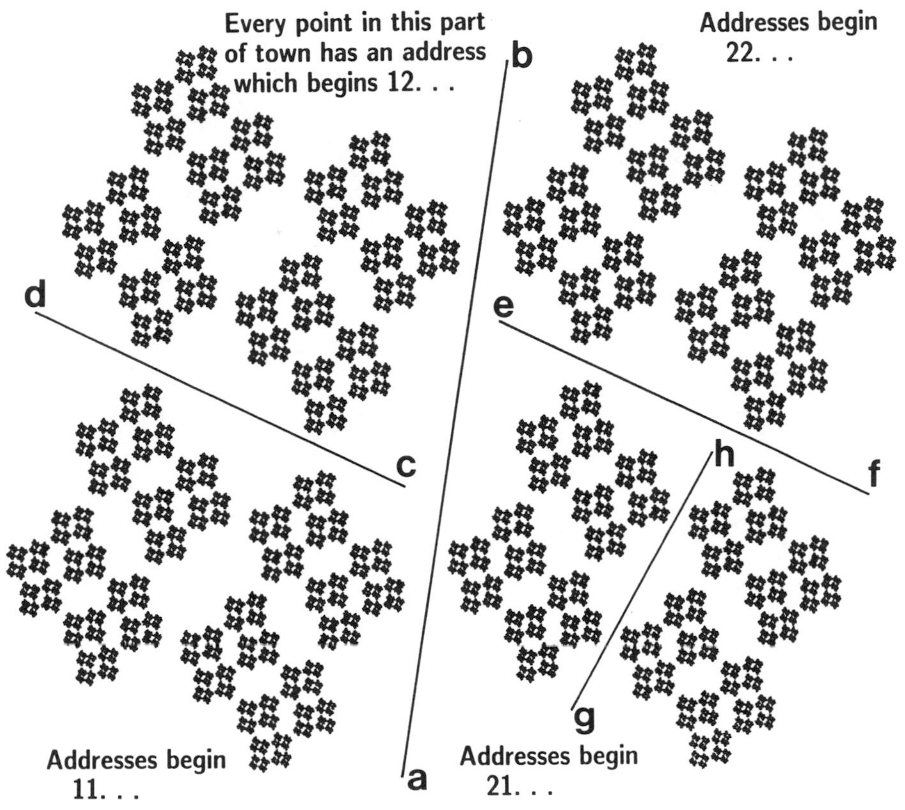
Every point in this part
of town has an address
which begins 12. . .
b
Addresses begin
22. . .
d
e
c
h
f
g
Addresses begin
11. . .
a
Addresses begin
21. . .

Figure 4.1.1
Addresses of points on an attractor. The lines ab, cd, ef, and gh are not part of the attractor.

assign a unique address to every point of A. In such cases we say that the IFS is "totally disconnected."

Here is a different type of example. Consider the IFS

$$\{C;\, w_1(z) = \tfrac{1}{2}z,\, w_2(z) = \tfrac{1}{2}z + \tfrac{1}{2},\, w_3(z) = \tfrac{1}{2}z + \tfrac{1}{2}i\}.$$

The attractor, A, of this IFS is a Sierpinski triangle with vertices at (0, 0), (1, 0), and (0, 1). Again we can address points on A according to the sequences of transformations which lead to them. This time there are at least three points in A which have two addresses, because there is a point in each of the sets $w_1(A) \cap w_2(A)$, $w_2(A) \cap w_3(A)$, and $w_3(A) \cap w_1(A)$, as illustrated in Figure 4.1.2.

On the other hand, some points on the Sierpinski triangle have only one address, such as the three vertices (0, 0), (1, 0), and (0, 1). Although the attractor is connected, the proportion of points with multiple addresses is "small," in a sense which we do not yet make precise. In such cases as this we

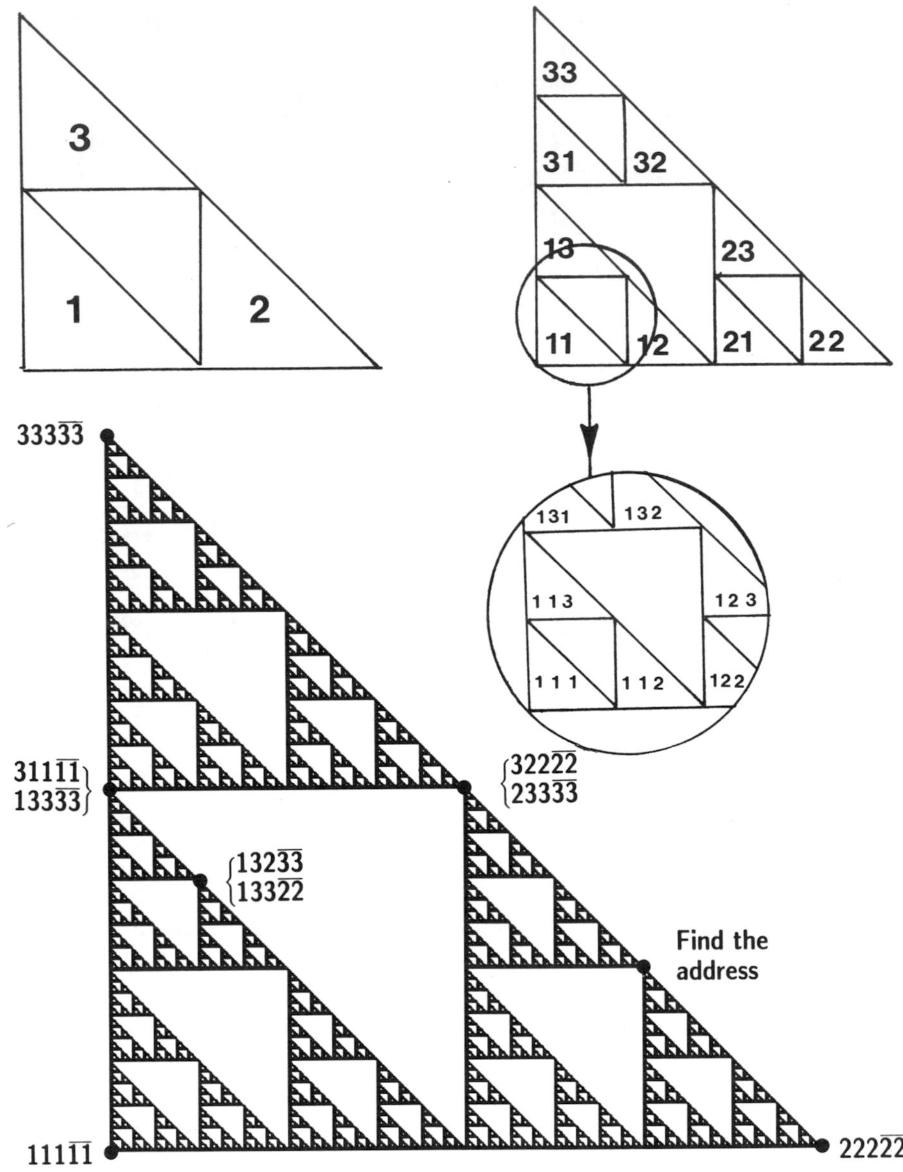

Figure 4.1.2 *Some points on this Sierpinski triangle have two addresses, while others have only one address. Overlining on the last symbols, in an expression such as* $311\overline{11}$, *means that the overlined symbols are repeated endlessly. For example,* $311\overline{11}$ = 31111111111111 111111..., *and* $311\overline{23}$ = 31123123123123....

say that the IFS is "just-touching." Notice that this terminology refers to the IFS itself rather than to its attractor.

Let us look at a third fundamentally different example. Consider the hyperbolic IFS

$$\{[0,1]; \tfrac{1}{2}x, \tfrac{3}{4}x + \tfrac{1}{4}\}.$$

The attractor is $A = [0, 1]$, but now

$$w_1(A) \cap w_2(A) = [0, \tfrac{1}{2}] \cap [\tfrac{1}{4}, 1] = [\tfrac{1}{4}, \tfrac{1}{2}];$$

so $w_1(A) \cap w_2(A)$ is a significant piece of the attractor. The attractor would look very different if this overlapping piece $[\frac{1}{4}, \frac{1}{2}]$ were missing. Now observe that every point in $[\frac{1}{4}, \frac{1}{2}]$ has at least two addresses. On the other hand, the points 0 and 1 have only one address each. Nonetheless, it appears that the proportion of points with multiple addresses is large. In such cases we say that the IFS is "overlapping."

The terminologies "totally disconnected," "just-touching," and "overlapping" refer to the IFS itself rather than to the attractor. The reason for this is that the same set may be the attractor of several different hyperbolic IFS's. Consider, for example, the two IFS

$$\{[0, 1]; w_1(x) = \tfrac{1}{2}x, w_2(x) = \tfrac{1}{2}x + \tfrac{1}{2}\}$$

and

$$\{[0, 1]; w_1(x) = \tfrac{1}{2}x, w_2(x) = -\tfrac{1}{2}x + 1\}.$$

The attractor of each one is the real interval $[0, 1]$. We can obtain two different addressing schemes for the points in $[0, 1]$, as illustrated in Figure 4.1.3. These two IFS are "just-touching." However the IFS

$$\{[0, 1]; w_1(x) = \tfrac{1}{2}x, w_2(x) = \tfrac{3}{4}x + \tfrac{1}{4}\}$$

is "overlapping" while its attractor is also $[0, 1]$.

Exercises & Examples

1.1. Figure 4.1.4 shows the attractor of an IFS of the form $\{\mathbb{R}^2; w_n, n = 1, 2, 3\}$ where each of the transformations $w_n\colon \mathbb{R}^2 \to \mathbb{R}^2$ is affine. The addresses of several points are given. Find the addresses of a, b, and c.

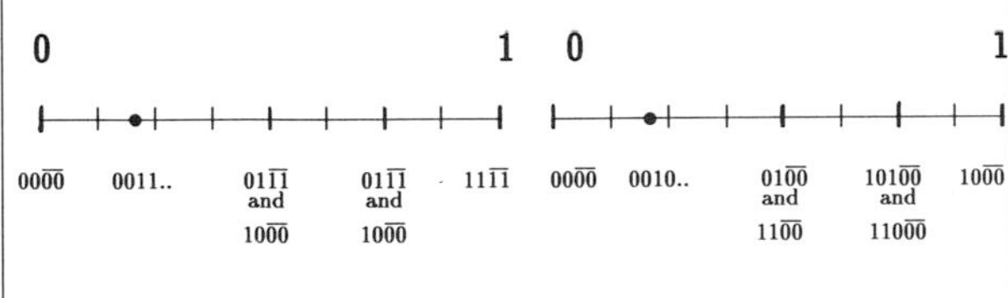

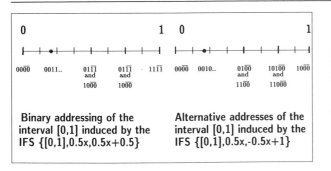

Figure 4.1.3 *Different IFS's with the same attractor provide different addressing schemes. Here the symbols $\{0, 1\}$ are used in place of $\{1, 2\}$ for obvious reasons.*

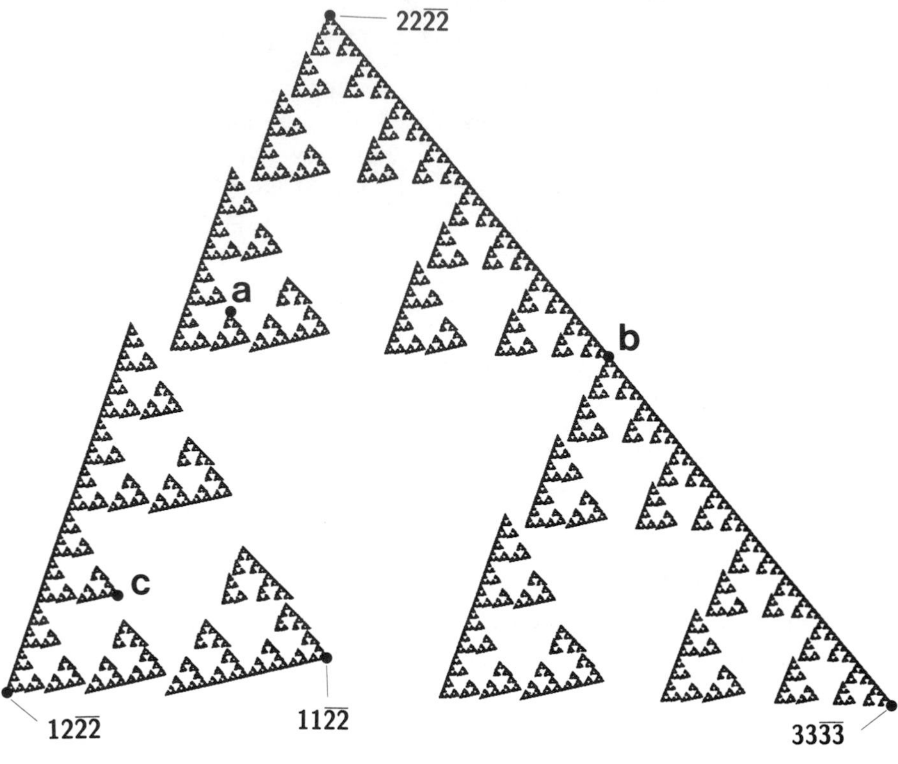

Figure 4.1.4
Can you find the addresses of a, b, and c?

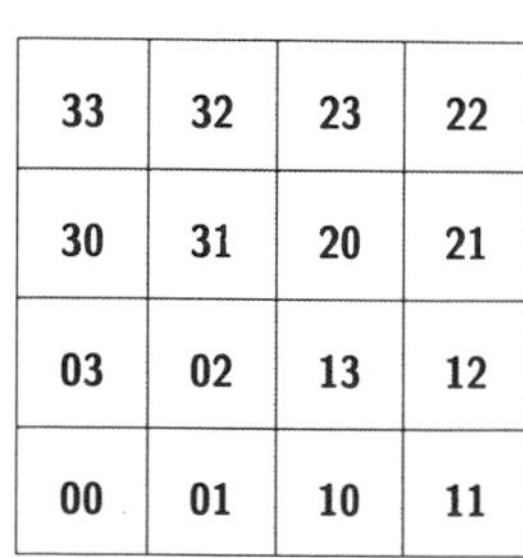

33	32	23	22
30	31	20	21
03	02	13	12
00	01	10	11

Figure 4.1.5
Addresses at depth two in a quadtree.

1.2. In Figure 4.1.4, locate the point whose address is $111\overline{11}$.

1.3. A *quadtree* is an addressing scheme used in computer science for addressing small squares in the unit square $\blacksquare = \{(x_1, x_2) \in \mathbb{R}^2 : 0 \leq x_1 \leq 1, 0 \leq x_2 \leq 1\}$ as follows. The square is broken into four quarters. Points in the first quarter have addresses which begin with 0, points in the second quarter have addresses which begin with 1, and so on, as illustrated in Figure 4.1.5. Find an IFS which gives rise to the addressing scheme suggested in Figure 4.1.5. Is this a "totally disconnected," "just-touching," or an "overlapping" IFS?

1.4. Addresses are assigned to the Sierpinski triangle, as in Figure 4.1.2. Characterize the addresses of the set of points which lie on the outermost boundary, the triangle with vertices $\overline{11}$, $\overline{22}$ and $\overline{33}$.

1.5. Characterize the addresses of points belonging to the boundary of the largest hole in Figure 4.1.6.

1.6. Consider a hyperbolic IFS with condensation set C. Suppose the condensation set is itself the attractor of another hyperbolic IFS. Design an

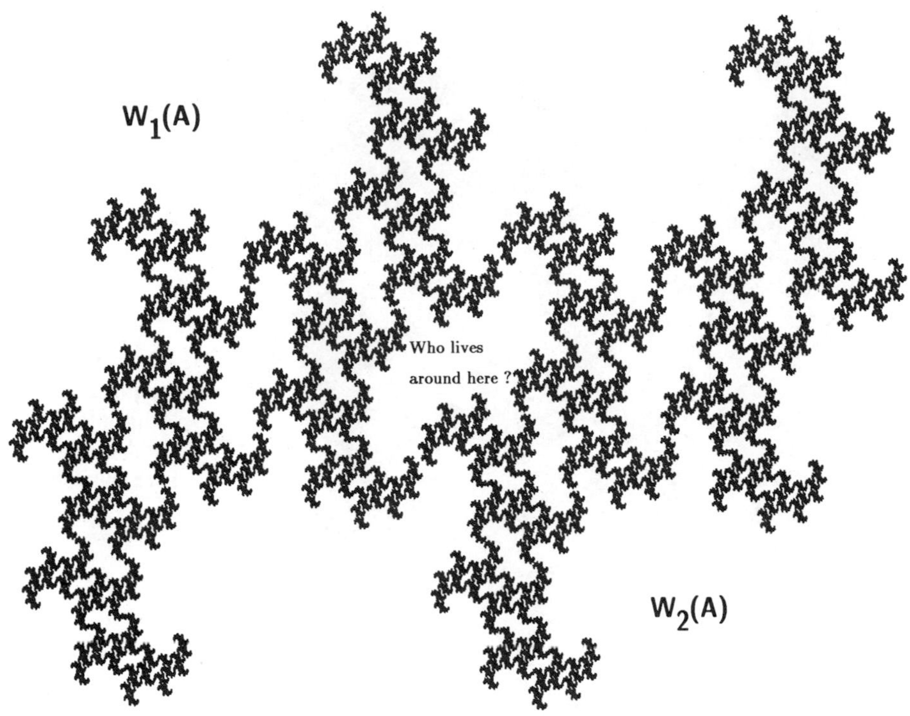
W1(A)
Who lives
around here ?
W2(A)

Figure 4.1.6
Can you describe the addresses of the points on the boundary of the central white region?

addressing scheme for the attractor of the IFS with condensation. Can all possible addresses occur?

1.7. Figure 4.1.7 shows an "overlapping" IFS attractor for two affine transformations in $\mathbb{R}^2$. Choose one point in each of the marked regions on the attractor. Find the first four numbers in two different addresses for each of these points. To remove ambiguities you should state a choice for how the two transformations act on the attractor.

1.8. ↣ Identify the set of addresses of points on the attractor, A, of a hyperbolic IFS with code space. Argue that nearby codes correspond to points on A which are nearby.

1.9. Address the real number 0.7513 in each of the two coding schemes given in Figure 4.1.3.

In thinking about the addresses of points on fractals, already we have been led to trying to compare "how many" points have a certain property to how many have another property. For example, in the case of the addressing scheme on the Sierpinski triangle described above, we wanted to compare the

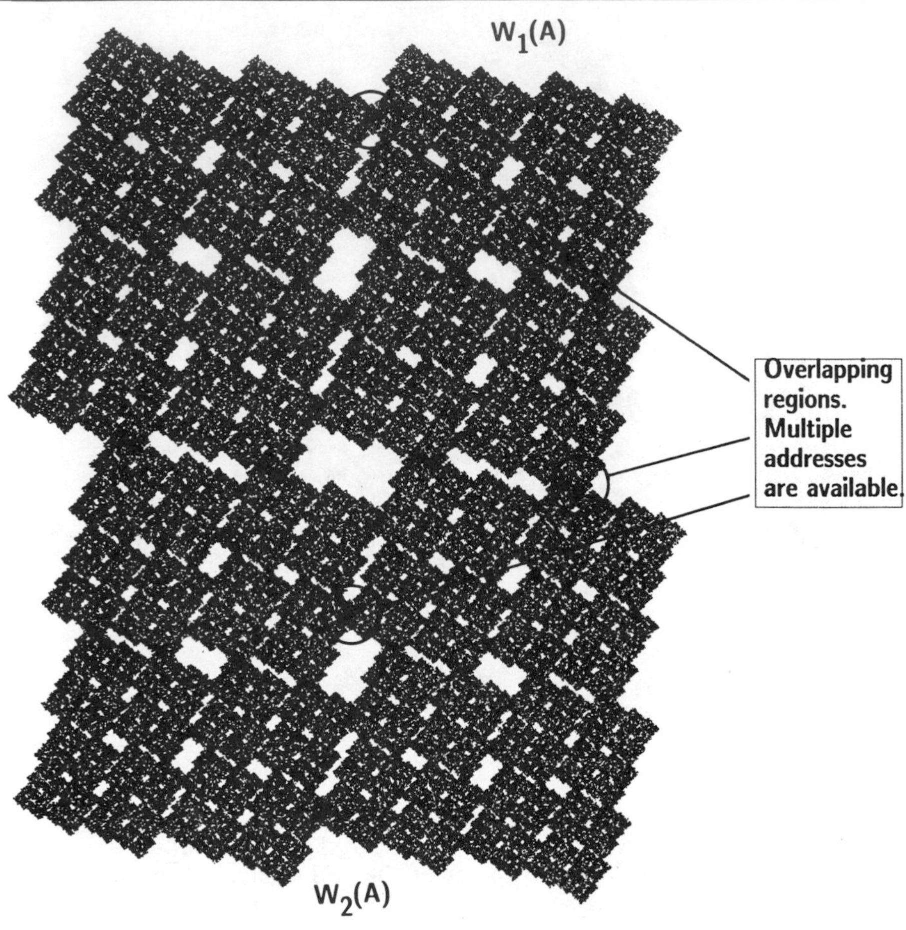
W$_1$(A)
Overlapping regions. Multiple addresses are available.
W$_2$(A)

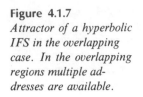

Figure 4.1.7 *Attractor of a hyperbolic IFS in the overlapping case. In the overlapping regions multiple addresses are available.*

number of points with multiple addresses to the number of points with single addresses. It turns out that both numbers are infinite. Yet still we want to compare their numbers. One way in which this may be done is through the concept of countability.

Definition 1. Let S be a set. S is *countable* if it is empty or if there is an onto transformation c: $I \to S$ where I is either: one of the sets $\{1\}$, $\{1,2\}$, $\{1,2,3\},\cdots,\{1,2,3,\ldots,n\},\cdots$ or: the positive integers $\{1,2,3,4,\ldots\}$. S is *uncountable* if it is not countable.

We think of an uncountable set as being larger than a countable set.

We are going to make fundamental use of code space to formalize the concept of addresses. How many points does code space contain?

Theorem 1. *Code space on two or more symbols is uncountable.*

Proof. The code space on the two symbols $\{1,2\}$ is proved here. Denote an element of code space Σ by $\omega = \omega_1\omega_2\omega_3\ldots$ where each $\omega_i \in \{1,2\}$. Define ρ: $\{1,2\} \to \{1,2\}$ by $\rho(1) = 2$ and $\rho(2) = 1$. Suppose code space is countable. Let the counting function be c: $\{1,2,3,\ldots\} \to \Sigma$. Consider the point $\sigma \in \Sigma$ defined by

$$\sigma = \sigma_1\sigma_2\sigma_3\ldots$$

where $\sigma_n = \rho((c(n))_n)$, and $(c(n))_n)$ means the nth symbol of $c(n)$. When does the counting function reach σ? Never! For example, $c(3) \neq \sigma$ because their third symbols are different! This completes the proof.

c(1) = 1/1
c(2) = 1/2
c(3) = 2/1
c(4) = 3/1
c(5) = 2/2
c(6) = 1/3
c(7) = 1/4
c(8) = 2/3
c(9) = 3/2
c(10) = 4/1
c(11) = 5/1
c(12) = 4/2
c(13) = 3/3
c(14) = 2/4
c(15) = 1/5
.
.

Figure 4.1.8
How to count the positive rational numbers. What is $c(24)$?

Exercises & Examples

1.10. The set of integers $\boldsymbol{Z} = \{0, \pm 1, \pm 2, \ldots,\}$ is countable. Define c: $\boldsymbol{N} \to \boldsymbol{Z}$ by $c(z) = (z-1)/2$ if z is odd, $c(z) = -z/2$ if z is even.

1.11. A countable union of countable sets is countable. An uncountable set, take away a countable set, is uncountable.

1.12. The *rational* numbers are countable. A rational number is one which can be written in the form p/q, where p and q are integers with $q \neq 0$. Figure 4.1.8 shows how to count the positive ones, some numbers being counted more than once. Make a rule which gets rid of the redundant countings. Also show how to include the negative rationals in the scheme.

1.13. Show that a Sierpinski triangle contains countably many triangles.

1.14. Let S be a perfect subset of a complete metric space. Suppose that S contains more than one point. Prove that S is uncountable.

1.15. Characterize the addresses of the missing pieces in Figure 4.1.9.

4.2 CONTINUOUS TRANSFORMATIONS FROM CODE SPACE TO FRACTALS

Definition 1. Let $\{\boldsymbol{X}; w_1, w_2, \ldots, w_N\}$ be a hyperbolic IFS. The *code space associated with the IFS*, (Σ, d_C), is defined to be the code space on N symbols $\{1,2,\ldots,N\}$, with the metric d_C given by

$$d_C(\omega,\sigma) = \sum_{n=1}^{\infty} \frac{|\omega_n - \sigma_n|}{(N+1)^n} \qquad \text{for all } \omega, \sigma \in \Sigma.$$

Our goal is to construct a continuous transformation ϕ from the code space associated with an IFS onto the attractor of the IFS. This will allow us to formalize our notion of addresses. In order to make this construction, we will need two lemmas. The first lemma tells us that if we have a hyperbolic IFS

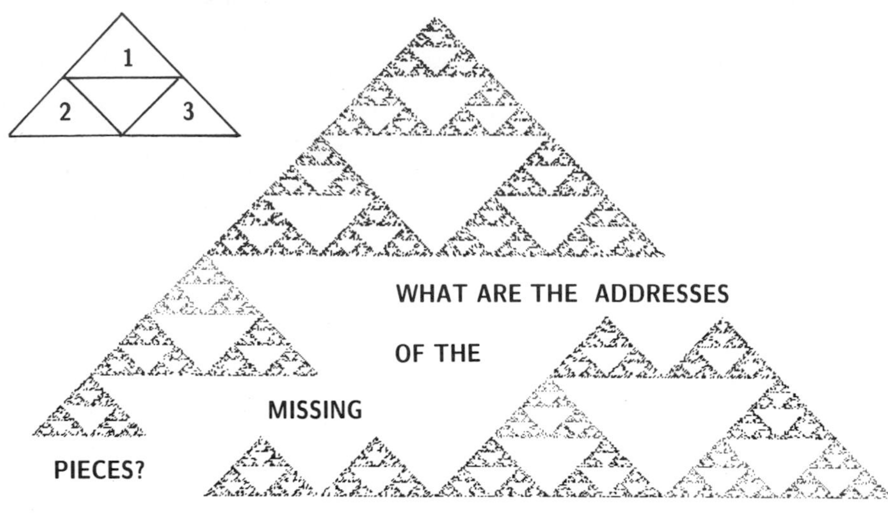

Figure 4.1.9 *Characterize the addresses of the missing pieces.*

acting on a complete metric space, but we are only interested in studying how the IFS acts in relation to a fixed compact subset of X, then we can treat the IFS as though it were defined on a compact metric space.

Lemma 1. *Let (X, d) be a complete metric space. Let $\{X;\ w_n\colon n = 1, 2, \ldots, N\}$ be a hyperbolic IFS. Let $K \in \mathscr{H}(X)$. Then there exists $\tilde{K} \in \mathscr{H}(X)$ such that $K \subset \tilde{K}$ and $w_n\colon \tilde{K} \to \tilde{K}$ for $n = 1, 2, \ldots, N$. In other words $\{\tilde{K};\ w_n\colon n = 1, 2, 3, \ldots, N\}$ is a hyperbolic IFS where the underlying space is compact.*

Proof. Define $W\colon \mathscr{H}(X) \to \mathscr{H}(X)$ by

$$W(B) = \bigcup_{n=1}^{N} w_n(B) \qquad \text{for all } B \in \mathscr{H}(\mathbf{X}).$$

To construct $\tilde{K}$ consider the IFS with condensation $\{X;\ w_n;\ n = 0, 1, 2, \ldots, N\}$ where the condensation map w_0 is associated with the condensation set K. By Theorem 3.7.1′ the attractor of this IFS belongs to $\mathscr{H}(X)$. By example 9.1 it can be written

$$\tilde{K} = (K \cup W^{\circ 1}(K) \cup W^{\circ 2}(K) \cup W^{\circ 3}(K) \cup W^{\circ 4}(K) \cdots \cup W^{\circ n}(K) \cup \cdots \cup)$$

It is readily seen that $K \subset \tilde{K}$ and that $W(\tilde{K}) \subset \tilde{K}$. This completes the proof.

The next lemma provides the first step in linking code space to IFS attractors, by introducing a certain transformation ϕ which maps the Cartesian product space $\Sigma \times N \times X$ into X. By taking appropriate limits, in

Theorem 1 below, we will eliminate the dependences on N and X to provide the desired connection between Σ and X.

Lemma 2. *Let (X, d) be a complete metric space. Let $\{X; w_n: n = 1, 2, \ldots, N\}$ be a hyperbolic IFS of contractivity s. Let (Σ, d_C) denote the code space associated with the IFS. For each $\sigma \in \Sigma$, $n \in N$, and $x \in X$, define*

$$\phi(\sigma, n, x) = w_{\sigma_1} \circ w_{\sigma_2} \circ \cdots \circ w_{\sigma_n}(x).$$

Let K denote a compact nonempty subset of X. Then there is a real constant D such that

$$d(\phi(\sigma, m, x_1), \phi(\sigma, n, x_2)) \leq Ds^{m \wedge n}$$

for all $\sigma \in \Sigma$, all $m, n \in \mathbf{N}$, and all $x_1, x_2 \in K$.

Proof. Let σ, m, n, x_1, and x_2 be as stated in the lemma. Construct $\tilde{K}$ as in Lemma 1. Without any loss of generality we can suppose that $m < n$. Then observe that

$$\phi(\sigma, n, x_2) = \phi(\sigma, m, \phi(\omega, n - m, x_2))$$

where $\omega = \sigma_{m+1}\sigma_{m+2} \cdots \sigma_n \cdots \in \Sigma$.

Let $x_3 = \phi(\omega, n - m, x_2)$. Then x_3 belongs to $\tilde{K}$. Hence we can write

$$\begin{aligned} d(\phi(\sigma, m, x_1), \phi(\sigma, n, x_2)) &= d(\phi(\sigma, m, x_1), \phi(\sigma, m, x_3)) \\ &\leq sd(w_{\sigma_2} \circ \cdots \circ w_{\sigma_m}(x_1), w_{\sigma_2} \circ \cdots \circ w_{\sigma_m}(x_3)) \\ &\leq s^2 d(w_{\sigma_3} \circ \cdots \circ w_{\sigma_m}(x_1), w_{\sigma_3} \circ \cdots \circ w_{\sigma_m}(x_3)) \\ &\leq s^m d(x_1, x_3) \leq s^m D, \end{aligned}$$

where $D = \text{Max}\{d(x_1, x_3): x_1, x_3 \in \tilde{K}\}$. D is finite because $\tilde{K}$ is compact. This completes the proof.

Theorem 1. *Let (X, d) be a complete metric space. Let $\{X; w_n: n = 1, 2, \ldots, N\}$ be a hyperbolic IFS. Let A denote the attractor of the IFS. Let (Σ, d_C) denote the code space associated with the IFS. For each $\sigma \in \Sigma$, $n \in N$, and $x \in X$, let*

$$\phi(\sigma, n, x) = w_{\sigma_1} \circ w_{\sigma_2} \circ \cdots \circ w_{\sigma_n}(x).$$

Then

$$\phi(\sigma) = \lim_{n \to \infty} \phi(\sigma, n, x)$$

exists, belongs to A, and is independent of $x \in X$. If K is a compact subset of X then the convergence is uniform over $x \in K$. The function $\phi: \Sigma \to A$ thus provided is continuous and onto.

Proof. Let $x \in X$. Let $K \in \mathscr{H}(X)$ be such that $x \in K$. Construct $\tilde{K}$ as in Lemma 1. Define W: $\mathscr{H}(X) \to \mathscr{H}(X)$ in the usual way. By Theorem 3.7.1, W is a contraction mapping on the metric space $(\mathscr{H}(X), h(d))$; and we have

$$A = \lim_{n \to \infty} \{W^{\circ n}(K)\}.$$

In particular $\{W^{\circ n}(K)\}$ is a Cauchy sequence in $(\mathscr{H}, h)$. Notice that $\phi(\sigma, n, x) \in W^{\circ n}(K)$. It follows from Theorem 2.7.1, that if $\text{Lim}_{n \to \infty}\phi(\sigma, n, x)$ exists, then it belongs to A.

That the latter limit does exist follows from the fact that, for fixed $\sigma \in \Sigma$, $\{\phi(\sigma, n, x)\}_{n=1}^{\infty}$ is a Cauchy sequence: by Lemma 2

$$d(\phi(\sigma, m, x), \phi(\sigma, n, x)) \leq Ds^{m \wedge n} \qquad \text{for all } x \in K,$$

and the right-hand side here tends to zero as m and n tend to infinity. The uniformity of the convergence follows from the fact that the constant D is independent of $x \in K$.

Next we prove that ϕ: $\Sigma \to A$ is continuous. Let $\epsilon > 0$ be given. Choose n so that $s^n D < \epsilon$, and let $\sigma, \omega \in \Sigma$ obey

$$d_C(\sigma, \omega) < \sum_{m=n+2}^{\infty} \frac{N}{(N+1)^m} = \frac{1}{(N+1)^{n+1}}.$$

Then one can verify that σ must agree with ω through n terms; that is, $\sigma_1 = \omega_1, \sigma_2 = \omega_2, \ldots, \sigma_n = \omega_n$. It follows that, for each $m \geq n$ we can write

$$d(\phi(\sigma, m, x), \phi(\omega, m, x)) = d(\phi(\sigma, n, x_1), \phi(\sigma, n, x_2)),$$

for some pair $x_1, x_2 \in \tilde{K}$. By Lemma 2 the right-hand side here is smaller than $s^n D$ which is smaller than ϵ. Taking the limit as $m \to \infty$ we find

$$d(\phi(\sigma), \phi(\omega)) < \epsilon.$$

Finally, we prove that ϕ is onto. Let $a \in A$. Then, since $A = \text{Lim}_{n \to \infty} W^{\circ n}(\{x\})$, it follows from Theorem 2.7.1 that there is a sequence $\{\omega^{(n)} \in \Sigma\colon n = 1, 2, 3, \ldots\}$ such that

$$\text{Lim}_{n \to \infty} \phi(\omega^{(n)}, n, x) = a.$$

Since (Σ, d_C) is compact, it follows that $\{\omega^{(n)}\colon n = 1, 2, 3, \ldots\}$ possesses a convergent subsequence with limit $\omega \in \Sigma$. Without loss of generality assume $\text{Lim}_{n \to \infty}\omega^{(n)} = \omega$. Then the number of successive initial agreements between the components of $\omega^{(n)}$ and ω increases without limit. That is, if

$$\alpha(n) = \text{number of elements in } \left\{ j \in N\colon \omega_k^{(n)} = \omega_k \text{ for } 1 \leq k \leq j \right\},$$

where $N = \{1, 2, 3, \ldots\}$, then $\alpha(n) \to \infty$ as $n \to \infty$. It follows that

$$d\big(\phi(\omega, n, x), \phi(\omega^{(n)}, n, x)\big) \leq s^{\alpha(n)} D.$$

By taking the limit on both sides as $n \to \infty$ we find $d(\phi(\omega), a) = 0$ which implies $\phi(\omega) = a$. Hence ϕ: $\Sigma \to A$ is onto. This completes the proof.

Definition 2. Let $\{X; w_n, n = 1, 2, 3, \dots, N\}$ be a hyperbolic IFS with associated code space Σ. Let $\phi: \Sigma \to A$ be the continuous function from code space onto the attractor of the IFS constructed in Theorem 1. An *address* of a point $a \in A$ is any member of the set

$$\phi^{-1}(a) = \{\omega \in \Sigma : \phi(\omega) = a\}.$$

This set is called the set of *addresses* of $a \in A$. The IFS is said to be *totally disconnected* if each point of its attractor possesses a unique address. The IFS is said to be *just-touching* if it is not totally disconnected yet its attractor contains an open set $\mathcal{O}$ such that

(i) $w_i(\mathcal{O}) \cap w_j(\mathcal{O}) = \varnothing \ \forall\, i, j \in \{1, 2, \dots, N\}$ with $i \neq j$;

(ii) $\bigcup_{i=1}^{N} w_i(\mathcal{O}) \subset \mathcal{O}$.

An IFS whose attractor obeys (i) and (ii) is said to obey the *open set* condition. The IFS is said to be *overlapping* if it is neither just-touching nor disconnected.

Theorem 2. *Let $\{X; w_n, n = 1, 2, \dots, N\}$ be a hyperbolic IFS with invertible maps and attractor A. The IFS is totally disconnected if and only if*

$$w_i(A) \cap w_j(A) = \varnothing \ \forall\, i, j \in \{1, 2, \dots, N\} \quad \text{with } i \neq j. \tag{4.2.1}$$

Proof. To demonstrate the first part, show that if Equation (4.2.1) is true, then every point possesses a unique address, and show that if a point has two addresses then Equation (4.2.1) cannot be true. To achieve the latter note that two different addresses must disagree first at some place: choose inverse images to get this place out front. This completes the proof.

Exercises & Examples

2.1. Show that the IFS $\{\mathbb{R}; \frac{1}{2}x, \frac{1}{2}x + \frac{1}{2}\}$ is just-touching, whilst the IFS $\{\mathbb{R}; \frac{1}{2}x, 1\}$ is overlapping.

2.2. Prove that the IFS $\{\mathbb{R}; \frac{1}{2}x, \frac{3}{4}x + \frac{1}{4}\}$ is overlapping.

2.3. Consider the IFS $\{[0, 1],\ w_n(x) = (n - 1)/10) + \frac{1}{10}x,\ n = 1, 2, 3, \dots, 10\}$ and for the associated code space use the symbols $\{0, 1, 2, \dots, 9\}$. Show that the attractor of the IFS is $[0, 1]$ and that it is just-touching. Identify the addresses of points with multiple addresses. Show that the address of a point is just its decimal representation. Comment on the fact that some numbers have two different representations.

2.4. Prove that the IFS $\{[0, 1];\ w_1(x) = \frac{1}{3}x,\ w_2(x) = \frac{1}{3}x + \frac{2}{3}\}$ is totally disconnected.

2.5. Prove that the IFS which generates the *Black Spleenwort* fern, given in Chapter 3, p. 103, is just-touching.

2.6. Show that the IFS $\{[0,1];\ w_1(x) = \frac{1}{2},\ w_2(x) = \frac{1}{2}\}$ is overlapping.

We need to understand the structure of code space. Theorem 1 told us that the code space on N symbols is the mother of all hyperbolic IFS consisting of N maps. We will use the following theorem to show that mother is metrically equivalent to a Classical Cantor Set.

Theorem 3. *Let Σ denote the code space of the N symbols, $\{1, 2, \ldots, N\}$ and define two different metrics on Σ by*

$$d_1(x, y) = \sum_{i=1}^{\infty} \frac{|x_i - y_i|}{(N+1)^i},\ d_2(x, y) = \left| \sum_{i=1}^{\infty} \frac{x_i - y_i}{(N+1)^i} \right|.$$

Then (Σ, d_1) and (Σ, d_2) are equivalent metric spaces.

Proof. We give the proof for the case $N = 10$. Let $x, y \in \Sigma$ be given. Clearly we have $d_2(x, y) \le d_1(x, y)$. We must show that there is a constant C so that $Cd_1(x, y) \le d_2(x, y)$ where C is independent of x and y. Here we pick $C = \frac{1}{19}$ and we show that it works.

We can suppose that for some $k \in \{1, 2, 3, \ldots\}$ $x_1 = y_1,\ x_2 = y_2, \ldots, x_{k-1} = y_{k-1},\ x_k \ne y_k$. Then

$$d_2(x, y) = \left| \sum_{i=k}^{\infty} \frac{x_i - y_i}{11^i} \right| \ge \frac{|x_k - y_k|}{11^k} - \sum_{i=k+1}^{\infty} \frac{|x_i - y_i|}{11^i}$$

$$\ge \frac{|x_k - y_k|}{11^k} - \sum_{i=k+1}^{\infty} \frac{9}{11^i} = \left(|x_k - y_k| - \frac{9}{10}\right)\frac{1}{11^k}$$

$$\ge \frac{1}{19}\left(|x_k - y_k| + \frac{9}{10}\right)\frac{1}{11^k},$$

(verify this by checking it for $|x_k - y_k| \in \{1, 2, \ldots, 9\}$,)

$$= \frac{1}{19}\left(\frac{|x_k - y_k|}{11^k} + \sum_{i=k+1}^{\infty} \frac{9}{11^i}\right) \ge \frac{1}{19}\left(\frac{|x_k - y_k|}{11^k} + \sum_{i=k+1}^{\infty} \frac{|x_i - y_i|}{11^i}\right)$$

$$\ge \frac{1}{19} \sum_{n=1}^{\infty} \frac{|x_i - y_i|}{11^i} = \frac{1}{19} d_1(x, y).$$

This completes the proof.

We now show that code space is metrically equivalent to a totally disconnected Cantor subset of $[0, 1]$. Define a hyperbolic IFS by $\{[0, 1];\ w_n(x) =$

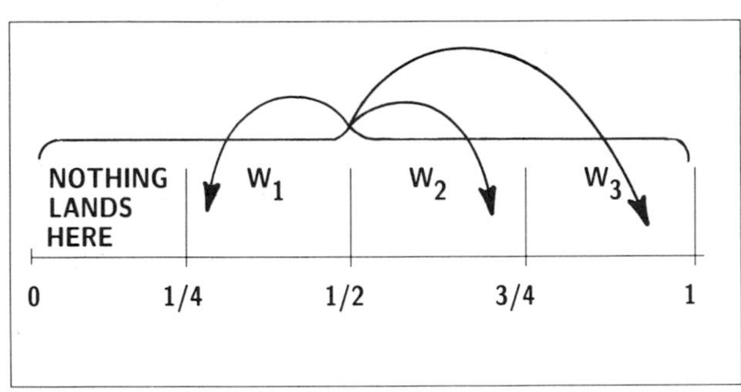

Figure 4.2.1 *Nothing lands here.*

$(1/(N+1))x + n/(N+1)$: $n = 1, 2, \ldots, N\}$. Thus

$$w_n([0,1]) = \left[\frac{n}{N+1}, \frac{n+1}{N+1}\right] \qquad \text{for } n = 1, 2, \ldots, N,$$

as illustrated for $N = 3$ in Figure 4.2.1.

The attractor for this IFS is totally disconnected, as illustrated in Figure 4.2.2 for $N = 3$.

In the case $N = 3$, the attractor is contained in $[\frac{1}{3}, 1]$. The fixed points of the three transformations $w_1(x) = \frac{1}{4}x + \frac{1}{4}$, $w_2(x) = \frac{1}{4}x + \frac{1}{2}$, $w_2(x) = \frac{1}{4}x + \frac{3}{4}$ are respectively $\frac{1}{3}$, $\frac{2}{3}$, and 1. Moreover, the address of any point on the attractor is exactly the same as the string of digits which represents it in base $N + 1$. What is happening here is this. At this zeroth level we begin with all numbers in $[0, 1]$ represented in base $(N + 1)$. We remove all those points whose first digit is 0. For example, in the case $N = 3$ this eliminates the interval $[0, \frac{1}{4}]$. At the second level we remove from the remaining points all

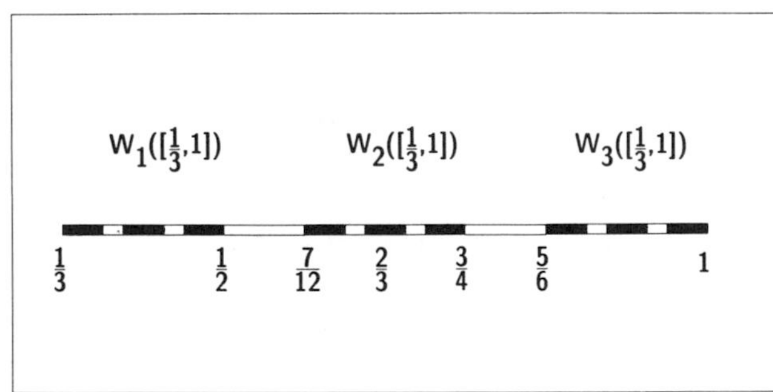

Figure 4.2.2 *A special ternary Cantor set in the making.*

those which have the digit 0 in the second place. And so on. We end up with those numbers whose expansion in base $(N + 1)$ does not contain the digit 0. Now consider the continuous transformation ϕ: $(\Sigma, d_C) \to (A, \text{Euclidean})$. It follows from Theorem 3 that the two metric spaces are equivalent. ϕ is the transformation which provides the equivalence. Thus, we have a realization, a way of picturing code space.

Exercises & Examples

2.7. Find the Figure analogous to Figure 4.2.2, corresponding to the case $N = 9$.

2.8. What is the smallest number in $[0, 1]$ whose decimal expansion contains no zeros?

We continue to discuss the relationship between the attractor A of a hyperbolic IFS $\{\mathbf{X};\ w_1, w_2, \ldots, w_N\}$ and its associated code space Σ. Let ϕ: $\Sigma \to \mathbf{X}$ be the code space map constructed in Theorem 1. Let $\omega = \omega_1\omega_2\omega_3\omega_4\ldots$ be an address of a point $x \in A$. Then

$$\tilde{\omega} = j\omega_1\omega_2\omega_3\omega_4\ldots$$

is an address of $w_j(x)$, for each $j \in \{1, 2, \ldots, N\}$.

Definition 3. Let A be the attractor of a hyperbolic IFS $\{\mathbf{X}, w_1, w_2, \ldots, w_N\}$. A point $a \in A$ is called a *periodic point* of the IFS if there is a finite sequence of numbers $\{\sigma(n) \in \{1, 2, \ldots, N\}\}_{n=1}^{P}$ such that

$$a = w_{\sigma(P)} \circ w_{\sigma(P-1)} \circ \cdots \circ w_{\sigma(1)}(a). \tag{4.2.2}$$

If $a \in A$ is periodic, then the smallest integer P such that the latter statement is true is called the *period* of a.

Thus, a point on an attractor is periodic if we can apply a sequence of w_n's to it, in such a way as to get back to exactly the same point after finitely many steps. Let $a \in A$ be a periodic point which obeys equation (4.2.2). Let σ be the point in the associated code space, defined by

$$\begin{aligned}\sigma &= \sigma(P)\sigma(P-1)\cdots\sigma(1)\sigma(P)\sigma(P-1)\cdots\sigma(1)\sigma(P)\sigma(P-1)\cdots \\ &= \overline{\sigma(P)\sigma(P-1)\cdots\sigma(1)}.\end{aligned} \tag{4.2.3}$$

Then, by considering $\text{Lim}_{n\to\infty}\phi(\sigma, n, a)$, we see that $\phi(\sigma) = a$.

Definition 4. A point in code space whose symbols are periodic, as in equation (4.2.3), is called a *periodic address*. A point in code space whose symbols are periodic after a finite initial set is omitted, is called *eventually periodic*.

Exercises & Examples

2.9. An example of a periodic address is:

$$121212121212121212121212121212121212121212121212121212\ldots$$

where (12) is repeated endlessly. An example of an eventually periodic address is:

$$112111111211111211111211111212221211212121212121212121212121\ldots$$

where (21) is repeated endlessly.

2.10. Prove the following theorem: "Let $\{X; w_1, w_2, \ldots, w_N\}$ be a hyperbolic IFS with attractor A. Then the following statements are equivalent:

(I) $x \in A$ is a periodic point;
(II) $x \in A$ possesses an address which is periodic;
(III) $x \in A$ is a fixed point of an element of the semigroup of transformations generated by $\{w_1, w_2, \ldots, w_N\}$."

2.11. Show that a point $x \in [0,1]$ is a periodic point of the IFS

$$\{[0,1]; \tfrac{1}{2}x, \tfrac{1}{2}x + \tfrac{1}{2}\}$$

if and only if it can be written $x = p/(2^N - 1)$ for some integer $0 \le p \le 2^N - 1$ and some integer $N \in \{1, 2, 3, \ldots\}$.

2.12. Let $\{X; w_1, w_2, \ldots, w_N\}$ denote a hyperbolic IFS with attractor A. Define $W(S) = \bigcup_{n=1}^{N} w_n(S)$ when S is a subset of X. Let P denote the set of eventually periodic points of the IFS. Show that $W(P) = P$.

2.13. Locate all the periodic points of period 3 for the IFS $\{\mathbb{R}^2; \tfrac{1}{2}z, \tfrac{1}{2}z + \tfrac{1}{2}, \tfrac{1}{2}z + \tfrac{i}{2}\}$. Mark the positions of these points on the attractor.

2.14. Locate all periodic points of the IFS $\{\mathbb{R}; w_1(x) = 0, w_2(x) = \tfrac{1}{2}x + \tfrac{1}{2}\}$.

Theorem 4. *The attractor of an IFS is the closure of its periodic points.*

Proof. Code space is the closure of the set of periodic codes. Lift this statement to A using the code space map $\phi\colon \Sigma \to A$. (ϕ is a continuous mapping from a metric space Σ onto a metric space A. If $S \subset \Sigma$ is such that its closure equals Σ, then the closure of $f(S)$ equals A.)

Exercises & Examples

2.15. Prove that the attractor of a totally disconnected hyperbolic IFS of two or more maps is uncountable.

2.16. Under what conditions does the attractor of a hyperbolic IFS contain uncountably many points with multiple addresses? Do not try to give a complete answer, just some conditions—and think about the problem.

2.17. Under what conditions do there exist points in the attractor of a hyperbolic IFS with uncountably many addresses? As in 2.16, do not try to give a full answer.

2.18. In the standard construction of the Classical Cantor Set $\mathscr{C}$, described in section 3.1 (example 1.5), a succession of open subintervals of $[0,1]$ are removed. The endpoints of each of these intervals belongs to $\mathscr{C}$. Show that the set of such interval endpoints is countable. Show that $\mathscr{C}$ itself is uncountable. $\mathscr{C}$ is the attractor of the IFS $\{[0,1]; \frac{1}{3}x, \frac{1}{3}x + \frac{2}{3}\}$. Characterize the addresses of the set of interval endpoints in $\mathscr{C}$.

4.3 INTRODUCTION TO DYNAMICAL SYSTEMS

We introduce the idea of a dynamical system, and some of the associated terminology.

Definition 1. A *dynamical system* is a transformation $f\colon \mathbf{X} \to \mathbf{X}$ on a metric space $(\mathbf{X}, d)$. It is denoted by $\{\mathbf{X}; f\}$. The *orbit* of a point $x \in \mathbf{X}$ is the sequence $\{f^{\circ n}(x)\}_{n=0}^{\infty}$.

As we will discover, dynamical systems are sources of deterministic fractals. The reasons for this are deeply intertwined with IFS theory, as we will see. Later we will introduce a special type of dynamical system, called a shift dynamical system, which can be associated with an IFS. By studying the orbits of these systems, we will learn more about fractals. One of our goals is to learn why the Random Iteration Algorithm, used in Program 3.8.2, successfully calculates the images of attractors of IFS. More information about the deep structure of attractors of IFS will be discovered.

Exercises & Examples

3.1. Define a function on code space, $f\colon \Sigma \to \Sigma$, by

$$f(x_1x_2x_3x_4\ \cdots) = x_2x_3x_4x_5\ \cdots$$

Then $\{\Sigma; f\}$ is a dynamical system.

3.2. $\{[0,1];\ f(x) = \lambda x(1-x)\}$ is a dynamical system for each $\lambda \in [0,4]$. We say that we have a *one-parameter family* of dynamical systems.

3.3. Let $w(x) = Ax + t$ be an affine transformation in $\mathbb{R}^2$. Then $\{\mathbb{R}^2; w\}$ is a dynamical system.

3.4. Define $T\colon C[0,1] \to C[0,1]$ by

$$(Tf)(x) = \tfrac{1}{2}f(\tfrac{1}{2}x) + \tfrac{1}{2}f(\tfrac{1}{2}x + \tfrac{1}{2}).$$

Then $\{C[0,1]; T\}$ is a dynamical system.

3.5. Let $w\colon \hat{\mathbb{C}} \to \hat{\mathbb{C}}$ be a Möbius transformation. That is $w(z) = (az + b)/$

$(cz + d)$, where $a, b, c, d \in \mathbb{C}$, and $(ad - bc) \neq 0$. Then $\{\hat{\mathbb{C}}; w(z)\}$ is a dynamical system.

3.6. $\{[0, 1]; 2x \bmod 1\}$ is a dynamical system. Here $2x \bmod 1 = 2x - [2x]$ where $[2x]$ denotes the greatest integer less than or equal to $2x$.

3.7. Define a transformation $f\colon \blacksquare \to \blacksquare$ as illustrated in Figure 4.3.1. $\{\blacksquare; f\}$ is a dynamical system.

In dynamical systems theory one is interested in what happens when one follows a typical orbit: is there some kind of attractor which usually occurs? Dynamical systems become interesting when the transformations involved are *not* contraction mappings, so that a single transformation suffices to produce interesting behavior. The orbit of a single point may be a geometrically complex set. Some thought about horizontal slices through Figure 4.3.2 will

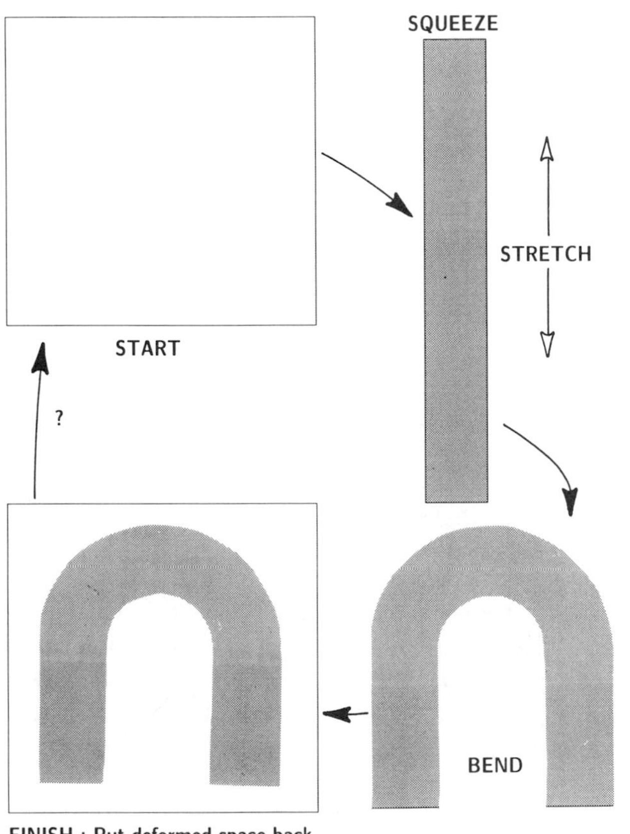

Figure 4.3.1
An example of a "Stretch, Squeeze, and Bend" dynamical system.

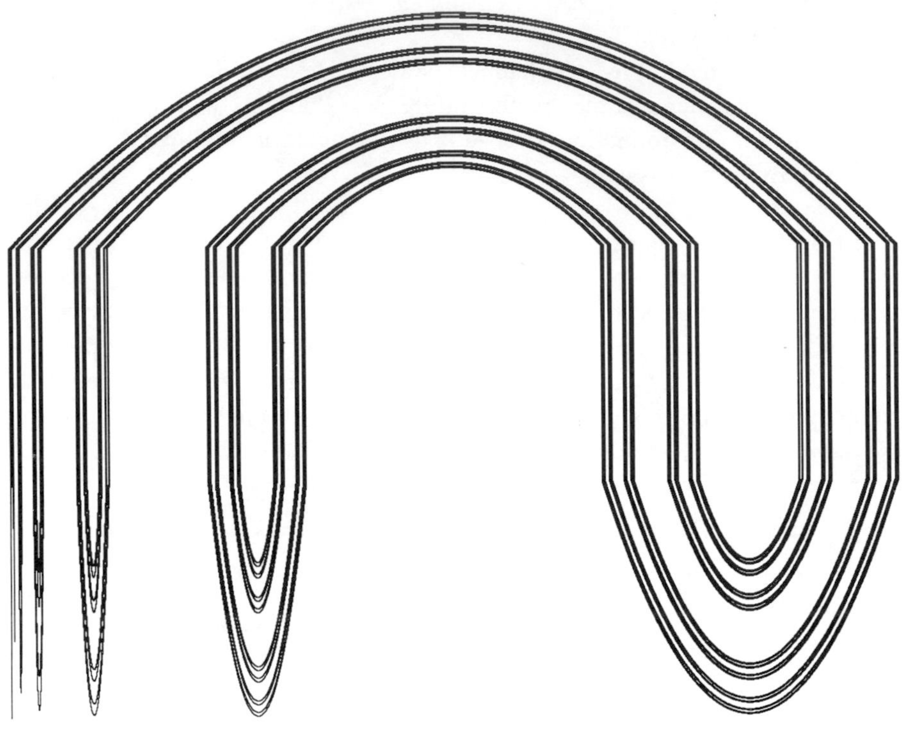

Figure 4.3.2
One million points of an orbit of a "Stretch, Squeeze, and Bend" dynamical system. Can you find a relationship to IFS theory?

quickly suggest to the inquisitive student that there is a close relationship between this noncontractive dynamical system and a hyperbolic IFS.

Definition 2. Let $\{X; f\}$ be a dynamical system. A *periodic point* of f is a point $x \in X$ such that $f^{\circ n}(x) = x$ for some $n \in \{1, 2, 3, \ldots\}$. If x is a periodic point of f then an integer n such that $f^{\circ n}(x) = x$, $n \in \{1, 2, 3, \ldots\}$ is called a period of x. The least such integer is called the *minimal period* of the periodic point x. The orbit of a periodic point of f is called a *cycle* of f. The minimal period of a cycle is the number of distinct points which it contains. A period of a cycle of f is a period of a point in the cycle.

Definition 3. Let $\{X; f\}$ be a dynamical system and let $x_f \in X$ be a fixed point of f. The point x_f is called an *attractive* fixed point of f if there is a number $\epsilon > 0$ so that f maps the ball $B(x_f, \epsilon)$ into itself, and moreover f is a contraction mapping on $B(x_f, \epsilon)$. Here $B(x_f, \epsilon) = \{y \in X\colon d(x_f, y) \leq \epsilon\}$. The point x_f is called a *repulsive* fixed point of f if there are numbers $\epsilon > 0$ and $C > 1$ such that

$$d(f(x_f), f(y)) \geq C d(x_f, y) \qquad \text{for all } y \in B(x_f, \epsilon).$$

A periodic point of f of period n is *attractive* if it is an attractive fixed point of $f^{\circ n}$. A cycle of period n is an *attractive cycle* of f if the cycle contains an attractive periodic point of f of period n. A periodic point of f of period n is *repulsive* if it is a repulsive fixed point of $f^{\circ n}$. A cycle of period n is a *repulsive cycle* of f if the cycle contains a repulsive periodic point of f of period n.

Definition 4. Let $\{X, f\}$ be a dynamical system. A point $x \in X$ is called an *eventually periodic* point of f if $f^{\circ m}(x)$ is periodic for some positive integer m.

Exercises & Examples

3.8. The point $x_f = 0$ is an attractive fixed point for the dynamical system $\{\mathbb{R}; \frac{1}{2}x\}$, and a repulsive fixed point for the dynamical system $\{\mathbb{R}; 2x\}$.

3.9. The point $z = 0$ is an attractive fixed point, and $z = \infty$ is a repulsive fixed point, for the dynamical system

$$\{\hat{\mathbb{C}}; (\cos 10° + i \sin 10°)(0.9)z\}.$$

A typical orbit, starting from near the Point at Infinity on the sphere, is shown in Figure 4.3.3 (a) and (b).

3.10. The point $x_f = 111\overline{111}$ is a repulsive fixed point for the dynamical system $\{\Sigma; f\}$ where $f\colon \Sigma \to \Sigma$ is defined by

$$f(x_1x_2x_3x_4x_5 \cdots) = x_2x_3x_4x_5 \cdots$$

Show that $x = 121\overline{212}$ is a repulsive periodic point of period 2, and that $\{1\overline{212}, 2\overline{121}\}$ is a repulsive cycle of period 2.

3.11. The dynamical system $\{[0,1]; 2x(1-x)\}$ possesses the attractive fixed point $x_f = \frac{1}{2}$. Can you find a repulsive fixed point for this system?

There is a delightful construction for representing orbits of a dynamical system of the special form $\{\mathbb{R}; f(x)\}$. It utilizes the graph of the function $f\colon \mathbb{R} \to \mathbb{R}$. We describe here how it is used to represent the orbit $\{x_n = f^{\circ n}(x_0)\}_{n=1}^{\infty}$ of a point $x_0 \in \mathbb{R}$.

For simplicity we suppose that $f\colon [0,1] \to [0,1]$. Draw the square $\{(x, y)\colon 0 \le x \le 1, 0 \le y \le 1\}$ and sketch the graphs of $y = f(x)$ and $y = x$ for $x \in [0,1]$. Start at the point (x_0, x_0) and connect it by a straight line segment to the point $(x_0, x_1 = f(x_0))$. Connect this point by a straight line segment to the point (x_1, x_1). Connect this point by a straight line segment to the point $(x_1, x_2 = f(x_1))$; and continue. The orbit itself shows up on the forty-five degree line $y = x$, as the sequence of points (x_0, x_0), (x_1, x_1), $(x_2, x_2), \ldots$. We call the result of this geometrical construction a *web diagram*.

It is straightforward to write computergraphical routines which plot web diagrams on the graphics display device of a microcomputer. The following program is written in BASIC. It runs without modification on an IBM PC with Color Graphics Adaptor and Turbobasic. On any line the words preceded by a ' are comments: they are not part of the program.

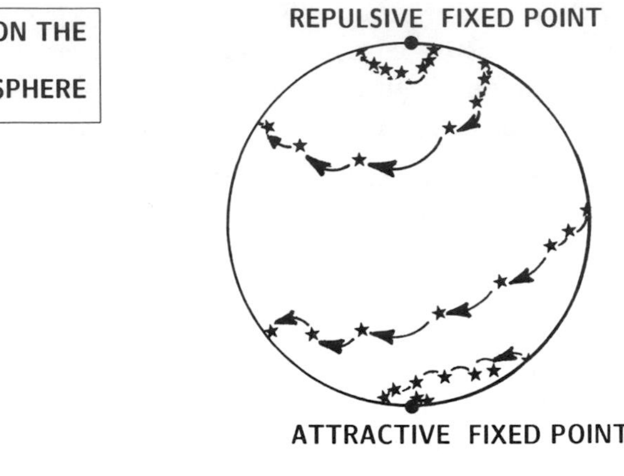

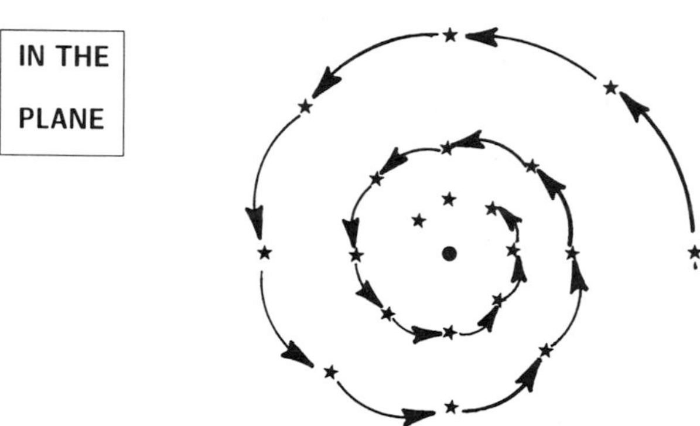

Figure 4.3.3(a) *The dynamics of a simple Möbius transformation. Points spiral away from one fixed point and they spiral in towards the other. What happens if the fixed points coincide?*

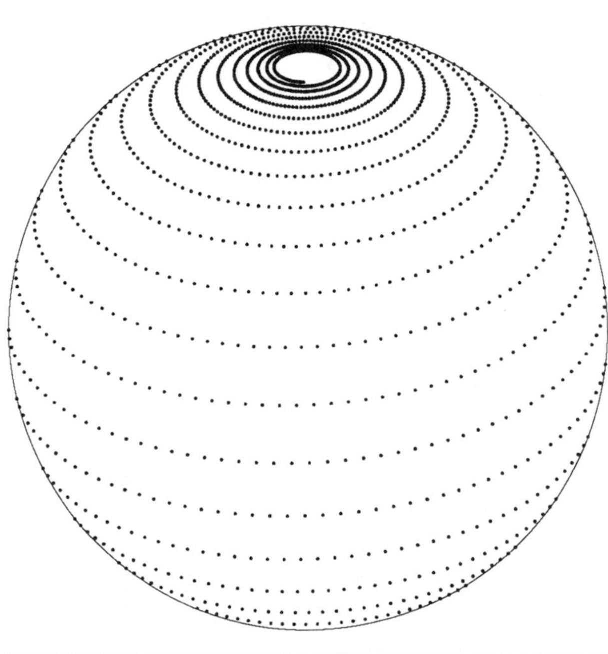

Figure 4.3.3(b) *Points belonging to an orbit of a Mobius transformation on a sphere.*

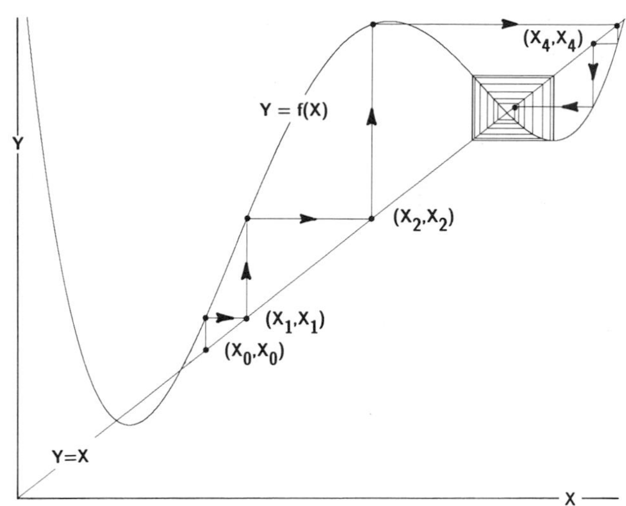

Figure 4.3.4 *This shows an example of a web diagram. A web diagram is a means for displaying and analysing the orbit of a point $x_0 \in \mathbb{R}$ for a dynamical system $\{\mathbb{R}, f\}$. The geometrical construction of a web diagram makes use of the graph of $f(x)$.*

PROGRAM 4.3.1

```
l = 3.79 : xn = 0.95          'parameter value 3.79, orbit starts
                              at 0.95

def fnf(xn) = l*xn*(1 - xn)   'change this function f(x) for other
                              dynamical systems

screen 1 : cls                'initialize computer graphics
window (0, 0) - (1, 1)        'set plotting window to 0 < x < 1,
                              0 < y < 1

for k = 1 to 400              'plot the graph of the f(x)
pset(k/400, fnf(k/400))
next k
do                            'the main computational loop
n = n + 1                     'increment the counter, n
y = fnf(xn)                   'compute the next point on the orbit
line (xn, xn) - (xn, y), n    'draw a linc from (xn, xn) to (xn, y)
                              in color n

line (xn, y) - (y, y), n      'draw a line segment from (xn, y)
                              to (y, y) in color n

xn = y                        'set xn to be the most recently
                              computed point on the orbit

loop until instat : end       'stop running if a key is pressed.
```

An example of some web diagrams computed using this program are shown in Figure 4.3.5. The dynamical system used in this case is $\{[0, 1]; f(x) = 3.79x(1 - x)\}$.

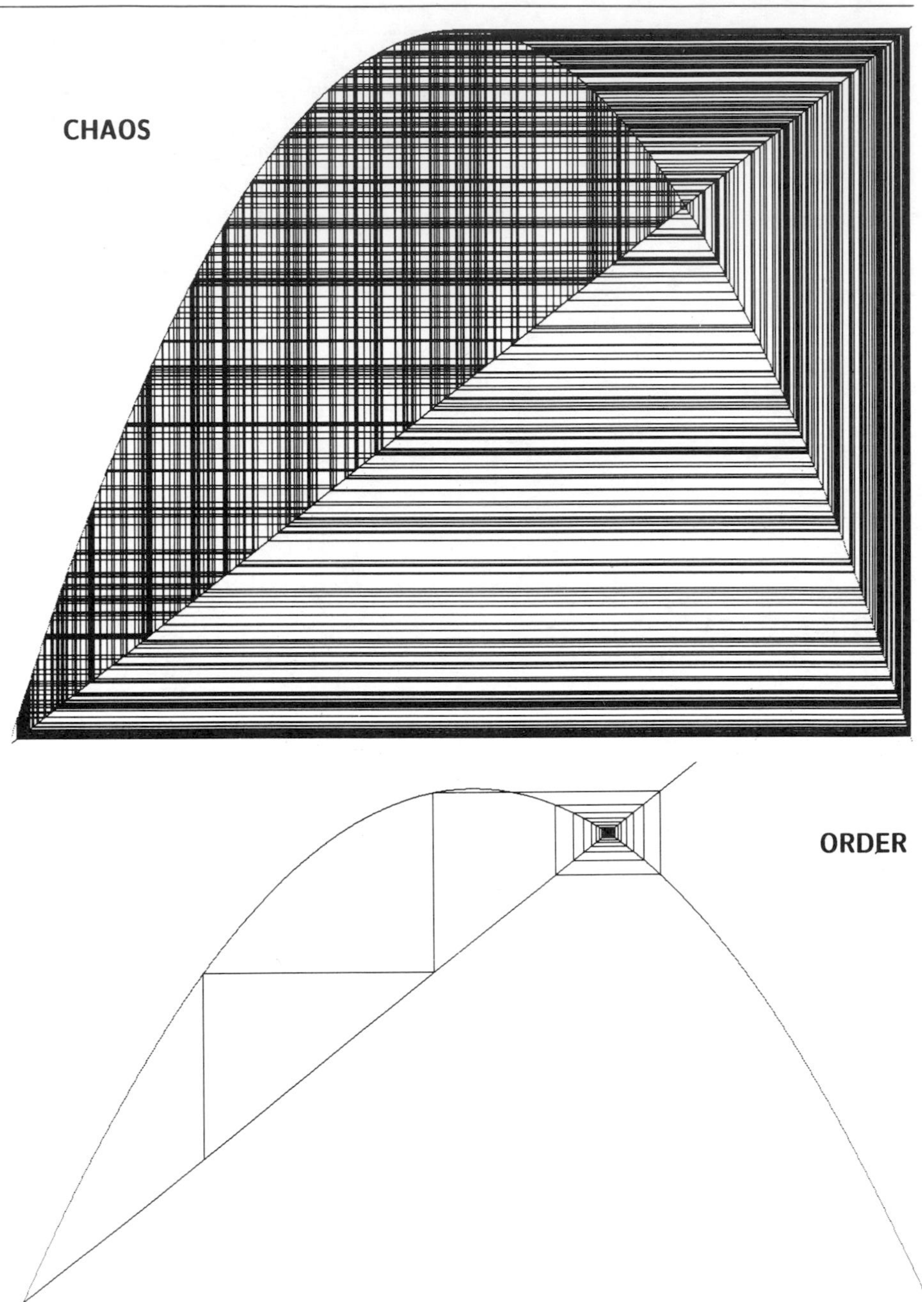

Figure 4.3.5 *Two examples of web diagrams computed using Program 4.3.1. The dynamical system in this case is* $\{[0,1]; f(x) = \lambda x(1-x)\}$, *for two different values of* $\lambda \in (0,4)$. *The system corresponding to the lower value of* λ *is orderly, the other is close to being chaotic.*

Exercises & Examples

3.12. Rewrite Program 1 in a form suitable for your own computer environment. Use the resulting system to study the dynamical systems $\{[0,1]; \lambda x(1-x)\}$ for $\lambda = 0.55$, 1.3, 2.225, 3.014, 3.794. Try to classify the various species of web diagrams that occur for this one-parameter family of dynamical systems.

3.13. Divide $[0,1]$ into sixteen subintervals $[0, \frac{1}{16}), [\frac{1}{16}, \frac{1}{16}), \ldots, [\frac{14}{16}, \frac{15}{16}), [\frac{15}{16}, 1]$. Let $f:[0,1] \to [0,1]$ be defined by $f(x) = \lambda x(1-x)$ where $\lambda \in [0,4]$ is a parameter. Compute $\{f^{\circ n}(\frac{1}{2}): n = 0, 1, 2, \ldots, 5000\}$ and keep track of the *frequency* with which $f^{\circ n}(\frac{1}{2})$ falls in the kth interval for $k = 1, 2, \ldots, 16$, and $\lambda = 0.55$, 1.3, 2.225, 3.014, 3.794. Make histograms of your results.

3.14. Describe the behavior for the one-parameter family of dynamical systems $\{\mathbb{R} \cup \{\infty\}; \lambda x\}$, where λ is a real parameter, in the cases (i) $\lambda = 0$; (ii) $0 < |\lambda| < 1$; (iii) $\lambda = -1$; (iv) $\lambda = 1$; (v) $1 < \lambda < \infty$.

3.15. Analyze possible behaviors of $\{\mathbb{R}^2; Ax + t\}$ where $Ax + t$ is an affine transformation.

3.16. Study possible behaviors of orbits for the dynamical system $\{\hat{C}$; Möbius transformation$\}$. You should make appropriate changes of coordinates to simplify the discussion.

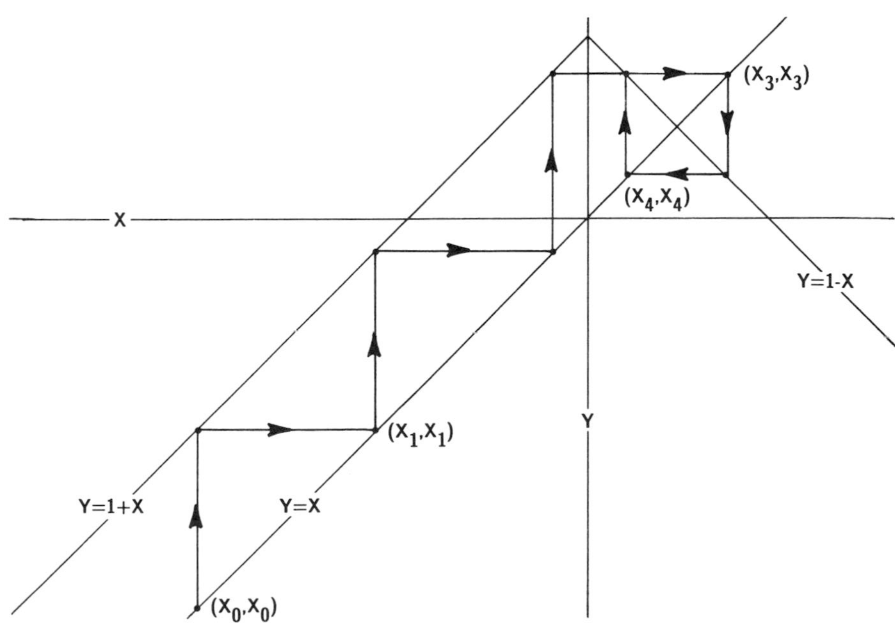

Figure 4.3.6
An orbit of the "Slide and Fold" dynamical system described in example 3.17. *Can you prove that all orbits are eventually periodic?*

Figure 4.3.7
A sign of things to come.

3.17. Show that all points are eventually periodic for the slide-and-fold dynamical system $\{\mathbb{R}; f\}$ where

$$f(x) = \begin{cases} x + 1 & \text{if } x \leq 0, \\ -x + 1 & \text{if } x \geq 0. \end{cases}$$

This system is illustrated in Figure 4.3.6.

3.18. Let $\{\boldsymbol{X}; w_1, w_2, \ldots, w_N\}$ be a hyperbolic IFS. Then $\{\mathscr{H}(\boldsymbol{X}); W\}$ is a dynamical system, where

$$W(B) = \bigcup_{n=1}^{N} w_n(B) \qquad \text{for all } B \in \mathscr{H}(\boldsymbol{X}).$$

Dynamical systems which act on sets in place of points are sometimes called *set dynamical systems*. Show that the attractor of the IFS is an attractive fixed point of the dynamical system $\{\mathscr{H}(\boldsymbol{X}); W\}$. You should quote appropriate results from earlier theorems.

4.4 DYNAMICS ON FRACTALS: OR HOW TO COMPUTE ORBITS BY LOOKING AT PICTURES

We continue with the main theme for this chapter, namely dynamical systems on fractals. We will need the following result.

Lemma 1. *Let* $\{\mathbf{X};\ w_n,\ n = 1, 2, \ldots, N\}$ *be a hyperbolic IFS with attractor* A. *If the IFS is totally disconnected, then for each* $n \in \{1, 2, \ldots, N\}$, *the transformation* $w_n \colon A \to A$ *is one-to-one.*

Proof. We use a code space argument. Suppose that there is an integer $n \in \{1, 2, \ldots, N\}$ and distinct points $a_1, a_2 \in A$ so that $w_n(a_1) = w_n(a_2) = a \in A$. If a_1 has address ω and a_2 has address σ, then a has the two addresses $n\omega$ and $n\sigma$. This is impossible because the IFS is totally disconnected. This completes the proof.

Lemma 1 shows that the following definition is good.

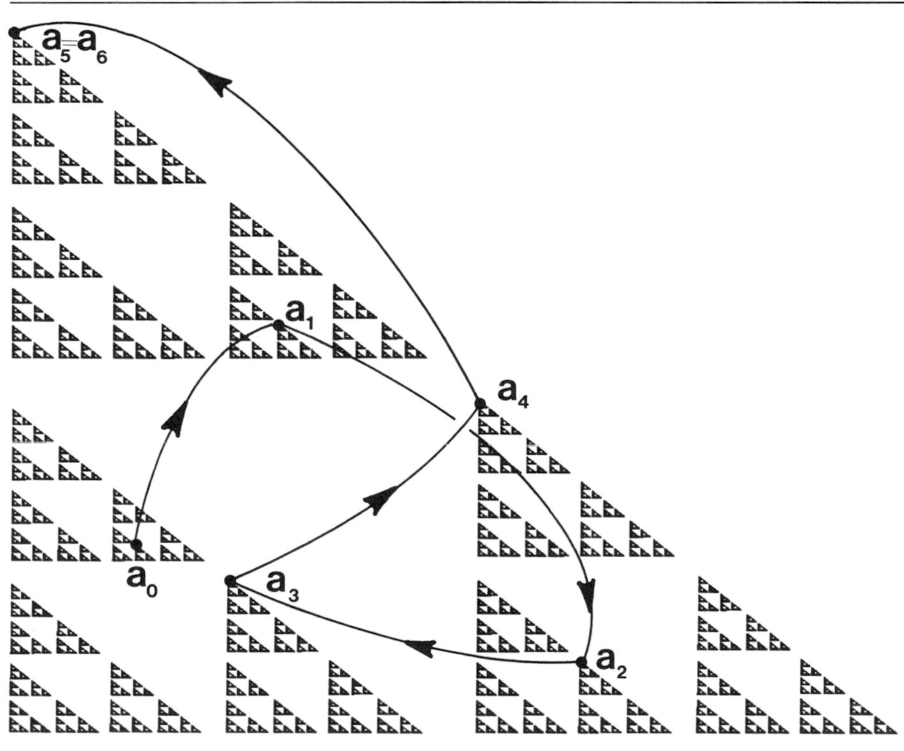

Figure 4.4.1
An orbit of a shift dynamical system on a fractal.

Definition 1. Let $\{X; w_n, n = 1, 2, \ldots, N\}$ be a totally disconnected hyperbolic IFS with attractor A. The associated *shift transformation* on A is the transformation $S\colon A \to A$ defined by

$$S(a) = w_n^{-1}(a) \qquad \text{for } a \in w_n(A),$$

where w_n is viewed as a transformation on A. The dynamical system $\{A; S\}$ is called the *shift dynamical system* associated with the IFS.

Exercises & Examples

4.1. Figure 4.4.1 shows the attractor of the IFS

$$\left\{\mathbb{R}^2; 0.47\begin{pmatrix} x_1 \\ x_2 \end{pmatrix}, 0.47\begin{pmatrix} x_1 \\ x_2 \end{pmatrix} + \begin{pmatrix} 0 \\ 1 \end{pmatrix}, 0.47\begin{pmatrix} x_1 \\ x_2 \end{pmatrix} + \begin{pmatrix} 1 \\ 0 \end{pmatrix}\right\}.$$

Figure 4.4.1 also shows an eventually periodic orbit $\{a_n = S^{\circ n}(a_0)\}_{n=0}^{\infty}$ for the associated shift dynamical system. This orbit actually ends up at the fixed point $\phi(22\overline{22})$. The orbit reads $a_0 = \phi(1231322\overline{22})$, $a_1 = \phi(231322\overline{22})$, $a_2 = \phi(31322\overline{22})$, $a_3 = \phi(1322\overline{22})$, $a_4 = \phi(322\overline{22})$, $a_5 = \phi(22\overline{22})$, where $\phi\colon \Sigma \to A$ is the associated code space map. The point $a_4 \in A$ is clearly a repulsive fixed point of the dynamical system.

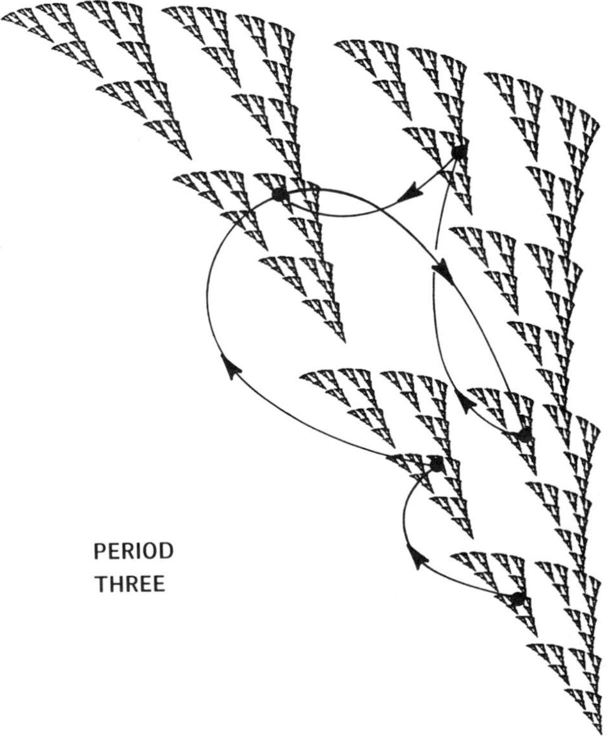

Figure 4.4.2 *This orbit ends up in a cycle of period three.*

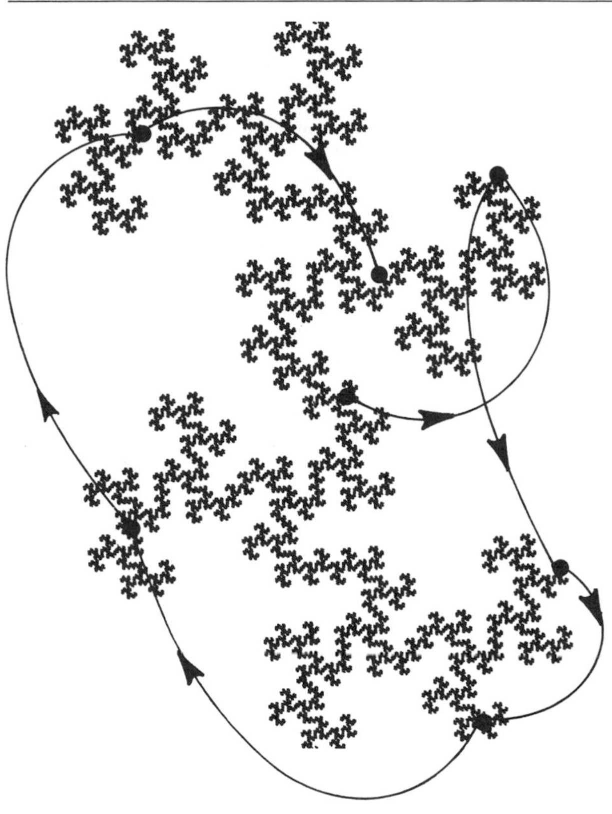

Figure 4.4.3
A chaotic orbit getting going. The shift dynamics are often wild. Why?

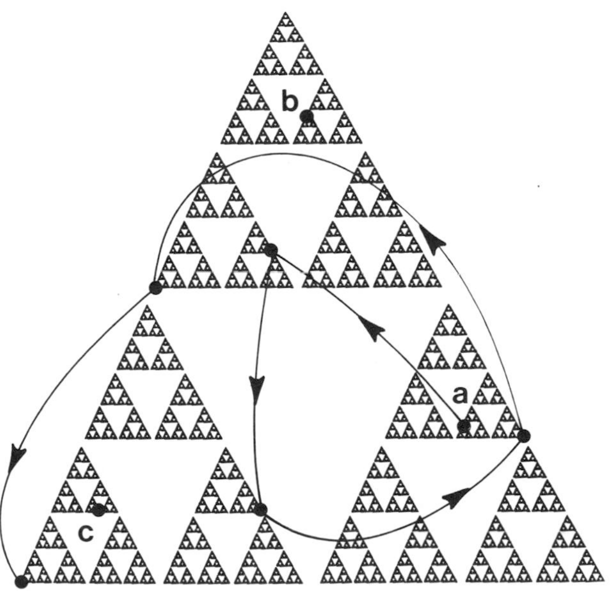

Figure 4.4.4
The orbit of the point a is shown. Can you plot the first few points of the orbits of b and c? Warning! The IFS here is not the usual one. See how the knowledge of some dynamics can imply some more!

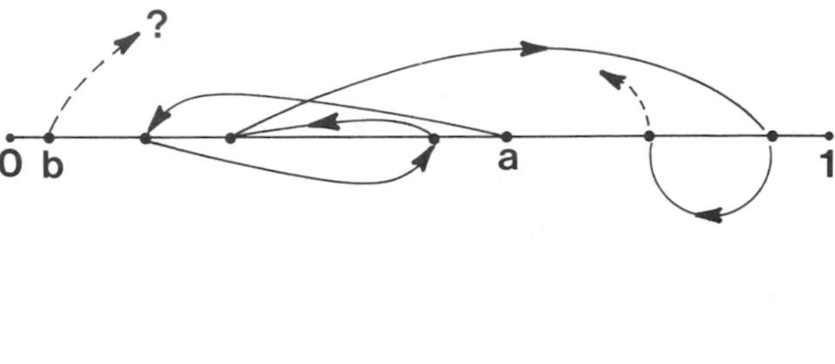

Figure 4.4.5 *This figure shows a sketch of part of an orbit of an IFS* $\{[0,1];\ w_1, w_2, w_3\}$ *on its attractor* $[0,1]$. *The transformation* $w_i:[0,1] \to [0,1]$ *is affine for* $i = 1,2,3$. *Sketch part of the orbit of* b.

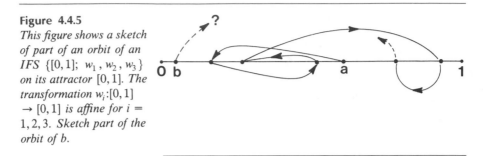

Notice how one can read off the orbit of the point a_0 from its address. Start from another point very close to a_0 and see what happens. Notice how the dynamics depend not only on A itself, but also on the IFS. A different IFS with the same attractor will in general lead to different shift dynamics.

4.2. Each of Figures 4.4.2 and 4.4.3 show attractors of IFS's. In each case the implied IFS is the obvious one. Give the addresses of the points $\{a_n = S^{\circ n}(a_0)\}_{n=0}^{\infty}$ of the eventually periodic orbit in Figure 4.4.2. Show that the cycle to which the orbit converges is a repulsive cycle of period 3. The orbit in Figure 4.4.3 is either very long or infinitely long: why is it hard for us to know which?

4.3. Figure 4.4.4 shows an orbit of a point under the shift dynamical system associated with a certain IFS $\{\mathbb{R}^2;\ w_1, w_2, w_3\}$ where w_1, w_2, and w_3 are affine transformations. Deduce the orbits of the points marked b and c in the figure.

4.4. Figure 4.4.5 shows the start of an orbit of a point under the shift dynamical system associated with a certain hyperbolic IFS. The IFS is of the form $\{\mathbb{R};\ w_1, w_2, w_3\}$ where the transformations $w_n\colon \mathbb{R} \to \mathbb{R}$ are affine, and the attractor is $[0,1]$. Sketch part of the orbit of the point labelled b in the Figure. (Notice that this IFS is actually just-touching: nonetheless, it is straightforward to define uniquely the associated shift dynamics on the open set $\mathcal{O}$ referred to in Definition 4.2.2.)

4.5 EQUIVALENT DYNAMICAL SYSTEMS

Definition 1. Two metric spaces $(\mathbf{X}_1, d_1)$ and $(\mathbf{X}_2, d_2)$ are said to be *topologically equivalent* if there is a homeomorphism $f\colon \mathbf{X}_1 \to \mathbf{X}_2$. Two subsets $S_1 \subset \mathbf{X}_1$ and $S_2 \subset \mathbf{X}_2$ are topologically equivalent, or *homeomorphic*, if the metric spaces (S_1, d_1) and (S_2, d_2) are topologically equivalent. S_1 and S_2 are *metrically equivalent* if the (S_1, d_1) and (S_2, d_2) are equivalent metric spaces.

The Cantor set and code space, discussed following Theorem 4.2.3, are metrically equivalent. Theorem 2.8.5 tells us that if $f\colon \mathbf{X}_1 \to \mathbf{X}_2$ is a continuous one-to-one mapping from a compact metric (X_1, d_1) onto a compact metric space (X_2, d_2), then f is a homeomorphism. So by means of the code space mapping $\phi: \Sigma \to A$ (Theorem 4.2.1) one readily establishes that the attractor of a totally disconnected hyperbolic IFS is topologically equivalent to a classical Cantor set.

Topological equivalence permits a great deal more "stretching and compression" to take place than is permitted by metric equivalence. Later we will define a quantity called the *fractal dimension*. The fractal dimension of a subset of a metric space such as $(\mathbb{R}^2$, Euclidean) provides a measure of the geometrical complexity of the set; it measures the wildness of the set, and it may be used to predict your excitement and wonder when you look at a picture of the set. We will show that two sets which are metrically equivalent have the same fractal dimension. If they are merely topologically equivalent, their fractal dimensions may be different.

With the naturally implied metrics, $[0, 1]$ is homeomorphic to $[0, 2]$. ■ is homeomorphic to ●, is homeomorphic to , and ——— is homeomorphic to .

In fractal geometry we are especially interested in the *geometry* of sets, and in the way they *look* when they are represented by pictures. Thus we use the restrictive condition of metric equivalence to start to define mathematically what we mean when we say that two sets are alike. However, in dynamical systems theory we are interested in *motion* itself, in the dynamics, in the way points move, in the existence of periodic orbits, in the asymptotic behavior of orbits, and so on. These structures are not damaged by homeomorphisms, as we will see, and hence we say that two dynamical systems are alike if they are related via a homeomorphism.

Definition 2. Two dynamical systems $\{X_1; f_1\}$ and $\{X_2; f_2\}$ are said to be ***equivalent***, or *topologically conjugate*, if there is a homeomorphism $\theta: X_1 \to X_2$ such that

$$f_1(x_1) = \theta^{-1} \circ f_2 \circ \theta(x_1) \qquad \text{for all } x_1 \in X_1,$$

$$f_2(x_2) = \theta \circ f_1 \circ \theta^{-1}(x_2) \qquad \text{for all } x_2 \in X_2.$$

In other words, the two dynamical systems are related by the commutative diagram shown in Figure 4.5.1.

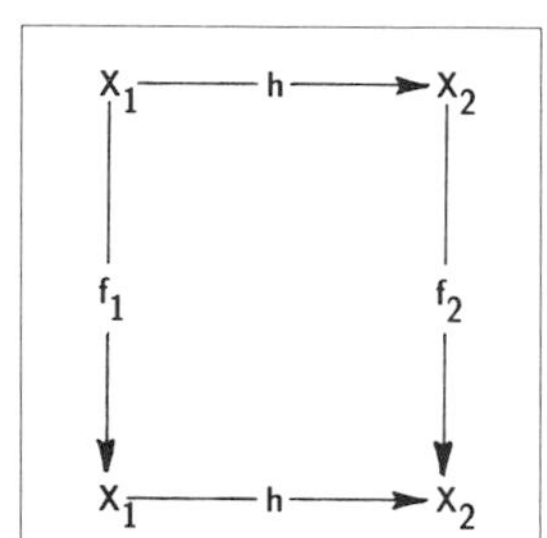

Figure 4.5.1 *Commutative diagram which establishes the equivalence between two dynamical systems $\{X_1; f_1\}$ and $\{X_2; f_2\}$. The function h: $X_1 \to X_2$ is a homeomorphism.*

The following Theorem expresses formally what already should be clear intuitively from our experience with shift dynamics on fractals.

Theorem 1. *Let $\{X; w_1, w_2, \ldots, w_N\}$ be a totally disconnected hyperbolic IFS and let $\{A; S\}$ be the associated shift dynamical system. Let Σ be the associated* *code space of N symbols and let* $T: \Sigma \to \Sigma$ *be defined by*

$$T(\sigma_1\sigma_2\sigma_3 \cdots) = \sigma_2\sigma_3\sigma_4 \cdots \quad \text{for all } \sigma = \sigma_1\sigma_2\sigma_3 \cdots \in \Sigma.$$

Then the two dynamical systems $\{A; S\}$ and $\{\Sigma; T\}$ are equivalent. The homeomorphism which provides this equivalence is $\phi: \Sigma \to A$, as defined in Theorem 4.2.1. Moreover $\{a_1, a_2, \ldots, a_p\}$ is a repulsive cycle of period p for S if, and only if, $\{\phi^{-1}(a_1), \phi^{-1}(a_2), \ldots, \phi^{-1}(a_p)\}$ is a repulsive cycle of period p for T.

Exercises & Examples

5.1. Let $\{X_1; f_1\}$ and $\{X_2; f_2\}$ be equivalent dynamical systems. Let a homeomorphism which provides this equivalence be denoted by $\theta: X_1 \to X_2$. Show that $\{x_1, x_2, \ldots, x_p\}$ is a cycle of period p for $\{X_1; f_1\}$ if and only if $\{\theta(x_1), \theta(x_2), \ldots, \theta(x_p)\}$ is a cycle of period p for $\{X_2; f_2\}$. Suppose that $\{x_1, x_2, \ldots, x_p\}$ is an attractive cycle for f_1. Show that this does not imply that $\{\theta(x_1), \ldots, \theta(x_p)\}$ is an attractive cycle for f_2.

5.2. Let $\{X_1; f_1\}$ and $\{X_2; f_2\}$ be equivalent dynamical systems. Let a homeomorphism which provides this equivalence be denoted by $\theta: X_1 \to X_2$. Let $\{f_1^{\circ n}(x)\}_{n=0}^{\infty}$ be an eventually periodic orbit of f_1. Show that $\{f_2^{\circ n}(\theta(x))\}_{n=0}^{\infty}$ is an eventually periodic orbit of f_2.

5.3. Let $\{X_1; f_1\}$ and $\{X_2; f_2\}$ be equivalent dynamical systems. Let a homeomorphism which provides this equivalence be denoted by $\theta: X_1 \to X_2$. Let this homeomorphism be such as to make the two spaces (X_1, d_1) and (X_2, d_2) metrically equivalent. Construct an example where $x_f \in X_1$ is a repulsive fixed point of the dynamical system $\{X_1, f_1\}$, yet $\theta(x_f)$ is not a repulsive fixed point of $\{X_2, f_2\}$.

5.4. Let $\{X_1; f_1\}$ and $\{X_2; f_2\}$ be equivalent metric spaces. Let a homeomorphism which provides their equivalence be denoted by $\theta: X_1 \to X_2$. Let $x_f \in X_1$ be a fixed point of f_1. Suppose there is an open set $\mathcal{O}$ which contains x_f and is such that $x \in \mathcal{O}$ implies $\text{Lim}_{n \to \infty} f_1^{\circ n}(x) = x_f$. Show that there is an open neighborhood of $\theta(x_f)$ in X_2 with a similar property.

5.5. Our definition of *attractive* and *repulsive* fixed points and cycles, Definition 4.3.4, has the feature that it depends heavily on the metric. It is motivated by the situation of analytic dynamics where small disks are almost mapped into disks. Show how one can use example 5.4 to make a definition of an attractive cycle in such a way that attractiveness of cycles is preserved under topological conjugacy.

5.6. Let $A \subset \mathbb{R}$. Then a function $f: A \to A$ is *differentiable* at a point $x_0 \in A$ if

$$\lim_{\substack{x \to x_0 \\ x \in A}} \left\{ \frac{f(x) - f(x_0)}{x - x_0} \right\}$$

exists. If this limit exists it is denoted by $f'(x_0)$. Let $\{\mathbb{R}; w_1, w_2, \ldots, w_N\}$ be a totally disconnected hyperbolic IFS acting on the metric space $(\mathbb{R}, \text{Euclidean})$. Suppose that, for each $n = 1, 2, \ldots, N$, $w_n(x)$ is differentiable, with $|w_n'(x)| > 0$ for all $x \in \mathbb{R}$. Show that the associated shift dynamical system $\{A; S\}$ is such that S is differentiable at each point $x_0 \in A$, and moreover $|S'(x_0)| > 1$ for all $x \in A$.

5.7. Let $\{\mathbb{R}; f\}$ and $\{\mathbb{R}; g\}$ be equivalent dynamical systems. Let a homeomorphism which provides their equivalence be denoted by $\theta: \mathbb{R} \to \mathbb{R}$. If $\theta(x)$ is infinitely differentiable for all $x \in \mathbb{R}$, then the dynamical systems are said to be *diffeomorphic*. Suppose that $\theta'(x) \neq 0$ for all $x \in \mathbb{R}$. Prove that a_1 is an attractive fixed point of f if and only if $\theta(a_1)$ is an attractive fixed point of g.

5.8. Let $\{\mathbb{R}; f\}$ be a dynamical system such that f is differentiable for all $x \in \mathbb{R}$. Consider the web diagrams associated with this system. Show that the fixed points of f are exactly the intersections of the line $y = x$ with the graph $y = f(x)$. Let a be a fixed point of f. Show that a is an attractive fixed point of f if and only if $|f'(a)| < 1$. Generalize this result to cycles. Note that if $\{a_1, a_2, \ldots, a_p\}$ is a cycle of period p then $(d/dx)(f^{\circ p}(x)|_{x=a_1} = f'(a_1)f'(a_2)\ldots f'(a_p)$. Assure yourself that the situation is correctly summarized in the web diagram shown in Figure 4.5.2.

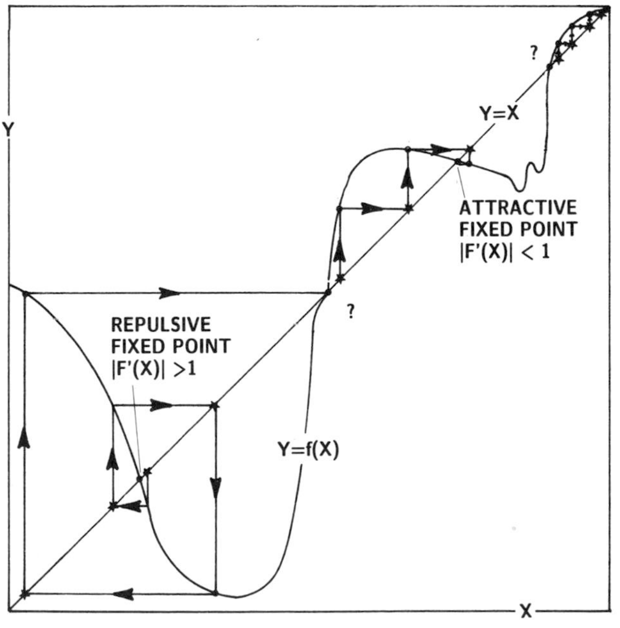

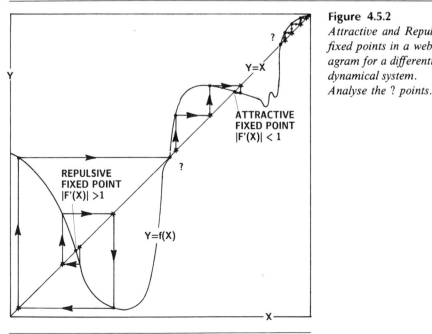

Figure 4.5.2 *Attractive and Repulsive fixed points in a web diagram for a differentiable dynamical system. Analyse the ? points.*

5.9. Consider the dynamical system $\{[0,1]; f(x)\}$ where

$$f(x) = \begin{cases} 1 - 2x & \text{when } x \in [0, \tfrac{1}{2}], \\ 2x - 1 & \text{when } x \in [\tfrac{1}{2}, 1]. \end{cases}$$

Consider also the just-touching IFS $\{[0,1], \frac{1}{2}x + \frac{1}{2}, -\frac{1}{2}x + \frac{1}{2}\}$. Show that it is possible to define a "shift transformation," S, on the attractor,

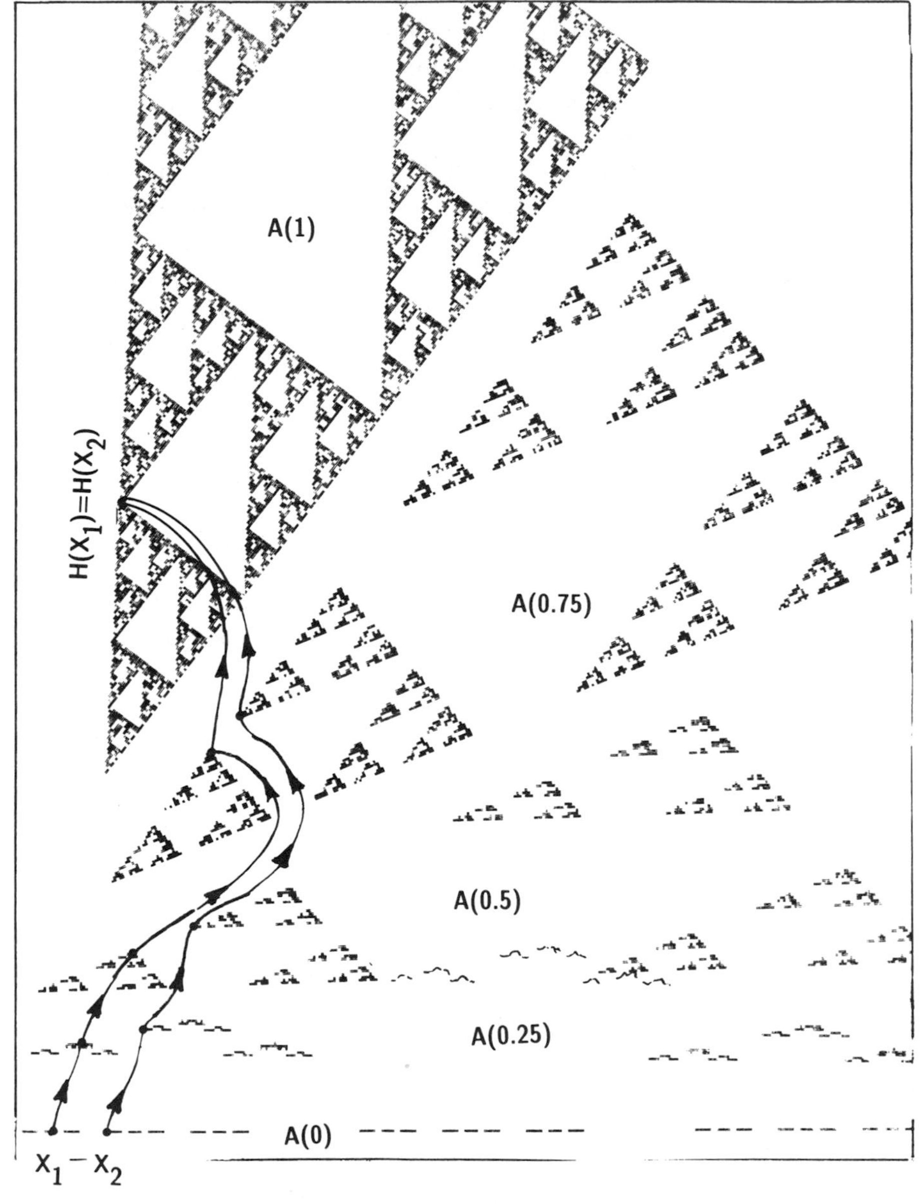

Figure 4.5.3 *Continuous transformation of a Cantor set into a Sierpinski triangle. The inverse transformation would involve some ripping.*

A, of this IFS in such a way that $\{[0,1]; S\}$ and $\{[0,1]; f(x)\}$ are equivalent dynamical systems. To do this you should define S: $A \to A$ in the obvious manner for points with unique addresses; and you should make a suitable definition for the action of S on points with multiple addresses.

5.10. Let $\{\mathbb{R}^2; w_1, w_2, w_3\}$ denote a one-parameter family of IFS, where

$$w_1\begin{pmatrix} x \\ y \end{pmatrix} = \begin{pmatrix} \left(\frac{1+p}{4}\right) & 0 \\ 0 & \left(\frac{1+p}{4}\right) \end{pmatrix}\begin{pmatrix} x \\ y \end{pmatrix};$$

$$w_2\begin{pmatrix} x \\ y \end{pmatrix} = \begin{pmatrix} \left(\frac{1+p}{4}\right) & 0 \\ 0 & \left(\frac{1+p}{4}\right) \end{pmatrix}\begin{pmatrix} x \\ y \end{pmatrix} + \begin{pmatrix} \frac{3+p}{8} \\ \frac{p}{2} \end{pmatrix},$$

$$w_3\begin{pmatrix} x \\ y \end{pmatrix} = \begin{pmatrix} \left(\frac{1+p}{4}\right) & 0 \\ 0 & \left(\frac{1+p}{4}\right) \end{pmatrix}\begin{pmatrix} x \\ y \end{pmatrix} + \begin{pmatrix} \frac{3-p}{4} \\ 0 \end{pmatrix} \text{ for } p \in [0,1].$$

Let the attractor of this IFS be denoted by $A(p)$. Show that $A(0)$ is a Cantor set and $A(1)$ is a Sierpinski triangle. Consider the associated family of code space maps $\phi(p)$: $\Sigma \to A(p)$. Show that $\phi(p)(\sigma)$ is continuous in p for fixed $\sigma \in \Sigma$; that is $\phi(\ )(\sigma)$: $[0,1] \to \mathbb{R}^2$ is a continuous path. Draw some of these paths, including ones which meet at $p = 1$. Interpret these observations in terms of the Cantor set becoming "joined to itself" at various points to make a Sierpinski triangle.

Since the IFS is totally disconnected when $p = 0$, $\phi(p = 0)$: $\Sigma \to A(0)$ is invertible. Hence we can define a continuous transformation θ: $A(0) \to A(1)$ by $\theta(x) = \phi(p = 1)(\phi^{-1}(p = 0)(x))$. Show that if we define a set $J(x) = \{y \in A(0)\text{: } \theta(y) = x\}$ for each $x \in A(1)$, then $J(x)$ is the set of points in $A(0)$ whose associated paths meet at $x \in A(1)$ when $p = 1$. Invent shift dynamics on paths.

4.6 THE SHADOW OF DETERMINISTIC DYNAMICS

Our goal in this section is to extend the definition of the shift dynamical system associated with a totally disconnected hyperbolic IFS to cover the just-touching and overlapping cases. This will lead us to the idea of a random shift dynamical system and to the discovery of a beautiful theorem. This theorem will be called the Shadow Theorem.

Let $\{\boldsymbol{X}; w_1, w_2, \ldots, w_N\}$ denote a hyperbolic IFS, and let A denote its attractor. Assume that w_n: $A \to A$ is invertible for each $n = 1, 2, \ldots, N$, but that the IFS is not totally disconnected. We want to define a dynamical system

$\{A; S\}$ which is analogous to the shift dynamical system defined earlier. Clearly we should define

$$S(x) = w_n^{-1}(x) \qquad \text{when } x \in w_n(A), \text{but } x \notin w_m(A) \text{ for } m \neq n,$$
$$\text{for each } n = 1, 2, \ldots, N.$$

However, at least one of the intersections $w_m(A) \cap w_n(A)$ is nonempty for some $m \neq n$. One idea is simply to make an assignment of which inverse map is to be applied in the overlapping region. For the case $N = 2$ we might define, for example,

$$S(x) = \begin{cases} w_1^{-1}(x) & \text{when } x \in w_1(A), \\ w_2^{-1}(x) & \text{when } x \in A \setminus w_1(A). \end{cases}$$

In the just-touching case the assignment of where S takes points which lie in the overlapping regions does not play a very important role: only a relatively small proportion of points will have somewhat arbitrarily specified orbits. We look at some examples, just to get the flavor.

Exercises & Examples

6.1. Consider the shift dynamical systems associated with the IFS

$$\{[0,1]; \tfrac{1}{2}x, \tfrac{1}{2}x + \tfrac{1}{2}\}.$$

We have $S(x) = 2x$ for $x \in [0, \frac{1}{2})$ – and $S(x) = 2x - 1$ for $x \in (\frac{1}{2}, 1]$. We can define the value of $S(\frac{1}{2})$ to be either 1 or 0. The two possible graphs for $S(x)$ are shown in Figure 4.6.1. The only points $x \in [0, 1] = A$ whose orbits are affected by the definition are those rational numbers whose binary expansions end $\ldots 01\overline{1}$ or $\ldots 10\overline{0}$, the dyadic rationals.

6.2. Show that, if we follow the ideas introduced above, there is only one dynamical system $\{A; S\}$ which can be associated with the just-touching IFS $\{[0,1]; -\frac{1}{2}x + \frac{1}{2}, \frac{1}{2}x\}$. The key here is that $w_1^{-1}(x) = w_2^{-1}(x)$ for all $x \in w_1(A) \cap w_2(A)$.

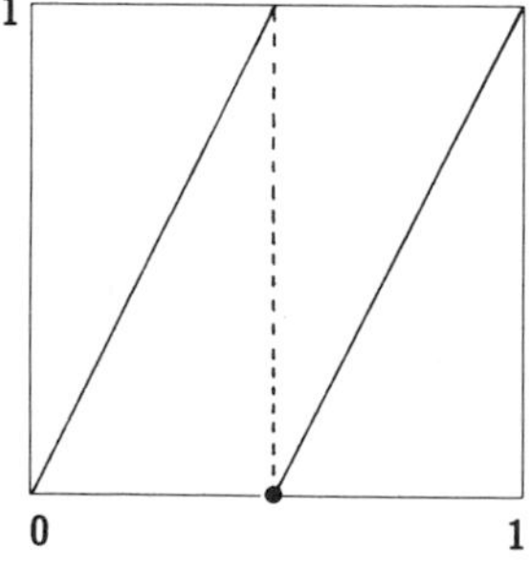

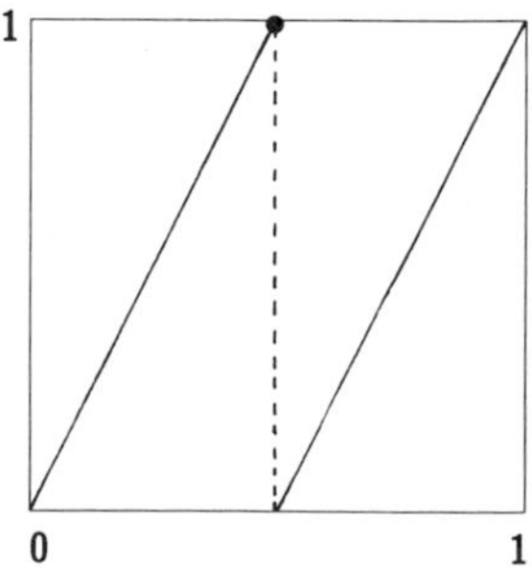

Figure 4.6.1 *The two possible shift dynamical systems associated with the just-touching IFS $\{[0,1]; \frac{1}{2}x, \frac{1}{2}x + \frac{1}{2}\}$ are represented by the two possible graphs of $S(x)$. "Most" orbits are unaffected by the difference between the two systems.*

6.3. Consider some possible "shift" dynamical systems $\{A; S\}$ which can be associated with the IFS

$$\{\mathbb{C}; \tfrac{1}{2}z, \tfrac{1}{2}z + \tfrac{1}{2}, \tfrac{1}{2}z + \tfrac{i}{2}\}.$$

The attractor, ▲, is overlapping at the three points $a = w_1($▲$) \cap w_2($▲$)$, $b = w_2($▲$) \cap w_3($▲$)$, and $c = w_3($▲$) \cap w_1($▲$)$. We might define $S(a) = w_1^{-1}(a)$ or $w_2^{-1}(a)$, $S(b) = w_2^{-1}(b)$ or $w_3^{-1}(b)$, and $S(c) = w_3^{-1}(c)$ or $w_1^{-1}(c)$. Show that regardless of which definition is made, the orbits of a, b, and c are eventually periodic.

6.4. Consider a just-touching IFS of the form $\{\mathbb{R}^2; w_1, w_2, w_3\}$ whose attractor is an equilateral Sierpinski triangle ▲. Assume that each of the maps is a similitude of scaling factor 0.5. Consider the possibility that each map involves a rotation through 0°, 120°, or 240°. The attractor, ▲, is overlapping at the three points $a = w_1($▲$) \cap w_2($▲$)$, $b = w_2($▲$) \cap w_3($▲$)$, and $c = w_3($▲$) \cap w_1($▲$)$. Show that it is possible to choose the maps so that $w_1^{-1}(a) = w_2^{-1}(a)$, $w_2^{-1}(b) = w_3^{-1}(b)$, and $w_3^{-1}(c) = w_1^{-1}(c)$.

6.5. Is code space on two symbols topologically equivalent to code space on three symbols? Yes! Construct a homeomorphism which establishes this equivalence.

6.6. Consider the hyperbolic IFS $\{\Sigma; t_1, t_2, \ldots, t_N\}$ where Σ is code space on N symbols $\{1, 2, \ldots, N\}$ and

$$t_n\sigma = n\sigma \qquad \text{for all } \sigma \in \Sigma.$$

Show that the associated shift dynamical system is exactly $\{\Sigma; T\}$ defined in Theorem 4.5.1. Can two such shift dynamical systems be equivalent for different values of N? To answer this question consider how many fixed points the dynamical system $\{\Sigma; T\}$ possesses for different values of N.

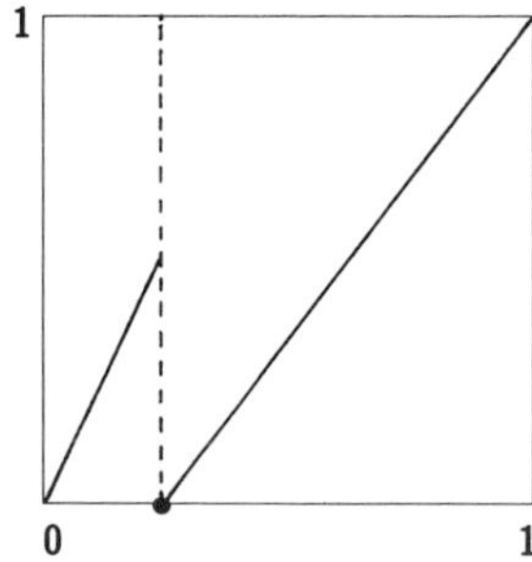

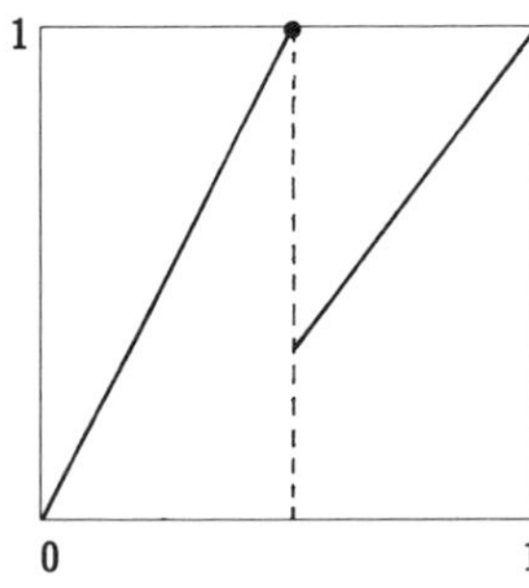

Figure 4.6.2 *Two possible shift dynamical systems which can be associated with the overlapping IFS* $\{[0,1]; \tfrac{1}{2}x, \tfrac{3}{4}x + \tfrac{1}{4}\}$. *In what ways are they alike?*

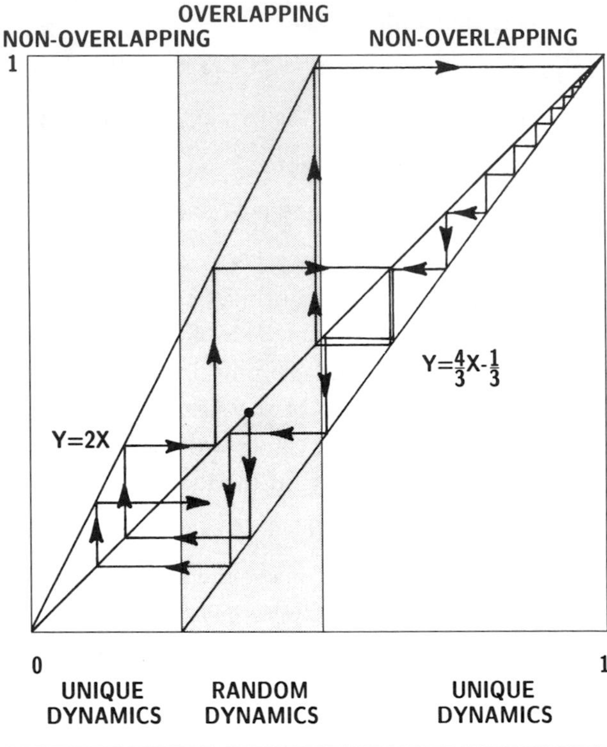

Figure 4.6.3 *A partially random and partially deterministic shift dynamical system associated with the IFS* $\{[0,1]; \frac{1}{2}x, \frac{3}{4}x + \frac{1}{4}\}$.

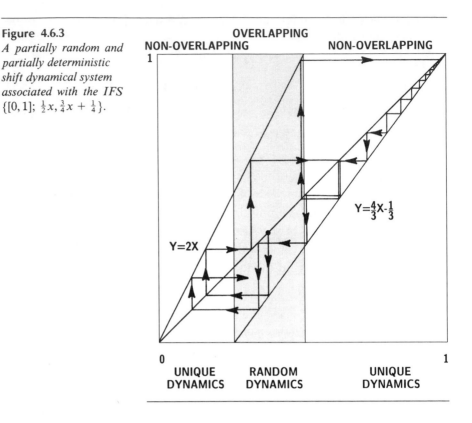

6.7. Consider the overlapping hyperbolic IFS $\{[0,1]; \frac{1}{2}x, \frac{3}{4}x + \frac{1}{4}\}$. Compare the two associated "shift" dynamical systems whose graphs are shown in Figure 4.6.2. What features do they share in common?

6.8. Demonstrate that code space on two symbols is not metrically equivalent to code space on three symbols.

In considering exercises such as 6.7 where two different dynamical systems are associated with an IFS in the overlapping case, we are tempted to entertain the idea that no particular definition of the shift dynamics in the overlapping regions is to be preferred. This suggests that we define the dynamics in overlapping regions in a somewhat random manner. Whenever a point on an orbit lands in an overlapping region, we should allow the possibility that the next point on the orbit is obtained by applying any one of the available inverse transformations. This idea is illustrated in Figure 4.6.3, which should be compared with Figure 4.6.2.

Definition 1. Let $\{X; w_1, w_2\}$ be a hyperbolic IFS. Let A denote the attractor of the IFS. Assume that both $w_1: A \to A$ and $w_2: A \to A$ are invertible. A sequence of points $\{x_n\}_{n=0}^{\infty}$ in A is called an orbit of the *random shift dynamical system* associated with the IFS if

$$x_{n+1} = \begin{cases} w_1^{-1}(x_n) & \text{when } x_n \in w_1(A), \text{ and } x_n \notin w_1(A) \cap w_2(A), \\ w_2^{-1}(x_n) & \text{when } x_n \in w_2(A), \text{ and } x_n \notin w_1(A) \cap w_2(A), \\ \text{one of } \{w_1^{-1}(x_n), w_2^{-1}(x_n)\}, & \text{when } x_n \in w_1(A) \cap w_2(A), \end{cases}$$

for each $n \in \{0, 1, 2, \dots\}$. We will use the notation $x_{n+1} = S(x_n)$ although there may be no well-defined transformation $S\colon A \to A$ which makes this true. Also we will write $\{A; S\}$ to denote the collection of possible orbits defined here, and we will call $\{A; S\}$ the random shift dynamical system associated with the IFS.

Notice that if $w_1(A) \cap w_2(A) = \varnothing$ then the IFS is totally disconnected and the orbits defined here are simply those of the shift dynamical system $\{A; S\}$ defined earlier.

We now show that there is a completely deterministic dynamical system acting on a higher dimensional space, whose projection into the original space $\mathbf{X}$, yields the "random dynamics" we have just described. Our random dynamics are seen as the shadow of deterministic dynamics. To achieve this we turn the IFS into a totally disconnected system by introducing an additional variable. To keep the notation succinct we restrict the following discussion to IFS's of two maps.

Definition 2. The *lifted* IFS associated with a hyperbolic IFS $\{\mathbf{X}; w_1, w_2\}$ is the hyperbolic IFS $\{\mathbf{X} \times \Sigma; \tilde{w}_1, \tilde{w}_2\}$ where Σ is the code space on two symbols $\{1, 2\}$, and

$$\tilde{w}_1(x, \sigma) = (w_1(x), 1\sigma) \qquad \text{for all } (x, \sigma) \in \mathbf{X} \times \Sigma;$$

$$\tilde{w}_2(x, \sigma) = (w_2(x), 2\sigma) \qquad \text{for all } (x, \sigma) \in \mathbf{X} \times \Sigma.$$

What is the nature of the attractor $\tilde{A} \subset \mathbf{X} \times \Sigma$ of the lifted IFS? Clearly $\tilde{A} = \{(\phi(\sigma), \sigma)\colon \sigma \in \Sigma\}$, the graph of the code space map ϕ; and

$$A = \{x \in \mathbf{X}\colon (x, \sigma) \in \tilde{A} \quad \text{for some } \sigma \in \Sigma\} = \phi(\Sigma).$$

In other words the projection of the attractor of the lifted IFS into the original space $\mathbf{X}$ is simply the attractor A of the original IFS. The projection of $\tilde{A}$ into Σ is Σ. Recall that Σ is equivalent to a classical Cantor set. This tells us that the attractor of the lifted IFS is totally disconnected, since the projection map from $\tilde{A}$ into Σ is one-to-one on Σ.

Lemma 1. *Let $\{\mathbf{X}; w_1, w_2\}$ be a hyperbolic IFS with attractor A. Let the two transformations $w_1\colon A \to A$ and $w_2\colon A \to A$ be invertible. Then the associated lifted IFS is hyperbolic and totally disconnected.*

Definition 3. Let $\{\mathbf{X}; w_1, w_2\}$ be a hyperbolic IFS. Let the two transformations $w_1: A \to A$ and $w_2: A \to A$ be invertible. Let $\tilde{A}$ denote the attractor of the associated lifted IFS. Then the shift dynamical system $\{\tilde{A}; \tilde{S}\}$ associated with the lifted IFS is called the *lifted shift dynamical system* associated with the IFS.

Notice that

$$\tilde{S}(x, \sigma) = \left(w_{\sigma_1}^{-1}(x), T(\sigma)\right) \qquad \text{for all } (x, \sigma) \in \tilde{A},$$

where $T(\sigma_1\sigma_2\sigma_3\sigma_4\ldots) = \sigma_2\sigma_3\sigma_4\sigma_5\ldots$ for all $\sigma = \sigma_1\sigma_2\sigma_3\sigma_4\ldots \in \Sigma$.

Theorem 1. (The Shadow Theorem.) *Let $\{\mathbf{X}; w_1, w_2\}$ be a hyperbolic IFS of invertible transformations w_1 and w_2 and attractor A. Let $\{x_n\}_{n=0}^{\infty}$ be any orbit of the associated random shift dynamical system $\{A; S\}$. Then there is an orbit $\{\tilde{x}_n\}_{n=0}^{\infty}$ of the lifted dynamical system $\{\tilde{A}; \tilde{S}\}$ such that the first component of $\tilde{x}_n$ is x_n for all n.*

We leave the proofs of Lemma 1 and Theorem 1 as exercises. It is fun however, and instructive, to look in a couple of different geometrical ways at what is going on here.

Examples

6.9. Consider the IFS $\{\mathbb{C}; w_1(z), w_2(z), w_3(z), w_4(z)\}$ where, in complex notation,

$$w_1(z) = (0.5)\left(\cos 45° - \sqrt{-1}\,\sin 45°\right)z + \left(0.4 - 0.2\sqrt{-1}\right),$$
$$w_2(z) = (0.5)\left(\cos 45° + \sqrt{-1}\,\sin 45°\right)z - \left(0.4 + 0.2\sqrt{-1}\right),$$
$$w_3(z) = (0.5)z + \sqrt{-1}\,(0.3),$$
$$w_4(z) = (0.5)z - \sqrt{-1}\,(0.3).$$

A sketch of its attractor is included in Figure 4.6.4. It looks like a maple leaf.

The leaf is made of four overlapping leaflets, which we think of as separate entities, at different heights "above" the attractor. In turn, we think of each leaflet as consisting of four smaller leaflets, again at different heights. One quickly gets the idea: one ends up with a set of heights distributed on a Cantor set in such a way that the shadow of the whole collection of infinitesimal leaflets is the leaf attractor in the $\mathbb{C}$ plane. The Cantor set is essentially Σ. The lifted attractor is totally disconnected; and it supports deterministic shifts dynamics, as illustrated in Figure 4.6.5.

6.10. Consider the overlapping hyperbolic IFS $\{\mathbb{R}; \frac{1}{2}x, \frac{3}{4}x + \frac{1}{4}\}$. We can lift

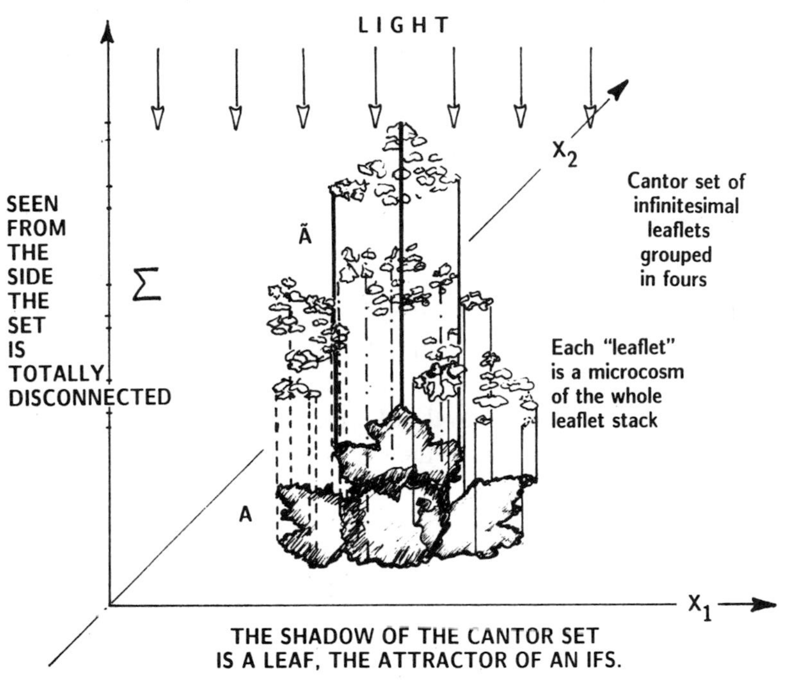

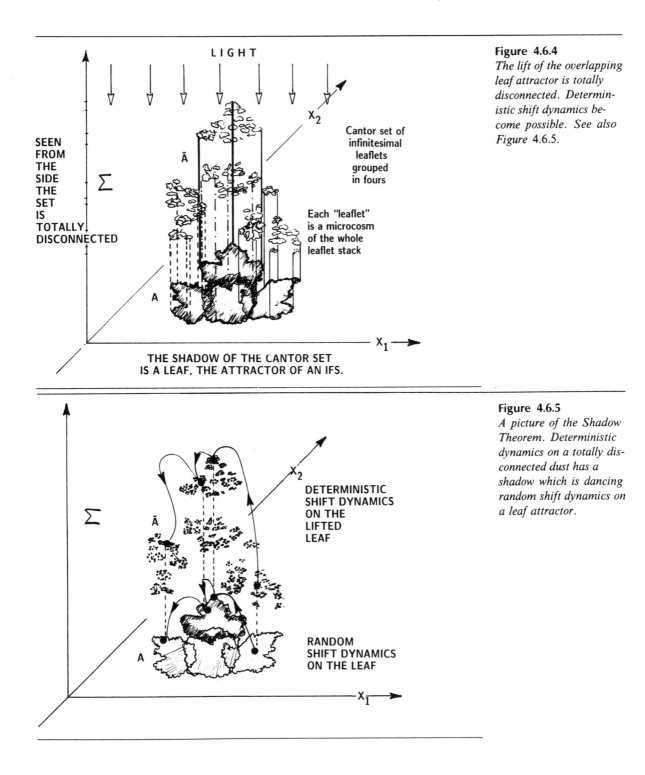

Figure 4.6.4
The lift of the overlapping leaf attractor is totally disconnected. Deterministic shift dynamics become possible. See also Figure 4.6.5.

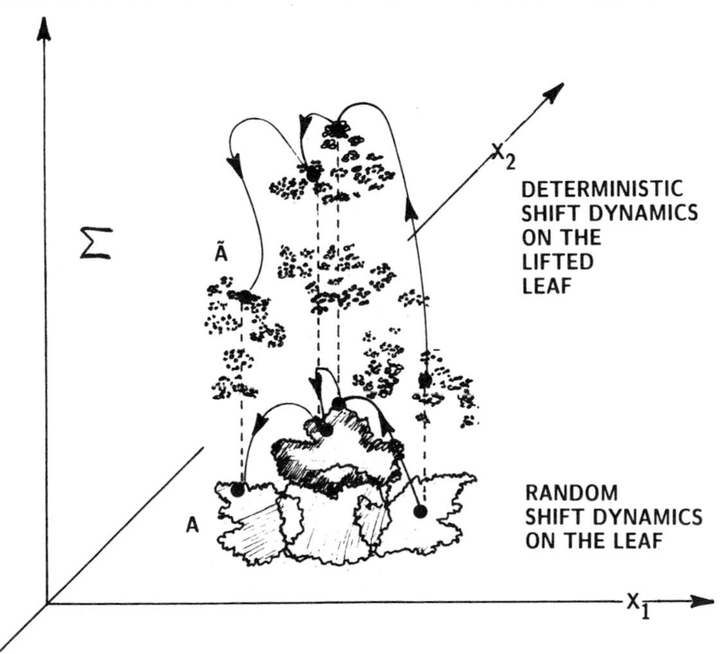

Figure 4.6.5
A picture of the Shadow Theorem. Deterministic dynamics on a totally disconnected dust has a shadow which is dancing random shift dynamics on a leaf attractor.

this to the hyperbolic IFS $\{\mathbb{R}^2; w_1(x), w_2(x)\}$ where

$$w_1\begin{pmatrix} x_1 \\ x_2 \end{pmatrix} = \begin{pmatrix} \frac{1}{2} & 0 \\ 0 & \frac{1}{3} \end{pmatrix}\begin{pmatrix} x_1 \\ x_2 \end{pmatrix};$$

$$w_2\begin{pmatrix} x_1 \\ x_2 \end{pmatrix} = \begin{pmatrix} \frac{3}{4} & 0 \\ 0 & \frac{1}{3} \end{pmatrix}\begin{pmatrix} x_1 \\ x_2 \end{pmatrix} + \begin{pmatrix} \frac{1}{4} \\ \frac{2}{3} \end{pmatrix}.$$

The attractor $\tilde{A}$ of this lifted system is shown in Figure 4.6.6, which also shows an orbit of the associated shift dynamical system. The shadow of this orbit is an apparently random orbit of the original system. The Shadow Theorem asserts that *any* orbit $\{x_n\}_{n=0}^{\infty}$ of a random shift dynamical system associated with the IFS $\{\mathbb{R}; \frac{1}{2}x, \frac{3}{4}x + \frac{1}{4}\}$ is the projection, or shadow, of some orbit for the shift dynamical system associated with the lifted IFS.

6.11. As a compelling illustration of the Shadow Theorem, consider the IFS $\{\mathbb{R}; \frac{1}{2}x, \frac{3}{4}x + \frac{1}{4}\}$. Let us look at the orbits $\{x_n\}_{n=0}^{\infty}$ of the "shift" dynamical system which is specified in the left-hand graph of Figure

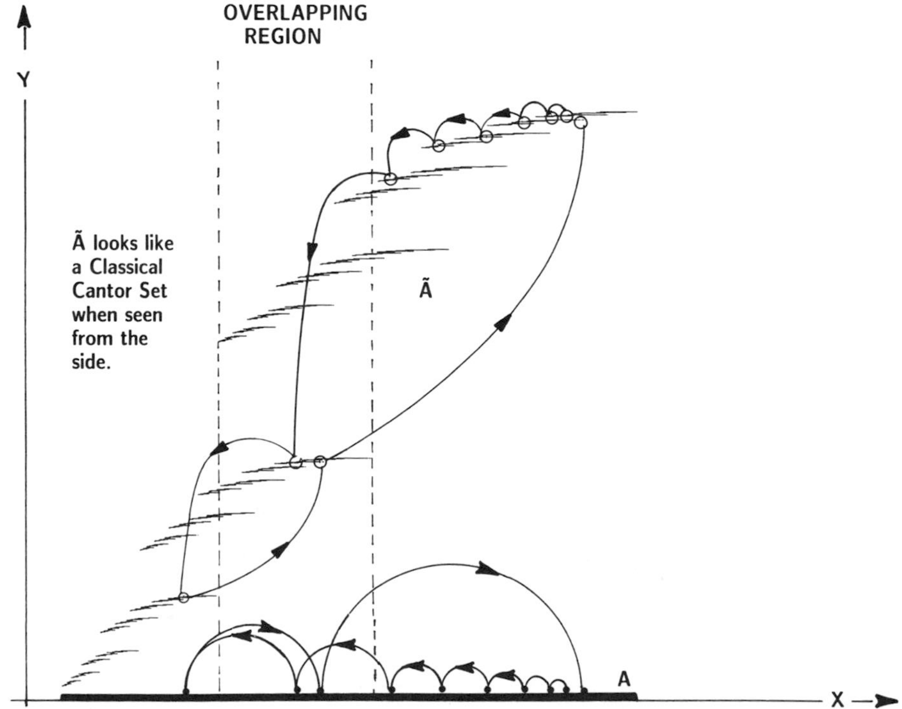

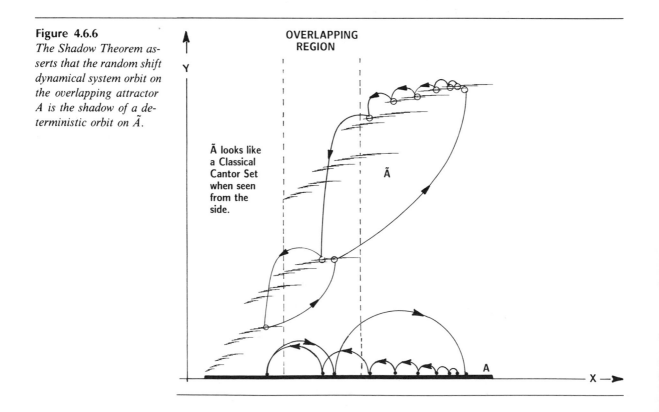

Figure 4.6.6 *The Shadow Theorem asserts that the random shift dynamical system orbit on the overlapping attractor A is the shadow of a deterministic orbit on $\tilde{A}$.*

4.6.2. In this case we always choose $S(x) = w_2^{-1}(x)$ in the overlapping region. What orbits $\{\tilde{x}_n\}_{n=0}^{\infty}$ of the lifted system, described in example 6.10, are these orbits the shadow of? Look again at Figure 4.6.6! Define the top of $\tilde{A}$ as

$$\tilde{A}_{\text{top}} = \{(x, y) \in \tilde{A}: (x, z) \in \tilde{A} \Rightarrow z \leq y, \quad \text{and } x \in [0,1]\}.$$

Notice that $\tilde{S}$: $\tilde{A}_{\text{top}} \to \tilde{A}_{\text{top}}$. It is easy to see that there is a one-to-one correspondence between orbits of the lifted system $\{\tilde{A}_{\text{top}}; \tilde{S}\}$ and orbits of the original system specified through the left-hand graph of Figure 4.6.2. Indeed,

$$\text{“}\{(x_n, y_n)\}_{n=0}^{\infty} \quad \text{is an orbit of the lifted system, and} \quad (x_0, y_0) \in \tilde{A}_{\text{top}}\text{”}$$

$$\Updownarrow$$

$$\text{“}\{x_n\}_{n=0}^{\infty} \quad \textit{is an orbit of the left-hand graph of Figure } 4.6.2.\text{”}$$

6.12. Draw some pictures to illustrate the Shadow Theorem in the case of the just-touching IFS $\{[0,1]; \frac{1}{2}x, \frac{1}{2}x + \frac{1}{2}\}$.

6.13. Illustrate the Shadow Theorem using the overlapping IFS $\{[0,1]; -\frac{3}{4}x + \frac{3}{4}, \frac{3}{4}x + \frac{1}{4}\}$. Can you find an orbit of period two whose lift has minimal period four? Do there exist periodic orbits whose lifts are not periodic?

6.14. Prove Lemma 4.6.1.

6.15. Prove Theorem 4.6.1.

4.7 THE MEANINGFULNESS OF INACCURATELY COMPUTED ORBITS IS ESTABLISHED BY MEANS OF A SHADOWING THEOREM

Let $\{X; w_1, w_2, \ldots, w_N\}$ be a hyperbolic IFS of contractivity of $0 < s < 1$. Let A denote the attractor of the IFS, and assume that w_n: $A \to A$ is invertible for each $n = 1, 2, \ldots N$. If the IFS is totally disconnected let $\{A; S\}$ denote the associated shift dynamical system; otherwise let $\{A; S\}$ denote the associated random shift dynamical system. Consider the following model for the inaccurate calculation of an orbit of a point $x_0 \in A$. This model will surely describe the reader's experiences in computing shift dynamics directly on pictures of fractals. Moreover it is a reasonable model for the occurrence of numerical errors when machine computation is used to compute an orbit.

Let an exact orbit of the point $x_0 \in A$ be denoted by $\{x_n\}_{n=0}^{\infty}$ where $x_n = S^{\circ n}(x_0)$ for each n. Let an approximate orbit of the point $x_0 \in A$ be denoted by $\{\tilde{x}_n\}_{n=0}^{\infty}$ where $\tilde{x}_0 = x_0$. Then we suppose that at each step there is made an error of at most θ for some $0 \leq \theta < \infty$; that is,

$$d(\tilde{x}_{n+1}, S(\tilde{x}_n)) \leq \theta, \qquad \text{for } n = 0, 1, 2, \ldots$$

We proceed to analyse this model. It is clear that the inaccurate orbit $\{\tilde{x}_n\}_{n=0}^{\infty}$ will usually start out by diverging from the exact orbit $\{x_n\}_{n=0}^{\infty}$ at an exponential rate. It may well occur "accidentally" that $d(x_n, \tilde{x}_n)$ is small for various large values of n, due to the compactness of A. But typically, if $d(x_n, \tilde{x}_n)$ is small enough then $d(x_{n+j}, \tilde{x}_{n+j})$, will again grow exponentially with increasing j. To be precise, suppose $d(\tilde{x}_1, S(\tilde{x}_0)) = \theta$ and that we make no further errors. Suppose also that for some integer M, and some integers $\sigma_1, \sigma_2, \ldots, \sigma_M \in \{1, 2, \ldots, N\}$ we have

$$\tilde{x}_n \text{ and } x_n \in w_{\sigma_n}(A), \qquad \text{for } n = 0, 1, 2, \ldots, M.$$

Moreover, suppose that

$$x_{n+1} = w_{\sigma_n}^{-1}(x_n) \quad \text{and } \tilde{x}_{n+1} = w_{\sigma_n}^{-1}(\tilde{x}_n), \qquad \text{for } n = 1, 2, \ldots, M.$$

Then we have

$$d(x_{n+1}, \tilde{x}_{n+1}) \geq s^{-n}\theta, \qquad \text{for } n = 0, 1, 2, \ldots, M.$$

For some integer $J > M$ it is likely to be the case that

$$x_{J+1} = w_{\sigma_J}^{-1}(x_J) \quad \text{and } \tilde{x}_{J+1} = w_{\tilde{\sigma}_J}^{-1}(\tilde{x}_J), \qquad \text{for some } \sigma_J \neq \tilde{\sigma}_J.$$

Then, without further assumptions, we cannot say anything more about the correlation between the exact orbit and the approximate orbit. Of course, we always have the error bound

$$d(x_n, \tilde{x}_n) \leq \text{diam}(A) = \text{Max}\{d(x, y): x \in A,\ y \in A\},$$
$$\text{for all } n = 1, 2, 3, \ldots .$$

Do the above comments make the situation hopeless? Are all of the calculations of shift dynamics which we have done in this chapter without point because they are so hopelessly riddled with errors? No! The following wonderful theorem tells us that, however many errors we make, there is an exact orbit which lies at every step within a small distance of our errorful one. This orbit shadows the errorful orbit. Here we use the word "shadows" in the sense of a secret agent who shadows a spy. The agent is always just out of sight, not too far away, usually not too close, but forever he follows the spy.

Theorem 1. [The Shadow*ing* Theorem] *Let $\{\mathbf{X}; w_1, w_2, \ldots, w_N\}$ be a hyperbolic IFS of contractivity s, where $0 < s < 1$. Let A denote the attractor of the IFS and suppose that each of the transformations w_n: $A \to A$ is invertible. Let $\{A; S\}$ denote the associated shift dynamical system in the case that the IFS is totally disconnected; otherwise let $\{A; S\}$ denote the associated random shift dynamical system. Let $\{\tilde{x}_n\}_{n=0}^{\infty} \subset A$ be an approximate orbit of S, such that*

$$d(\tilde{x}_{n+1}, S(\tilde{x}_n)) \leq \theta \qquad \text{for all } n = 0, 1, 2, 3, \ldots$$

for some fixed constant θ with $0 \leq \theta \leq$ diam(A). Then there is an exact orbit $\{x_n = S^{\circ n}(x_0)\}_{n=0}^{\infty}$ for some $x_0 \in A$, such that

$$d(\tilde{x}_{n+1}, x_{n+1}) \leq \frac{s\theta}{(1 - s)} \qquad \text{for all } n = 0, 1, 2, \ldots .$$

Proof. As usual we exploit code space! For $n = 1, 2, 3, \ldots,$ let $\sigma_n \in \{1, 2, \ldots, N\}$ be chosen so that $w_{\sigma_1}^{-1}, w_{\sigma_2}^{-1}, w_{\sigma_3}^{-1}, \ldots,$ is the actual sequence of inverse maps used to compute $S(\tilde{x}_0), S(\tilde{x}_1), S(\tilde{x}_2), \ldots$. Let $\phi\colon \Sigma \to A$ denote the code space map associated with the IFS. Then define

$$x_0 = \phi(\sigma_1\sigma_2\sigma_3\ldots).$$

Then we compare the exact orbit of the point x_0,

$$\{x_n = S^{\circ n}(x_0) = \phi(\sigma_{n+1}\sigma_{n+2}\cdots)\}_{n=0}^{\infty}$$

with the errorful orbit $\{\tilde{x}_n\}_{n=0}^{\infty}$.

Let M be a large positive integer. Then, since x_M and $S(\tilde{x}_{M-1})$ both belong to A, we have

$$d(S(x_{M-1}), S(\tilde{x}_{M-1})) \leq \operatorname{diam}(A) < \infty.$$

Since $S(x_{M-1})$ and $S(\tilde{x}_{M-1})$ are both computed with the same inverse map $w_{\sigma_M}^{-1}$ it follows that

$$d(x_{M-1}, \tilde{x}_{M-1}) \leq s \operatorname{diam}(A).$$

Hence

$$\begin{aligned} d(S(x_{M-2}), S(\tilde{x}_{M-2})) &= d(x_{M-1}, S(\tilde{x}_{M-2})) \\ &\leq d(x_{M-1}, \tilde{x}_{M-1}) + d(\tilde{x}_{M-1}, S(\tilde{x}_{M-2})) \\ &\leq \theta + s \operatorname{diam}(A); \end{aligned}$$

and repeating the argument used above we now find

$$d(x_{M-2}, \tilde{x}_{M-2}) \leq s(\theta + s \operatorname{diam}(A))$$

Repeating the same argument k times we arrive at

$$d(x_{M-k}, \tilde{x}_{M-k}) \leq s\theta + s^2\theta + \cdots + s^{k-1}\theta + s^k \operatorname{diam}(A).$$

Hence for any positive integer M and any integer n such that $0 < n < M$ we have

$$d(x_n, \tilde{x}_n) \leq s\theta + s^2\theta + \cdots + s^{M-n-1}\theta + s^{M-n} \operatorname{diam}(A).$$

Now take the limit of both sides of this equation as $M \to \infty$ to obtain

$$d(x_n, \tilde{x}_n) \leq s\theta(1 + s + s^2 + \cdots) = \frac{s\theta}{(1-s)}, \qquad \text{for all } n = 1, 2, \ldots.$$

This completes the proof.

Exercises & Examples

7.1. Let us apply the Shadowing Theorem to an orbit on the Sierpinski triangle, using the random shift dynamical system associated with the IFS

$$\{C; \tfrac{1}{2}z, \tfrac{1}{2}z + \tfrac{1}{2}, \tfrac{1}{2}z + \tfrac{i}{2}\}$$

Since the system is just-touching, we must assign values to the shift transformation applied to the just-touching points. We do this by defining

$$S(x_1 + ix_2) = 2x_1 \bmod 1 + i(2x_2 \bmod 1).$$

We consider the orbit of the point $\tilde{x}_0 = (0.2147, 0.0353)$. We compute the first eleven points on the exact orbit of this point, and compare it to the results obtained when a deliberate error $\theta = 0.0001$ is introduced at each step. We obtain:

ERRORFUL	**EXACT**
$\tilde{x}_0 = (0.2174, 0.0353)$	$S^{\circ 0}(\tilde{x}_0) = (0.2147, 0.0353)$
$\tilde{x}_1 = (0.4295, 0.0705)$	$S^{\circ 1}(\tilde{x}_0) = (0.4294, 0.0706)$
$\tilde{x}_2 = (0.8591, 0.1409)$	$S^{\circ 2}(\tilde{x}_0) = (0.8588, 0.1412)$
$\tilde{x}_3 = (0.7183, 0.2817)$	$S^{\circ 3}(\tilde{x}_0) = (0.7176, 0.2824)$
$\tilde{x}_4 = (0.4365, 0.5635)$	$S^{\circ 4}(\tilde{x}_0) = (0.4352, 0.5648)$
$\tilde{x}_5 = (0.8731, 0.1269)$	$S^{\circ 5}(\tilde{x}_0) = (0.8704, 0.1296)$
$\tilde{x}_6 = (0.7463, 0.2537)$	$S^{\circ 6}(\tilde{x}_0) = (0.7408, 0.2592)$
$\tilde{x}_7 = (0.4927, 0.5073)$	$S^{\circ 7}(\tilde{x}_0) = (0.4816, 0.5184)$
$\tilde{x}_8 = (0.9855, 0.0145)$	$S^{\circ 8}(\tilde{x}_0) = (0.9632, 0.0368)$
$\tilde{x}_9 = (0.9711, 0.0289)$	$S^{\circ 9}(\tilde{x}_0) = (0.9264, 0.0736)$
$\tilde{x}_{10} = (0.9423, 0.0577)$	$S^{\circ 10}(\tilde{x}_0) = (0.8528, 0.1472)$

Notice how the orbit with errors diverges from the exact orbit of $\tilde{x}_0$. Nonetheless, the Shadowing Theorem asserts that there is an *exact* orbit $\{x_n\}$ such that

$$d(x_n, \tilde{x}_n) \leq \frac{\frac{1}{2}}{1 - \frac{1}{2}}(0.0001) = 0.0001,$$

where $d(\cdot, \cdot)$ denotes the Manhattan metric. This really *seems* unlikely; but it must be true! Here's an example of such a shadowing orbit, also computed exactly.

EXACT SHADOWING ORBIT $x_n = S^{\circ n}(x_0)$	$d(x_n, \tilde{x}_n) \leq 0.0001$
$x_0 = (0.21478740234375, 0.03521259765625)$	0.00009
$x_1 = (0.4295748046875, 0.0704251953125)$	0.00008
$x_2 = (0.8591496093750, 0.1408503906250)$	0.00005
$x_3 = (0.7182992187500, 0.2817007812500)$	0.000001
$x_4 = (0.4365984375000, 0.5634015625000)$	0.0001
$x_5 = (0.8731968750000, 0.1268031250000)$	0.0001
$x_6 = (0.7463937500000, 0.2536062500000)$	0.0001
$x_7 = (0.4927875000000, 0.5072125000000)$	0.00009
$x_8 = (0.9855750000000, 0.0144250000000)$	0.00008
$x_9 = (0.9711500000000, 0.0288500000000)$	0.00005
$x_{10} = (0.9423000000000, 0.0577000000000)$	0.000000

Figure 4.7.1 illustrates the idea.

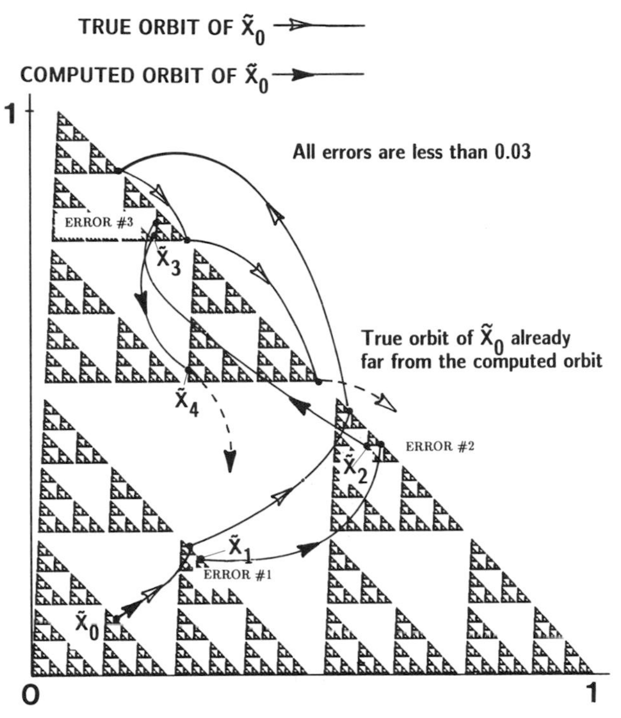

Figure 4.7.1 *The Shadowing Theorem tells us there is an exact orbit which is closer to* $\{\tilde{x}_n\}$ *than* 0.03 *for all n.*

7.2. Consider the shift dynamical system $\{\Sigma; T\}$ on the code space of two symbols $\{1, 2\}$. Show that the sequence of points $\{\tilde{x}_n\}$ given by

$$\tilde{x}_0 = 21\bar{2}, \quad \text{and } \tilde{x}_n = 1\bar{2} \qquad \text{for all } n = 1, 2, 3, \ldots .$$

is an errorful orbit for the system. Illustrate the divergence of $T^{\circ n}\tilde{x}_0$ from $\tilde{x}_n$. Find a shadowing orbit $\{x_n\}_{n=0}^{\infty}$; and verify the error estimate provided by the Shadowing Theorem.

7.3. Illustrate the Shadowing Theorem by constructing an erroneous orbit, and an orbit which shadows it, for the shift dynamical system $\{[0,1]; \frac{1}{3}x, \frac{1}{2}x + \frac{1}{2}\}$.

7.4. Compute an orbit for a random shift dynamical system associated with the overlapping IFS $\{[0,1]; \frac{3}{4}x, \frac{1}{2}x + \frac{1}{2}\}$.

7.5. An orbit of the shift dynamical system associated with the IFS

$$\left\{\mathbb{R}^2; \tfrac{1}{2}\begin{pmatrix} x \\ y \end{pmatrix}, \tfrac{3}{4}\begin{pmatrix} x \\ y \end{pmatrix} + \tfrac{1}{4}\begin{pmatrix} 1 \\ 1 \end{pmatrix}, \tfrac{1}{2}\begin{pmatrix} x \\ y \end{pmatrix} + \begin{pmatrix} 2 \\ 0 \end{pmatrix}, \tfrac{1}{8}\begin{pmatrix} x \\ y \end{pmatrix} + \begin{pmatrix} 0 \\ 7 \end{pmatrix}\right\},$$

is computed to accuracy 0.0005. How close a shadowing orbit does there exist? Use the Manhattan metric.

7.6. In Figure 4.7.2 an orbit of the random shift dynamical system associated with the overlapping IFS $\{[0,1], w_1(x), w_2(x)\}$ is computed by drawing a

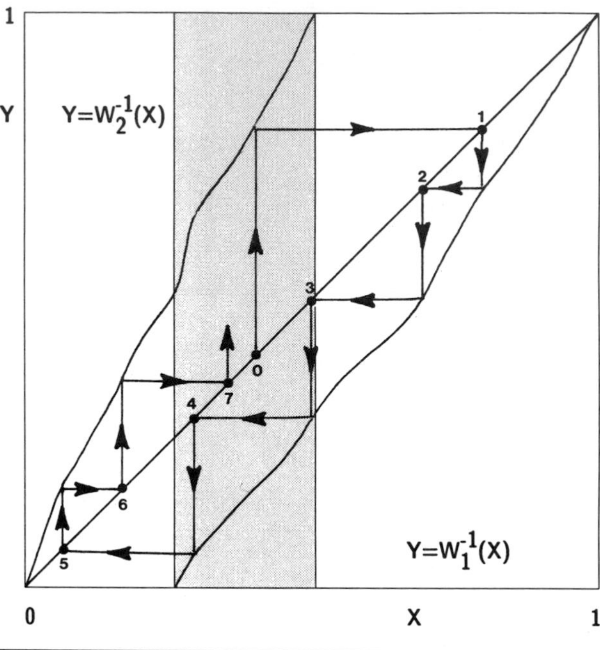

Figure 4.7.2 *An exact orbit shadows the orbit "computed" by "drawing," in this web diagram for a random shift dynamical system.*

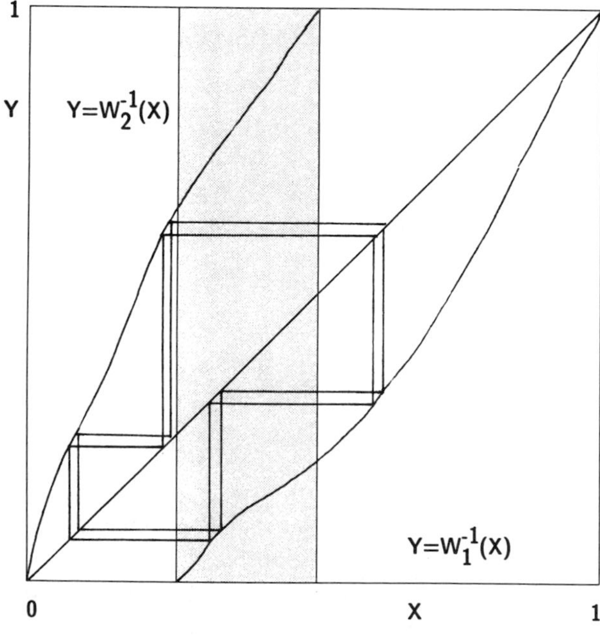

Figure 4.7.3 *Only the* Shadow *knows. Inside the "orbit tube" there is an* exact *orbit* $\{x_n\}_{n=0}^{\infty}$ *of the random shift dynamical system associated with the IFS.*

web diagram. The computer in this case consists of a pencil and a drafting table. Estimate the errors in the drawing and then deduce how closely an exact orbit shadows the plotted one. You will need to estimate the contractivity of the IFS. Also draw a tube around the plotted orbit, within which an exact orbit lies.

7.7. Figure 4.7.3 shows an orbit $\{x_n\}$ of the random shift dynamical system associated with the IFS $\{[0,1];\, w_1(x),\, w_2(x)\}$. It was obtained by defining $S(x) = w_2^{-1}(x)$ for $x \in w_1(A) \cap w_2(A)$. A contractivity factor for the IFS is readily estimated from the drawing to be $\frac{3}{5}$. Hence if the web diagram is accurate to within 1 mm at each iteration, that is

$$d(\tilde{x}_{n+1}, S(\tilde{x}_n)) \leq 1 \text{ mm},$$

then there is an exact orbit $\{x_n = S^{\circ n}(x_0)\}_{n=0}^{\infty}$ such that

$$d(x_n, \tilde{x}_n) \leq \frac{(\frac{3}{5})}{(\frac{2}{5})} = 1.5 \text{ mm}.$$

Thus there is an actual orbit which remains within the shaded tube shown in Figure 4.7.3.

4.8 CHAOTIC DYNAMICS ON FRACTALS

The shift dynamical system $\{A; S\}$ associated with a totally disconnected hyperbolic IFS is equivalent to the shift dynamical system $\{\Sigma, T\}$, where Σ is the code space associated with the IFS. As we have seen, this equivalence means that the two systems have a number of properties in common; for example, the two systems have the same number of cycles of minimal period seven. A particularly important property which they share is that they are both "chaotic" dynamical systems, a concept which we explain in this section. First, however, we want to emphasize that the two systems are deeply different from the point of view of the interplay of their dynamics with the geometry of the underlying spaces.

Consider the case of an IFS of three transformations. Let Σ denote the code space of the three symbols $\{1, 2, 3\}$, and look at the orbit of the point $\sigma \in \Sigma$ given by

$$\begin{aligned}\sigma = {} & 1231112132122233132331111211312112212313113\\ & 2133211212213221222223231232233311312313321 3\\ & 2232333133233311111121113112111221123113111\\ & 3211331211121212131221122212231231123212331 3\\ & 111312121 2 \ldots\ldots\ldots\ldots \text{FOREVER}.\end{aligned}$$

This orbit $\{T^{\circ n}\sigma\}_{n=0}^{\infty}$ may be plotted on a Cantor set of three symbols, as sketched in Figure 4.8.1. This can be compared with the orbit $\{S^{\circ n}(\phi(\sigma))\}_{n=0}^{\infty}$

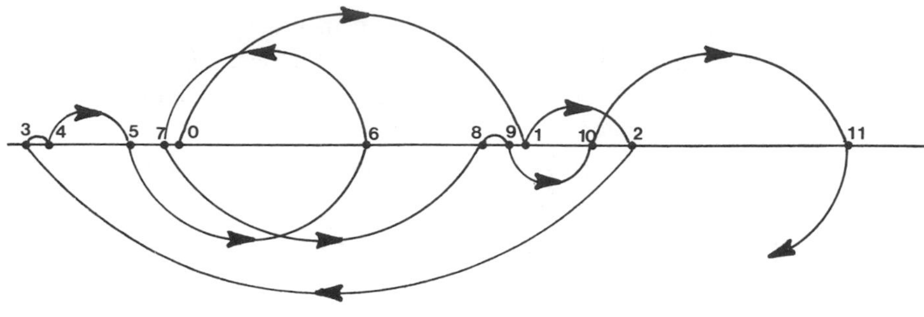

Figure 4.8.1
The start of a chaotic orbit on a Ternary Cantor Set.

of the shift dynamical system $\{A, S\}$ associated with an IFS of three maps, as plotted in Figure 4.8.2. Figure 4.8.3 shows an equivalent orbit, but this time for the just-touching IFS $\{[0, 1]; \frac{1}{3}x, \frac{1}{3}x + \frac{1}{3}, \frac{1}{3}x + \frac{2}{3}\}$, and it is displayed using a web diagram.

In each case the "same" dynamics look entirely different. The qualities of beauty and harmony present in the observed orbits are different. This is not surprising: the equivalence of the dynamical systems is a topological equivalence. It does not provide much information about the interplay of the dynamics with the geometries of the spaces on which they act. This interplay is an open area for research. For example, what are the special conserved properties of two metrically equivalent dynamical systems? Can you quantify the grace and delicacy of a dancing orbit on a fractal?

Figure 4.8.2
The start of an orbit of a deterministic shift dynamical system. This orbit is chaotic. It will visit the part of the attractor inside each of these little circles infinitely many times.

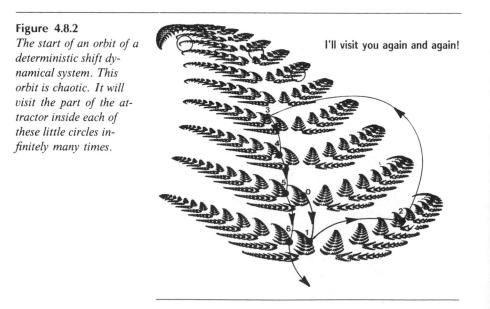

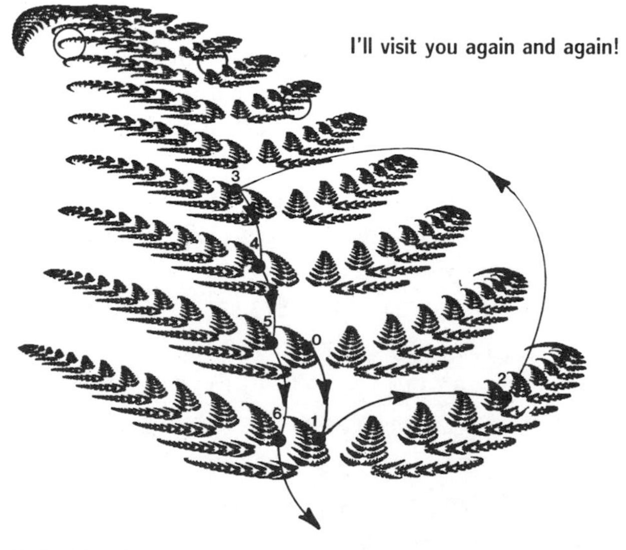

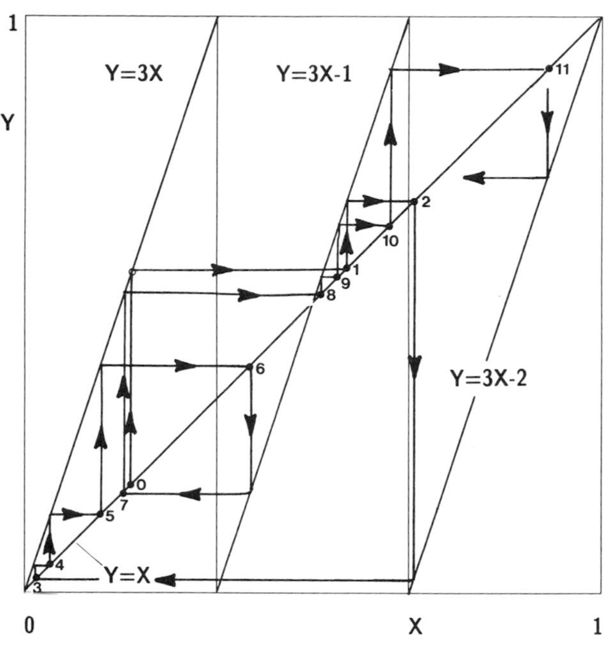

Figure 4.8.3 *Equivalent orbit to the one in Figures* 4.8.1 *and* 4.8.2, *this time plotted using a web diagram. The starting point has address* 12311121321222331. . . . *This manifestation of an orbit which goes arbitrarily close to any point, takes place on a just-touching attractor.*

This said, we turn our attention back to an important collection of properties which are shared by all shift dynamical systems. For simplicity we formalize the discussion for the case of the shift dynamical system $\{A, S\}$ associated with a totally disconnected hyperbolic IFS.

Definition 1. Let $(\mathbf{X}, d)$ be a metric space. A subset $B \subset \mathbf{X}$ is said to be *dense* in $\mathbf{X}$ if the closure of B equals $\mathbf{X}$. A sequence $\{x_n\}_{n=0}^{\infty}$ of points in $\mathbf{X}$ is said to be dense in $\mathbf{X}$ if, for each point $a \in A$, there is a subsequence $\{x_{\sigma_n}\}_{n=0}^{\infty}$ which converges to a. In particular, an orbit $\{x_n\}_{n=0}^{\infty}$ of a dynamical system $\{\mathbf{X}, f\}$ is said to be dense in $\mathbf{X}$ if the sequence $\{x_n\}_{n=0}^{\infty}$ is dense in $\mathbf{X}$.

By now you will have had some experience with using the Random Iteration Algorithm, Program 3.8.2, for computing images of the attractor A of IFS in $\mathbb{R}^2$. If you run the algorithm starting from a point $x_0 \in A$, then all of the computed points lie on A. Apparently, the sequences of points which we plot are examples of sequences which are dense in the metric space (A, d).

The property of being dense is invariant under homeomorphism: if B is dense in a metric space $(\mathbf{X}, d)$ and if θ: $\mathbf{X} \to \mathbf{Y}$ is a homeomorphism, then $\theta(B)$ is dense in $\mathbf{Y}$. If $\{\mathbf{X}; f\}$ and $\{\mathbf{Y}; g\}$ are equivalent dynamical systems under θ; and if $\{x_n\}$ is an orbit of f which is dense in $\mathbf{X}$, then $\{\theta(x_n)\}$ is an orbit of g which is dense in $\mathbf{Y}$.

Definition 2. A dynamical system $\{X, f\}$ is *transitive* if, whenever $\mathscr{U}$ and $\mathscr{V}$ are open subsets of the metric space (X, d), there exists a finite integer n such that

$$\mathscr{U} \cap f^{\circ n}(\mathscr{V}) \neq \varnothing .$$

The dynamical system $\{[0,1]; f(x) = \text{Min}\{2x, 2-2x\}\}$ is topologically transitive. To verify this just let $\mathscr{U}$ and $\mathscr{V}$ be any pair of open intervals in the metric space ([0, 1], Euclidean). Clearly each application of the transformation increases the length of the interval $\mathscr{V}$ in such a way that it eventually overlaps $\mathscr{U}$.

Definition 3. The dynamical system $\{X; f\}$ is *sensitive to initial conditions* if there exists $\delta > 0$ such that, for any $x \in X$ and any ball $B(x, \epsilon)$ with radius $\epsilon > 0$ there is $y \in B(x, \epsilon)$ and an integer $n \geq 0$ such that $d(f^{\circ n}(x), f^{\circ n}(y)) > \delta$.

Roughly, orbits which begin close together get pushed apart by the action of the dynamical system. For example, the dynamical system $\{[0,1]; 2x \bmod 1\}$ is sensitive to initial conditions.

Exercises & Examples

8.1. Show that the rational numbers are dense in the metric space $(\mathbb{R}$, Euclidean).

8.2. Let $r(n)$ be a counting function which counts all of the rational numbers which lie in the interval $[0,1]$. Prove that the sequence of real numbers $\{r(n) \in [0,1]\colon\ n = 1, 2, 3, \ldots\}$, is dense in the metric space ([0, 1], Euclidean).

8.3. Consider the dynamical system $\{[0,1];\ f(x) = 2x \bmod 1\}$. Find a point $x_0 \in [0,1]$ whose orbit is dense in $[0,1]$.

8.4. Show that the dynamical system $\{[0,\infty) : f(x) = 2x\}$ is sensitive to initial conditions; but that the dynamical system $\{[0,\infty) : f(x) = (0.5)x\}$ is not.

8.5. Show that the shift dynamical system $\{\Sigma; T\}$, where Σ is the code space of two symbols, is transitive and sensitive to initial conditions.

8.6. Let $\{X, f\}$ and $\{Y, g\}$ be equivalent dynamical systems. Show that $\{X, f\}$ is transitive if and only if $\{Y, g\}$ is transitive. In other words, the property of being transitive is preserved between equivalent dynamical systems.

Definition 4. A dynamical system $\{X, f\}$ is *chaotic* if

(i) it is transitive;
(ii) it is sensitive to initial conditions;
(iii) the set of periodic orbits of f is dense in X.

Theorem 1. *The shift dynamical system associated with a totally disconnected hyperbolic IFS of two or more transformations is chaotic.*

Sketch of Proof. First one establishes that the shift dynamical system $\{\Sigma; T\}$ is chaotic where Σ is the code space of N symbols, with $N \geq 2$. One then uses the code space map $\phi\colon \Sigma \to A$ to carry the results over to the equivalent dynamical system $\{A; S\}$.

Theorem 1 applies to the lifted IFS associated with a hyperbolic IFS. Hence the lifted shift dynamical system associated with an IFS of two or more transformations is chaotic. In turn, this implies certain characteristics to the behaviour of the projection of a lifted shift dynamical system, namely a random shift dynamical system.

Let us consider now why the Random Iteration Algorithm works, from an intuitive point of view. Consider the hyperbolic IFS $\{\mathbb{R}^2; w_1, w_2\}$. Let $a \in A$; suppose that the address of a is $\sigma \in \Sigma$, the associated code space. That is

$$a = \phi(\sigma).$$

With the aid of a random number generator a sequence of one million 1's and 2's, is selected. For example, suppose that the actual sequence produced is the following one, *which has been written from right to left.*

21...121211211212111211121111112112111211112112121222 11

By this we mean that the first number chosen is a 1, then a 1, then three 2's, and so on. Then the following sequence of points on the attractor is computed:

$$\begin{aligned}
&a = \phi(\sigma)\\
&w_1(a) = \phi(1\sigma)\\
&w_1 \circ w_1(a) = \phi(11\sigma)\\
&w_2 \circ w_1 \circ w_1(a) = \phi(211\sigma)\\
&w_2 \circ w_2 \circ w_1 \circ w_1(a) = \phi(2211\sigma)\\
&w_2 \circ w_2 \circ w_2 \circ w_1 \circ w_1(a) = \phi(22211\sigma)\\
&w_1 \circ w_2 \circ w_2 \circ w_2 \circ w_1 \circ w_1(a) = \phi(122211\sigma)\\
&w_2 \circ w_1 \circ w_2 \circ w_2 \circ w_2 \circ w_1 \circ w_1(a) = \phi(2122211\sigma)\\
&w_1 \circ w_2 \circ w_1 \circ w_2 \circ w_2 \circ w_2 \circ w_1 \circ w_1(a) = \phi(12122211\sigma)\\
&w_2 \circ w_1 \circ w_2 \circ w_1 \circ w_2 \circ w_2 \circ w_2 \circ w_1 \circ w_1(a) = \phi(212122211\sigma)\\
&w_1 \circ w_2 \circ w_1 \circ w_2 \circ w_1 \circ w_2 \circ w_2 \circ w_2 \circ w_1 \circ w_1(a) = \phi(1212122211\sigma)\\
&w_1 \circ w_1 \circ w_2 \circ w_1 \circ w_2 \circ w_1 \circ w_2 \circ w_2 \circ w_2 \circ w_1 \circ w_1(a) = \phi(11212122211\sigma)\\
&\ldots\\
&w_2 \circ w_1 \circ \cdots\ w_1 \circ w_1 \circ w_2 \circ w_1 \circ w_2 \circ w_1 \circ w_2 \circ w_2 \circ w_2 \circ w_1 \circ w_1(a)\\
&\qquad = \phi(21\ldots11212122211\sigma)
\end{aligned}$$

We imagine that instead of plotting the points as they are computed, we keep a list of the one million computed points. This done, we plot the points in the reverse order from the order in which they were computed. That is, we begin by plotting the point $\phi(21\ldots 11212122211\sigma)$ and we finish by plotting the point $\phi(\sigma)$. What will we see? We will see a million points on the orbit of the shift dynamical system $\{A; S\}$, namely $\{S^{\circ n}(\phi(21\ldots 11212122211\sigma))\}_{n=0}^{1,000,000}$.

Now from our experience with shift dynamics and from our theoretical knowledge and intuitions, what do we expect of such an orbit? We expect it to be *chaotic* and to visit a widely distributed collection of points on the attractor. We are looking at part of a "randomly chosen" orbit of the shift dynamical system; and we expect it to be dense in the attractor.

For example, suppose that you are doing shift dynamics on a picture of a totally disconnected fractal, or a fern. You should be convinced that by making sly adjustments in the orbit at each step, as in the Shadowing Theorem, you can most easily coerce an orbit into visiting, to within a distance $\epsilon > 0$, each point in the image. But then the Shadow*ing* Theorem ensures that there is an actual orbit close to our artificial one, and it too goes close to every point on the fractal, say to within a distance 2ϵ of each point on the image. This suggests that "most" orbits of the shift dynamical system are dense in the attractor.

Exercises & Examples

8.7. Make experiments on a picture of the attractor of a totally disconnected hyperbolic IFS to verify assertion in the last paragraph, that "by making sly adjustments in an orbit... you can most easily coerce the orbit into visiting to within a distance $\epsilon > 0$ of each point in the image." Can you make some experimental estimates of how many orbits go to within a distance $\epsilon > 0$, for several values of ϵ, of every point in the picture? One way to do this might be to work with a discretized image and to try to count the number of available orbits.

8.8. Run the Random Iteration Algorithm, Program 3.8.2, to produce an image of a fractal, for example a fern without a stem as used in Figure 4.8.2. As the points are calculated and plotted, keep a list of them. Then plot the points over again in reverse order, this time making them flash on and off on the picture of the attractor on the screen, so that you can see where they land. This way you will see the interplay of the geometry with the shift dynamics on the attractor. See if the orbit is beautiful. If you think that it is, try to make your impression objective.

We want to begin to formulate the idea that "most" orbits of the shift dynamical system associated with a totally disconnected IFS are dense in the attractor. The following lemma counts the number of cycles of minimal period p.

Lemma 1. *Let $\{A; S\}$ be the shift dynamical system associated with a totally disconnected hyperbolic IFS $\{\mathbf{X}; w_1, w_2, \dots, w_N\}$. Let $\mathcal{N}(p)$ denote the number of distinct cycles of minimal period p, for $p \in \{1, 2, 3, \dots\}$. Then*

$$\mathcal{N}(p) = \left(N^p - \sum_{\substack{k=1 \\ k \text{ divides } p}}^{p-1} k\mathcal{N}(k) \right) \Big/ p \qquad \text{for } p = 1, 2, 3, \dots .$$

Proof. It suffices to restrict attention to code space, and to give the main idea, consider only the case $N = 2$. For $p = 1$, the cycles of period 1 are the fixed points of T. The equation

$$T\sigma = \sigma \qquad \sigma \in \Sigma$$

implies $\sigma = \overline{1111}$ or $\sigma = \overline{2222}$. Thus $\mathcal{N}(1) = 2$. For $p = 2$, any point which lies on a cycle of period 2 must be a fixed point of $T^{\circ 2}$, namely

$$T^{\circ 2}\sigma = \sigma,$$

whence $\sigma = \overline{11}, \overline{12}, \overline{21}$, or $\overline{22}$. The only cycles here which are not of minimal period two must have minimal period one. Furthermore, there are two distinct points on a cycle of minimal period two, so

$$\mathcal{N}(2) = (2^2 - \mathcal{N}(1))/2 = 2/2 = 1.$$

One quickly gets the idea. Mathematical induction on p completes the proof for $N = 2$.

For $N = 2$, we find, for example, $\mathcal{N}(2) = 1$, $\mathcal{N}(3) = 2$, $\mathcal{N}(4) = 3$, $\mathcal{N}(5) = 6$, $\mathcal{N}(6) = 9$, $\mathcal{N}(7) = 18$, $\mathcal{N}(8) = 30$, $\mathcal{N}(9) = 56$, $\mathcal{N}(10) = 99$, $\mathcal{N}(11) = 186$, $\mathcal{N}(12) = 335$, $\mathcal{N}(13) = 630$, $\mathcal{N}(14) = 1161$, $\mathcal{N}(15) = 2182$, $\mathcal{N}(16) = 4080$, $\mathcal{N}(17) = 7710$, $\mathcal{N}(18) = 14532$, $\mathcal{N}(19) = 27594$, $\mathcal{N}(20) = 52377$. In particular, 99.8 percent of all points lying on cycles of period 20 lie on cycles of minimal period 20.

Here is the idea we are getting at. We know that the set of periodic cycles are dense in the attractor of a hyperbolic IFS. It follows that we may approximate the attractor by the set of all cycles of some finite period, say period twelve billion. Thus we replace the attractor A by such an approximation $\tilde{A}$, which consists of $2^{12{,}000{,}000{,}000}$ points. Suppose we pick one of these points at random. Then this point is extremely likely to lie on a cycle of *minimal* period twelve billion. Hence the orbit of a point chosen "at random" on the approximate attractor $\tilde{A}$ is extremely likely to consist of twelve billion *distinct* points on $\tilde{A}$.

In fact one can show that a statistically random sequence of symbols contains every possible finite subsequence. So we expect that the set of twelve billion distinct points on A is likely to contain at least one representative from each part of the attractor!

5 Fractal Dimension

5.1 FRACTAL DIMENSION

How big is a fractal? When are two fractals similar to one another in some sense? What experimental measurements might we make to tell if two different fractals may be metrically equivalent? What is it, which is the same about the two fractals in Figure 5.1.1?

There are various numbers associated with fractals which can be used to compare them. They are generally referred to as *fractal dimensions*. They are attempts to quantify a subjective feeling which we have about how densely the fractal occupies the metric space in which it lies. Fractal dimensions provide an objective means for comparing fractals.

Fractal dimensions are important because they can be defined in connection with real-world data, and they can be measured approximately by means of experiments. For example, one can measure "the fractal dimension" of the coastline of Great Britain; its value is about 1.2. Fractal dimensions can be attached to clouds, trees, coastlines, feathers, networks of neurons in the body, dust in the air at an instant in time, the clothes you are wearing, the distribution of frequencies of light reflected by a flower, the colors emitted by

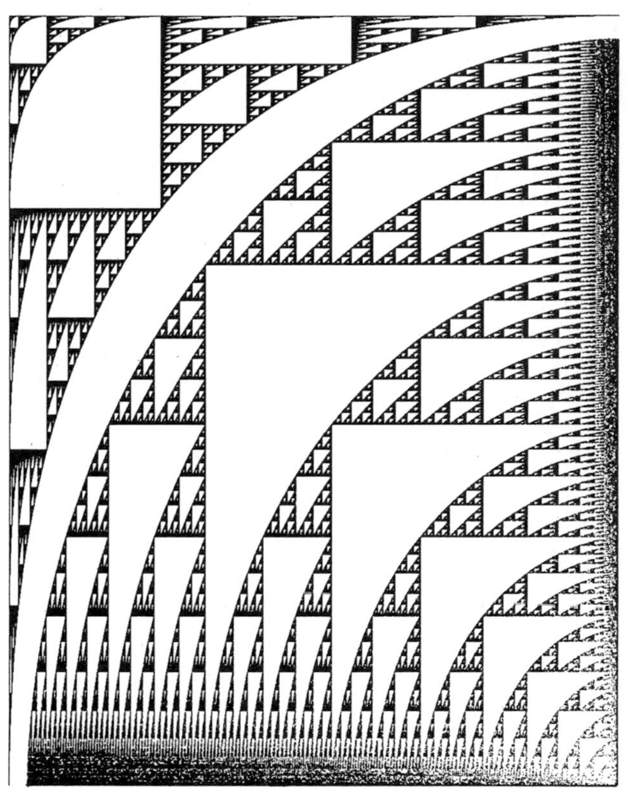

Figure 5.1.1 *Do the two implied fractals have the same dimension?*

the sun, and the wrinkled surface of the sea during a storm. These numbers allow us to compare sets in the real world with the laboratory fractals, such as attractors of IFS.

We restrict attention to compact subsets of metric spaces. This fits well with the idea of modelling the real physical world by subsets of metric spaces. Suppose that an experimentalist is studying a physical entity, and he wishes to model this entity by means of a subset of $\mathbb{R}^3$. Then he can use a compact set for his model. For example, he can assume that the distances which he measures are Euclidean distances, and he can assume that the universe is bounded. He can assume that any Cauchy sequence of points in his model set converges to a point in his model set, because he cannot experimentally invalidate this assumption. Although mathematically we can distinguish between a set and its closure, we cannot make the same distinction between their physical counterparts. The assumption of compactness will allow the model to be handled theoretically with relative ease.

Let $(\mathbb{X}, d)$ denote a complete metric space. Let $A \in \mathscr{H}(\mathbb{X})$ be a nonempty compact subset of $\mathbb{X}$. Let $\epsilon > 0$. Let $B(x, \epsilon)$ denote the closed ball of radius ϵ

and center at a point $x \in \mathbb{X}$. We wish to define an integer, $\mathcal{N}(A, \epsilon)$ to be the least number of closed balls of radius ϵ needed to cover the set A. That is

$$\mathcal{N}(A, \epsilon) = \text{smallest positive integer } M \text{ such that } A \subset \bigcup_{n=1}^{M} B(x_n, \epsilon),$$

for some set of distinct points $\{x_n\colon n = 1, 2, \ldots, M\} \subset \mathbb{X}$. How do we know that there is such a number $\mathcal{N}(A, \epsilon)$? Easy! The logic is this: surround every point $x \in A$ by an *open* ball of radius $\epsilon > 0$ to provide a cover of A by open sets. Because A is compact, this cover possesses a finite subcover, consisting of an integer number, say $\hat{M}$, of open balls. By taking the closure of each ball, we obtain a cover consisting of $\hat{M}$ *closed* balls. Let C denote the set of covers of A by at most $\hat{M}$ closed balls of radius ϵ. Then C contains at least one element. Let $f\colon C \to \{1, 2, 3, \ldots, \hat{M}\}$ be defined by $f(c) =$ number of balls in the cover $c \in C$. Then $\{f(c)\colon c \in C\}$ is a finite set of positive integers. It follows that it contains a least integer, $\mathcal{N}(A, \epsilon)$.

The intuitive idea behind fractal dimension is that a set A has fractal dimension D if:

$$\mathcal{N}(A, \epsilon) \approx C\epsilon^{-D} \qquad \textit{for some positive constant } C.$$

Here we use the notation "$\approx$" as follows. Let $f(\epsilon)$ and $g(\epsilon)$ be real valued functions of the positive real variable ϵ. Then $f(\epsilon) \approx g(\epsilon)$ means that $\mathrm{Lim}_{\epsilon \to 0}\{\mathrm{Ln}(f(\epsilon))/\mathrm{Ln}(g(\epsilon))\} = 1$.

If we "solve" for D we find that

$$D \approx \frac{\mathrm{Ln}\,\mathcal{N}(A, \epsilon) - \mathrm{Ln}\,C}{\mathrm{Ln}(1/\epsilon)}$$

We use the notation $\mathrm{Ln}(x)$ to denote the logarithm to the base e of the positive real number x. Now notice that the term $\mathrm{Ln}\,C/\mathrm{Ln}(1/\epsilon)$ approaches zero as $\epsilon \to 0$. This leads us to the following definition.

Definition 1. Let $A \in \mathcal{H}(\mathbb{X})$ where $(\mathbb{X}, d)$ is a metric space. For each $\epsilon > 0$ let $\mathcal{N}(A, \epsilon)$ denote the smallest number of closed balls of radius $\epsilon > 0$ needed to cover A. If

$$D = \lim_{\epsilon \to 0}\left\{\frac{\mathrm{Ln}(\mathcal{N}(A, \epsilon))}{\mathrm{Ln}(1/\epsilon)}\right\}$$

exists, then D is called the *fractal dimension* of A. We will also use the notation $D = D(A)$, and will say "A has fractal dimension D."

Exercises & Examples

1.1. This example takes place in the metric space ($\mathbb{R}^2$, Euclidean). Let $a \in \mathbb{X}$ and let $A = \{a\}$. A consists of a single point in the space. For each $\epsilon > 0$, $\mathcal{N}(A, \epsilon) = 1$. It follows that $D(A) = 0$.

1.2. This example takes place in the metric space ($\mathbb{R}^2$, Manhattan). Let A denote the line segment $[0,1]$. Let $\epsilon > 0$. Then it is quite easy to see that $\mathcal{N}(A,\epsilon) = -[-1/\epsilon]$, where $[x]$ denotes the integer part of the real number x. In Figure 5.1.2 we have plotted the graph of $\text{Ln}(\mathcal{N}(A,\epsilon))$ as a function of $\text{Ln}(1/\epsilon)$. Despite a rough start, it appears clear that

$$\underset{\epsilon\to 0}{\text{Lim}}\left\{\frac{\text{Ln}(\mathcal{N}(A,\epsilon))}{\text{Ln}(1/\epsilon)}\right\} = 1.$$

In fact, for $0 < \epsilon < 1$

$$\frac{\text{Ln}(1/\epsilon)}{\text{Ln}(1/\epsilon)} \leq \frac{\text{Ln}(-[-1/\epsilon])}{\text{Ln}(1/\epsilon)} = \frac{\text{Ln}(\mathcal{N}(A,\epsilon))}{\text{Ln}(1/\epsilon)}$$

$$\leq \frac{\text{Ln}(1/\epsilon + 1)}{\text{Ln}(1/\epsilon)} = \frac{\text{Ln}(1+\epsilon) + \text{Ln}(1/\epsilon)}{\text{Ln}(1/\epsilon)}.$$

Both sides here converge to one as $\epsilon \to 0$. Hence the quantity in the middle also converges to one. We conclude that the fractal dimension of a closed line segment is one. We would have obtained the same result if we had used the Euclidean metric.

1.3. Let $(\mathbf{X}, d)$ be a metric space. Let $a, b, c \in \mathbf{X}$ and let $A = \{a, b, c\}$. Prove that $D(A) = 0$.

The following two theorems simplify the process of calculating the fractal dimension. They allow one to replace the continuous variable ϵ by a discrete variable.

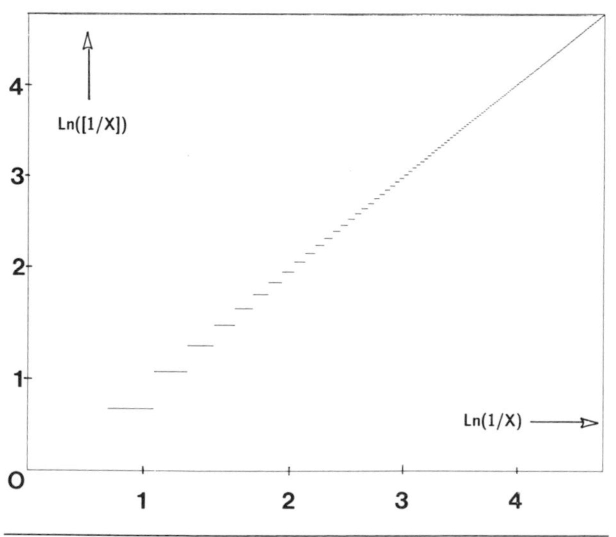

Figure 5.1.2
Plot of $Ln([1/x])$ as a function of $Ln(1/x)$. This illustrates that in the computation of the fractal dimension one usually evaluates the limiting "slope" of a discontinuous function. In the present example this slope is one.

Theorem 1. *Let $A \in \mathcal{H}(\mathbb{X})$ where $(\mathbb{X}, d)$ is a metric space. Let $\epsilon_n = Cr^n$ for real numbers $0 < r < 1$ and $C > 0$, and integers $n = 1, 2, 3, \ldots$. If*

$$D = \lim_{n \to \infty} \left\{ \frac{\operatorname{Ln}(\mathcal{N}(A, \epsilon_n))}{\operatorname{Ln}(1/\epsilon_n)} \right\},$$

then A has fractal dimension D.

Proof. Let the real numbers r and C, and the sequence of numbers $E = \{\epsilon_n: n = 1, 2, 3, \ldots\}$ be as defined in the statement of the theorem. Define $f(\epsilon) = \operatorname{Max}\{\epsilon_n \in E: \epsilon_n \leq \epsilon\}$. Assume that $\epsilon \leq r$. Then

$$f(\epsilon) \leq \epsilon \leq f(\epsilon)/r \quad \text{and} \quad \mathcal{N}(A, f(\epsilon)) \geq \mathcal{N}(A, \epsilon) \geq \mathcal{N}(A, f(\epsilon)/r).$$

Since $\operatorname{Ln}(x)$ is an increasing positive function of x for $x \geq 1$, it follows that

$$\left\{ \frac{\operatorname{Ln}(\mathcal{N}(A, f(\epsilon)/r))}{\operatorname{Ln}(1/f(\epsilon))} \right\} \leq \left\{ \frac{\operatorname{Ln}(\mathcal{N}(A, \epsilon))}{\operatorname{Ln}(1/\epsilon)} \right\}$$
$$\leq \left\{ \frac{\operatorname{Ln}(\mathcal{N}(A, f(\epsilon)))}{\operatorname{Ln}(r/f(\epsilon))} \right\}. \tag{5.1.1}$$

Assume that $\mathcal{N}(A; \epsilon) \to \infty$ as $\epsilon \to 0$; if not then the theorem is true. The right-hand-side of (5.1.1) obeys

$$\lim_{\epsilon \to 0} \left\{ \frac{\operatorname{Ln}(\mathcal{N}(A, f(\epsilon)))}{\operatorname{Ln}(r/f(\epsilon))} \right\} = \lim_{n \to \infty} \left\{ \frac{\operatorname{Ln}(\mathcal{N}(A, \epsilon_n))}{\operatorname{Ln}(r/\epsilon_n)} \right\}$$
$$= \lim_{n \to \infty} \left\{ \frac{\operatorname{Ln}(\mathcal{N}(A, \epsilon_n))}{\operatorname{Ln}(r) + \operatorname{Ln}(1/\epsilon_n)} \right\}$$
$$= \lim_{n \to \infty} \left\{ \frac{\operatorname{Ln}(\mathcal{N}(A, \epsilon_n))}{\operatorname{Ln}(1/\epsilon_n)} \right\}.$$

The left-hand-side of (5.1.1) obeys

$$\lim_{\epsilon \to 0} \left\{ \frac{\operatorname{Ln}(\mathcal{N}(A, f(\epsilon)/r))}{\operatorname{Ln}(1/f(\epsilon))} \right\} = \lim_{n \to \infty} \left\{ \frac{\operatorname{Ln}(\mathcal{N}(A, \epsilon_{n-1}))}{\operatorname{Ln}(1/\epsilon_n)} \right\}$$
$$= \lim_{n \to \infty} \left\{ \frac{\operatorname{Ln}(\mathcal{N}(A, \epsilon_{n-1}))}{\operatorname{Ln}(1/r) + \operatorname{Ln}(1/\epsilon_{n-1}))} \right\}$$
$$= \lim_{n \to \infty} \left\{ \frac{\operatorname{Ln}(\mathcal{N}(A, \epsilon_n))}{\operatorname{Ln}(1/\epsilon_n)} \right\}.$$

So as $\epsilon \to 0$ both the left-hand-side and the right-hand-side of equation (5.1.1) approach the same value, claimed in the theorem. By the Sandwich Theorem of Calculus, the limit as $\epsilon \to 0$ of the quantity in the middle of (5.1.1) also exists, and it equals the same value. This completes the proof of the theorem.

Theorem 2. (The Box Counting Theorem) *Let $A \in \mathcal{H}(\mathbb{R}^m)$, where the Euclidean metric is used. Cover $\mathbb{R}^m$ by closed just-touching square boxes of side* *length* $(1/2^n)$, *as exemplified in Figure* 5.1.3 *for* $n = 2$ *and* $m = 2$. *Let* $\mathcal{N}_n(A)$ *denote the number of boxes of side length* $(1/2^n)$ *which intersect the attractor. If*

$$D = \lim_{n \to \infty} \left\{ \frac{\text{Ln}(\mathcal{N}_n(A))}{\text{Ln}(2^n)} \right\},$$

then A has fractal dimension D.

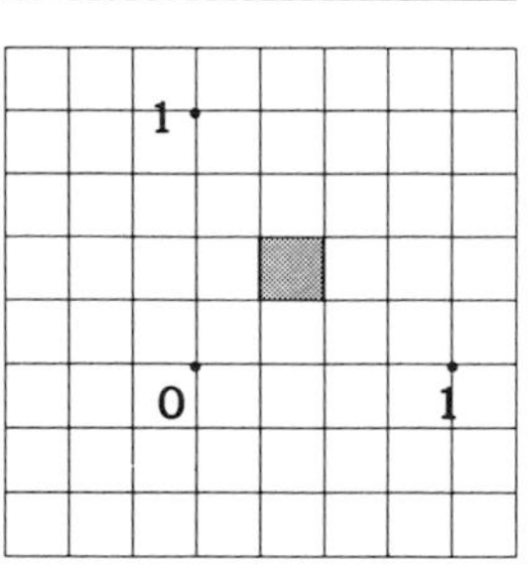

Figure 5.1.3
Closed boxes of side $(1/2^n)$ *cover* $\mathbb{R}^2$. *Here* $n = 2$. *See Theorem* 2.

Proof. We observe that for $m = 1, 2, 3, \ldots$,

$$2^{-m}\mathcal{N}_{n-1} \le \mathcal{N}(A, 1/2^n) \le \mathcal{N}_{k(n)} \qquad \text{for all } n = 1, 2, 3, \ldots$$

where $k(n)$ is the smallest integer k satisfying $k \ge n - 1 + 1/2 \log_2 m$. The first inequality holds because a ball of radius $1/2^n$ can intersect at most 2^m "on-grid" boxes of side $1/2^{n-1}$. The second follows from the fact that a box of side s can fit inside a ball of radius r provided $r^2 \ge (s/2)^2 + (s/2)^2 + \cdots + (s/2)^2 = m(s/2)^2$ by the theorem of Pythagoras. Now

$$\lim_{n \to \infty} \left\{ \frac{\text{Ln}(\mathcal{N}_{k(n)})}{\text{Ln}(2^n)} \right\} = \lim_{n \to \infty} \left\{ \frac{\text{Ln}(2^{k(n)})}{\text{Ln}(2^n)} \frac{\text{Ln}(\mathcal{N}_{k(n)})}{\text{Ln}(2^{k(n)})} \right\} = D,$$

since $k(n)/n \to 1$. Since also

$$\lim_{n \to \infty} \left\{ \frac{\text{Ln}\, 2^{-m}\mathcal{N}_{n-1}}{\text{Ln}(2^n)} \right\} = \lim_{n \to \infty} \left\{ \frac{\text{Ln}\, \mathcal{N}_{n-1}}{\text{Ln}(2^{n-1})} \right\} = D,$$

Theorem 5.1.1 with $r = 1/2$ completes the proof.

There is nothing magical about using boxes of side $(1/2)^n$ in Theorem 2. One can equally well use boxes of side Cr^n, where $C > 0$ and $0 < r < 1$ are fixed real numbers.

Exercises & Examples

1.4. Consider the $\blacksquare \subset \mathbb{R}^2$. It is easy to see that $\mathcal{N}_1(\blacksquare) = 4$, $\mathcal{N}_2(\blacksquare) = 16$, $\mathcal{N}_3(\blacksquare) = 64$, $\mathcal{N}_4(\blacksquare) = 256$, and in general that $\mathcal{N}_n(\blacksquare) = 4^n$ for $n = 1, 2, 3, \ldots$ (see Figure 5.1.4).

Theorem 5.1.2 implies that

$$D(\blacksquare) = \lim_{n \to \infty} \left\{ \frac{\text{Ln}(\mathcal{N}_n(\blacksquare))}{\text{Ln}(2^n)} \right\} = \lim_{n \to \infty} \left\{ \frac{\text{Ln}(4^n)}{\text{Ln}(2^n)} \right\} = 2.$$

1.5. Consider the Sierpinski triangle △, in Figure 5.1.5, as a subset of ($\mathbb{R}^2$, Euclidean).

We see that $\mathcal{N}_1(△) = 3$, $\mathcal{N}_2(△) = 9$, $\mathcal{N}_3(△) = 27$, $\mathcal{N}_4(△) = 81$, and in general $\mathcal{N}_n(△) = 3^n$ for $n = 1, 2, 3, \ldots$.

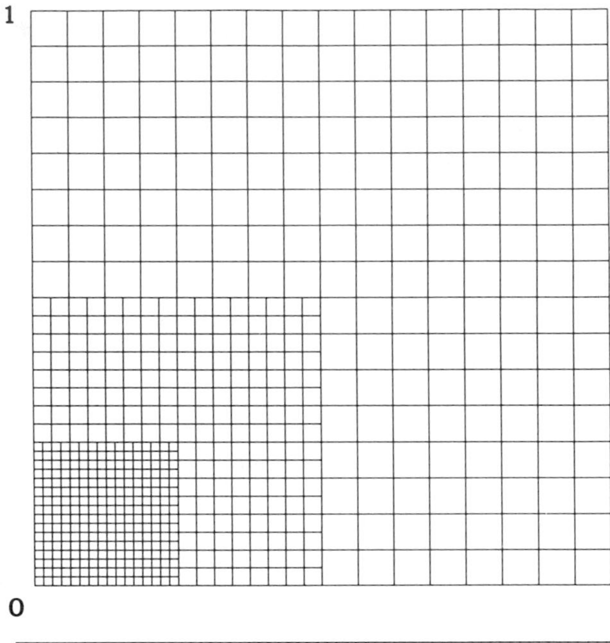

Figure 5.1.4
It requires $(1/2^n)^{-2}$ *boxes of side* $(1/2^n)$ *to cover* ■ $\subset \mathbb{R}^2$. *We deduce, with a feeling of relief, that the fractal dimension of* ■ *is* 2. *Which collage is this image related to*?

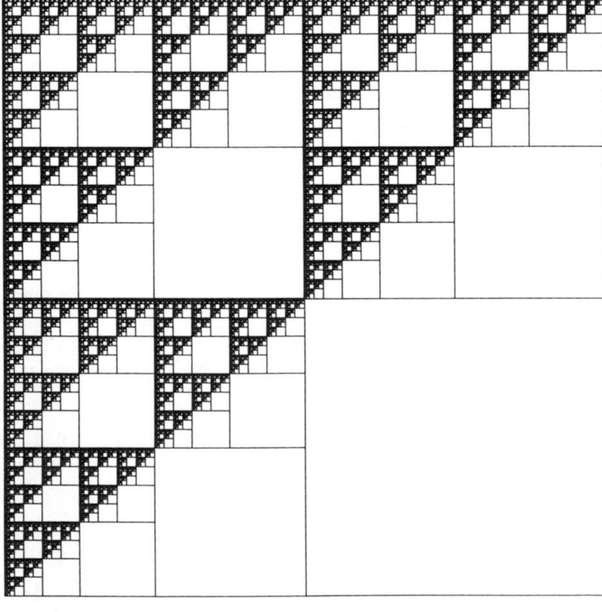

Figure 5.1.5
It requires 3^n *closed boxes of side* $(1/2)^n$ *to cover the Sierpinski triangle* $\subset \mathbb{R}^2$. *We deduce that its fractal dimension is* $Ln(3)/Ln(2)$.

Theorem 5.1.2 implies that

$$D(\triangle) = \lim_{n\to\infty}\left\{\frac{\mathrm{Ln}(\mathcal{N}_n(\triangle))}{\mathrm{Ln}(2^n)}\right\} = \lim_{n\to\infty}\left\{\frac{\mathrm{Ln}(3^n)}{\mathrm{Ln}(2^n)}\right\} = \frac{\mathrm{Ln}(3)}{\mathrm{Ln}(2)}.$$

1.6. Use the Box Counting Theorem, but with boxes of side length $(1/3)^n$, to calculate the fractal dimension of the classical Cantor set $\mathscr{C}$ described in Section 3.1.1 (example 5).

1.7. Use the Box Counting Theorem to estimate the fractal dimension of the fractal subset of $\mathbb{R}^2$ shown in Figure 5.1.6. You will need to take as your first box the obvious one suggested by the figure. You should then find that there appears to be a pattern to the sequence of numbers $\mathcal{N}_1, \mathcal{N}_2, \mathcal{N}_3, \ldots$.

1.8. The same problem as 1.7, this time applied to Figure 5.1.7. By making the right choice of Cartesian coordinate system, you will make this problem easy.

What happens to the fractal dimension of a set if we deform it "with bounded distortion?" The following theorem tells us that metrically equivalent sets have the same fractal dimension. For example, the two fractals in Figure 5.1.1 have the same fractal dimension!

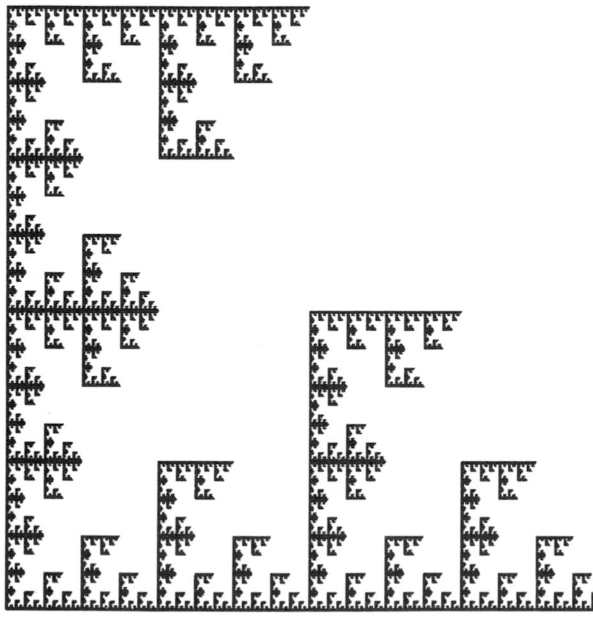

Figure 5.1.6 *Use the Box Counting Theorem to estimate the fractal dimension of the subset of* ($\mathbb{R}^2$, *Euclidean*) *shown here. What other well-known fractal has the same fractal dimension?*

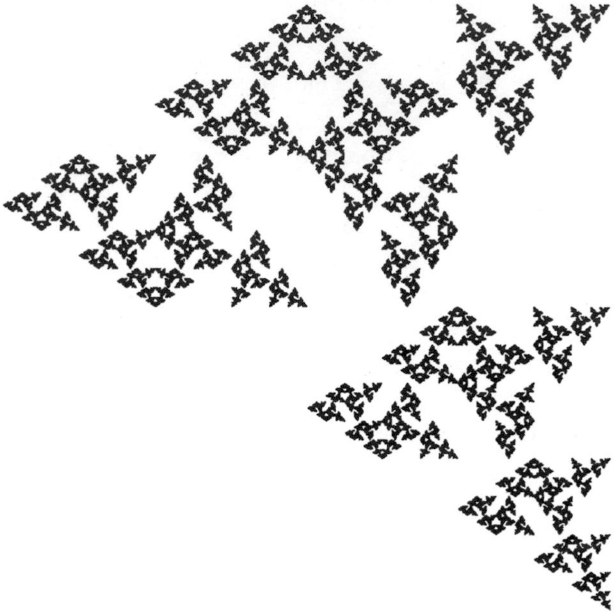

Figure 5.1.7
If you choose the "first" box just right, the fractal dimension of this fractal is easily estimated. Count the number $\mathcal{N}_n$ of boxes of side $1/2^n$ which intersect the set, for $n = 1, 2, 3, \ldots$ and apply the Box Counting Theorem.

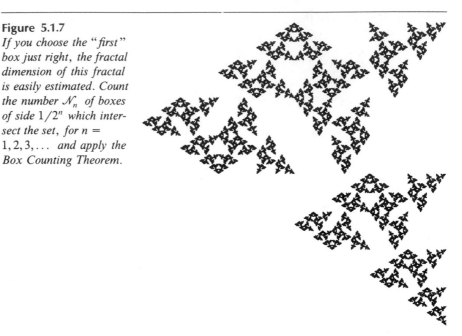

Theorem 3. *Let $(\mathbb{X}_1, d_1)$ and $(\mathbb{X}_2, d_2)$ be metrically equivalent metric spaces. Let $\theta\colon \mathbb{X}_1 \to \mathbb{X}_2$ be a transformation which provides the equivalence of the spaces. Let $A_1 \in \mathscr{H}(\mathbb{X}_1)$ have fractal dimension D. Then $A_2 = \theta(A_1)$ has fractal dimension D. That is*

$$D(A_1) = D(\theta(A_1)).$$

Proof. This proof makes use of the concepts of the Limsup and Liminf of a function.

Since the two spaces $(\mathbb{X}_1, d_1)$ and $(\mathbb{X}_2, d_2)$ are equivalent under θ, there exist positive constants e_1 and e_2 such that

$$e_1 d_2(\theta(x), \theta(y)) \le d_1(x, y) \le e_2 d_2(\theta(x), \theta(y)) \quad \text{for all } x, y \in \mathbb{X}_1. \tag{5.1.2}$$

Without loss of generality we assume that $e_1 < 1 < e_2$. Equation (5.1.2) implies

$$d_2(\theta(x), \theta(y)) \le \frac{d_1(x, y)}{e_1} \qquad \text{for all } x, y \in \mathbb{X}_1.$$

This implies

$$\theta(B(x, \epsilon)) \subset B(\theta(x), \epsilon/e_1) \qquad \text{for all } x \in \mathbb{X}_1. \tag{5.1.3}$$

Now, from the definition of $\mathcal{N}(A_1, \epsilon)$, we know that there is a set of points

$\{x_1, x_2, \ldots, x_{\mathcal{N}}\} \subset \mathbf{X}_1$, where $\mathcal{N} = \mathcal{N}(A_1, \epsilon)$, such that the set of closed balls $\{B(x_n, \epsilon): n = 1, 2, \ldots, \mathcal{N}(A_1, \epsilon)\}$ provides a cover of A_1. It follows that $\{\theta(B(x_n, \epsilon)): n = 1, 2, \ldots, \mathcal{N}(A_1, \epsilon)\}$ provides a cover of A_2. Equation (5.1.3) now implies that $\{B(\theta(x_n), \epsilon/e_1)): n = 1, 2, \ldots, \mathcal{N}(A_1, \epsilon)\}$ provides a cover of A_2. Hence

$$\mathcal{N}(A_2, \epsilon/e_1) \leq \mathcal{N}(A_1, \epsilon).$$

Hence, when $\epsilon < 1$,

$$\frac{\mathrm{Ln}(\mathcal{N}(A_2, \epsilon/e_1))}{\mathrm{Ln}(1/\epsilon)} \leq \frac{\mathrm{Ln}(\mathcal{N}(A_1, \epsilon))}{\mathrm{Ln}(1/\epsilon)}.$$

It follows that

$$\text{(5.1.4)} \quad \operatorname*{Limsup}_{\epsilon \to 0} \left\{ \frac{\mathrm{Ln}(\mathcal{N}(A_2, \epsilon))}{\mathrm{Ln}(1/\epsilon)} \right\} = \operatorname*{Limsup}_{\epsilon \to 0} \left\{ \frac{\mathrm{Ln}(\mathcal{N}(A_2, \epsilon/e_1))}{\mathrm{Ln}(1/\epsilon)} \right\}$$

$$\leq \operatorname*{Lim}_{\epsilon \to 0} \left\{ \frac{\mathrm{Ln}(\mathcal{N}(A_1, \epsilon))}{\mathrm{Ln}(1/\epsilon)} \right\} = D(A_1).$$

We now seek an inequality in the opposite direction. Equation (5.1.2) implies that

$$d_1(\theta^{-1}(x), \theta^{-1}(y)) \leq e_2 d_2(x, y) \qquad \text{for all } x, y \in \mathbf{X}_2.$$

This tells us that

$$\theta^{-1}(B(x, \epsilon)) \subset B(\theta^{-1}(x), e_2\epsilon) \qquad \text{for all } x \in \mathbf{X}_2,$$

and this in turn implies

$$\mathcal{N}(A_1, e_2\epsilon) \leq \mathcal{N}(A_2, \epsilon).$$

Hence, when $\epsilon < 1$,

$$\frac{\mathrm{Ln}(\mathcal{N}(A_1, e_2\epsilon))}{\mathrm{Ln}(1/\epsilon)} \leq \frac{\mathrm{Ln}(\mathcal{N}(A_2, \epsilon))}{\mathrm{Ln}(1/\epsilon)}.$$

It follows that

$$\text{(5.1.5)} \qquad D(A_1) = \operatorname*{Lim}_{\epsilon \to 0} \left\{ \frac{\mathrm{Ln}(\mathcal{N}(A_1, \epsilon))}{\mathrm{Ln}(1/\epsilon)} \right\}$$

$$= \operatorname*{Lim}_{\epsilon \to 0} \left\{ \frac{\mathrm{Ln}(\mathcal{N}(A_1, e_2\epsilon))}{\mathrm{Ln}(1/\epsilon)} \right\}$$

$$\leq \operatorname*{Liminf}_{\epsilon \to 0} \left\{ \frac{\mathrm{Ln}(\mathcal{N}(A_2, \epsilon))}{\mathrm{Ln}(1/\epsilon)} \right\}.$$

By combining (5.1.4) and (5.1.5) we obtain

$$\operatorname*{Liminf}_{\epsilon \to 0} \left\{ \frac{\mathrm{Ln}(\mathcal{N}(A_2, \epsilon))}{\mathrm{Ln}(1/\epsilon)} \right\} = D(A_1) = \operatorname*{Limsup}_{\epsilon \to 0} \left\{ \frac{\mathrm{Ln}(\mathcal{N}(A_2, \epsilon))}{\mathrm{Ln}(1/\epsilon)} \right\}.$$

From this it follows that

$$D(A_2) = \operatorname*{Lim}_{\epsilon \to 0} \left\{ \frac{\mathrm{Ln}(\mathcal{N}(A_2, \epsilon))}{\mathrm{Ln}(1/\epsilon)} \right\} = D(A_1).$$

This completes the proof.

Exercises & Examples

1.9. Let $\mathscr{C}$ denote the Classical Cantor Set, living in $[0,1]$ and obtained by omitting "middle thirds." Let $\tilde{\mathscr{C}}$ denote the Cantor set which is obtained by starting from the closed interval $[0,3]$ and omitting "middle thirds." Use Theorem 5.1.3 to show that they have the same fractal dimension. Verify the conclusion by means of a box counting argument.

1.10. Let A be a compact nonempty subset of $\mathbb{R}^2$. Suppose that A has fractal dimension D_1 when evaluated using the Euclidean metric and fractal dimension D_2 when evaluated using the Manhattan metric. Show that $D_1 = D_2$.

1.11. This example takes place in the metric space ($\mathbb{R}^2$, Manhattan). Let A_1 and A_2 denote the attractors of the following two hyperbolic IFS:

$$\left\{\mathbb{R}^2; \begin{pmatrix} \frac{1}{2} & 0 \\ 0 & \frac{1}{2} \end{pmatrix}\begin{pmatrix} x \\ y \end{pmatrix} + \begin{pmatrix} 1 \\ 0 \end{pmatrix}, \begin{pmatrix} \frac{1}{2} & 0 \\ 0 & \frac{1}{2} \end{pmatrix}\begin{pmatrix} x \\ y \end{pmatrix}, \begin{pmatrix} \frac{1}{2} & 0 \\ 0 & \frac{1}{2} \end{pmatrix}\begin{pmatrix} x \\ y \end{pmatrix} + \begin{pmatrix} 0 \\ 1 \end{pmatrix}\right\},$$

and

$$\left\{\mathbb{R}^2; \begin{pmatrix} \frac{1}{2} & 0 \\ 0 & \frac{1}{2} \end{pmatrix}\begin{pmatrix} x \\ y \end{pmatrix} + \begin{pmatrix} 2 \\ 0 \end{pmatrix}, \begin{pmatrix} \frac{1}{2} & 0 \\ 0 & \frac{1}{2} \end{pmatrix}\begin{pmatrix} x \\ y \end{pmatrix}, \begin{pmatrix} \frac{1}{2} & 0 \\ 0 & \frac{1}{2} \end{pmatrix}\begin{pmatrix} x \\ y \end{pmatrix} + \begin{pmatrix} 1 \\ 1 \end{pmatrix}\right\}.$$

By finding a suitable change of coordinates, show A_1 and A_2 have the same fractal dimensions.

5.2 THE THEORETICAL DETERMINATION OF THE FRACTAL DIMENSION

The following definition extends Definition 5.2.1. It provides a value for the fractal dimension for a wider collection of sets.

Definition 1. Let $(\mathbb{X}, d)$ be a complete metric space. Let $A \in \mathscr{H}(\mathbb{X})$. Let $\mathscr{N}(\epsilon)$ denote the minimum number of balls of radius ϵ needed to cover A. If

$$D = \lim_{\epsilon \to 0}\left\{\operatorname{Sup}\left\{\frac{\operatorname{Ln} \mathscr{N}(\tilde{\epsilon})}{\operatorname{Ln}(1/\tilde{\epsilon})} : \tilde{\epsilon} \in (0, \epsilon)\right\}\right\}$$

exists, then D is called the *fractal dimension* of A. We will also use the notation $D = D(A)$, and will say "A has fractal dimension D."

In stating this definition we have "spelled out" the Limsup. For any function $f(\epsilon)$, defined for $0 < \epsilon < 1$ for example, we have

$$\operatorname*{Limsup}_{\epsilon \to 0} f(\epsilon) = \lim_{\epsilon \to 0}\{\operatorname{Sup}\{f(\epsilon) : \tilde{\epsilon} \in (0, \epsilon)\}\}.$$

It can be proved that Definition 5.2.1 is consistent with Definition 5.1.1: if a set has fractal dimension D according to Definition 5.1.1 then it has the same dimension according to Definition 5.2.1. Also, all of the theorems in this book apply with either definition. The broader definition provides a fractal dimension in some cases where the previous definition makes no assertion.

Theorem 1. *Let m be a positive integer; and consider the metric space ($\mathbb{R}^m$, Euclidean). The fractal dimension $D(A)$ exists for all $A \in \mathscr{H}(\mathbb{R}^m)$. Let $B \in \mathscr{H}(\mathbb{R}^m)$ be such that $A \subset B$; and let $D(B)$ denote the fractal dimension of B. Then $D(A) \leq D(B)$. In particular,*

$$0 \leq D(A) \leq m.$$

Proof. We prove the theorem for the case $m = 2$. Without loss of generality we can suppose that $A \subset \blacksquare$. It follows that $\mathscr{N}(A, \epsilon) \leq \mathscr{N}(\blacksquare, \epsilon)$ for all $\epsilon > 0$. Hence for all ϵ such that $0 < \epsilon < 1$ we have

$$0 \leq \frac{\operatorname{Ln}(\mathscr{N}(A, \epsilon))}{\operatorname{Ln}(1/\epsilon)} \leq \frac{\operatorname{Ln}(\mathscr{N}(\blacksquare, \epsilon))}{\operatorname{Ln}(1/\epsilon)}.$$

It follows that

$$\operatorname*{Limsup}_{\epsilon \to 0} \left\{ \frac{\operatorname{Ln}(\mathscr{N}(A, \epsilon))}{\operatorname{Ln}(1/\epsilon)} \right\} \leq \operatorname*{Limsup}_{\epsilon \to 0} \left\{ \frac{\operatorname{Ln}(\mathscr{N}(\blacksquare, \epsilon))}{\operatorname{Ln}(1/\epsilon)} \right\}.$$

The Limsup on the right-hand side exists and has value 2. It follows that the Limsup on the left-hand side exists and is bounded above by 2. Hence the fractal dimension $D(A)$ is defined and bounded above by 2. Also $D(A)$ is non-negative.

If $A, B \in \mathscr{H}(\mathbb{R}^2)$ with $A \subset B$, then the fractal dimensions of A and B are defined. The above argument wherein $\blacksquare$ is replaced by B shows that $D(A) \leq D(B)$. This completes the proof.

The following theorem helps us to calculate the fractal dimension of the union of two sets.

Theorem 2. *Let m be a positive integer; and consider the metric space ($\mathbb{R}^m$, Euclidean). Let A and B belong to $\mathscr{H}(\mathbb{R}^m)$. Let A be such that its fractal dimension is given by*

$$D(A) = \operatorname*{Lim}_{\epsilon \to 0} \left\{ \frac{\operatorname{Ln}(\mathscr{N}(A, \epsilon))}{\operatorname{Ln}(1/\epsilon)} \right\}.$$

Let $D(B)$ and $D(A \cup B)$ denote the fractal dimensions of B and $A \cup B$ respectively. Suppose that $D(B) \leq D(A)$. Then

$$D(A \cup B) = D(A).$$

Proof. Assume for simplicity $D(B) < D(A)$. From Theorem 1 it follows that $D(A \cup B) \geq D(A)$. We want to show that $D(A \cup B) \leq D(A)$. We begin by observing that, for all $\epsilon > 0$,

$$\mathscr{N}(A \cup B, \epsilon) \leq \mathscr{N}(A, \epsilon) + \mathscr{N}(B, \epsilon).$$

It follows that

$$\begin{aligned} D(A \cup B) &= \operatorname*{Limsup}_{\epsilon \to 0} \left\{ \frac{\operatorname{Ln}(\mathcal{N}(A \cup B, \epsilon))}{\operatorname{Ln}(1/\epsilon)} \right\} \\ &\leq \operatorname*{Limsup}_{\epsilon \to 0} \left\{ \frac{\operatorname{Ln}(\mathcal{N}(A, \epsilon) + \mathcal{N}(B, \epsilon))}{\operatorname{Ln}(1/\epsilon)} \right\} \\ &\leq \operatorname*{Limsup}_{\epsilon \to 0} \left\{ \frac{\operatorname{Ln}(\mathcal{N}(A, \epsilon))}{\operatorname{Ln}(1/\epsilon)} \right\} \\ &\quad + \operatorname*{Limsup}_{\epsilon \to 0} \left\{ \frac{\operatorname{Ln}(1 + \mathcal{N}(B, \epsilon)/\mathcal{N}(A, \epsilon))}{\operatorname{Ln}(1/\epsilon)} \right\}. \end{aligned}$$

The proof is completed by showing that $\mathcal{N}(B, \epsilon)/\mathcal{N}(A, \epsilon)$ is less than one when ϵ is sufficiently small. This would imply that the second limit on the right here is equal to zero. The first limit on the right converges to $D(A)$.

Notice that

$$\operatorname{Sup}\left\{ \frac{\operatorname{Ln}(\mathcal{N}(B, \tilde{\epsilon}))}{\operatorname{Ln}(1/\tilde{\epsilon})} : \tilde{\epsilon} < \epsilon \right\}$$

is a decreasing function of the positive variable ϵ. It follows that

$$\frac{\operatorname{Ln}(\mathcal{N}(B, \epsilon))}{\operatorname{Ln}(1/\epsilon)} < D(A) \qquad \text{for all sufficiently small } \epsilon > 0.$$

Because the limit which is explicitly stated in the theorem exists, it follows that

$$\frac{\operatorname{Ln}(\mathcal{N}(B, \epsilon))}{\operatorname{Ln}(1/\epsilon)} < \frac{\operatorname{Ln}(\mathcal{N}(A, \epsilon))}{\operatorname{Ln}(1/\epsilon)} \qquad \text{for all sufficiently small } \epsilon > 0.$$

This allows us to conclude that

$$\frac{\mathcal{N}(B, \epsilon)}{\mathcal{N}(A, \epsilon)} < 1 \qquad \text{for all sufficiently small } \epsilon > 0.$$

This completes the proof. Slightly more care is needed when $D(A) = D(B)$.

Exercises & Examples

2.1. The fractal dimension of the hairy set $A \subset \mathbb{R}^2$, suggested in Figure 5.2.1 is 2. The contribution from the hairs to $\mathcal{N}(A, \epsilon)$ becomes exponentially small compared to the contribution from ■, as $\epsilon \to 0$.

We now give you a wonderful theorem which provides the fractal dimension of the attractor of an important class of IFS. It will allow you to estimate fractal dimensions "on the fly," simply from inspection of pictures of fractals, once you get used to it.

Theorem 3. *Let* $\{\mathbb{R}^m; w_1, w_2, \ldots, w_N\}$ *be a hyperbolic IFS, and let* A *denote its attractor. Suppose* w_n *is a similitude of scaling factor* s_n *for each* $n \in$

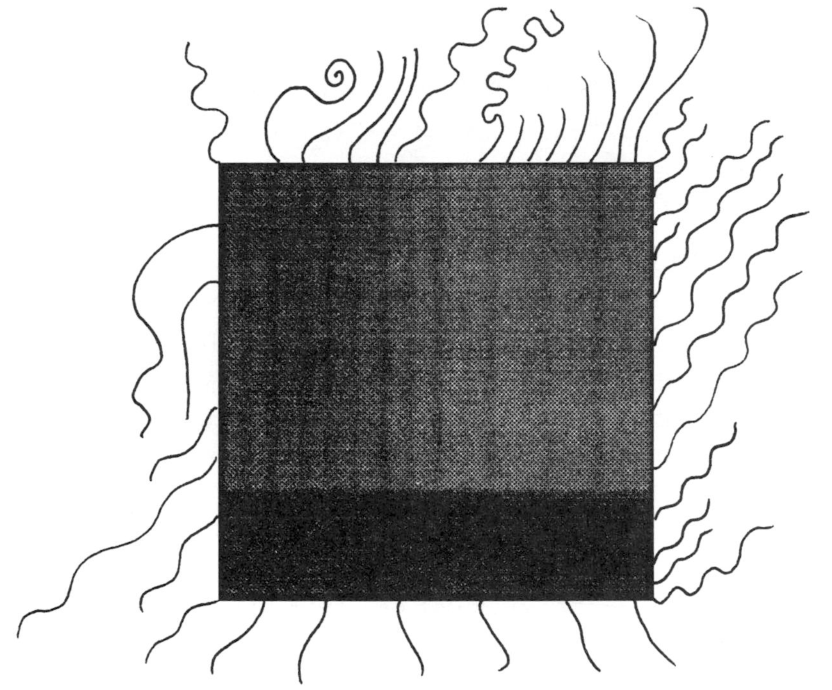

Figure 5.2.1
Picture of a hairy box. The fractal dimension of the subset of $\mathbb{R}^2$ suggested here is the same as the fractal dimension of the box. The hairs are overpowered.

$\{1, 2, 3, \ldots, N\}$. *If the IFS is totally disconnected or just-touching then the attractor has fractal dimension $D(A)$, which is given by unique solution of*

$$\sum_{n=1}^{N} |s_n|^{D(A)} = 1,\ D(A) \in [0, m].$$

If the IFS is overlapping then $\overline{D} \geq D(A)$ where $\overline{D}$ is the solution of

$$\sum_{n=1}^{N} |s_n|^{\overline{D}} = 1,\ \overline{D} \in [0, \infty).$$

Sketch of Proof. The full proof can be found in [Bedford 1986], [Hardin 1985], [Hutchinson 1981], and [Reuter 1987]. The following argument gives a valuable insight into the fractal dimension. We restrict attention to the case where the IFS $\{\mathbb{R}^m; w_1, w_2, \ldots, w_N\}$ is totally disconnected. We suppose that the scaling factor s_i associated with the similitude w_i is nonzero for each $i \in \{1, 2, \ldots, N\}$. Let $\epsilon > 0$. We begin by making two observations.

Observation (i). Let $i \in \{1, 2, \ldots, N\}$. Since w_i is a similitude of scaling factor s_i it maps closed balls onto closed balls, according to

$$w_i(B(x, \epsilon)) = B(w_i(x), |s_i|\epsilon).$$

Assume that $s_i \neq 0$. Then w_i is invertible, and we obtain

$$w_i^{-1}(B(x,\epsilon)) = B\left(w_i^{-1}(x), |s_i|^{-1}\epsilon\right).$$

The latter two relations allow us to establish that for all $\epsilon > 0$,

$$\mathcal{N}(A,\epsilon) = \mathcal{N}\left(w_i(A), |s_i|\epsilon\right);$$

which is equivalent to

$$\mathcal{N}\left(w_i(A),\epsilon\right) = \mathcal{N}\left(A, |s_i|^{-1}\epsilon\right). \tag{5.2.6}$$

This applies for each $i \in \{1,2,3,\ldots,N\}$.

Observation (ii). The attractor A of the IFS is the disjoint union

$$A = w_1(A) \cup w_2(A) \cup \cdots \cup w_N(A),$$

where each of the sets $w_n(A)$ is compact. Hence we can choose the positive number ϵ so small that if, for some point $x \in \mathbb{R}^2$ and some integer $i \in \{1,2,\ldots,N\}$, we have $B(x,\epsilon) \cap w_i(A) \neq \varnothing$, then $B(x,\epsilon) \cap w_j(A) = \varnothing$ for all $j \in \{1,2,\ldots,N\}$ with $j \neq i$. It follows that if the number ϵ is sufficiently small we have

$$\mathcal{N}(A,\epsilon) = \mathcal{N}\left(w_1(A),\epsilon\right) + \mathcal{N}\left(w_2(A),\epsilon\right) + \mathcal{N}\left(w_3(A),\epsilon\right) + \cdots + \mathcal{N}\left(w_N(A),\epsilon\right).$$

We put our two observations together. Substitute from equation (5.2.6) into the last equation to obtain

$$\mathcal{N}(A,\epsilon) = \mathcal{N}\left(A, |s_1|^{-1}\epsilon\right) + \mathcal{N}\left(A, |s_2|^{-1}\epsilon\right) + \mathcal{N}\left(A, |s_3|^{-1}\epsilon\right) + \cdots + \mathcal{N}\left(A, |s_N|^{-1}\epsilon\right). \tag{5.2.7}$$

This functional equation is true for all positive numbers ϵ which are sufficiently small. The proof is completed by showing formally that this implies the assertion in the theorem.

Here we demonstrate the reasonableness of the last step. Let us make the assumption $\mathcal{N}(A,\epsilon) \sim C\epsilon^{-D}$. Then substituting into (5.2.7) we obtain the equation:

$$C\epsilon^{-D} \approx C|s_1|^D\epsilon^{-D} + C|s_2|^D\epsilon^{-D} + C|s_3|^D\epsilon^{-D} + \cdots + C|s_N|^D\epsilon^{-D}.$$

From this we deduce that

$$1 = |s_1|^D + |s_2|^D + |s_3|^D + \cdots + |s_N|^D.$$

This completes our sketch of the proof of Theorem 5.2.3.

Exercises & Examples

2.1. This example takes place in the metric space ($\mathbb{R}^2$, Euclidean). A Sierpinski triangle is the attractor of a just-touching IFS of three similitudes, each with scaling factor 0.5. Hence the fractal dimension is the solution D of the equation

$$(0.5)^D + (0.5)^D + (0.5)^D = 1$$

from which we find

$$D = \frac{\mathrm{Ln}(1/3)}{\mathrm{Ln}(0.5)} = \frac{\mathrm{Ln}(3)}{\mathrm{Ln}(2)}.$$

2.2. Find a just-touching IFS of similitudes in $\mathbb{R}^2$ whose attractor is ■. Verify that Theorem 5.2.3 yields the correct value for the fractal dimension of ■.

2.3. The Classical Cantor set is the attractor of the hyperbolic IFS

$$\{[0,1];\ w_1(x) = \tfrac{1}{3}x;\ w_2(x) = \tfrac{1}{3}x + \tfrac{2}{3}\}.$$

Use Theorem 5.2.3 to calculate its fractal dimension.

2.4. The attractor of a just-touching hyperbolic IFS $\{\mathbb{R}^2;\ w_i(x),\ i = 1, 2, 3, 4\}$ is represented in Figure 5.2.2. The affine transformations w_i: $\mathbb{R}^2 \to \mathbb{R}^2$ are similitudes, and are given in tabular form in Table 5.2.1. Use Theorem 5.2.3 to calculate the fractal dimension of the attractor.

2.5. The attractor of a just-touching hyperbolic IFS $\{\mathbb{R}^2;\ w_i(x),\ i = 1, 2, 3\}$ is represented in Figure 5.2.3. The affine transformations w_i: $\mathbb{R}^2 \to \mathbb{R}^2$ are

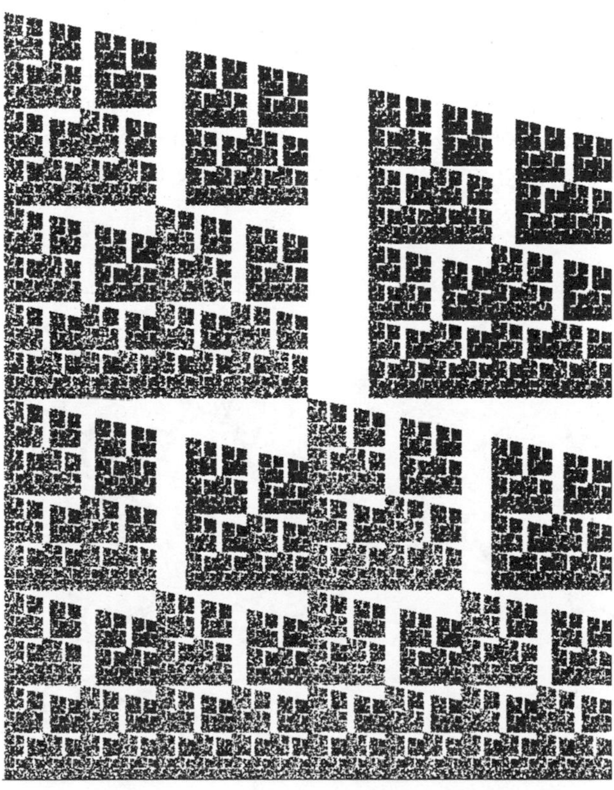

Figure 5.2.2 *The Castle fractal. This is an example of a self-similar fractal, and its fractal dimension may be calculated with the aid of Theorem* 5.2.3. *The associated IFS code is given in Table* 5.2.1.

Table 5.2.1
IFS code for a Castle.

w	a	b	c	d	e	f	p
1	0.5	0	0	0.5	0	0	0.25
2	0.5	0	0	0.5	2	0	0.25
3	0.4	0	0	0.4	0	1	0.25
4	0.5	0	0	0.5	2	1	0.25

similitudes. Use the Collage Theorem to find the similitudes, and then use Theorem 5.2.3 to calculate the fractal dimension of the attractor.

2.6. Figure 5.2.4 represents the attractor of an overlapping hyperbolic IFS $\{\mathbb{R}^2; w_i(x), i = 1,2,3,4\}$. Use the Collage Theorem and Theorem 5.2.3 to obtain an upper bound to the fractal dimension of the attractor.

2.7. Calculate the fractal dimension of the subset of $\mathbb{R}^2$ represented by Figure 5.2.5.

2.8. Consider the attractor A of a totally disconnected hyperbolic IFS $\{\mathbb{R}^7; w_i(x), i = 1,2\}$ where the two maps $w_1: \mathbb{R}^7 \to \mathbb{R}^7$ and $w_2: \mathbb{R}^7 \to \mathbb{R}^7$ are similitudes, of scaling factors s_1 and s_2 respectively. Show that A is also the attractor of the totally disconnected hyperbolic IFS $\{\mathbb{R}^7; v_i(x), i = 1,2,3,4\}$ where $v_1 = w_1 \circ w_1$, $v_2 = w_1 \circ w_2$, $v_3 = w_2 \circ w_1$, and $v_4 = w_2 \circ w_2$. Show that $v_i(x)$ is a similitude, and find its scaling factor, for $i = 1,2,3,4$. Now apply Theorem 5.2.3 to yield two apparently different equations for the fractal dimension of A. Prove that these two equations have the same solution.

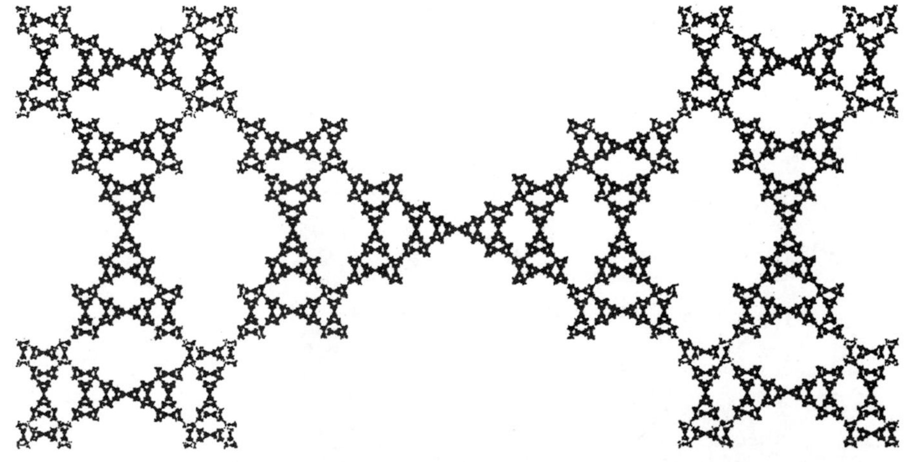

Figure 5.2.3
To calculate the fractal dimension of the subset of $\mathbb{R}^2$ represented here, first apply the Collage Theorem to find a corresponding set of similitudes. Then use Theorem 5.2.3.

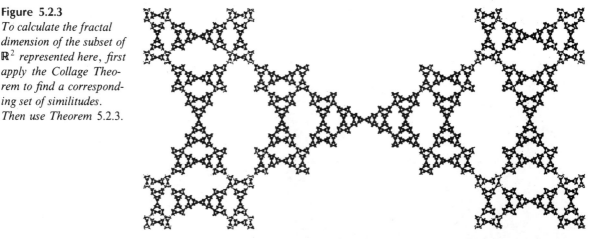

Figure 5.2.4
An upper bound to the fractal dimension of attractor of an overlapping IFS, corresponding to this picture, can be computed with the aid of Theorem 5.2.3.

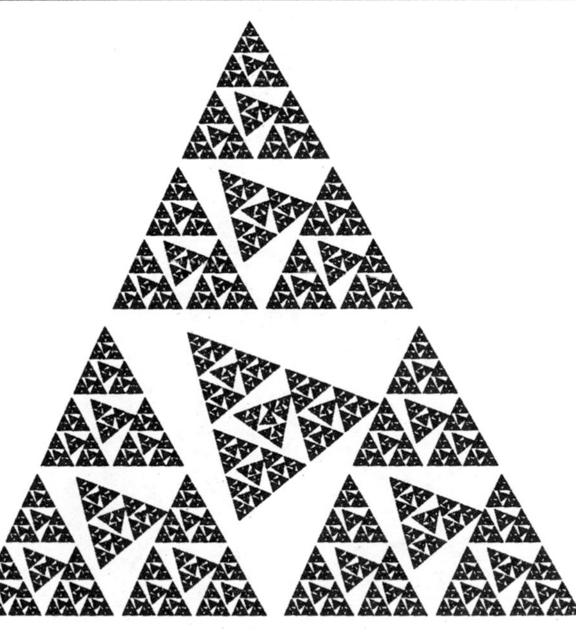

Figure 5.2.5
Calculate the fractal dimension of the subset of $\mathbb{R}^2$ represented by this image.

5.3 THE EXPERIMENTAL DETERMINATION OF THE FRACTAL DIMENSION

In this section we consider the experimental determination of the fractal dimension of sets in the physical world. We model them, as best we can, as subsets of ($\mathbb{R}^2$, Euclidean) or ($\mathbb{R}^3$, Euclidean). Then, based on the definition of the fractal dimension, and sometimes in addition to one or other of the preceding theorems, such as the Box Counting Theorem, we analyse the model to provide a fractal dimension for the real-world set.

In the following examples we emphasize that when the fractal dimension of a physical set is quoted, some indication of how it was calculated must also be provided. There is not yet a broadly accepted unique way of associating a fractal dimension with a set of experimental data.

Example

3.1. There is a curious cloud of dots in the woodcut in Figure 5.3.1. Let us try to estimate its fractal dimension by direct appeal to Definition 5.1.1.

We begin by covering the cloud of points by disks of radius ϵ for a range of ϵ-values from $\epsilon = 3$ cm down to $\epsilon = 0.3$ cm; and in each case we count the number of disks needed. This provides the set of approximate values for $\mathcal{N}(A, \epsilon)$ given in Table 5.3.1. The data is redisplayed in log-log format in Table 5.3.2. The data in Table 5.3.2 is plotted in Figure 5.3.2. A straight line

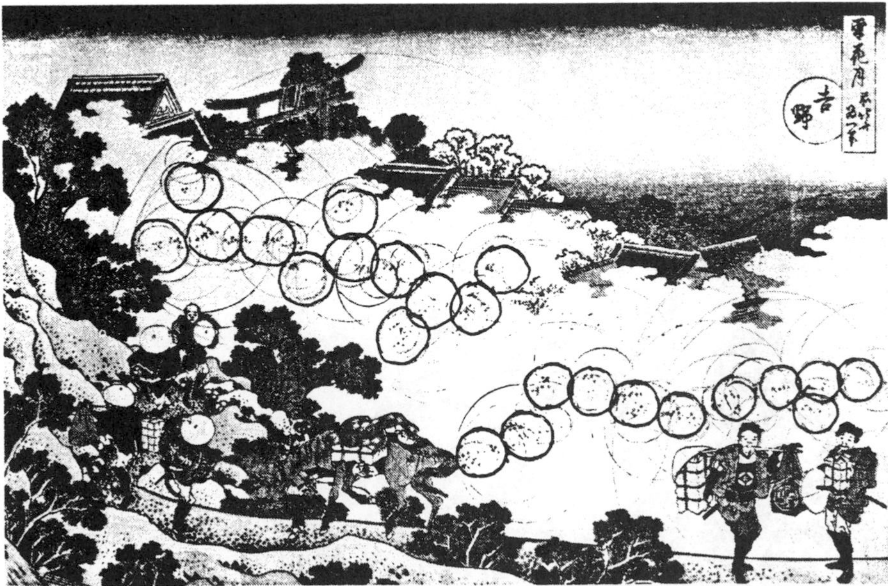

Figure 5.3.1 *Covering a cloud of dots in a woodcut by balls of radius $\epsilon > 0$.*

Table 5.3.1
Minimal numbers of balls, of various radii, needed to cover a "dust" in a woodcut.

ϵ	$\mathcal{N}(A, \epsilon)$
3 cm	2
2 cm	3
1.5 cm	4
1.2 cm	6
1 cm	7
0.75 cm	10
0.5 cm	16
0.4 cm	23
0.3 cm	31
0.015 cm	267

Table 5.3.2
The data in Table 5.3.1 is tabulated in log-log form. These values are used to obtain the fractal dimension.

$\ln(1/\epsilon)$	$\ln(\mathcal{N}(A, \epsilon))$
− 1.1	0.69
−0.69	1.09
−0.405	1.39
−0.182	1.79
0	1.95
0.29	2.30
0.693	2.77
0.916	3.13
1.204	3.43
4.2	5.59

which approximately passes through the points is drawn. The slope of this straight line is our approximation to the fractal dimension of the cloud of points.

The experimental number $\mathcal{N}(A, 0.015 \text{ cm})$ is not very accurate. It is a very rough estimate based on the size of the dots themselves, and is not included in the plot in Figure 5.3.2. The slope of the straight line in Figure 5.3.2 gives

$$D(A) \simeq 1.2, \quad \text{over the range } 0.5 \text{ cm to } 5 \text{ cm}, \tag{5.3.8}$$

where A denotes the set of points whose dimension we are approximating.

The straight line in Figure 5.3.2 was drawn "by eye." Thus if one was to repeat the experiment, a different value for $D(A)$ may be obtained. In order to make the results consistent from experiment to experiment, the straight line should be estimated by a least squares method.

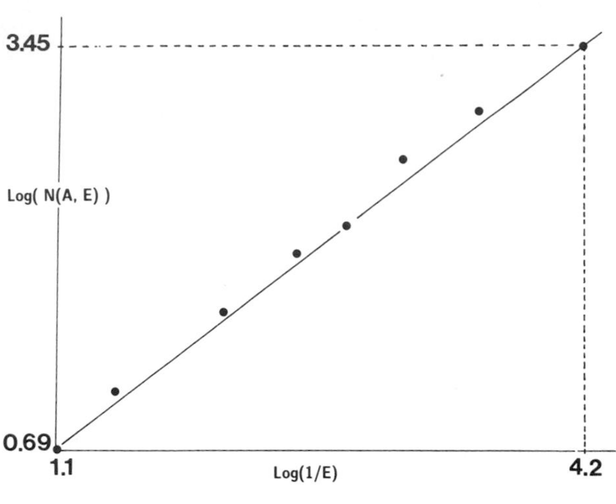

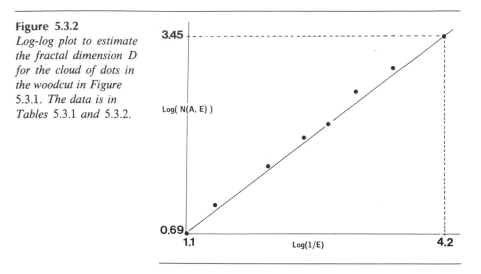

Figure 5.3.2 *Log-log plot to estimate the fractal dimension D for the cloud of dots in the woodcut in Figure* 5.3.1. *The data is in Tables* 5.3.1 *and* 5.3.2.

In proceeding by direct appeal to Definition 5.1.1, the estimates of $\mathcal{N}(A, \epsilon)$ need to be made very carefully. One needs to be quite sure that $\mathcal{N}(A, \epsilon)$ is indeed the *least* number of balls of radius ϵ needed. For large sets of data this could be very time consuming.

It is clearly important to state the range of scales used: we have no idea or definition concerning the structure of the dots in Figure 5.3.1 at higher resolutions than, say, 0.015 cm. Moreover, regardless of how much experimental data we have, and regardless of how many scales of observation are available to us, we will always end up estimating the slope of a straight line corresponding to a finite range of scales. If we include the data point (0.015 cm, 267) in the above estimation we obtain

(5.3.9) $\quad D(A) \simeq 0.9,$ over the range of scales 0.05 to 5 cm.

We comment on the difference between the estimates (5.3.8) and (5.3.9). If we restrict ourselves to the range of scales in (5.3.8), there is little information present in the data to distinguish the cloud of points from a very irregular curve. However the data used to obtain (5.3.9) contains values for $\mathcal{N}(A, \epsilon)$ for several values of ϵ such that the corresponding coverings of A are disconnected. The data is "aware" that A is disconnected. This lowers the experimentally determined value of D.

Example

3.2. In this example we consider the physical set labelled A in Figure 5.3.3. A is actually an approximation to a classical Cantor set. In this case we make an experimental estimate of the fractal dimension, based on the Box

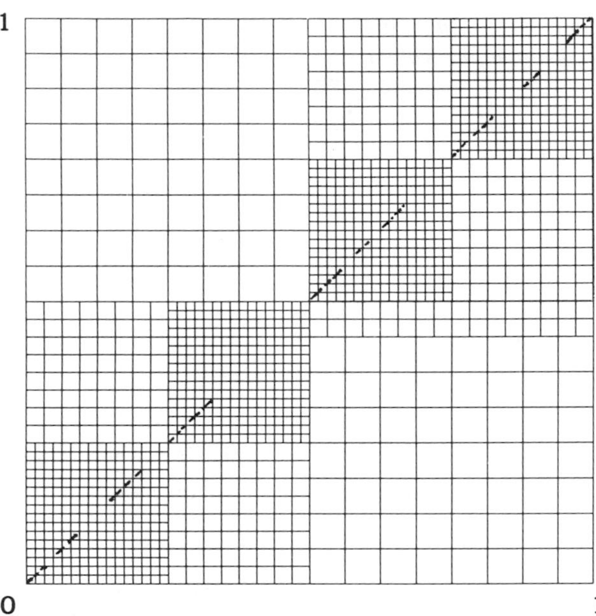

Figure 5.3.3 *Successive subdivision of overlaying grid to obtain the box counts needed for the application of Theorem 5.1.2 to estimate the fractal dimension of the Cantor Set A. The counts are presented in Table 5.3.3.*

Counting Theorem. A Cartesian coordinate system is set up as shown and we attempt to count the number of square boxes $\mathcal{N}_n(A)$ of side $(1/2^n)$ which intersect A. We are able to obtain fairly accurate values of $\mathcal{N}_n(A)$ for $n = 0, 1, 2, 3, 4, 5$, and 6. These values are presented in Table 5.3.3. We note that these values depend on the choice of coordinate system. Nonetheless, the values of $\mathcal{N}_n(A)$ are much easier to measure than the values of $\mathcal{N}(A, \epsilon)$ used in example 5.3.1.

The analysis of the data proceeds just as in example 5.3.1. It is represented in Table 5.3.3 and Figure 5.3.4. We obtain

$$D(A) \simeq 0.8, \qquad \text{over the range } \tfrac{1}{8} \text{ inch to 8 inches.}$$

Example

3.3. In this example we show how a good experimentalist [Strahle 1987] overcomes the inherent difficulties with the experimental determination of fractal dimensions. In so doing he obtains a major scientific result. The idea is to compare two sets of experimental data, obtained by different means, on the same physical system. The physical system is a laboratory jet flame. The data are time series for the temperature and velocity at two different points in the jet. The idea is to apply the same procedure to the analysis of the two sets of data, to obtain a value for the fractal

Table 5.3.3
The data determined from Figure 5.3.3, in the experimental calculation of the fractal dimension of a physical set A.

n	$\mathcal{N}_n(A)$	$\ln \mathcal{N}_n(A)$	$n \ln 2$
0	1	0	0
1	3	1.10	0.69
2	7	1.95	1.38
3	10	2.30	2.08
4	19	2.94	2.77
5	33	3.50	3.46
6	58	4.06	4.16

dimension. The two values are same. Instead of drawing the conclusion that the two sets of data "have the same fractal dimension," he deduces that the two sets of data have a common source. That common source is physical, real world, chaos.

The experimental setup is as follows. A flame is probed by (a) a laser beam and (b) a very thin wire. These two probes, coupled with appropriate measuring devices, allow measurements to be made of the temperature and velocity in the jet at two different points, as a function of time. In (a) the light bounces off the fast moving molecules in the exhaust and a receiver measures the characteristics of the bounced light. The output from the receiver is a voltage. This voltage, suitably rescaled, gives the temperature of the jet as a function of time. In (b) a constant temperature is maintained through a wire in the flame. The voltage required to hold the temperature constant is recorded. This voltage, suitably rescaled, gives the velocity of the jet as a function of time. In

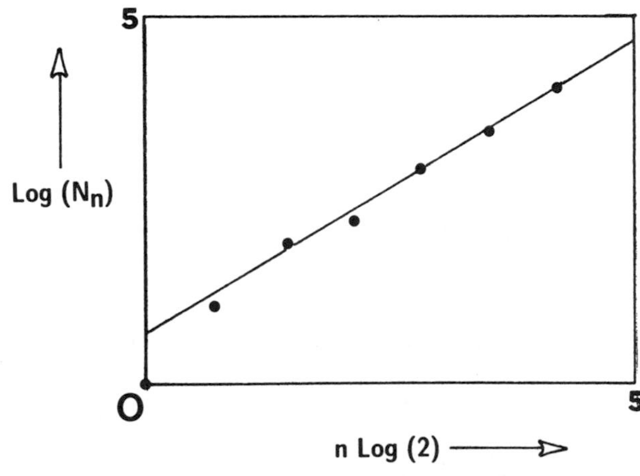

Figure 5.3.4
Slope of the plot of the data in Table 5.3.3 *gives an approximation to the fractal dimension of the Set A in Figure* 5.3.3.

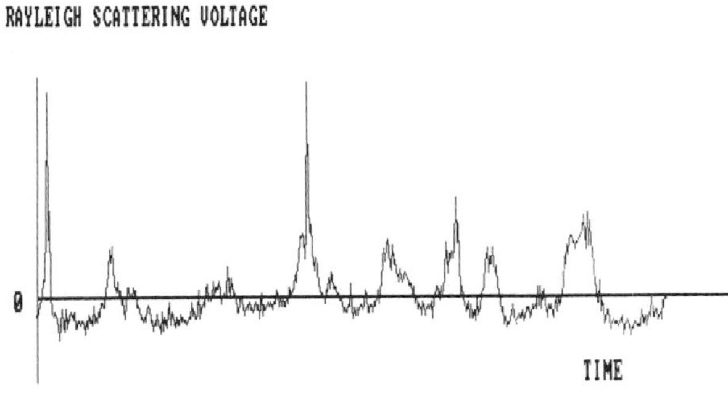

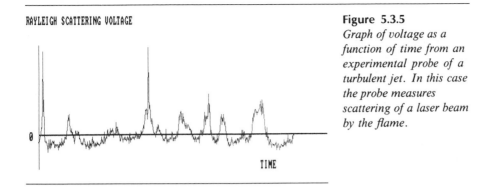

Figure 5.3.5
Graph of voltage as a function of time from an experimental probe of a turbulent jet. In this case the probe measures scattering of a laser beam by the flame.

this way we obtain two independent readings of two different, but related, quantities.

Of course the experimental apparatus is much more sophisticated than it sounds from the above description. What is important is that the measuring devices are of very high resolution, accuracy, and sensitivity. A reading of the velocity can be made once every microsecond. Vast amounts of data can be obtained. A sample of the experimental output from (a) is shown in Figure 5.3.5, where it is represented as the graph of voltage against time. It is a very complex curve. If one "magnifies up" the curve, one finds that its geometrical complexity in the curve continues to be present. It is just the sort of thing we fractal geometers like to analyse.

A sample of the experimental output from (b) is shown in Figure 5.3.6, again represented as a graph of voltage against time. You should compare Figures 5.3.5 and 5.3.6. They look different. Is there a relationship between

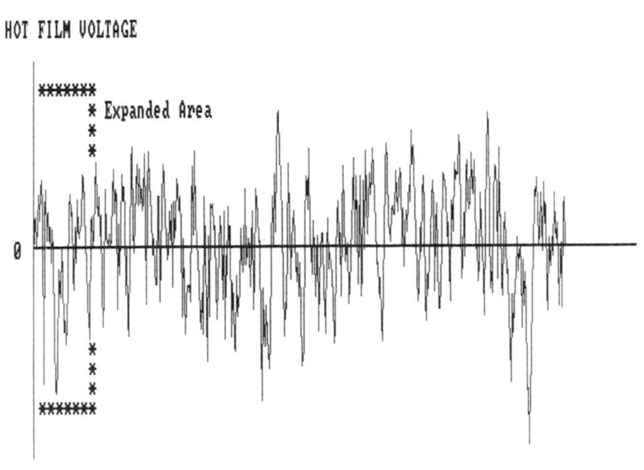

Figure 5.3.6
Graph of voltage as a function of time from an experimental probe of a turbulent jet. In this case the probe measures the voltage across a wire in the flame. This data has a definite fractal character, as demonstrated by the expanded piece shown in Figure 5.3.7.

them? There should be: they both probe the same burning gas and they are in the same units.

In order to bring out the fractal character in the data, an expanded piece of the data in Figure 5.3.6 is shown in Figure 5.3.7.

The fractal dimensions of the graphs of the two time series, obtained from (a) and (b), is calculated using a method based on the Box Counting Theorem. Exactly the same method is applied to both sets of data, over the same range of scales. Figure 5.3.8 shows the graphical analysis of the resulting box counts. Both experiments yield the same value

$$D \approx 1.5 \qquad \text{over the range of scales } 2^6 \times 10^{-5} \text{ to } 2^{13} \times 10^{-5} \text{ sec.}$$

This suggests that, despite the different appearances of their graphs, there is a common source for the data.

We believe that this common source is chaotic dynamics of a certain special flavor and character, present in the jet exhaust. If so, then fractal dimension provides an experimentally measurable parameter which can be used to quantify chaos.

Exercises & Examples

3.4. Use a method based on the direct application of Definition 5.1.1 to make an experimental determination of the fractal dimension of the physical set defined by the black ink in Figure 5.3.9. Give the range of scales to which your result applies.

3.5. Use a method based on the Box Counting Theorem, as in example 5.3.2, to estimate the fractal dimension of the "random dendrite" given in Figure 5.3.10. State the range of scales over which your estimate applies.

Figure 5.3.7 *A blow-up of a piece of the graph in Figure* 5.3.6.

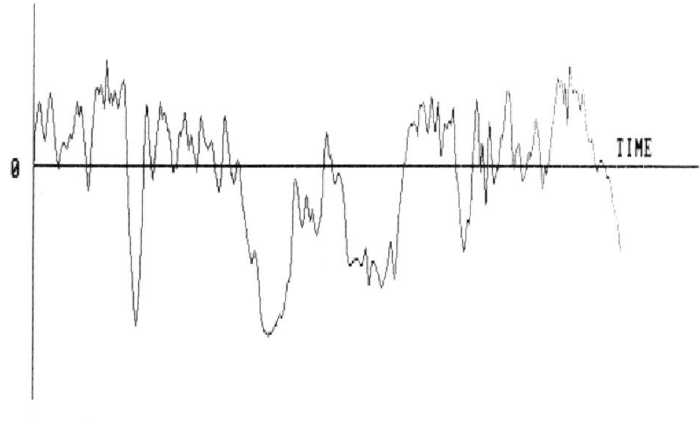

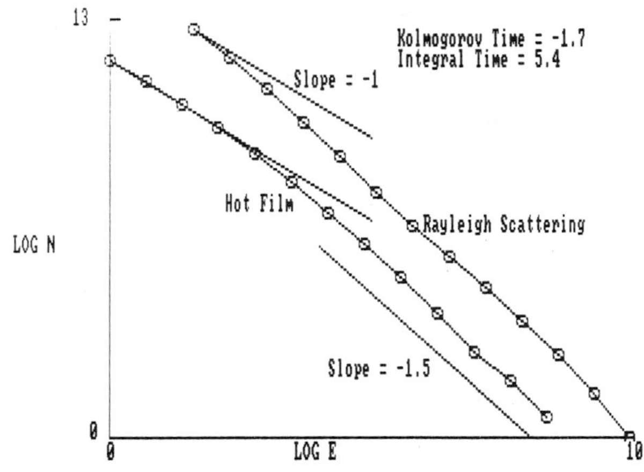

Figure 5.3.8 *Graphical analysis of Box Counts associated with experiments (a) and (b). The data analysed is illustrated in Figures 5.3.5 and 5.3.6. The two data streams, which come from probes of a single turbulent system, when analysed in exactly the same way, yield the same value $D = 1.5$ for the fractal dimension. This suggests that, despite the different appearances of their graphs, there is a common source for the data. This source is chaotic dynamics of a certain special flavor and character. The fractal dimension provides a measurable symptom of the brand of chaos.*

Make several complete experiments to obtain some idea of the accuracy of your result.

3.6. Make an experimental estimate of the fractal dimension of the dendrite shown in Figure 5.3.11. Note that a grid of boxes of size $(1/12)^{\text{th}}$ inch by $(1/12)^{\text{th}}$ inch has been printed on top of the dendrite. Compare the result you obtain with the result of example 5.3.5. It is important that you follow exactly the same procedure in both experiments.

3.7. Make an experimental determination of the fractal dimension of the set in Figure 5.2.5. Compare your result with a theoretical estimate based on Theorem 5.2.3, as in example 5.2.7.

3.8. Obtain maps of Great Britain of various sizes. Make an experimental determination of the fractal dimension of the coastline, over as wide a range of scales as possible.

3.9. Obtain data showing the variations of a Stock Market index, at several different time scales, for example: hourly, daily, monthly, and yearly. Make an experimental determination of the fractal dimension. Find a second economic indicator for the same system, and analyse its fractal dimension. Compare the results.

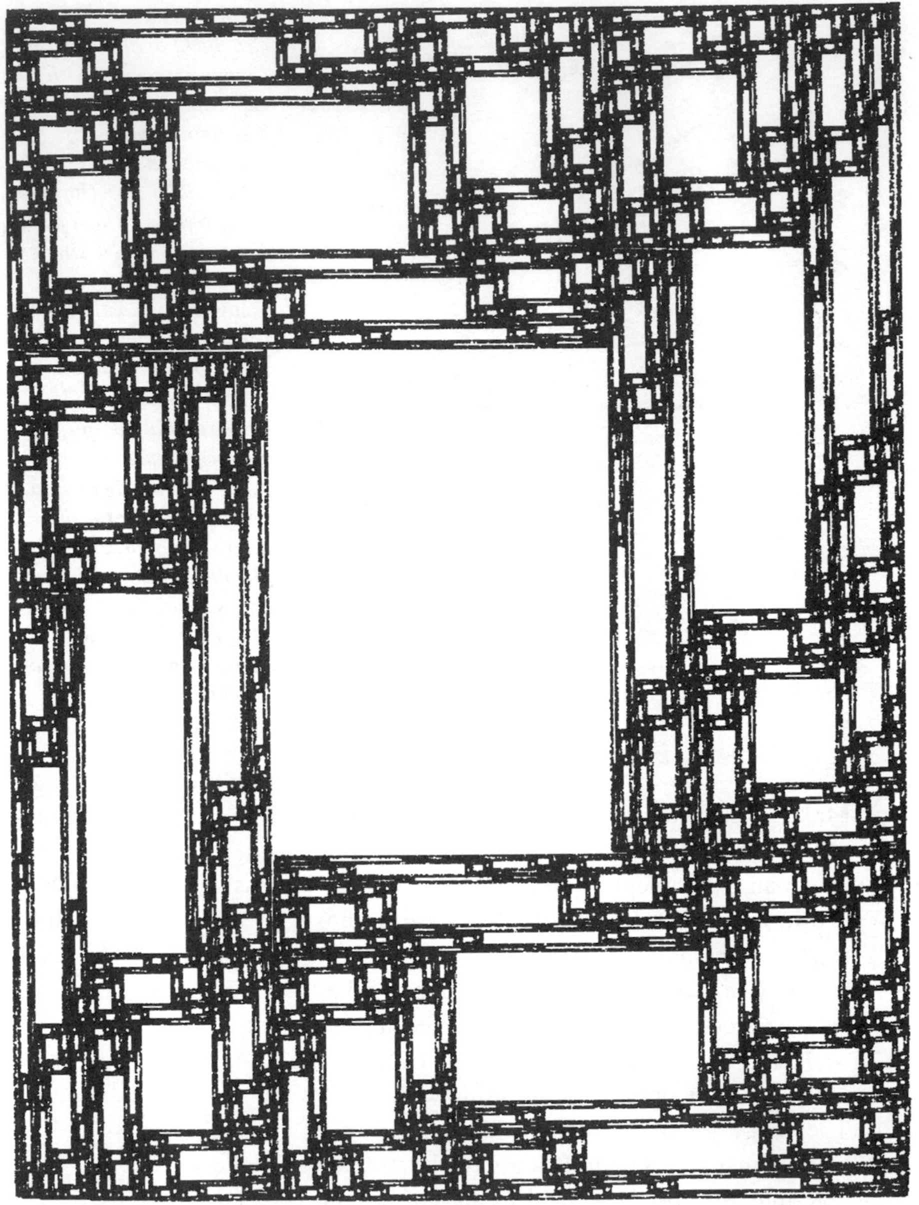

Figure 5.3.9 *Make an experimental estimate of the fractal dimension of the set A of black ink, above, over the range of scales 5 inches 0.1 inches. Base your experimental method directly on the Definition 5.1.1.*

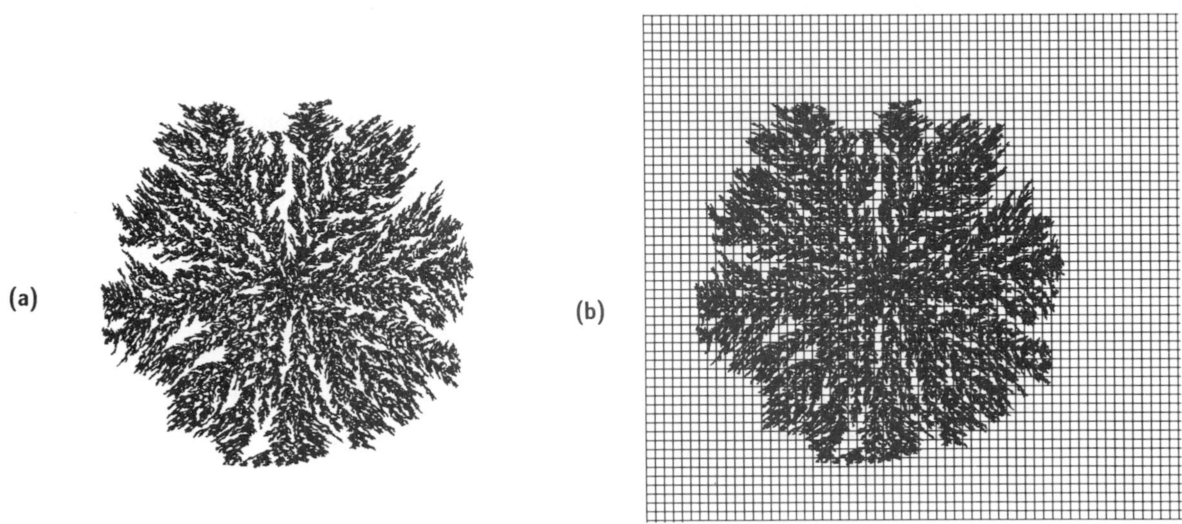

Figure 5.3.10
Make an experimental estimate of the fractal dimension of this set in (a), over the range of scales 5 inches to $(1/12)^{th}$ *inch, basing your method on the Box Counting Theorem and graphical analysis. In order to help you with your work, in (b) we have overlayed on the set a grid of boxes of size* $(1/12)^{th}$ *inch by* $(1/12)^{th}$ *inch.*

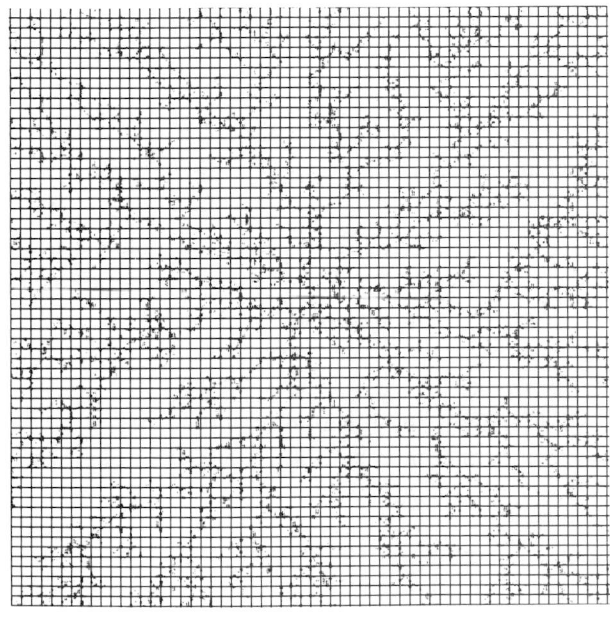

Figure 5.3.11
Make an experimental estimate of the fractal dimension of the random dendrite shown here. Note that a grid of boxes of size $(1/12)^{th}$ *inch by* $(1/12)^{th}$ *inch has been printed on top of the dendrite. Compare the experimental fractal dimension here with that of the dendrite in Figure 5.3.10. In advance, which one do you expect will have the lower fractal dimension?*

5.4 THE HAUSDORFF-BESICOVITCH FRACTAL DIMENSION

The Hausdorff-Besicovitch dimension of bounded subset of $\mathbb{R}^m$ is another real number which can be used to characterize the geometrical complexity of bounded subsets of $\mathbb{R}^m$. Its definition is more complex and subtle than that of the fractal dimension. One of the reasons for its importance is that it is associated with a method for comparing the "sizes" of sets whose fractal dimensions are the same. It is harder to work with than the fractal dimension, and its definition is not usually used as the basis of experimental procedures for the determination of fractal dimensions of physical sets.

Throughout we work in the metric space $(\mathbb{R}^m, d)$, where m is a positive integer and d denotes the Euclidean metric. Let $A \subset \mathbb{R}^m$ be bounded. Then we use the notation

$$\operatorname{diam}(A) = \inf\{d(x, y): x, y \in A\}.$$

Let $0 < \epsilon < \infty$, and $0 \le p < \infty$. Let $\mathscr{A}$ denote the set of sequences of subsets $\{A_i \subset A\}$, such that $A = \cup_{i=1}^{\infty} A_i$. Then we define

$$\mathscr{M}(A, p, \epsilon) = \inf\left\{ \sum_{i=1}^{\infty} (\operatorname{diam}(A_i))^p : \{A_i\} \in \mathscr{A}, \text{ and } \operatorname{diam}(A_i) < \epsilon \right.$$
$$\left. \text{for } i = 1, 2, 3, \ldots \right\}.$$

Here we use the convention that $(\operatorname{diam}(A_i))^0 = 0$ when A_i is empty. $\mathscr{M}(A, p, \epsilon)$ is a number in the range $[0, \infty]$; its value may be zero, finite or infinite. You should verify that it is a nonincreasing function of ϵ. We now define

$$\mathscr{M}(A, p) = \sup\{\mathscr{M}(A, p, \epsilon): \epsilon > 0\}.$$

Then for each $p \in [0, \infty]$ we have $\mathscr{M}(A, p) \in [0, \infty]$.

Definition 1. Let m be a positive integer and let A be a bounded subset of the metric space ($\mathbb{R}^m$, Euclidean). For each $p \in [0, \infty)$ the quantity $\mathscr{M}(A, p)$ described above is called the *Hausdorff p-dimensional measure* of A.

Exercises

4.1. Show that $\mathscr{M}(A, p)$ is a nonincreasing function of $p \in [0, \infty]$.

4.2. Let A denote a set of seven distinct points in ($\mathbb{R}^2$, Euclidean). Show that $\mathscr{M}(A, 0) = 7$ and $\mathscr{M}(A, p) = 0$ for $p > 0$.

4.3. Let A denote a countable infinite set of distinct points in ($\mathbb{R}^2$, Euclidean). Show that $\mathscr{M}(A, 0) = \infty$ and $\mathscr{M}(A, p) = 0$ for $p > 0$.

4.4. Let $\mathscr{C}$ denote the Classical Cantor set in $[0, 1]$. Show that $\mathscr{M}(\mathscr{C}, 0) = \infty$ and $\mathscr{M}(\mathscr{C}, 1) = 0$.

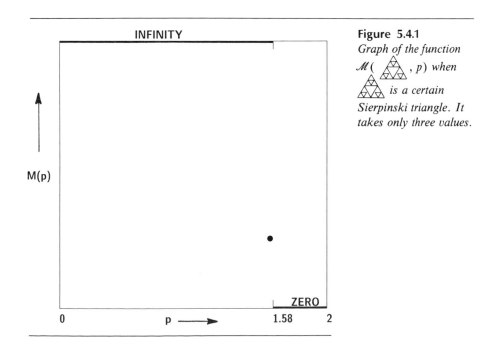

4.5. Let △ denote a convenient Sierpinski triangle. Show that $\mathcal{M}(△, 1) = \infty$ and $\mathcal{M}(△, 2) = 0$. Can you evaluate $\mathcal{M}(△, \ln(3)/\ln(2))$? At least try to argue why this might be an interesting number.

The Hausdorff p-dimensional measure $\mathcal{M}(A, p)$, as a function of $p \in [0, \infty]$, behaves in a remarkable manner. Its range consists of only one, two, or three values! The possible values are zero, a finite number, and infinity. In Figure 5.4.1 we illustrate this behaviour when A is a certain Sierpinski triangle.

Theorem 1. *Let m be a positive integer. Let A be a bounded subset of the metric space $(\mathbb{R}^m$, Euclidean). Let $\mathcal{M}(A, p)$ denote the function of $p \in [0, \infty)$ defined above. Then there is a unique real number $D_H \in [0, m]$ such that*

$$\mathcal{M}(A, p) = \begin{cases} \infty & \text{if } p < D_H \text{ and } p \in [0, \infty), \\ 0 & \text{if } p > D_H \text{ and } p \in [0, \infty). \end{cases}$$

Proof. This can be found, for example, in [Federer 1969, section 2.10.3].

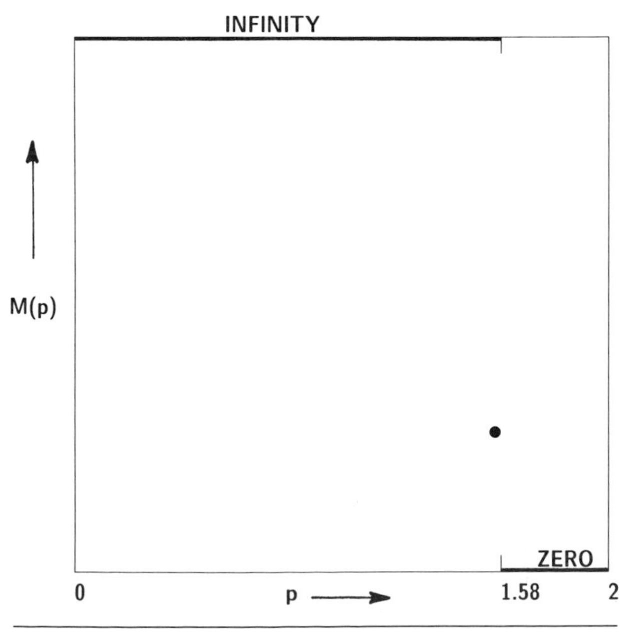

Figure 5.4.1 *Graph of the function $\mathcal{M}(△, p)$ when △ is a certain Sierpinski triangle. It takes only three values.*

Definition 2. Let m be a positive integer and let A be a bounded subset of the metric space ($\mathbb{R}^m$, Euclidean). The corresponding real number D_H, occurring in Theorem 5.4.1, is called the *Hausdorff-Besicovitch dimension* of the set A. This number will also be denoted by $D_H(A)$.

Theorem 2. *Let m be a positive integer and let A be a subset of the metric space ($\mathbb{R}^m$, Euclidean). Let $D(A)$ denote the fractal dimension of A and let $D_H(A)$ denote the Hausdorff-Besicovitch dimension of A. Then*

$$0 \leq D_H(A) \leq D(A) \leq m.$$

Exercises

4.6. Describe a situation where you would expect $D_H(A) < D(A)$.

4.7. Prove Theorem 5.4.2.

Theorem 3. *Let m be a positive integer. Let $\{\mathbb{R}^m; w_1, w_2, \ldots, w_N\}$ be a hyperbolic IFS, and let A denote its attractor. Let w_n be a similitude of scaling factor s_n for each $n \in \{1, 2, 3, \ldots, N\}$. If the IFS is totally disconnected or just-touching, then the Hausdorff-Besicovitch dimension $D_H(A)$ and the fractal dimension $D(A)$ are equal. In fact, $D(A) = D_H(A) = D$ where D is the unique solution of*

$$\sum_{n=1}^{N} |s_n|^D = 1, \quad D \in [0, m].$$

Moreover, if D is positive then the Hausdorff D-dimensional measure $\mathcal{M}(A, D_H(A))$ is a positive real number.

Proof. This can be found in [Hutchinson 1981].

In the situation referred to in Theorem 5.4.3, the Hausdorff $D_H(A)$-dimensional measure can be used to compare the "sizes" of fractals which have the same fractional dimension. The larger the value of $\mathcal{M}(A, D_H(A))$, the "larger" the fractal. Of course, if two fractals have different fractal dimensions, then we say that the one with the higher fractal dimension is the "larger" one.

Exercises & Examples

4.8. Here we provide some intuition about the functions $\mathcal{M}(A, p, \epsilon)$ and $\mathcal{M}(A, p)$, and the "sizes" of fractals. We illustrate how these quantities can be estimated. The type of procedure we use can often be followed, for attractors of just-touching and totally disconnected IFS whose maps are all similitudes, and should lead to correct values. Formal justification is tedious, and follows the lines suggested in [Hutchinson 1981].

Consider the Sierpinski triangle $\triangle$ with vertices at $(0,0)$, $(0,1)$ and $(1,0)$. We work in $\mathbb{R}^2$ with the Euclidean metric. We begin by estimating the number $\mathcal{M}(\triangle, p, \epsilon)$ for $p \in [0, \infty)$ for various values of ϵ. The values of ϵ we consider are $\epsilon = \sqrt{2}(1/2)^n$ for $n = 0, 1, 2, 3, \ldots$. Now notice that $\triangle$ can be covered very efficiently by 3^n closed disks of diameter $\sqrt{2}(1/2)^n$. We guess that this covering is one for which the infimum in the definition of $\mathcal{M}(\triangle, p, \epsilon = \sqrt{2}(1/2)^n)$ is actually achieved. We obtain the estimate

$$\mathcal{M}\left(\triangle, p, \sqrt{2}(1/2)^n\right) = 3^n(\sqrt{2})^p(1/2)^{np} \qquad \text{for } n = 1, 2, 3, \ldots.$$

The supremum in the definition of $\mathcal{M}(\triangle, p)$ can be replaced by a limit; so we obtain

$$\mathcal{M}\left(\triangle, p\right) = \lim_{n\to\infty} \left\{3^n(\sqrt{2})^p(1/2)^{np}\right\}$$

$$= \begin{cases} \infty \text{ if } p < \ln(3)/\ln(2), \\ (\sqrt{2})^{\ln(3)/\ln(2)} \text{ if } p = \ln(3)/\ln(2), \\ 0 \text{ if } p > \ln(3)/\ln(2). \end{cases}$$

This tells us that $D_H(\triangle) = \ln(3)/\ln(2)$, which we already know from Theorem 5.4.3. It also tells us that $\mathcal{M}(\triangle, D_H(\triangle)) = (\sqrt{2})^{\ln(3)/\ln(2)}$. This is our estimate of the "size" of the particular Sierpinski triangle under consideration.

If one repeats the above steps for the Sierpinski triangle $\tilde{\triangle}$ with vertices at $(0,0)$, $(0,1/\sqrt{2})$ and $(1/\sqrt{2},0)$ one finds $\mathcal{M}(\tilde{\triangle}, D_H(\tilde{\triangle})) = 1$. Thus $\tilde{\triangle}$ is "smaller" than $\triangle$.

Similar estimates can be made for pairs of attractors of totally disconnected or just-touching IFS whose maps are similitudes, and whose fractal dimensions are equal. The comparison of "sizes" becomes exciting when the two attractors are not metrically equivalent.

4.9. Estimate the "sizes" of the two fractals represented in Figure 5.4.2. Which one is "largest?" Does the computed estimate agree with your subjective feeling about which one is largest?

4.10. Prove that the Hausdorff-Besicovitch dimension of two metrically equivalent bounded subsets of ($\mathbb{R}^m$, Euclidean) is the same.

4.11. Let d denote a metric on $\mathbb{R}^2$ which is equivalent to the Euclidean metric. Let A denote a bounded subset of $\mathbb{R}^2$. Suppose that d is used in place of the Euclidean metric to calculate a "Hausdorff-Besicovitch" dimension of A, denoted by $\tilde{D}_H(A)$. Prove that $D_H(A) = \tilde{D}_H(A)$. Show, however,

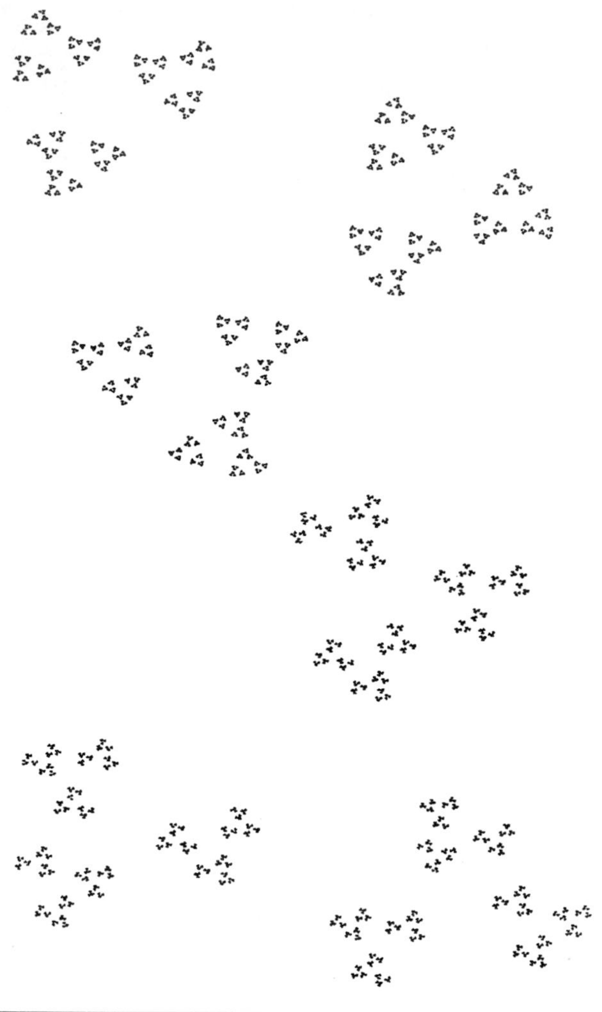

Figure 5.4.2 *The two images here represent the attractors of two different IFS of the form* $\{\mathbb{R}^2; w_1, w_2, w_3\}$ *where all of the maps are similitudes of scaling factor* 0.4. *Both sets have the same fractal dimension* $Ln(3)/Ln(2.5)$. *So which one is the "largest"? Compare their "sizes" by estimating their Hausdorff* $Ln(3)/Ln(2.5)$*-dimensional measures.*

that the "size" of the set, $\mathcal{M}(A, D_H(A))$, may be different when computed using d in place of the Euclidean metric.

4.12. If distance in $\mathbb{R}^2$ is measured in inches, and a subset A of $\mathbb{R}^2$ has fractal dimension 1.391, what are the units of $\mathcal{M}(A, 1.391)$?

4.13. The image in Figure 5.4.3 represents the attractor A of a certain hyperbolic IFS.

(a) Explain, with support from appropriate theorems, why the fractal dimension D and the Hausdorff-Besicovitch dimension D_H of the attractor of the IFS are equal.

(b) Evaluate D.

(c) Using inches as the unit, compare the Hausdorff-Besicovitch D-

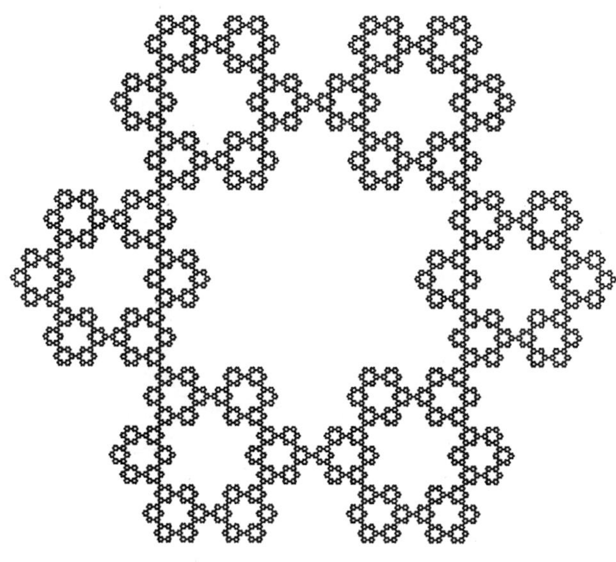

Figure 5.4.3
Why are the fractal dimension and the Hausdorff-Besicovitch dimension of the attractor of the IFS represented by this image equal?

dimensional measures of A and $w(A)$, where $w(A)$ denotes one of the small "first generation" copies of A.

4.14. By any means you like, estimate the Hausdorff-Besicovitch dimension of the coastline of Baron von Koch's Island, shown in Figure 5.4.4. It is recommended that theoreticians try to make an experimental estimate, and that experimentalists try to make a theoretical estimate.

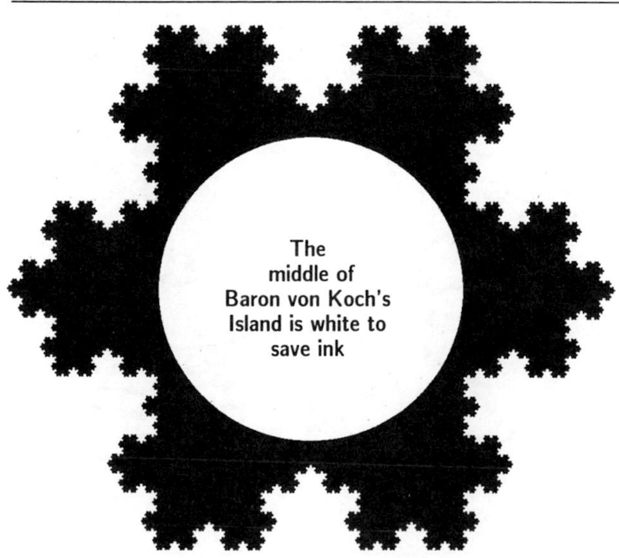

The
middle of
Baron von Koch's
Island is white to
save ink

Figure 5.4.4
By any means you like, estimate the Hausdorff-Besicovitch dimension of the coastline of Baron von Koch's Island.

4.15. Does the work of some artists have a characteristic fractal dimension? Make a comparison of the empirical fractal dimensions of Romeo and Juliet in Figure 5.4.5 over an appropriate range of scales.

Figure 5.4.5
Does the work of some artists have a characteristic fractal dimension? Make a comparison of the empirical fractal dimensions of Romeo and Juliet, over an appropriate range of scales.

Fractal Interpolation

6

6.1 INTRODUCTION: APPLICATIONS FOR FRACTAL FUNCTIONS

Euclidean geometry, trigonometry, and calculus have taught us to think about modelling the shapes we see in the real world in terms of straight lines, circles, parabolas, and other simple curves. Consequences of this way of thinking are abundant in our everyday lives. They include the design of household objects; the common usage of drafting tables, straight-edges, and compasses; and the "applications" which accompany introductory calculus courses. We note in particular the provision of functions for drawing points, lines, polygons, and circles in computer graphics software such as MacPaint and Turbobasic. Most computer graphics hardware is designed specifically to provide rapid computation and display of classical geometrical shapes.

Euclidean geometry and elementary functions, such as the sine, cosine, and polynomials, are the basis of the traditional method for analysing experimental data. Consider an experiment which measures values of a real-valued function $F(x)$ as a function of a real variable x. For example, $F(x)$ may denote a voltage as a function of time, as in the experiments on the jet-engine exhaust described in Section 5.3, example 3.3. The experiment may be a numerical experiment on a computer. In any case the result of the experiment will be a collection of data of the form:

$$\{(x_1, F_i): i = 0, 1, 2, \ldots, N\}.$$

Here N is a positive integer, $F_i = F(x_i)$, and the x_i's are real numbers such that

$$x_0 < x_1 < x_2 < x_3 < \cdots < x_N.$$

The traditional method for analysing this data begins by representing it graphically as a subset of $\mathbb{R}^2$. That is, the data points are plotted on graph paper. Next the graphical data is analysed geometrically. For example, one may seek a straight line segment which is a good approximation to the graph of the data. Or else, one might construct a polynomial of as low degree as possible, whose graph is a good fit to the data over the interval $[x_0, x_N]$. In place of a polynomial, a linear combination of elementary functions might be used. The goal is always the same: to represent the data, viewed as a subset of $\mathbb{R}^2$, by a classical geometrical entity. This entity is represented by a simple formula, one that can be communicated easily to someone else. The process is illustrated in Figure 6.1.1.

Elementary functions, such as trigonometric functions and rational functions, have their roots in Euclidean geometry. They share the feature that when their graphs are "magnified" sufficiently, locally they "look like" straight lines. That is, the tangent line approximation can be used effectively in the vicinity of most points. Moreover, the fractal dimension of the graphs of these functions is always one. These elementary "Euclidean" functions are useful not only because of their geometrical content, but because they can be expressed by simple formulas. We can use them to pass information easily from one person to another. They provide a common language for our scientific work. Moreover, elementary functions are used extensively in scientific computation, computer-aided design, and data analysis because they can be stored in small files and computed by fast algorithms.

Graphics systems founded on traditional geometry are effective for making pictures of man-made objects, such as bricks, wheels, roads, buildings and cogs. This is not surprising, since these objects were in the first place designed using Euclidean geometry. However, it is desirable for graphics systems to be able to deal with a wider range of problems.

In this chapter we introduce fractal interpolation functions. The graphs of these functions can be used to approximate image components such as the profiles of mountain ranges, the tops of clouds, stalactite-hung roofs of caves and horizons over forests, as illustrated in Figure 6.1.2. Rather than treating the image component as arising from a random assemblage of objects, such as individual mountains, cloudlets, stalactites, or tree tops, one models the image

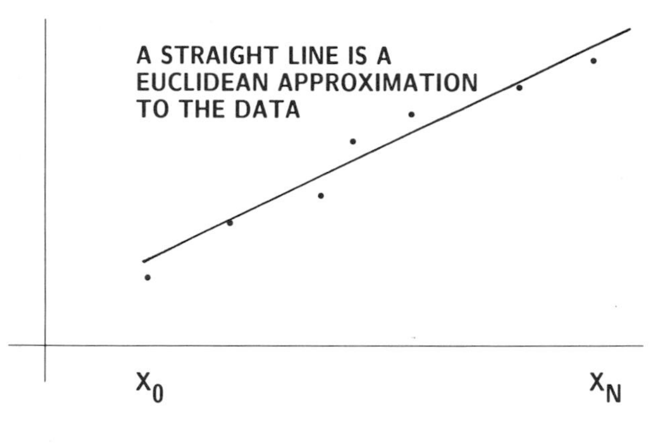

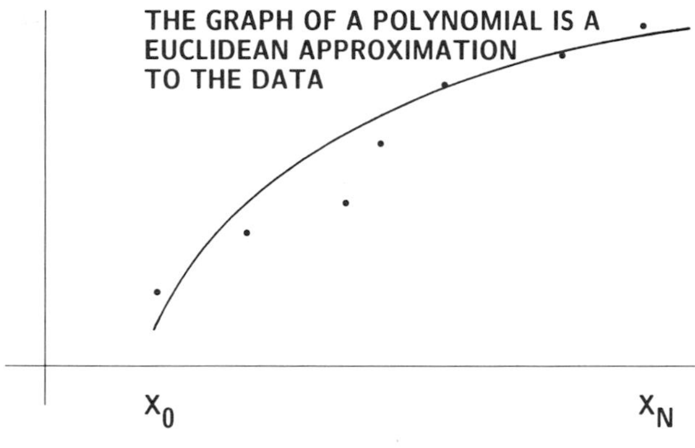

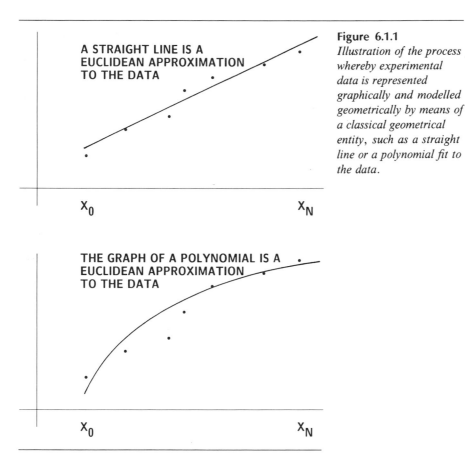

Figure 6.1.1
Illustration of the process whereby experimental data is represented graphically and modelled geometrically by means of a classical geometrical entity, such as a straight line or a polynomial fit to the data.

component as an interrelated single system. Such components are not well described by elementary functions or Euclidean graphics functions.

Fractal interpolation functions also provide a new means for fitting experimental data. Clearly it does not suffice to make a polynomial "least-squares" fit to the wild experimental data of Strahle for the temperature in a jet exhaust as a function of time, as illustrated in Figure 5.3.5. Nor would classical geometry be a good tool for the analysis of voltages at a point in the human brain as read by an electroencephalograph. However, fractal interpolation functions can be used to "fit" such experimental data: that is, the graph of the fractal interpolation function can be made close, in the Hausdorff metric, to the data. Moreover, one can ensure that the fractal dimension of the graph of the fractal interpolation function agrees with that of the data, over an appropriate range of scales. This idea is illustrated in Figure 6.1.3.

Fractal interpolation functions are like elementary functions in that they are of a geometrical character, that they can be represented succinctly by

Figure 6.1.2
The fractal interpolation functions introduced in this chapter may be used in computer graphics software packages to provide a simple means for rendering profiles of mountain ranges, the tops of clouds, and horizons over forests.

"formulas," and that they can be computed rapidly. The main difference is their fractal character. For example, they can have a noninteger fractal dimension. They are easy to work with—once one is accustomed to working with sets rather than points, and with IFS theory using affine maps. If we start to pass them from one to another, fractal functions will become part of the common language of science. So read on!

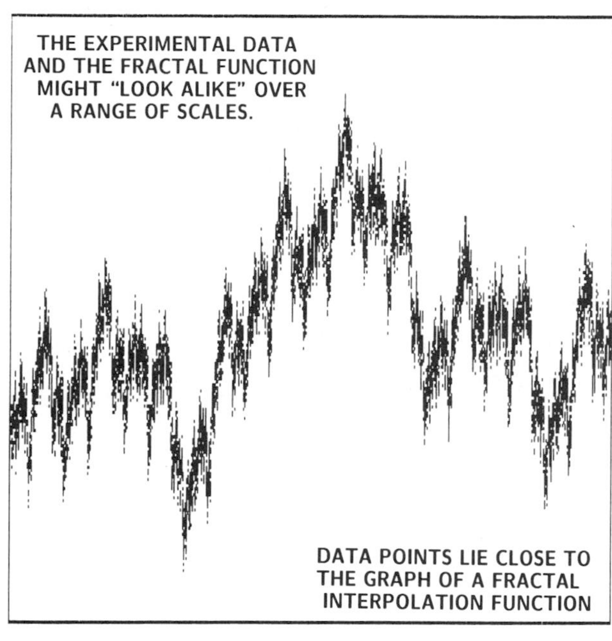

THE EXPERIMENTAL DATA
AND THE FRACTAL FUNCTION
MIGHT "LOOK ALIKE" OVER
A RANGE OF SCALES.
DATA POINTS LIE CLOSE TO
THE GRAPH OF A FRACTAL
INTERPOLATION FUNCTION

Figure 6.1.3
This figure illustrates the idea of using a fractal interpolation function to fit experimental data. The graph of the interpolation function may be close, in the Hausdorff metric, to the graph of the experimental data. The fractal dimension of the interpolation function may agree with that of the data over an appropriate range of scales.

Exercises & Examples

1.1. Write an essay on the influences of Euclidean geometry on the way in which we view the physical world. How does fractal geometry change that view?

1.2. Find the linear approximation $l(x)$ to the function $f(x) = \sin(x)$, about the point $x = 0$. Let $\epsilon > 0$. Find the linear change of coordinates $(x', y') = \theta(x, y)$ in $\mathbb{R}^2$, such that $\theta([0, \epsilon] \times [0, \epsilon]) = [0,1] \times [0,1]$. Let $l'(x')$ denote the function $l(x)$ represented in the new coordinate system. Let $f'(x')$ denote the function $f(x)$ in the new coordinate system. Let L denote the graph of $l'(x')$ for $x' \in [0,1]$ and let G denote the graph of $f'(x')$ for $x' \in [0,1]$. How small must ϵ be chosen to ensure that the Hausdorff distance from L to G is less than 0.01? The Hausdorff distance should be computed with respect to the Manhattan metric in $\mathbb{R}^2$.

6.2 FRACTAL INTERPOLATION FUNCTIONS

Definition 1. A set of *data* is a set of points of the form $\{(x_i, F_i) \in \mathbb{R}^2: i = 0,1,2,\ldots, N\}$, where

$$x_0 < x_1 < x_2 < x_3 < \cdots < x_N.$$

An *interpolation function* corresponding to this set of data is a continuous function $f\colon [x_0, x_N] \to \mathbb{R}$ such that

$$f(x_i) = F_i \qquad \text{for } i = 0, 1, 2, \ldots, N.$$

The points $(x_i, F_i) \in \mathbb{R}^2$ are called the *interpolation points*. We say that the function f *interpolates* the data; and that (the graph of) f *passes through* the interpolation points.

Exercises & Examples

2.1. The function $f(x) = 1 + x$ is an interpolation function for the set of data $\{(0,1),(1,2)\}$. Consider the hyperbolic IFS $\{\mathbb{R}^2; w_1, w_2\}$, where

$$w_1\begin{pmatrix} x \\ y \end{pmatrix} = \begin{pmatrix} 0.5 & 0 \\ 0 & 0.5 \end{pmatrix}\begin{pmatrix} x \\ y \end{pmatrix} + \begin{pmatrix} 0 \\ 0.5 \end{pmatrix}; \quad \text{and}$$

$$w_2\begin{pmatrix} x \\ y \end{pmatrix} = \begin{pmatrix} 0.5 & 0 \\ 0 & 0.5 \end{pmatrix}\begin{pmatrix} x \\ y \end{pmatrix} + \begin{pmatrix} 0.5 \\ 1 \end{pmatrix}.$$

Let G denote the attractor of the IFS. Then it is readily verified that G is the straight line segment which connects the pair of points $(0,1)$ and $(1,2)$. In other words, G is the graph of the interpolation function $f(x)$ over the interval $[0,1]$.

2.2. Let $\{(x_i, F_i)\colon i = 0, 1, 2, \ldots, N\}$ denote a set of data. Let $f\colon [x_0, x_N] \to \mathbb{R}$ denote the unique continuous function which passes through the interpolation points and which is linear on each of the subintervals $[x_{i-1}, x_i]$. That is

$$f(x) = F_{i-1} + \frac{(x - x_{i-1})}{(x_i - x_{i-1})}(F_i - F_{i-1}) \qquad \text{for } x \in [x_{i-1}, x_i],\ i = 1, 2, \ldots, N.$$

The function $f(x)$ is called a piecewise linear interpolation function. The graph of $f(x)$ is illustrated in Figure 6.2.1. This graph, G, is also the

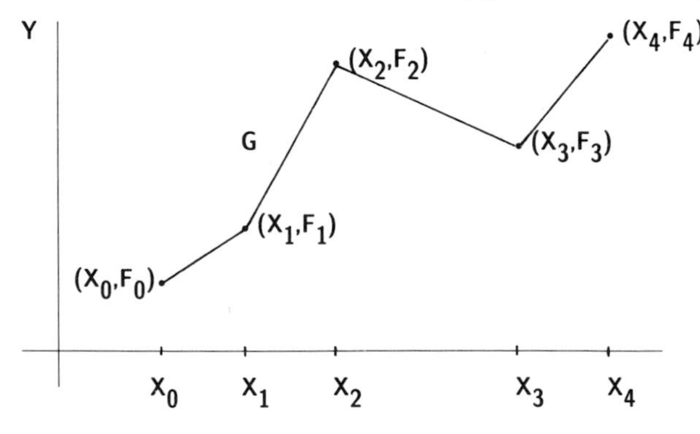

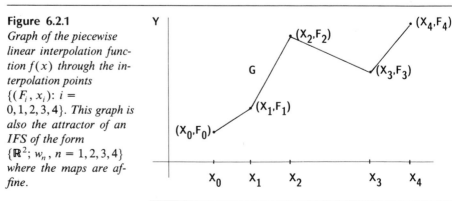

Figure 6.2.1 *Graph of the piecewise linear interpolation function $f(x)$ through the interpolation points $\{(F_i, x_i)\colon i = 0, 1, 2, 3, 4\}$. This graph is also the attractor of an IFS of the form $\{\mathbb{R}^2; w_n, n = 1, 2, 3, 4\}$ where the maps are affine.*

attractor of an IFS of the form $\{\mathbb{R}^2; w_n, n = 1, 2, \ldots, N\}$ where the maps are affine. In fact

$$w_n\begin{pmatrix} x \\ y \end{pmatrix} = \begin{pmatrix} a_n & 0 \\ c_n & 0 \end{pmatrix}\begin{pmatrix} x \\ y \end{pmatrix} + \begin{pmatrix} e_n \\ f_n \end{pmatrix},$$

where

$$a_n = \frac{(x_n - x_{n-1})}{(x_N - x_0)}, \quad e_n = \frac{(x_N x_{n-1} - x_0 x_n)}{(x_N - x_0)},$$

$$c_n = \frac{(F_n - F_{n-1})}{(x_N - x_0)}, \quad f_n = \frac{(x_N F_{n-1} - x_0 F_n)}{(x_N - x_0)}, \qquad \text{for } n = 1, 2, \ldots, N.$$

Notice that the IFS may not be hyperbolic with respect to the Euclidean metric in $\mathbb{R}^2$. Can you prove that, nonetheless, G is the unique non-empty compact subset of $\mathbb{R}^2$ such that

$$G = \bigcup_{n=1}^{N} w_n(G)?$$

2.3. Verify the claims in example 6.2.2 in the case of the data set $\{(0,0), (1,3), (2,0)\}$ by applying either the Deterministic Algorithm, Program 3.8.1, or the Random Iteration Algorithm, Program 3.8.2. You will need to modify the programs slightly.

2.4. The parabola defined by $f(x) = 2x - x^2$ on the interval $[0, 2]$ is an interpolation function for the set of data $\{(0,0), (1,1), (2,0)\}$. Let G denote the graph of $f(x)$. That is

$$G = \{(x, 2x - x^2): x \in [0, 2]\}.$$

Then we claim that G is the attractor of the hyperbolic IFS $\{\mathbb{R}^2; w_1, w_2\}$, where

$$w_1\begin{pmatrix} x \\ y \end{pmatrix} = \begin{pmatrix} 0.5 & 0 \\ 0.5 & 0.25 \end{pmatrix}\begin{pmatrix} x \\ y \end{pmatrix}; \quad \text{and} \quad w_2\begin{pmatrix} x \\ y \end{pmatrix} = \begin{pmatrix} 0.5 & 0 \\ -0.5 & 0.25 \end{pmatrix}\begin{pmatrix} x \\ y \end{pmatrix} + \begin{pmatrix} 1 \\ 1 \end{pmatrix}.$$

We verify this claim directly. We simply note that for all $x \in [0, 2]$

$$w_1\begin{pmatrix} x \\ f(x) \end{pmatrix} = \begin{pmatrix} \frac{1}{2}x \\ 2(\frac{1}{2}x) - (\frac{1}{2}x)^2 \end{pmatrix} = \begin{pmatrix} \frac{1}{2}x \\ f(\frac{1}{2}x) \end{pmatrix},$$

$$w_2\begin{pmatrix} x \\ f(x) \end{pmatrix} = \begin{pmatrix} 1 + \frac{1}{2}x \\ 2(1 + \frac{1}{2}x) - (1 + \frac{1}{2}x)^2 \end{pmatrix} = \begin{pmatrix} 1 + \frac{1}{2}x \\ f(1 + \frac{1}{2}x) \end{pmatrix}.$$

As x varies over $[0, 2]$, the right-hand side of the first equation yields the part of the graph of $f(x)$ lying over the interval $[0, 1]$, while the right-hand side of the second equation yields the part of the graph of $f(x)$ lying over the interval $[1, 2]$. Hence $G = w_1(G) \cup w_2(G)$. Since $G \in \mathscr{H}(\mathbb{R}^2)$, we conclude that it is the attractor of the IFS. Notice that the IFS is just-touching.

2.5. Find a hyperbolic IFS of the form $\{\mathbb{R}^2; w_1, w_2\}$, where w_1 and w_2 are affine transformations in $\mathbb{R}^2$, whose attractor is the graph of the quadratic function which interpolates the data $\{(0,0),(1,1),(2,4)\}$.

Let a set of data $\{(x_i, F_i)\colon i = 0, 1, 2, \ldots, N\}$ be given. We explain how one can construct an IFS in $\mathbb{R}^2$ such that its attractor, which we denote by G, is the graph of a continuous function $f\colon [x_0, x_N] \to \mathbb{R}$ which interpolates the data. Throughout we will restrict our attention to affine transformations. The usage of more general transformations is discussed in [Barnsley 1986a], [Barnsley 1986c], [Hardin 1985], [Massopust 1986].

We consider an IFS of the form $\{\mathbb{R}^2; w_n, n = 1, 2, \ldots, N\}$ where the maps are affine transformations of the special structure

$$w_n\begin{pmatrix} x \\ y \end{pmatrix} = \begin{pmatrix} a_n & 0 \\ c_n & d_n \end{pmatrix}\begin{pmatrix} x \\ y \end{pmatrix} + \begin{pmatrix} e_n \\ f_n \end{pmatrix}$$

The transformations are constrained by the data according to

$$w_n\begin{pmatrix} x_0 \\ F_0 \end{pmatrix} = \begin{pmatrix} x_{n-1} \\ F_{n-1} \end{pmatrix} \quad \text{and} \quad w_n\begin{pmatrix} x_N \\ F_N \end{pmatrix} = \begin{pmatrix} x_n \\ F_n \end{pmatrix} \qquad \text{for } n = 1, 2, \ldots, N.$$

The situation is summarized in Figure 6.2.2.

Let $n \in \{1, 2, 3, \ldots, N\}$. The transformation w_n is specified by the five real numbers a_n, c_n, d_n, e_n, and f_n, which must obey the four linear equations

$$\begin{aligned}
&a_n x_0 + e_n = x_{n-1}\\
&a_n x_N + e_n = x_n\\
&c_n x_0 + d_n F_0 + f_n = F_{n-1}\\
&c_n x_N + d_n F_N + f_n = F_n
\end{aligned}$$

It follows that there is effectively one free parameter in each transformation. We choose this parameter to be d_n for the following reason: The transformation w_n is a *shear* transformation: it maps lines parallel to the y-axis into lines parallel to the y-axis. Let L denote a line segment parallel to the y-axis. Then $w_n(L)$ is also a line segment parallel to the y-axis. The ratio of the length of $w_n(L)$ to the length of L is $|d_n|$. We call d_n the *vertical scaling factor* in the transformation w_n. By choosing d_n to be the free parameter, we are able to specify the vertical scaling produced by the transformation. With $d_n = 0$, $n = 1, 2, \ldots, N$, one recovers the piecewise-linear interpolation function. In this section we will show that these parameters determine the fractal dimension of the attractor of the IFS.

Let d_n be any real number. We demonstrate that we can always solve the above equations for a_n, c_n, e_n, and f_n in terms of the data and d_n. We find

(6.2.1) $a_n = (x_n - x_{n-1})/(x_N - x_0)$,
(6.2.2) $e_n = (x_N x_{n-1} - x_0 x_n)/(x_N - x_0)$,
(6.2.3) $c_n = (F_n - F_{n-1})/(x_N - x_0) - d_n(F_N - F_0)/(x_N - x_0)$,
(6.2.4) $f_n = (x_N F_{n-1} - x_0 F_n)/(x_N - x_0) - d_n(x_N F_0 - x_0 F_N)/(x_N - x_0)$.

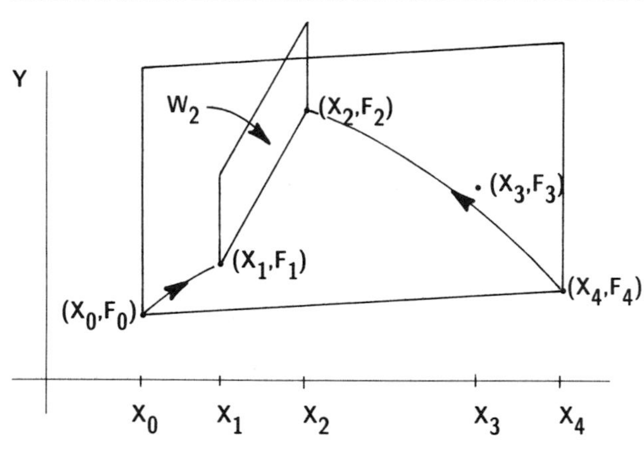

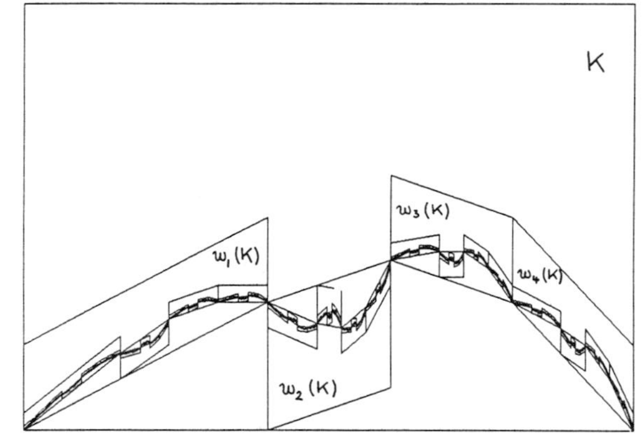

Figure 6.2.2 *Two illustrations showing how an IFS of shear transformations is used to construct a fractal interpolation function. (Produced by Peter Massopust.)*

Now let $\{\mathbb{R}^2; w_n, n = 1, 2, \ldots, N\}$ denote the IFS defined above. Let the vertical scaling factor d_n obey $0 \leq d_n < 1$ for $n = 1, 2, \ldots, N$. Even with this condition, the IFS is not in general hyperbolic on the metric space ($\mathbb{R}^2$, Euclidean). Despite this, let us see what happens if we apply the Random Iteration Algorithm to the IFS.

Here we present Program 3.8.2 modified so that the input data consists of the interpolation points and vertical scaling factors. It is written for $N = 3$ and the data set

$$\{(0,0),(30,50),(60,40),(100,10)\}.$$

The vertical scaling factors are input by the user during execution of the code. The program calculates the coefficients in the shear transformations from the data, and then applies the Random Iteration Algorithm to the resulting IFS. The program is written in BASIC. It runs without modification on an IBM PC with enhanced graphics adaptor and Turbobasic. On any line the words preceded by a ' are comments: they are not part of the program.

PROGRAM 6.2.1

```
x[0] = 0: x[1] = 30: x[2] = 60: x[3] = 100   'Data set
F[0] = 0: F[1] = 50: F[2] = 40: F[3] = 10
input "enter scaling factors d(1), d(2), and d(3)", d(1), d(2), d(3)
   'Vertical Scaling Factors
for n = 1 to 3   'Calculate the shear transformations from the Data and
   Vertical Scaling Factors
      b = x[3] - x[0]: a[n] = (x[n] - x[n - 1])/b: e[n] = (x[3]*x[n - 1]
   - x[0]*x[n])/b
      c[n] = (F[n] - F[n - 1] - d[n]*(F[3] - F[0]))/b
      ff[n] = (x[3]*F[n - 1] - x[0]*F[n] - d[n]*(x[3]*F[0] - x[0]*F[3]))/b
next
screen 2: cls   'initialize graphics
window (0,0) - (100,100)   'change this to zoom and/or pan
x = 0: y = 0   'initial point from which the random iteration begins
for n = 1 to 1000   'Random Iteration Algorithm
      k = int(3*rnd - 0.0001) + 1
      newx = a[k]*x + e[k]
      newy = c[k]*x + d[k]*y + ff[k]
      x = newx: y = newy
      pset(x, y)   'plot the most recently computed point on the screen
next
end
```

The result of running an adaptation of this program on a Masscomp workstation and then printing the contents of the graphics screen, is presented in Figure 6.2.3. In this case $d_1 = 0.5$, $d_2 = -0.5$, and $d_3 = 0.23$. Notice that if the size of the plotting window is decreased, for example by replacing the window call by WINDOW $(0,0) - (50,50)$, then a portion of the image is plotted at a higher resolution. The number of iterations can be increased to improve the quality of the computed image.

Exercises & Examples

2.6. Rewrite Program 6.2.1 in a form suitable for your own computer environment, then run it and obtain hardcopy of the output.

2.7. Vary the data used by Program 6.2.1. Verify, by means of computer-graphical experiments, that the corresponding IFS always seems to have a unique attractor, provided that the vertical scaling factors are less than one in norm. Verify that, provided sufficiently many points are plotted, the attractor always contains the data points, and that it looks like the graph of a function.

2.8. Show that the shear transformations w_n, described above, need not be

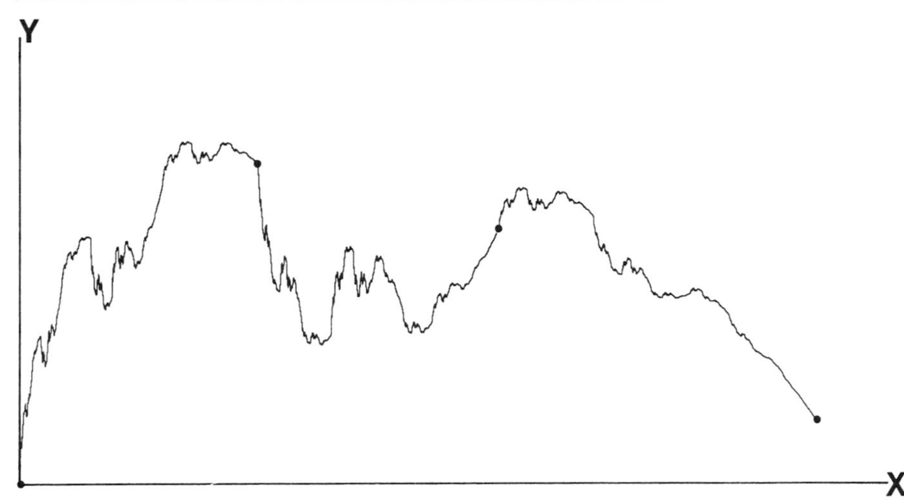

Figure 6.2.3 *The result of running program 6.2.1 with vertical scaling factors* 0.5, −0.5, *and* 0.23. *It appears that the corresponding IFS possesses a unique attractor which is the graph of a function which passes through the interpolation points* $\{(0,0),(30,50),(60,40),(100,10)\}$. *Is there a metric such that the IFS is hyperbolic?*

contractions in the Euclidean metric, even though the magnitudes of the vertical scaling factors are less than one. Once you have found such an example, use Program 6.2.1, suitably modified, to obtain graphical evidence concerning the possible existence of an attractor. You are supposed to discover that even though the IFS is not hyperbolic in the Euclidean metric, it appears to possess an attractor.

2.9. Use Program 6.2.1 to verify that the attractor of the IFS in example 6.2.4 is a parabola.

We now give the theoretical basis for our experimental observations.

Theorem 1. *Let N be a positive integer greater than one. Let* $\{\mathbb{R}^2; w_n, n = 1,2,\ldots,N\}$ *denote the IFS defined above, associated with the data set* $\{(x_n, F_n): n = 1,2,\ldots,N\}$. *Let the vertical scaling factor* d_n *obey* $0 \le d_n < 1$ *for* $n = 1,2,\ldots,N$. *Then there is a metric d on* $\mathbb{R}^2$, *equivalent to the Euclidean metric, such that the IFS is hyperbolic with respect to d. In particular, there is a unique nonempty compact set* $G \subset \mathbb{R}^2$ *such that*

$$G = \bigcup_{n=1}^{N} w_n(G).$$

Proof. We define a metric d on $\mathbb{R}^2$ by

$$d((x_1, y_1),(x_2, y_2)) = |x_1 - x_2| + \theta|y_1 - y_2|$$

where θ is a positive real number which we specify below. We leave it as an exercise to the reader to prove that this metric is equivalent to the Euclidean metric on $\mathbb{R}^2$. Let $n \in \{1, 2, \ldots, N\}$. Let the numbers a_n, c_n, e_n, f_n, be defined by equations (6.2.1), (6.2.2), (6.2.3), and (6.2.4). Then we have

$$\begin{aligned} d(w_n(x_1, y_1), w_n(x_2, y_2)) &= d((a_n x_1 + e_n, c_n x_1 + d_n y_1 + f_n), \\ &\qquad (a_n x_2 + e_n, c_n x_2 + d_n y_2 + f_n)) \\ &= a_n|x_1 - x_2| + \theta|c_n(x_1 - x_2) + d_n(y_1 - y_2)| \\ &\le (|a_n| + \theta|c_n|)|x_1 - x_2| + \theta|d_n||y_1 - y_2|. \end{aligned}$$

Now notice that $|a_n| = |x_n - x_{n-1}|/|x_N - x_0| < 1$ because $N \ge 2$. If $c_1 = c_2 \cdots = c_n = 0$ then we choose $\theta = 1$. Otherwise we choose

$$\theta = \frac{\text{Min}\{1 - |a_n|: n = 1, 2, \ldots, N\}}{\text{Max}\{2|c_n|: n = 1, 2, \ldots, N\}}.$$

Then it follows that

$$\begin{aligned} d(w_n(x_1, y_1), w_n(x_2, y_2)) &\le (|a_n| + \theta|c_n|)|x_1 - x_2| + \theta|d_n||y_1 - y_2| \\ &\le a|x_1 - x_2| + \theta\delta|y_1 - y_2| \\ &\le \text{Max}\{a, \delta\} d((x_1, y_1), (x_2, y_2)), \end{aligned}$$

where

$$a = \left(1/2 + \frac{\text{Max}\{|a_n|: n = 1, 2, \ldots, N\}}{2}\right) < 1 \text{ and}$$

$$\delta = \text{Max}\{|d_n|: n = 1, 2, \ldots, N\} < 1.$$

This completes the proof.

Theorem 2. *Let N be a positive integer greater than one. Let $\{\mathbb{R}^2; w_n, n = 1, 2, \ldots, N\}$ denote the IFS defined above, associated with the data set $\{(x_n, F_n): n = 1, 2, \ldots, N\}$. Let the vertical scaling factor d_n obey $0 \le d_n < 1$ for $n = 1, 2, \ldots, N$, so that the IFS is hyperbolic. Let G denote the attractor of the IFS. Then G is the graph of a continuous function $f: [x_0, x_N] \to \mathbb{R}$ which interpolates the data $\{(x_i, F_i): i = 1, 2, \ldots, N\}$. That is*

$$G = \{(x, f(x)): x \in [x_0, x_N]\},$$

where

$$f(x_i) = F_i \qquad \text{for } i = 0, 1, 2, 3, \ldots, N.$$

Proof. Let $\mathscr{F}$ denote the set of continuous functions $f: [x_0, x_1] \to \mathbb{R}$ such that $f(x_0) = F_0$ and $f(x_N) = F_N$. We define a metric d on $\mathscr{F}$ by

$$d(f, g) = \text{Max}\{|f(x) - g(x)|: x \in [x_0, x_N]\} \qquad \text{for all } f, g \text{ in } \mathscr{F}.$$

Then $(\mathscr{F}, d)$ is a complete metric space; see for example Rudin [1966], or prove it yourself.

Let the real numbers a_n, c_n, e_n, f_n, be defined by equations (6.2.1), (6.2.2),

(6.2.3), and (6.2.4). Define a mapping T: $\mathscr{F} \to \mathscr{F}$ by

$$(Tf)(x) = c_n l_n^{-1}(x) + d_n f(l_n^{-1}(x)) + f_n \text{ for } x \in [x_{n-1}, x_n],$$
$$\text{for } n = 1, 2, \ldots, N,$$

where l_n: $[x_0, x_N] \to [x_{n-1}, x_n]$ is the invertible transformation

$$l_n(x) = a_n x + e_n.$$

We verify that T does indeed take $\mathscr{F}$ into itself. Let $f \in \mathscr{F}$. Then the function $(Tf)(x)$ obeys the endpoint conditions because

$$\begin{aligned}(Tf)(x_0) &= c_1 l_1^{-1}(x_0) + d_1 f(l_1^{-1}(x_0)) + f_1 = c_1 x_0 + d_1 f(x_0) + f_1 \\ &= c_1 x_0 + d_1 F_0 + f_1 = F_0\end{aligned}$$

and

$$\begin{aligned}(Tf)(x_N) &= c_N l_N^{-1}(x_N) + d_N f(l_N^{-1}(x_N)) + f_N = c_N x_N + d_N f(x_N) + f_N \\ &= c_N x_N + d_N F_N + f_N = F_N.\end{aligned}$$

The reader can prove that $(Tf)(x)$ is continuous on the interval $[x_{n-1}, x_n]$ for $n = 1, 2, \ldots, N$. Then it remains to demonstrate that $(Tf)(x)$ is continuous at each of the points $x_1, x_2, x_3, \ldots, x_{N-1}$. At each of these points the value of $(Tf)(x)$ is apparently defined in two different ways. For $n \in \{1, 2, \ldots, N-1\}$ we have

$$\begin{aligned}(Tf)(x_n) &= c_{n+1} l_{n+1}^{-1}(x_n) + d_{n+1} f(l_{n+1}^{-1}(x_n)) + f_{n+1} \\ &= c_{n+1} x_0 + d_{n+1} f(x_0) + f_{n+1} = F_n,\end{aligned}$$

and also

$$(Tf)(x_n) = c_n l_n^{-1}(x_n) + d_n f(l_n^{-1}(x_n)) + f_n = c_n x_N + d_n f(x_N) + f_n = F_n$$

so both methods of evaluation lead to the same result. We conclude that T does indeed take $\mathscr{F}$ into $\mathscr{F}$.

Now we show that T is a contraction mapping on the metric space $(\mathscr{F}, d)$. Let $f, g \in \mathscr{F}$. Let $n \in \{1, 2, \ldots, N\}$ and $x \in [x_{n-1}, x_n]$. Then

$$|(Tf)(x) - (Tg)(x)| = |d_n| |f(l_n^{-1}(x)) - g(l_n^{-1}(x))| \le |d_n| d(f, g).$$

It follows that

$$d(Tf, Tg) \le \delta\, d(f, g) \qquad \text{where } \delta = \text{Max}\{|d_n|: n = 1, 2, \ldots, N\} < 1.$$

We conclude that T: $\mathscr{F} \to \mathscr{F}$ is a contraction mapping. The Contraction Mapping Theorem implies that T possesses a unique fixed point in $\mathscr{F}$. That is, there exists a function $f \in \mathscr{F}$ such that

$$(Tf)(x) = f(x) \qquad \text{for all } x \in [x_0, x_N].$$

The reader should convince himself that f passes through the interpolation points.

Let $\tilde{G}$ denote the graph of f. Notice that the equations which define T can be rewritten

$$(Tf)(a_n x + e_n) = c_n x + d_n f(x) + f_n \quad \text{for } x \in [x_0, x_N], \quad \text{for } n = 1, 2, \ldots, N,$$

which implies that

$$\tilde{G} = \bigcup_{n=1}^{N} w_n(\tilde{G}).$$

But $\tilde{G}$ is a nonempty compact subset of $\mathbb{R}^2$. By Theorem 6.2.1 there is only one nonempty compact set G, the attractor of the IFS, which obeys the latter equation. It follows that $G = \tilde{G}$. This completes the proof.

Definition 1. The function $f(x)$ whose graph is the attractor of an IFS as described in Theorems 6.2.1 and 6.2.2 above, is called a *fractal interpolation function* corresponding to the data $\{(x_i, F_i): i = 1, 2, \ldots, N\}$.

Figure 6.2.4 shows an example of a sequence of iterates $\{T^{\circ n} f_0 : n = 0, 1, 2, 3, \ldots\}$ obtained by repeated application of the contraction mapping T, introduced in the proof of Theorem 6.2.2. The initial function $f_0(x)$ is linear. The sequence converges to the fractal interpolation function f which is the fixed point of T. Notice that the whole image can be interpreted as the attractor of an IFS with condensation, where the condensation set is the graph of the function $f_0(x)$.

The reader may wonder, in view of the proof of Theorem 6.2.2, why we go to the trouble of establishing that there is a metric such that the IFS is contractive. After all, we could simply use T to construct fractal interpolation functions. The answer has two parts, (a) and (b). (a) We can now apply the

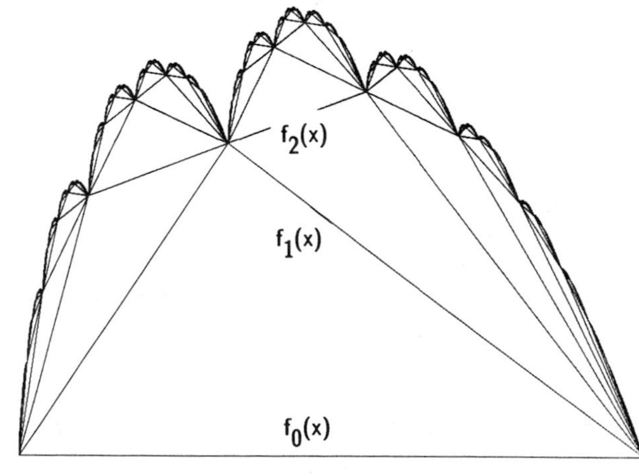

Figure 6.2.4
A sequence of functions $\{f_{n+1}(x) = (Tf_n)(x)\}$ *converging to the fixed point of the mapping* T: $\mathscr{F} \to \mathscr{F}$ *used in the proof of Theorem* 6.2.4. *This is another example of a contraction mapping doing its work.*

theory of hyperbolic IFS to fractal interpolation functions. Of especial importance, this means that we can use IFS algorithms to compute fractal interpolation functions, that the Collage Theorem can be used as an aid to finding fractal interpolation functions which approximate given data, and that we can use the Hausdorff metric to discuss the accuracy of approximation of experimental data by a fractal interpolation function. (b) By treating fractal interpolation functions as attractors of IFS of affine transformations we provide a common language for the description of an important class of functions and sets: the same type of formula, namely an IFS code, can be used in all cases.

One consequence of the fact that the IFS $\{\mathbb{R}; w_n, n = 1, 2, \ldots, N\}$ associated with a set of data $\{(x_n, F_n): n = 1, 2, \ldots, N\}$ is hyperbolic is that any set $A_0 \in \mathscr{H}(\mathbb{R}^2)$ leads to a Cauchy sequence of sets $\{A_n\}$ which converges to G in the Hausdorff metric. In the usual way we define $W: \mathscr{H}(\mathbb{R}^2) \to \mathscr{H}(\mathbb{R}^2)$ by

$$W(B) = \bigcup_{n=1}^{N} w_n(B) \qquad \text{for all } B \in \mathscr{H}(\mathbb{R}^2).$$

Then $\{A_n = W^{\circ n}(A_0)\}$ is a Cauchy sequence of sets which converges to G in the Hausdorff metric. This idea is illustrated in Figures 6.2.5(a) and (b). Notice that if A_0 is the graph of a function $f_0 \in \mathscr{F}$ then A_n is the graph of $T^{\circ n} f_0$.

Exercises & Examples

2.10. Prove that the metric on $\mathbb{R}^2$ introduced in the proof of Theorem 6.2.1 is equivalent to the Euclidean metric on $\mathbb{R}^2$.

2.11. Use the Collage Theorem to help you find a fractal interpolation function which approximates the function whose graph is shown in Figure 6.2.6.

2.12. Write a program which allows you to use the Deterministic Algorithm to compute fractal interpolation functions.

2.13. Explain why Theorems 6.2.1 and 6.2.2 have the restriction that N is greater than one.

2.14. Let a set of data $\{(x_i, F_i): i = 0, 1, 2, \ldots, N\}$ be given. Let the metric space $(\mathscr{F}, d)$ and the transformation $T: \mathscr{F} \to \mathscr{F}$ be defined as in the proof of Theorem 6.2.2. Prove that if $f \in \mathscr{F}$ then Tf is an interpolation function associated with the data. Deduce that if $f \in \mathscr{F}$ is a fixed point of T, then f is an interpolation function associated with the data.

2.15. Make a nonlinear generalization of the theory of fractal interpolation functions. For example, consider what happens if one uses an IFS made up of nonlinear transformations $w_n: \mathbb{R}^2 \to \mathbb{R}^2$ of the form

$$w_n(x, y) = \left(a_n x + e_n, c_n x + d_n y + g_n y^2 + f_n\right)$$

where a_n, e_n, c_n, d_n, g_n, and f_n are real constants. This example uses "quadratic scaling" in the vertical direction instead of linear scaling.

Figure 6.2.5(a) and (b). *Examples of the convergence of a sequence of sets* $\{A_n\}$ *in the Hausdorff metric, to the graph of a fractal interpolation function.*

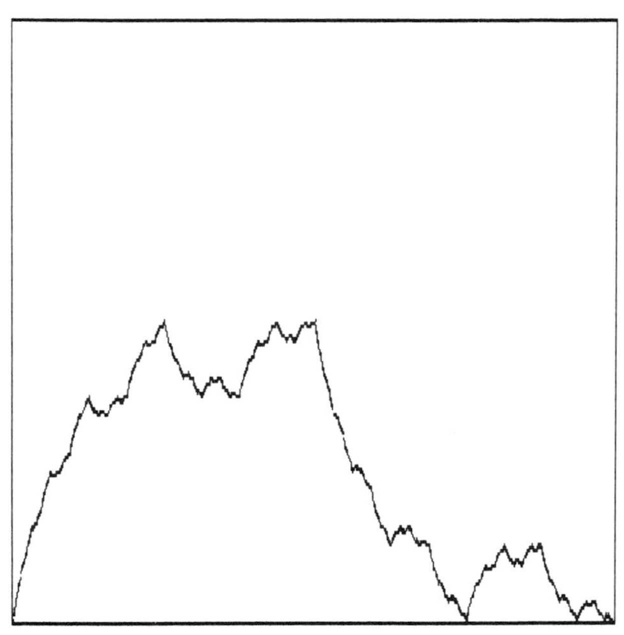

Figure 6.2.6 *Use the Collage Theorem to find an IFS* $\{\mathbb{R}^2; w_1, w_2\}$, *where* w_1 *and* w_2 *are shear transformations on* $\mathbb{R}^2$, *such that the attractor of the IFS is a good approximation to the graph of the function shown here.*

Determine sufficient conditions for the IFS to be hyperbolic, with an attractor which is the graph of a function which interpolates the data $\{(x_i, F_i): i = 0, 1, 2, \ldots, N\}$. Note that in certain circumstances the IFS generates the graph of a differentiable interpolation function.

2.16. Let $f(x)$ denote a fractal interpolation function associated with a set of data $\{(x_i, F_i): i = 0, 1, 2, \ldots, N\}$, where $N > 1$. Let the metric space $(\mathscr{F}, d)$ and the transformation $T: \mathscr{F} \to \mathscr{F}$ be defined as in the proof of Theorem 6.2.2. The functional equation $Tf = f$ can be used to evaluate various integrals of f. As an example, consider the problem of evaluating the integral

$$I = \int_{x_0}^{x_N} f(x)\, dx.$$

The integral is well-defined because $f(x)$ is continuous. We have

$$I = \int_{x_0}^{x_N} (Tf)(x)\, dx = \sum_{n=1}^{N} \int_{x_{n-1}}^{x_n} (Tf)(x)\, dx$$

$$= \sum_{n=1}^{N} \int_{x_0}^{x_N} (c_n x + d_n f(x) + f_n)\, d(a_n x + e_n) = \alpha I + \beta$$

where

$$\alpha = \left(\sum_{n=1}^{N} a_n d_n \right) \quad \text{and} \quad \beta = \sum_{n=1}^{N} a_n \int_{x_0}^{x_N} (c_n x + f_n)\, dx.$$

Show that, under the standard assumptions, $|\alpha| < 1$. Show also that

$$\beta = \int_{x_0}^{x_N} f_0(x)\,dx,$$

where $f_0(x)$ is the piecewise linear interpolation function associated with the data. Conclude that

$$\int_{x_0}^{x_N} f(x)\,dx = \frac{\beta}{(1-\alpha)}.$$

Check this result for the case of the parabola, described in exercise 6.2.4. In Figure 6.2.7 we illustrate a geometrical way of thinking about the integration of a fractal interpolation function.

2.17. Let $f(x)$ denote a fractal interpolation function associated with a set of data $\{(x_i, F_i)\colon i = 0, 1, 2, \ldots, N\}$, where $N > 1$. By following similar steps to those in example 2.16, find a formula for the integral

$$I_1 = \int_{x_0}^{x_N} xf(x)\,dx.$$

Check your formula by applying it to the parabola which is described in exercise 6.2.4.

2.18. Figure 6.2.8 shows a fractal interpolation function together with a zoom.

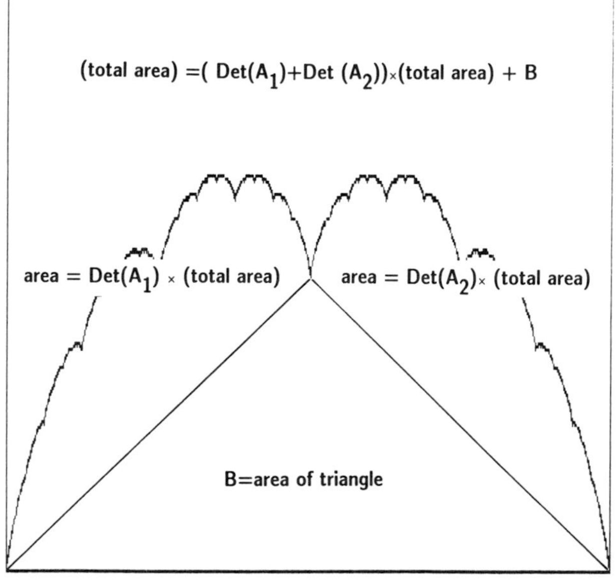

(total area) =(Det(A1)+Det (A2))×(total area) + B
area = Det(A1) × (total area)
area = Det(A2)× (total area)
B=area of triangle

Figure 6.2.7 *Illustration of the geometrical viewpoint concerning the integration of fractal interpolation functions.*

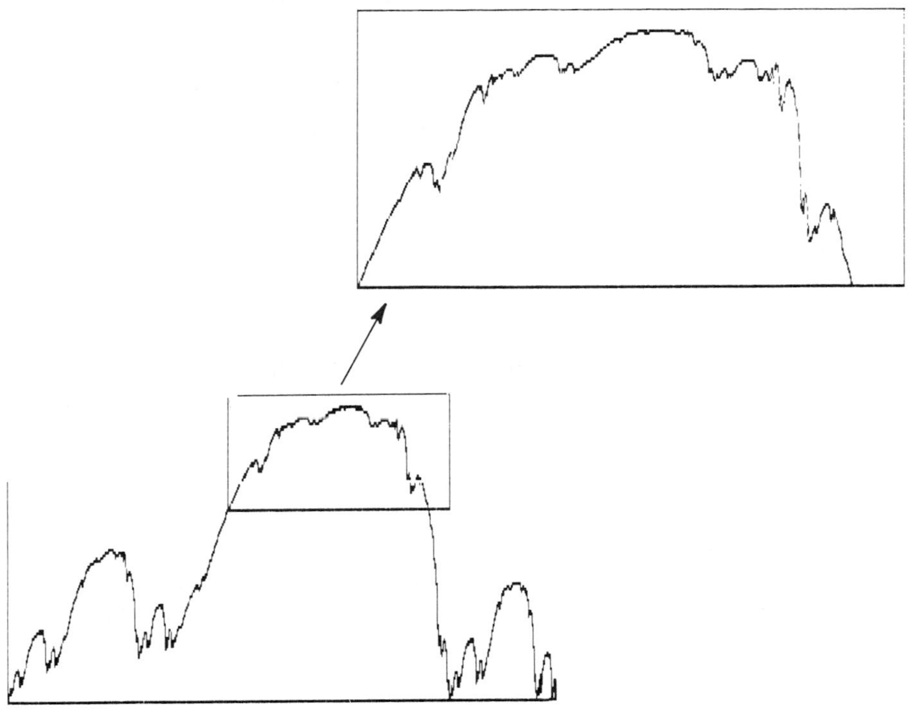

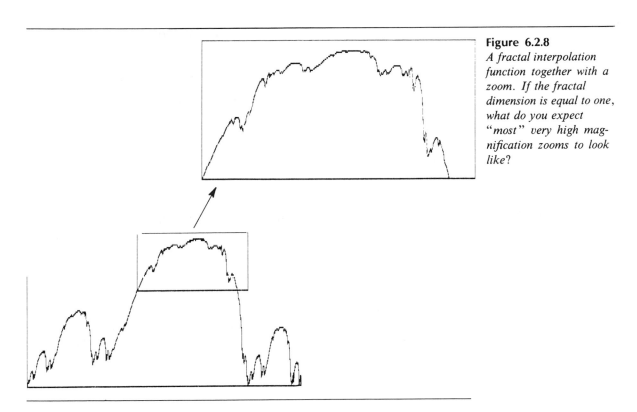

Figure 6.2.8 *A fractal interpolation function together with a zoom. If the fractal dimension is equal to one, what do you expect "most" very high magnification zooms to look like?*

Can you reproduce these images, and then make a further zoom? What do you expect a very high magnification zoom to look like?

6.3 THE FRACTAL DIMENSION OF FRACTAL INTERPOLATION FUNCTIONS

The following excellent theorem tells us the fractal dimension of fractal interpolation functions.

Theorem 1. *Let N be a positive integer greater than one. Let $\{(x_n, F_n) \in \mathbb{R}^2: n = 1, 2, \ldots, N\}$ be a set of data. Let $\{\mathbb{R}^2; w_n, n = 1, 2, \ldots, N\}$ be an IFS associated with the data, where*

$$w_n\begin{pmatrix} x \\ y \end{pmatrix} = \begin{pmatrix} a_n & 0 \\ c_n & d_n \end{pmatrix}\begin{pmatrix} x \\ y \end{pmatrix} + \begin{pmatrix} e_n \\ f_n \end{pmatrix} \qquad \text{for } n = 1, 2, \ldots, N.$$

The vertical scaling factors d_n obey $0 \leq d_n < 1$; and the constants a_n, c_n, e_n, and f_n, are given by Equations (6.2.1), (6.2.2), (6.2.3) and (6.2.4), for $n = 1, 2, \ldots, N$. Let G denote the attractor of the IFS, so that G is the graph of a *fractal interpolation function associated with the data. If*

$$\sum_{n=1}^{N} |d_n| > 1, \tag{6.3.1}$$

and the interpolation points do not all lie on a single straight line, then the fractal dimension of G is the unique real solution D of

$$\sum_{n=1}^{N} |d_n| a_n^{D-1} = 1.$$

Otherwise the fractal dimension of G is one.

Informal Demonstration: The formal proof of this theorem can be found in [Barnsley 1986c]. Here we give an informal argument for why it is true. We use the notation in the statement of the theorem.

Let $\epsilon > 0$. We consider G to be superimposed on a grid of closed square boxes of side length ϵ, as illustrated in Figure 6.3.1.

Let $\mathcal{N}(\epsilon)$ denote the number of square boxes of side length ϵ which intersect G. These boxes are similar to the ones used in the Box Counting Theorem, Theorem 5.1.2, except that their sizes are arbitrary. On the basis of the intuitive idea introduced in Chapter 5, Section 1, we suppose that G has

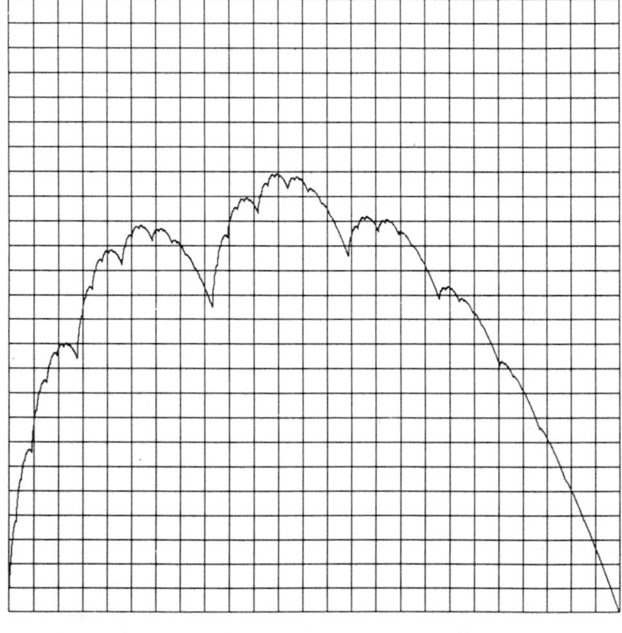

Figure 6.3.1
The graph G of a fractal interpolation function is superimposed on a grid of closed square boxes of side length ϵ. $\mathcal{N}(\epsilon)$ is used to denote the number of boxes which intersect G. What is the value of $\mathcal{N}(\epsilon)$?

fractal dimension D, where

$$\mathcal{N}(\epsilon) \approx \text{constant} \cdot \epsilon^{-D} \qquad \text{as } \epsilon \to 0.$$

We want to estimate the value of D on the basis of this assumption.

Let $n \in \{1, 2, \ldots, N\}$. Let $\mathcal{N}_n(\epsilon)$ denote the number of boxes of side length ϵ which intersect $w_n(G)$ for $n = 1, 2, \ldots, N$. We suppose that ϵ is very small compared to $|x_N - x_0|$. Then, because the IFS is just-touching, it is reasonable to make the approximation

$$\mathcal{N}(\epsilon) \approx \mathcal{N}_1(\epsilon) + \mathcal{N}_2(\epsilon) + \mathcal{N}_3(\epsilon) + \cdots + \mathcal{N}_N(\epsilon). \tag{6.3.2}$$

We now look for a relationship between $\mathcal{N}(\epsilon)$ and $\mathcal{N}_n(\epsilon)$. The boxes which intersect G can be thought of as being organized into columns, as illustrated in Figure 6.3.2.

Let the set of columns of boxes of side length ϵ which intersect G be denoted by $\{c_j(\epsilon): j = 1, 2, \ldots, \mathcal{K}(\epsilon)\}$, where $\mathcal{K}(\epsilon)$ denotes the number of columns. Under the conditions in Equation (6.3.1), in the statement of the theorem, one can prove that the minimum number of boxes in a column increases without limit as ϵ approaches zero. To simplify the discussion we assume that

$$|d_n| > a_n \qquad \text{for } n = 1, 2, \ldots, N.$$

(Notice that

$$\sum_{n=1}^{N} a_n = \sum_{n=1}^{N} \frac{(x_n - x_{n-1})}{(x_N - x_0)} = 1,$$

which tells us that this assumption is stronger than the assumption $\sum_{n=1}^{N} |d_n| >$

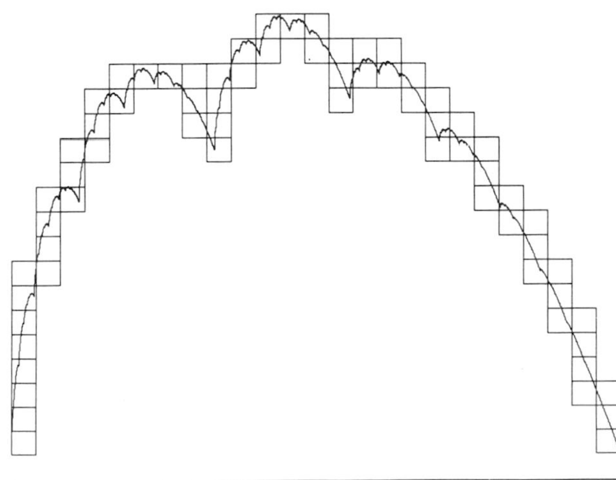

Figure 6.3.2
The boxes which intersect G can be thought of as organized in columns. The set of columns of boxes of side length ϵ which intersect G is denoted by $\{c_j(\epsilon): j = 1, 2, \ldots, \mathcal{K}(\epsilon)\}$, where $\mathcal{K}(\epsilon)$ denotes the number of columns. What is the value of $\mathcal{K}(\epsilon)$ and how many boxes are there in $c_2(\epsilon)$, in this illustration?

1.) Then consider what happens to a column of boxes $c_j(\epsilon)$ of side length ϵ, when we apply the affine transformation w_n to it. It becomes a column of parallelograms. The width of the column is $a_n\epsilon$ and the height of the column is $|d_n|$ times the height of the column before transformation. Let $\mathcal{N}(c_j(\epsilon))$ denote the number of boxes in the column $c_j(\epsilon)$. Then the column $w_n(c_j(\epsilon))$ can be thought of as being made up of square boxes of side length $a_n\epsilon$, each of which intersects $w_n(G)$. How many boxes of side length $a_n\epsilon$ are there in this column? Approximately $|d_n|\mathcal{N}(c_j(\epsilon))/a_n$. Adding up the contribution to $\mathcal{N}_n(a_n\epsilon)$ from each column we obtain

$$\mathcal{N}_n(a_n\epsilon) \approx \sum_{j=1}^{\mathcal{K}(\epsilon)} \frac{|d_n|\mathcal{N}(c_j(\epsilon))}{a_n} = \frac{|d_n|}{a_n}\sum_{j=1}^{\mathcal{K}(\epsilon)} \mathcal{N}(c_j(\epsilon)) = \frac{|d_n|}{a_n}\mathcal{N}(\epsilon).$$

The situation is illustrated in Figure 6.3.3.

From the last equation we deduce that when ϵ is very small compared to $[x_0, x_N]$,

$$\mathcal{N}_n(\epsilon) \approx \frac{|d_n|}{a_n}\mathcal{N}\left(\frac{\epsilon}{a_n}\right) \quad \text{for } n = 1, 2, \ldots, N. \tag{6.3.3}$$

We now substitute from (6.3.3) into (6.3.2) to obtain the functional equation

$$\mathcal{N}(\epsilon) \approx \frac{|d_1|}{a_1}\mathcal{N}\left(\frac{\epsilon}{a_1}\right) + \frac{|d_2|}{a_2}\mathcal{N}\left(\frac{\epsilon}{a_2}\right) + \frac{|d_3|}{a_3}\mathcal{N}\left(\frac{\epsilon}{a_3}\right) + \cdots + \frac{|d_N|}{a_N}\mathcal{N}\left(\frac{\epsilon}{a_N}\right).$$

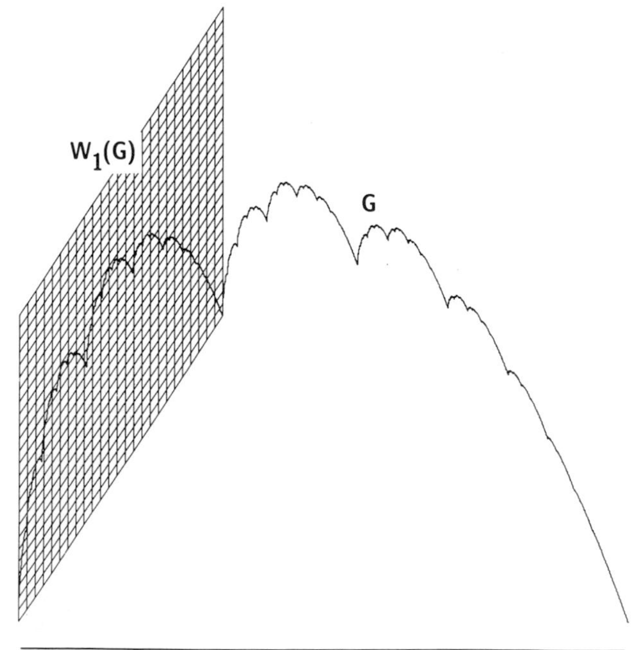

Figure 6.3.3 *When the shear transformation w_1 is applied to the columns of boxes which cover the graph, G, the result is a set of thinner columns, of width $a_1\epsilon$, which cover $w_1(G)$. The new columns are made up of small parallelograms, but the number of square boxes of side length $a_1\epsilon$ which they contain is readily estimated.*

Into this equation we substitute our assumption $\mathcal{N}(\epsilon) \approx \text{constant} \cdot \epsilon^{-D}$ to obtain the equation

$$\epsilon^{-D} \approx |d_1|a_1^{D-1}\epsilon^{-D} + |d_2|a_2^{D-1}\epsilon^{-D} + |d_3|a_3^{D-1}\epsilon^{-D} + \cdots + |d_N|a_N^{D-1}\epsilon^{-D}.$$

The main formula in the statement of the theorem follows at once.

If the interpolation points are collinear, then the attractor of the IFS is the line segment which connects the point (x_0, F_0) to the point (x_N, F_N), and this has fractal dimension one. If $\Sigma_{n=1}^{N}|d_n| \leq 1$ then one can show that $\mathcal{N}(\epsilon)$ behaves like a constant times ϵ^{-1}, whence the fractal dimension is one. This completes our informal demonstration of the theorem.

Exercises & Examples

3.1. We consider the fractal dimension of a fractal interpolation function in the case where the interpolation points are equally spaced. Let $x_i = x_0 + (i/N)(x_N - x_0)$ for $i = 0, 1, 2, \ldots, N$. Then it follows that $a_n = 1/N$ for $n = 1, 2, \ldots, N$. Hence if condition (1) in Theorem 6.3.1 holds, then the fractal dimension D of the interpolation function obeys

$$\sum_{n=1}^{N} |d_n|\left(\frac{1}{N}\right)^{D-1} = \left(\frac{1}{N}\right)^{D-1} \sum_{n=1}^{N} |d_n| = 1.$$

It follows that

$$D = 1 + \frac{\operatorname{Log}\left(\sum_{n=1}^{N} |d_n|\right)}{\operatorname{Log}(N)}.$$

This is a delightful formula for reasons of two types, (a) and (b). (a) This formula confirms our understanding of the fractal dimension of fractal interpolation functions. For example, notice that $\Sigma_{n=1}^{N}|d_n| < N$. Hence the dimension of a fractal interpolation function is less than two: however we can make it arbitrarily close to 2. Also, under the assumption that $\Sigma_{n-1}^{N}|d_n| > 1$, the fractal dimension is greater than 1: however we can vary it smoothly down to one. (b) It is remarkable that the fractal dimension does not depend on the values $\{F_i\colon i = 0, 1, 2, \ldots, N\}$, aside from the constraint that the interpolation points be noncollinear. Hence it is easy to explore a collection of fractal interpolation functions, all of which have the same fractal dimension, by imposing the following simple constraint on the vertical scaling factors:

$$\sum_{n=1}^{N} |d_n| = N^{D-1}.$$

Figures 6.3.4(a)–(c) illustrate some members of the family of fractal

Figure 6.3.4(a)–(c) *Members of the family of fractal interpolation functions corresponding to the set of data* $\{(0,0),(1,1),(2,1),(3,2)\}$, *such that the fractal dimension of each member of the family is* $D = 1.3$.

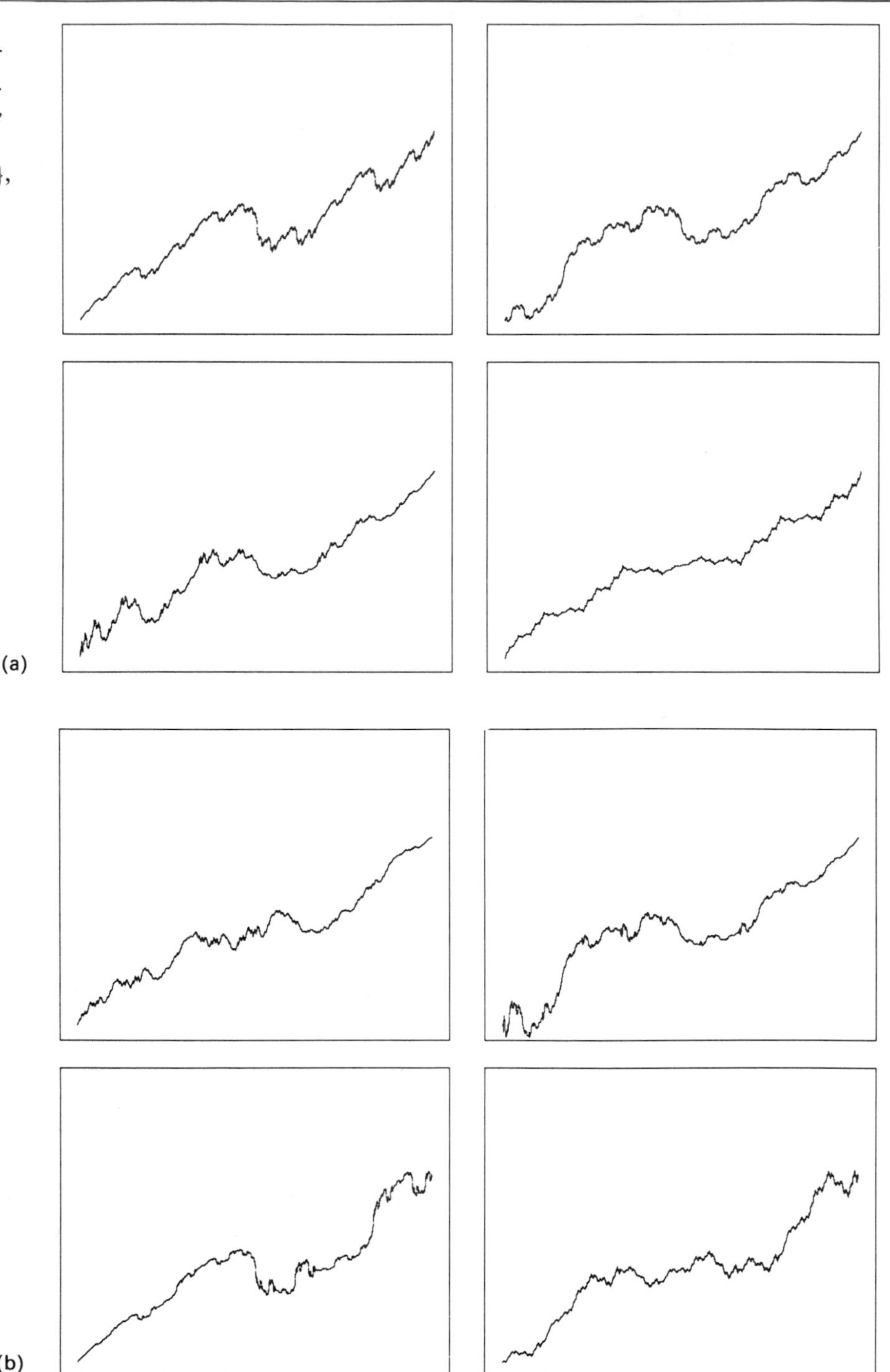

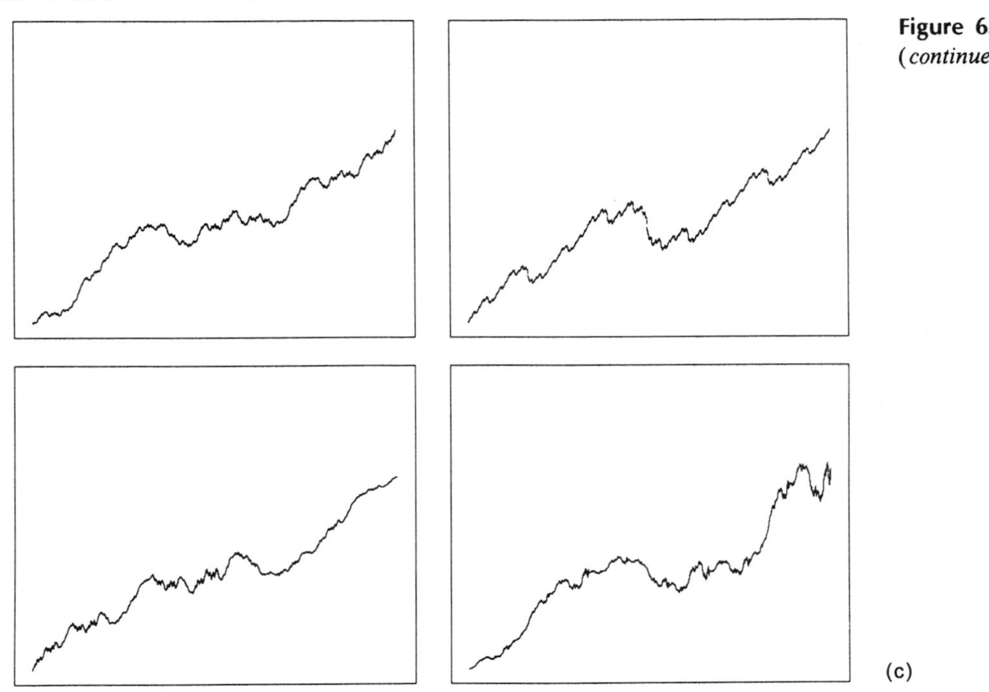

Figure 6.3.4 (*continued*).

(c)

interpolation functions corresponding to the set of data $\{(0,0),(1,1),(2,1),(3,2)$, such that the fractal dimension of each member of the family is $D = 1.3$.

Figures 6.3.5(a) and (b) illustrate members of a family of fractal interpolation functions parameterized by the fractal dimension D. Each function interpolates the same set of data.

3.2. Make an experimental estimate of the fractal dimension of the graphical data in Figure 6.3.6. Find a fractal interpolation function associated with the data $\{(0,0),(50,50),(100,0)\}$, which has the same fractal dimension, and which has two equal vertical scaling factors. Compare the graph of the fractal interpolation function with the graphical data.

3.3. Find a fractal interpolation function which approximates the experimental data shown in Figure 5.3.5.

3.4. Figure 6.3.7 shows the graphs of functions belonging to various one-parameter families of fractal interpolation functions. Each graph is the attractor of an IFS consisting of two affine transformations. Find the IFS associated with one of the families.

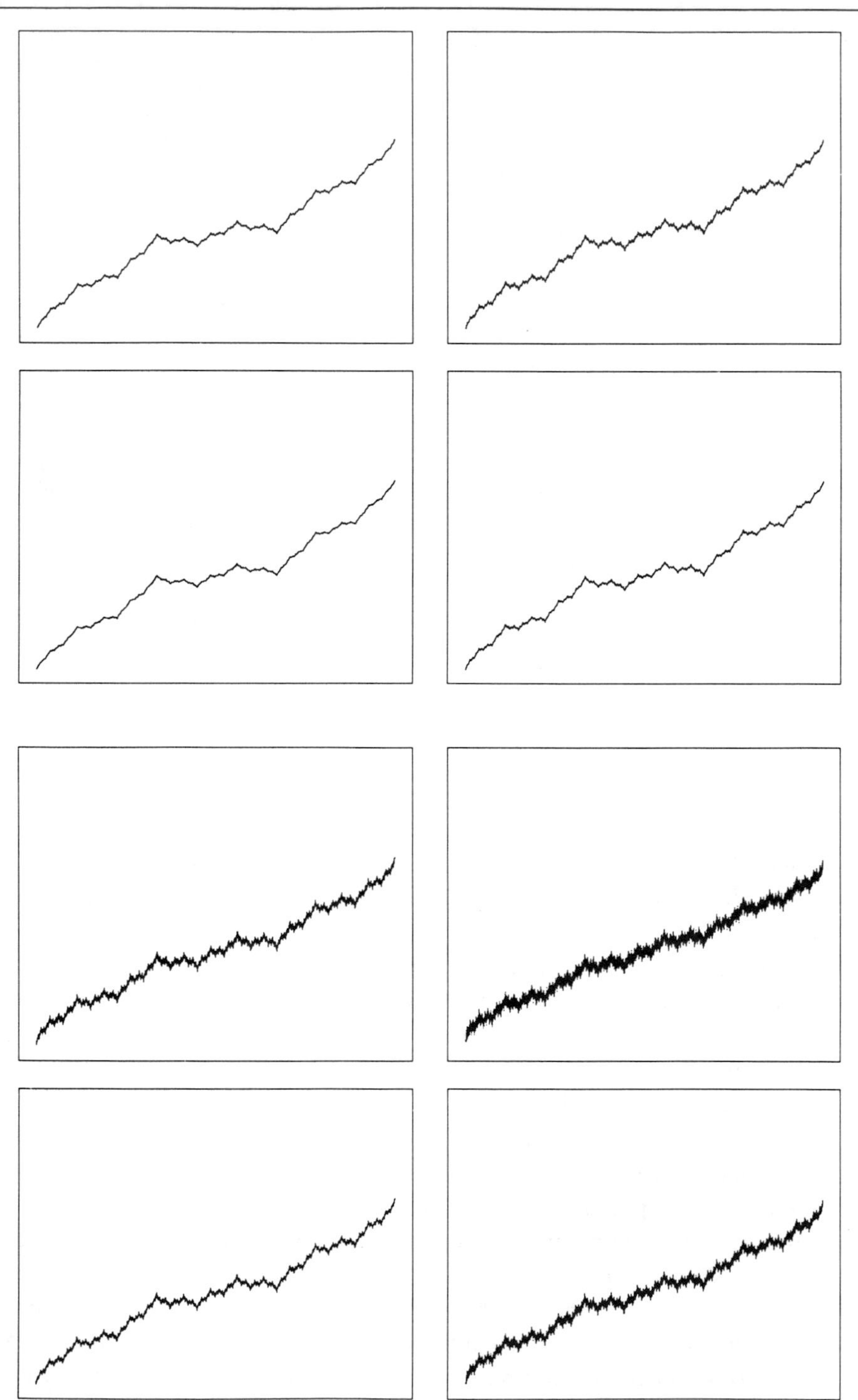

Figure 6.3.5(a) and (b) *Members of a one-parameter family of fractal interpolation functions. They correspond to the set of data* $\{(0,0),(1,1),(2,1),(3,2)\}$ *with vertical scaling factors* $d_1 = -d_2 = d_3 = 3^{D-2}$ *for* $D = 1, 1.1, 1.2,$ *and* $1.3, 1.4, 1.5, 1.6,$ *and* 1.7. *D is the fractal dimension of the fractal interpolation function.*

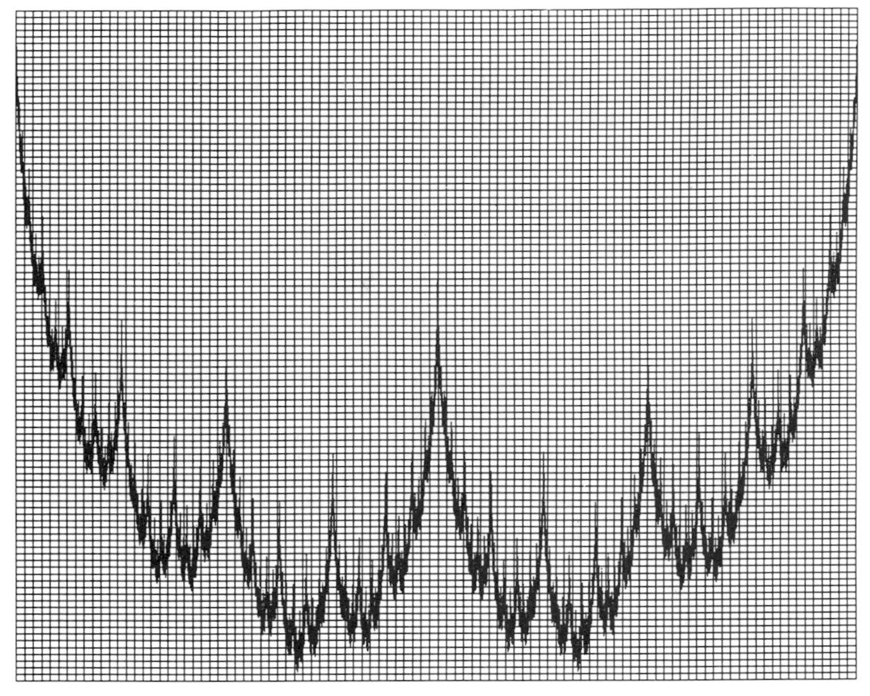

Figure 6.3.6
Make an experimental estimate of the fractal dimension of the graphical data shown here. Find a fractal interpolation function associated with the data $\{(0,0),(50,-50),(100,0)\}$, *which has the same fractal dimension, and which has two equal vertical scaling factors. Compare the graph of the fractal interpolation function with the graphical data.*

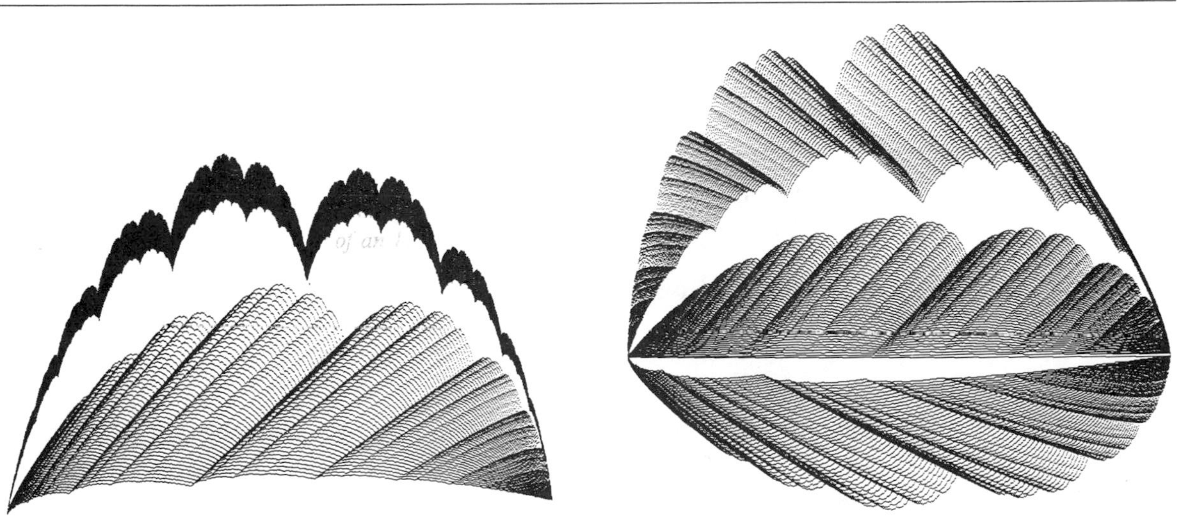

Figure 6.3.7
This figure shows graphs of various one-parameter families of fractal interpolation functions. Each graph is the attractor of an IFS consisting of two affine transformations. Can you find the families?

6.4 HIDDEN VARIABLE FRACTAL INTERPOLATION

We begin by generalizing the results of Section 6.2. Throughout this section, let $(\mathbb{Y}, d_Y)$ denote a complete metric space.

Definition 1. Let $I \subset \mathbb{R}$. Let $f\colon I \to \mathbb{Y}$ be a function. The *graph* of f is the set of points

$$G = \{(x, f(x)) \in \mathbb{R} \times \mathbb{Y} : x \in I\}.$$

Definition 2. A set of *generalized data* is a set of points of the form $\{(x_i, F_i) \in \mathbb{R} \times \mathbb{Y}: i = 0, 1, 2, \ldots, N\}$, where

$$x_0 < x_1 < x_2 < x_3 < \cdots < x_N.$$

An *interpolation function* corresponding to this set of data is a continuous function $f\colon [x_0, x_N] \to \mathbb{Y}$ such that

$$f(x_i) = F_i \qquad \text{for } i = 1, 2, \ldots, N.$$

The points $(x_i, F_i) \in \mathbb{R} \times \mathbb{Y}$ are called the *interpolation points*. We say that the function f *interpolates* the data and that (the graph of) f *passes through* the interpolation points.

Let $\boldsymbol{X}$ denote the Cartesian product space $\mathbb{R} \times \mathbb{Y}$. Let θ denote a positive number. Define a metric d on $\boldsymbol{X}$ by

$$d(X_1, X_2) = |x_1 - x_2| + \theta d_Y(y_1, y_2), \tag{6.4.1}$$

for all points $X_1 = (x_1, y_1)$ and $X_2 = (x_2, y_2)$ in $\boldsymbol{X}$. Then $(\boldsymbol{X}, d)$ is a complete metric space.

Let N be an integer greater than one. Let a set of generalized data $\{(x_i, F_i) \in \boldsymbol{X}: i = 0, 1, 2, \ldots, N\}$ be given. Let $n \in \{1, 2, \ldots, N\}$. Define $L_n\colon \mathbb{R} \to \mathbb{R}$ by

$$L_n(x) = a_n x + e_n \quad \text{where } a_n = \frac{(x_n - x_{n-1})}{(x_N - x_0)} \quad \text{and } e_n = \frac{(x_N x_{n-1} - x_0 x_n)}{(x_N - x_0)} \tag{6.4.2}$$

so that $L_n([x_0, x_N]) = [x_{n-1}, x_n]$. Let c and s be real numbers, with $0 \le s < 1$ and $c > 0$. For each $n \in \{1, 2, \ldots, N\}$ let $M_n\colon \boldsymbol{X} \to \mathbb{Y}$ be a function which obeys

$$d_Y(M_n(a, y), M_n(b, y)) \le c|a - b| \qquad \text{for all } a, b \in \mathbb{R},\ y \in \mathbb{Y}, \tag{6.4.3}$$

and

$$d_Y(M_n(x, a), M_n(x, b)) \le s d_Y(a, b) \qquad \text{for all } a, b \in \mathbb{Y},\ x \in \mathbb{R}. \tag{6.4.4}$$

Define a transformation w_n: $\mathbf{X} \to \mathbf{X}$ by

$$w_n(x, y) = (L_n(x), M_n(x, y)) \qquad \text{for all } (x, y) \in \mathbf{X},\ n = 1, 2, \ldots, N.$$

Theorem 1. *Let the IFS* $\{\mathbf{X}; w_n, 1, 2, \ldots, N\}$ *be defined as above. In particular, assume that there are real constants c and s such that* $0 \leq s < 1$, $0 < c$, *and conditions* (6.4.3), *and* (6.4.4) *are obeyed. Let the constant* θ *in the definition of the metric d in Equation* (6.4.1) *be defined by*

$$\theta = \frac{(1-a)}{2c} \qquad \textit{where } a = \mathrm{Max}\{a_i : i = 1, 2, \ldots, N\}.$$

Then the IFS $\{\mathbf{X}; w_n, n = 1, 2, \ldots, N\}$ *is hyperbolic with respect to the metric d.*

Proof. This follows very similar lines to the proof of Theorem 6.2.1. We leave it as an exercise for enthusiastic readers. The proof can also be found in [Barnsley, 1986c].

We now constrain the hyperbolic IFS $\{\mathbf{X}; w_n, n = 1, 2, \ldots, N\}$, defined above, to ensure that its attractor includes the set of generalized data. We assume that

$$(6.4.5) \quad M_n(x_0, F_0) = F_{n-1} \ \text{ and } \ M_n(x_N, F_N) = F_n \qquad \text{for } n = 1, 2, \ldots, N.$$

Then it follows that

$$w_n(x_0, F_0) = (x_{n-1}, F_{n-1}) \ \text{ and } \ w_n(x_N, F_N) = (x_n, F_n) \qquad \text{for } n = 1, 2, \ldots, N.$$

Theorem 2. *Let N be a positive integer greater than one. Let* $\{\mathbf{X}; w_n, n = 1, 2, \ldots, N\}$ *denote the IFS defined above, associated with the generalized data set* $\{(x_i, F_i) \in \mathbb{R} \times \mathbb{Y} : i = 1, 2, \ldots, N\}$. *In particular, assume that there are real constants c and s such that* $0 \leq s < 1$, $0 < c$, *and conditions* (6.4.3), (6.4.4), *and* (6.4.5) *are obeyed. Let* $G \in \mathscr{H}(\mathbf{X})$ *denote the attractor of the IFS. Then G is the graph of a continuous function* f: $[x_0, x_N] \to \mathbb{Y}$ *which interpolates the data* $\{(x_i, F_i): i = 1, 2, \ldots, N\}$. *That is*

$$G = \{(x, f(x)) : x \in [x_0, x_N]\},$$

where

$$f(x_i) = F_i \qquad \textit{for } i = 0, 1, 2, 3, \ldots, N.$$

Proof. Again we refer to [Barnsley 1986c]. The proof is analogous to the proof of Theorem 6.2.2.

Definition 3. The function whose graph is the attractor of an IFS, as described in Theorems 6.4.1 and 6.4.2, above, is called a *generalized fractal* *interpolation function*, corresponding to the generalized data $\{(x_i, F_i): i = 1, 2, \ldots, N\}$.

We now show how to use the idea of generalized fractal interpolation functions to produce interpolation functions which are more flexible than heretofore. The idea is to construct a generalized fractal interpolation function, using affine transformations acting on $\mathbb{R}^3$, and to project its graph into $\mathbb{R}^2$. This can be done in such a way that the projection is the graph of a function which interpolates a set of data $\{(x_i, F_i) \in \mathbb{R}^2: i = 1, 2, \ldots, N\}$. The extra degrees of freedom provided by working in $\mathbb{R}^3$ give us "hidden" variables. These variables can be used to adjust the shape and fractal dimension of the interpolation functions. The benefits of working with affine transformations are kept.

Let N be an integer greater than one. Let a set of data $\{(x_i, F_i) \in \mathbb{R}^2: i = 0, 1, 2, \ldots, N\}$ be given. Introduce a set of real parameters $\{H_i: i = 0, 1, 2, \ldots, N\}$. For the moment let us suppose that these parameters are fixed. Then we define a generalized set of data to be $\{(x_i, F_i, H_i) \in \mathbb{R} \times \mathbb{R}^2: i = 0, 1, 2, \ldots, N\}$. In the present application of Theorem 6.4.2 we take $(\mathbb{Y}, d_Y)$ to be ($\mathbb{R}^2$, Euclidean). We consider an IFS $\{\mathbb{R}^3; w_n, n = 1, 2, \ldots, N\}$ where for $n \in \{1, 2, \ldots, N\}$ the map $w_n: \mathbb{R}^3 \to \mathbb{R}^3$ is an affine transformation is of the special structure:

$$w_n \begin{bmatrix} x \\ y \\ z \end{bmatrix} = \begin{bmatrix} a_n & 0 & 0 \\ c_n & d_n & h_n \\ k_n & l_n & m_n \end{bmatrix} \begin{bmatrix} x \\ y \\ z \end{bmatrix} + \begin{bmatrix} e_n \\ f_n \\ g_n \end{bmatrix}.$$

Here a_n, c_n, d_n, e_n, f_n, g_n, h_n, k_n, l_n, and m_n, are real numbers. We assume that they obey the constraints

$$w_n \begin{bmatrix} x_0 \\ F_0 \\ H_0 \end{bmatrix} = \begin{bmatrix} x_{n-1} \\ F_{n-1} \\ H_{n-1} \end{bmatrix},$$

and

$$w_n \begin{bmatrix} x_N \\ F_N \\ H_N \end{bmatrix} = \begin{bmatrix} x_n \\ F_n \\ H_n \end{bmatrix}, \qquad \text{for } n = 1, 2, \ldots, N.$$

Then we can write

$$w_n(x, y, z) = (L_n(x), M_n(x, y, z)) \qquad \text{for all } (x, y, z) \in \mathbb{R}^3, n = 1, 2, \ldots, N,$$

where $L_n(x)$ is defined in Equation (6.4.2) and $M_n: \mathbb{R}^3 \to \mathbb{R}^2$ is defined by

$$M_n \begin{bmatrix} x \\ y \\ z \end{bmatrix} = A_n \begin{bmatrix} y \\ z \end{bmatrix} + \begin{bmatrix} f_n + c_n x \\ g_n + k_n x \end{bmatrix},$$

where

$$A_n = \begin{bmatrix} d_n & h_n \\ l_n & m_n \end{bmatrix} \quad \text{for } n = 1, 2, \ldots, N. \tag{6.4.6}$$

Let us replace F_n in condition (6.4.5) by (F_n, H_n). Then M_n obeys condition (6.4.5). Let us define

$$c = \text{Max}\{\text{Max}\{c_i, k_i\} : i = 1, 2, \ldots, N\}.$$

Then condition (6.4.3) is true. Lastly, assume that the linear transformations A_n: $\mathbb{R}^2 \to \mathbb{R}^2$ are contractive with contractivity factor s with $0 \leq s < 1$. Then condition (6.4.4) is true. We conclude that, under the conditions given in this paragraph, the IFS $\{\mathbb{R}^3; w_n, n = 1, 2, \ldots, N\}$ satisfies the conditions of Theorem 6.4.2. It follows that the attractor of the IFS is the graph of a continuous function f: $[x_0, x_N] \to \mathbb{R}^2$ such that

$$f(x_i) = (F_i, H_i) \qquad \text{for } 1, 2, \ldots, N.$$

Now write

$$f(x) = (f_1(x), f_2(x)).$$

Then f_1: $[x_0, x_N] \to \mathbb{R}$ is a continuous function such that

$$f_1(x_i) = F_i \qquad \text{for } i = 1, 2, \ldots, N.$$

Definition 4. The function f_1: $[x_0, x_N] \to \mathbb{R}^2$ constructed in the previous paragraph is called a *hidden variable* fractal interpolation function, associated with the set of data $\{(x_i, F_i) \in \mathbb{R}^2 : i = 1, 2, \ldots, N\}$.

The easiest method for computing the graph of a hidden variable fractal interpolation function is with the aid of the Random Iteration Algorithm. Here we present an adaptation of Program 6.2.1. It computes points on the graph of a hidden variable fractal interpolation function and displays them on a graphics monitor. It is written for $N = 3$ and the data set

$$\{(0,0), (30,50), (60,40), (100,10)\}.$$

The "hidden" variables, namely the entries of the matrices A_n and the number H_n for $n = 1, 2, 3$, are input by the user during execution of the code. The program calculates the coefficients in the three-dimensional affine transformations from the data, and then applies the Random Iteration Algorithm to the resulting IFS. The first two coordinates of each successively computed point, which has three coordinates, is plotted on the screen of the graphics monitor. The program is written in BASIC. It runs without modification on an IBM PC with Enhanced Graphics Adaptor and Turbobasic. On any line the words preceded by a ' are comments: they are not part of the program.

PROGRAM 6.4.1

```
x[0] = 0: x[1] = 30: x[2] = 60: x[3] = 100   'Data set
F[0] = 0: F[1] = 50: F[2] = 40: F[3] = 10
input "enter the hidden variables H[0], H[1], H[2] and H[3]", H[0], H[1], H[2],
    H[3]   'Hidden Variables
for n = 1 to 3: print "for n = ", n
input "enter the hidden variables d, h, l, m", d[n], hh[n], l[n], m[n]   'More
    Hidden Variables
next
for n = 1 to 3   'Calculate the affine transformations from the Data and the
    Hidden Variables
  p = F[n-1]-d[n]*F[0]-hh[n]*H[0]: q = h[n-1]-l[n]*F[0]-m[n]*H[0]
  r = F[n]-d[n]*F[3]-hh[n]*H[3]: s = H[n]-l[n]*F[3]-m[n]*H[3]
    b = x[3]-x[0]: c[n] = (r-p)/b: k[n] = (s-q)/b
  a[n] = (x[n]-x[n-1])/b: e[n] = (x[3]*x[n-1]-x[0]*x[n])/b
  ff[n] = p-c[n]*x[0]: g[n] = q-k[n]*x[0]
next
screen 2: cls   'initialize graphics
window (0, 0)-(100, 100)   'change this to zoom and/or pan
x = 0: y = 0: z = h[0]   'initial point from which the random iteration
    begins
for n = 1 to 1000   'Random Iteration Algorithm
  kk = int(3*rnd-0.0001) + 1
  newx = a[kk]*x + e[kk]
  newy = c[kk]*x + d[kk]*y + hh[kk]*z + ff[kk]
  newz = k[kk]*x + l[kk]*y + m[kk]*z + g[kk]
  x = newx: y = newy: z = newz
  pset (x, y), z   'plot the most recently computed point, in color z, on
    the screen
next
end
```

The result of running an adaptation of this program on a Masscomp workstation, and then printing the contents of the graphics screen, is presented in Figure 6.4.1. In this case $H[0] = 0$, $H[1] = 30$, $H[2] = 60$, $H[3] = 100$, $d(1) = d(2) = d(3) = 0.3$, $h(1) = h(2) = 0.2$, $h(3) = 0.1$, $l(1) = l(2) = l(3) = -0.1$, $m(1) = 0.3$, $m(2) = 0$, $m(3) = -0.1$. Remember that the linear transformation A_n must be contractive, so certainly do not enter values of magnitude larger than one for any of the numbers $d(n)$, $h(n)$, $l(n)$, and $m(n)$. The program renders each point in a color which depends on its z-coordinate. This helps the user to visualize the "hidden" three-dimensional character of the curve.

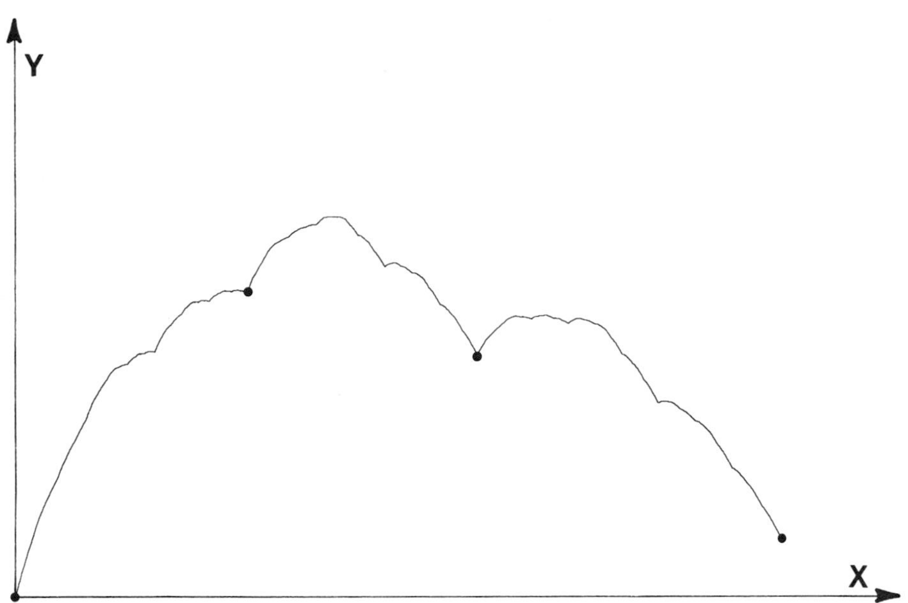

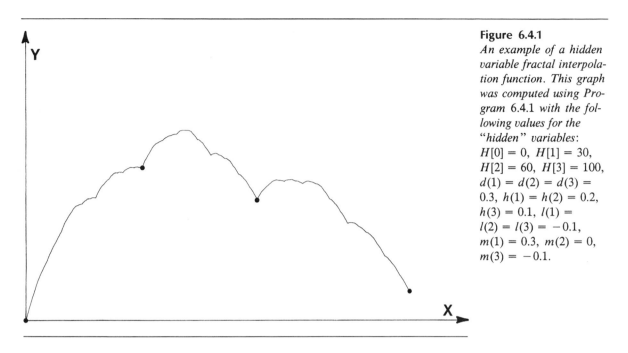

Figure 6.4.1
An example of a hidden variable fractal interpolation function. This graph was computed using Program 6.4.1 with the following values for the "hidden" variables: $H[0] = 0$, $H[1] = 30$, $H[2] = 60$, $H[3] = 100$, $d(1) = d(2) = d(3) = 0.3$, $h(1) = h(2) = 0.2$, $h(3) = 0.1$, $l(1) = l(2) = l(3) = -0.1$, $m(1) = 0.3$, $m(2) = 0$, $m(3) = -0.1$.

The important point about hidden variable fractal interpolation is this: Although the attractor of the IFS is a union of affine transformations applied to the attractor, this is not the case in general when we replace the word "attractor" by the phrase "projection of the attractor." The graph of the hidden variable fractal interpolation function $f_1(x)$ is not self-similar, or self-affine, or self-anything!

The idea of hidden variable fractal interpolation functions can be developed using any number of "hidden" dimensions. As the number of dimensions is increased, the process of specifying the function becomes more and more onerous, and the function itself, seen by us in flatland, becomes more and more random. One would never guess, from looking at pictures of them, that they are generated by *deterministic* fractal geometry.

Exercises & Examples

4.1. Generalize the proof of Theorem 6.2.1 to obtain a proof of Theorem 6.4.1.

4.2. Let $\mathscr{F}$ denote the set of continuous functions: $f\colon [x_0, x_N] \to \mathbb{Y}$ such that $f(x_0) = F_0$ and $f(x_N) = F_N$. Define a metric d on $\mathscr{F}$ by $d(f, g) = \text{Max}\{d_Y(f(x), g(x))\colon x \in [x_0, x_N]\}$. Then $(\mathscr{F}, d)$ is a complete metric space; see for example [F[illegible]in, 1966]. Use this fact to help you generalize the proof of Theorem 6.[illegible]o provide a proof of Theorem 6.4.2.

4.3. Rewrite Program 6.4.1 in a form suitable for your own computer environment, then run it and obtain hardcopy of the output.

4.4. Modify your version of Program 6.4.1 so that you can adjust one of the "hidden" variables while it is running. In this way, make a picture which shows a one-parameter family of hidden variable fractal interpolation functions.

4.5. Modify your version of Program 6.4.1 so that you can see the projection of the attractor of the IFS into the (y, z)-plane. To do this, simply plot (y, z) in place of (x, y). Make hardcopy of the output.

4.6. Figure 6.4.2(a) shows three projections of the graph G of a generalized fractal interpolation function f: $[0, 1] \to \mathbb{R}^2$. The projections are: (i) into the (x, y) plane, (ii) into the (x, z) plane, and (iii) into the (y, z) plane. G is the attractor of an IFS of the form $\{\mathbb{R}^3; w_1, w_2\}$ where w_1 and w_2 are affine transformations. Find w_1 and w_2. See also Figure 6.4.2(b).

4.7. Use a hidden-variable fractal interpolation function to fit the experimental data in Figure 5.3.5. Here is one way to proceed. (a) Modify your version of Program 6.4.1 so that you can adjust the "hidden" variables from the keyboard. (b) Trace the data in Figure 5.3.5 onto a sheet of flexible transparent material, such as a viewgraph. (c) Attach the tracing to the screen of your graphics monitor using clear sticky tape. (d) Interactively adjust the "hidden" variables to provide a good visual fit to the data.

4.8. $\rightarrowtail$ Show that, with hidden variables, one can use affine transformations to construct graphs of polynomials of any degree.

6.5 SPACE-FILLING CURVES

Here we make a delightful application of Theorem 6.4.2. Let A denote a nonempty pathwise connected compact subset of $\mathbb{R}^2$. We show how to construct a continuous function f: $[0, 1] \to \mathbb{R}^2$ such that $f([0, 1]) = A$.

Let $(\mathbb{Y}, d_Y)$ denote the metric space $(\mathbb{R}^2$, Euclidean). We represent points in $\mathbb{Y}$ using a Cartesian coordinate system defined by a y-axis and a z-axis. Thus, (y, z) may represent a point in $\mathbb{Y}$. To motivate the development we take $A = \blacksquare \subset \mathbb{Y}$. Consider the just-touching IFS $\{\mathbb{Y};\ w_1, w_2, w_3, w_4\}$ where the maps are similitudes of scaling factor 0.5, corresponding to the collage in Figure 6.5.1.

Let

$$(F_0, H_0) = (0,0), (F_1, H_1) = (0,0.5), (F_2, H_2) = (0.5,0.5),$$
$$(F_3, H_3) = (1,0.5), \text{ and } (F_4, H_4) = (1,0).$$

The maps are chosen so that

$$w_n(F_0, H_0) = (F_{n-1}, H_{n-1}) \quad \text{and} \quad w_n(F_4, H_4) = (F_n, H_n) \qquad \text{for } n = 1,2,3,4.$$

The IFS code for this IFS is given in Table 6.5.1.

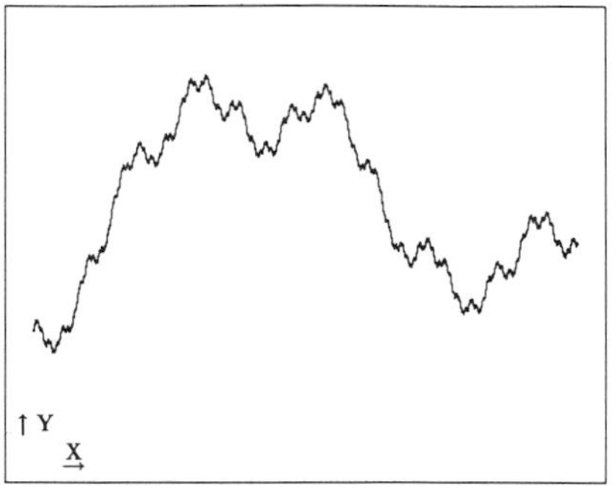

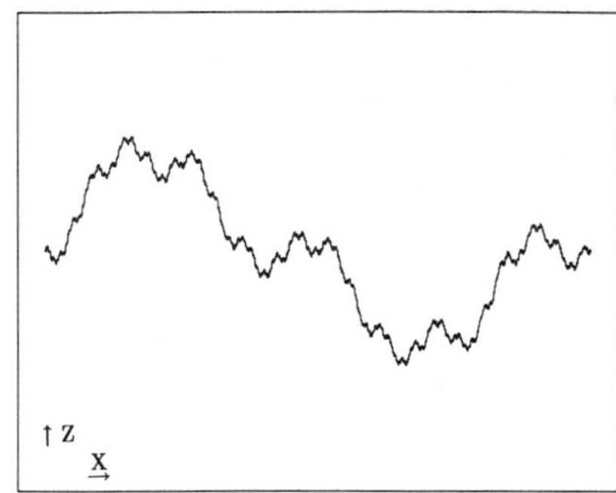

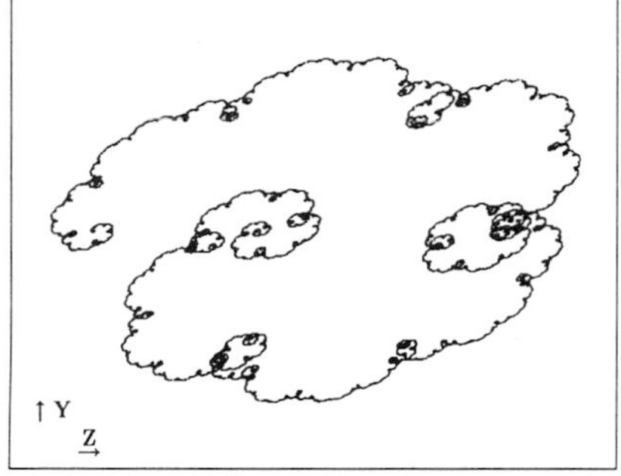

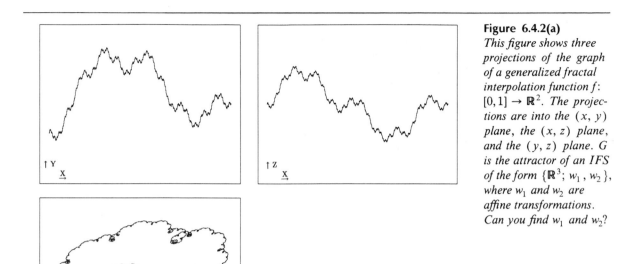

Figure 6.4.2(a)
This figure shows three projections of the graph of a generalized fractal interpolation function $f: [0,1] \rightarrow \mathbb{R}^2$. *The projections are into the* (x, y) *plane, the* (x, z) *plane, and the* (y, z) *plane. G is the attractor of an IFS of the form* $\{\mathbb{R}^3; w_1, w_2\}$, *where* w_1 *and* w_2 *are affine transformations. Can you find* w_1 *and* w_2?

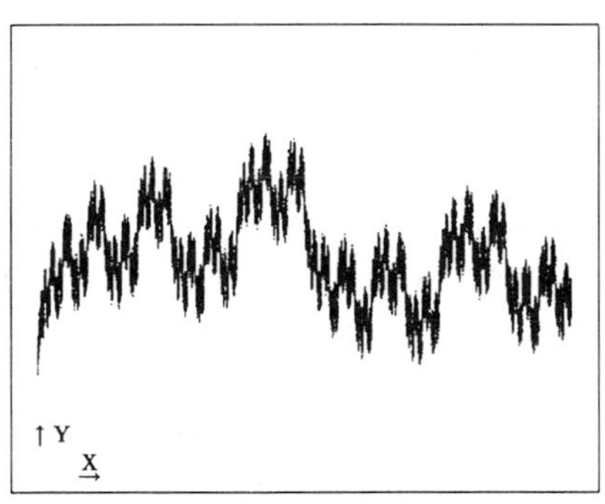

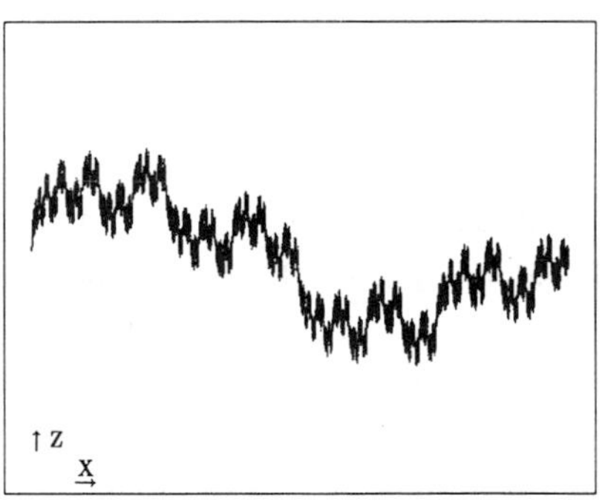

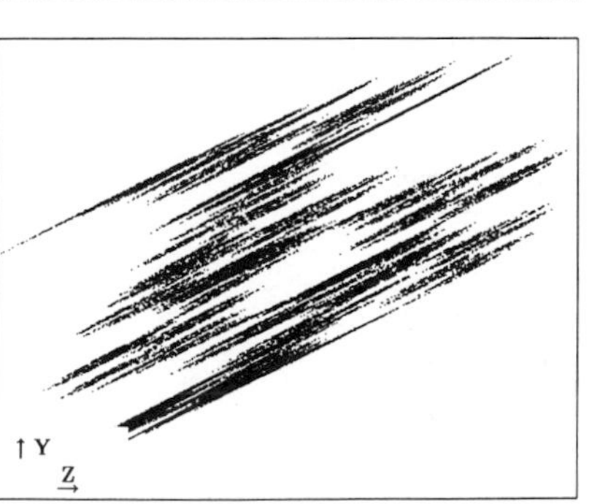

Figure 6.4.2(b)
Three orthogonal projections of the graph of a generalized fractal interpolation function. The fractal dimension here is higher than that for Figure 6.4.2(a).

Figure 6.5.1
Collage of ■ using four similitudes of scaling factor 0.5. The map w_n is chosen so that $w_n(F_0, H_0) = (F_{n-1}, H_{n-1})$ and $w_n(F_4, H_4) = (F_n, H_n)$ for $n = 1, 2, 3, 4$.

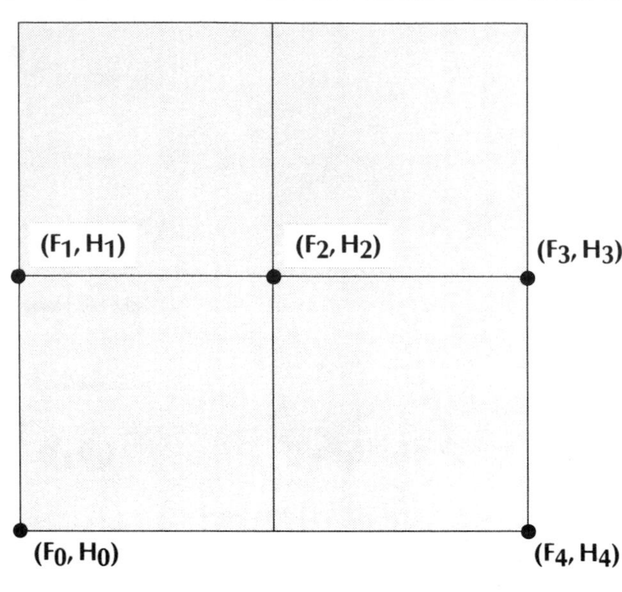

Let $A_0 \in \mathscr{H}(\blacksquare)$ denote a simple curve which connects the point (F_0, H_0) to the point (F_4, H_4), such that $A_0 \cap \partial\blacksquare = \{(F_0, H_0), (F_4, H_4)\}$. This last condition says that the curve lies in the interior of the unit square box, except for the two endpoints of the curve. Consider the sequence of sets $\{A_n = W^{\circ n}(A_0)\}_{n=0}^{\infty}$ where $W: \mathscr{H}(\blacksquare) \to \mathscr{H}(\blacksquare)$ is defined by

$$W(B) = \bigcup_{n=1}^{4} w_n(B) \qquad \text{for all } B \in \mathscr{H}(\blacksquare).$$

It follows from Theorem 3.7.1 that the sequence converges to ■ in the Hausdorff metric. The reader should verify that for each $n = 1, 2, \ldots,$ A_n is a simple curve which connects the points (F_0, H_0) to the point (F_4, H_4). Sequences of such curves are illustrated in Figures 6.5.2(a)–(d).

We use the IFS defined in the previous paragraph to construct a continuous function f: $[0, 1] \to \blacksquare$ such that $f([0, 1]) = \blacksquare$. We achieve this by exploiting a hidden variable fractal function constructed in a special way. We use

Table 6.5.1
IFS code for ■, constrained to yield a space-filling curve.

w	a	b	c	d	e	f	p
1	0	0.5	0.5	0	0	0	0.25
2	0.5	0	0	0.5	0	0.5	0.25
3	0.5	0	0	0.5	0.5	0.5	0.25
4	0	−0.5	−0.5	0	1	0.5	0.25

Figure 6.5.2(a) *A sequence of curves "converging to" a space-filling curve. These are obtained by application of the Deterministic Algorithm to the IFS code in Table* 6.5.1, *starting from a curve* A_0 *which connects* $(0,0)$ *to* $(1,0)$ *and lies in* ■.

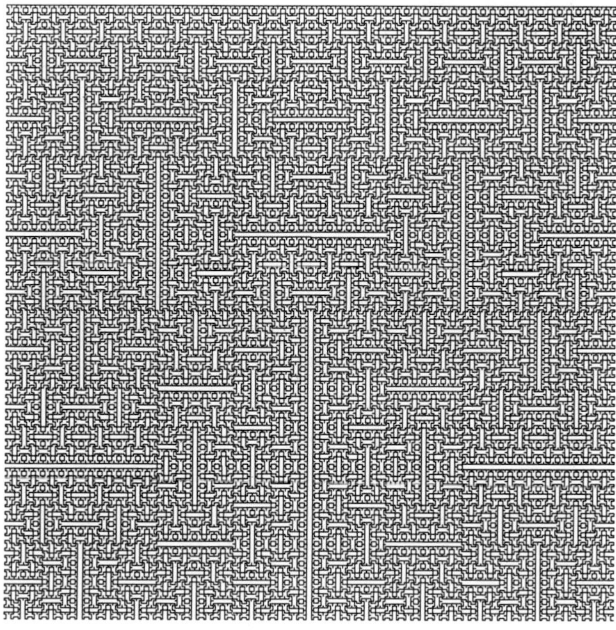

Figure 6.5.2(b) *A higher resolution view of one of the panels in Figure* 6.5.2(*a*). *How long is the shortest path from the lower left corner to the lower right corner?*

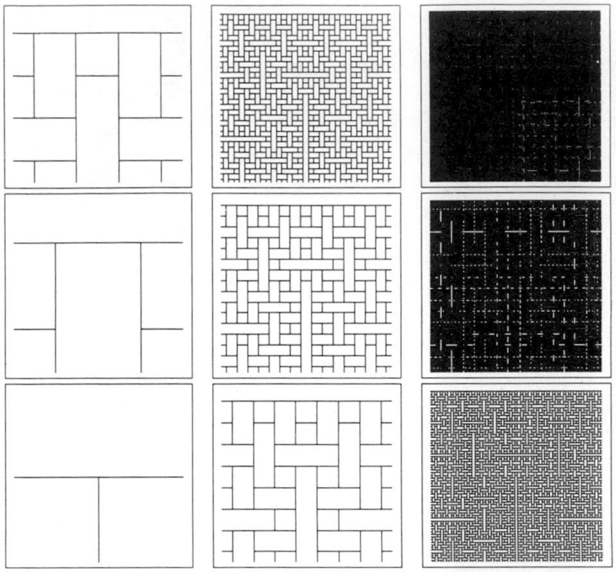

Figure 6.5.2(c)
A sequence of sets "converging to" a ■. These are obtained by application of the Deterministic Algorithm to the IFS code in Table 6.5.1, *starting from the set A_0 in the lower left panel. How fascinating they are!*

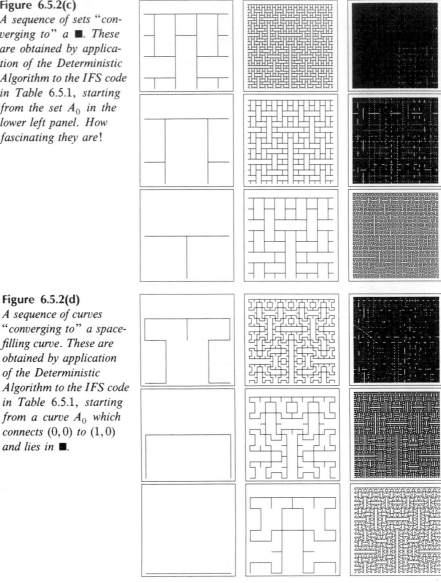

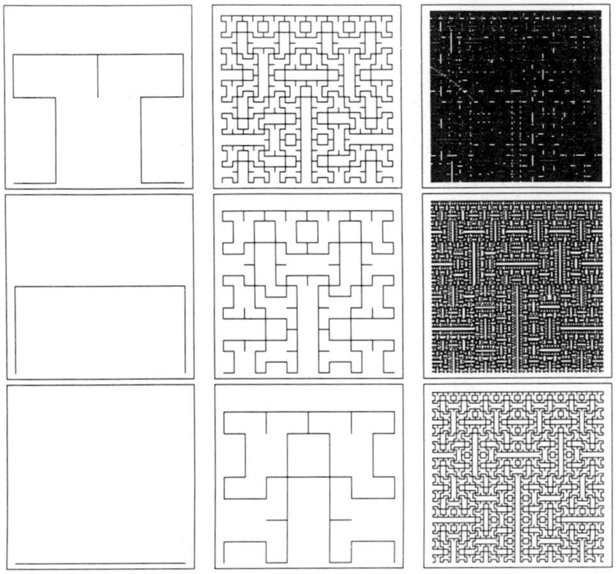

Figure 6.5.2(d)
A sequence of curves "converging to" a space-filling curve. These are obtained by application of the Deterministic Algorithm to the IFS code in Table 6.5.1, *starting from a curve A_0 which connects* (0,0) *to* (1,0) *and lies in* ■.

ideas presented in Chapter 6, Section 4. Consider the IFS $\{\mathbb{R}^3; w_n, n = 1, 2, \ldots, 4\}$ where the map $w_n\colon \mathbb{R}^3 \to \mathbb{R}^3$ is the affine transformation

$$w_n \begin{bmatrix} x \\ y \\ z \end{bmatrix} = \begin{bmatrix} 0.25 & 0 & 0 \\ 0 & a_n & b_n \\ 0 & c_n & d_n \end{bmatrix} \begin{bmatrix} x \\ y \\ z \end{bmatrix} + \begin{bmatrix} (n-1)/4 \\ e_n \\ f_n \end{bmatrix} \quad \text{for } n \in \{1, 2, 3, 4\}.$$

The constants a_n, b_n, c_n, d_n, e_n, and f_n are defined in Table 6.5.1. This IFS

Plate 8.4.3
A map of a small piece of the parameter space for the family of dynamical systems in Example 8.4.2. *The map was computed using Algorithm* 8.4.1. *The grainy multicolored areas resemble the repelling sets of the dynamical system for the corresponding values of* λ.

Plate 8.4.4
An extreme close-up on a grainy region in Color Plate 8.4.3.

Plate 9.8.1
Chimneys in a landscape giving off smoke.

Plate 9.8.2
Zoom on the smoke coming out of a chimney.

Plate 9.8.3
A colorful leaf produced by rendering a Borel measure.

Plate 9.8.4
A leaf, showing its veins.

Plate 9.8.5
A sequence of frames of an IFS encoded cloud. This illustrates the continuous dependence of the invariant measure on parameters in the IFS code.

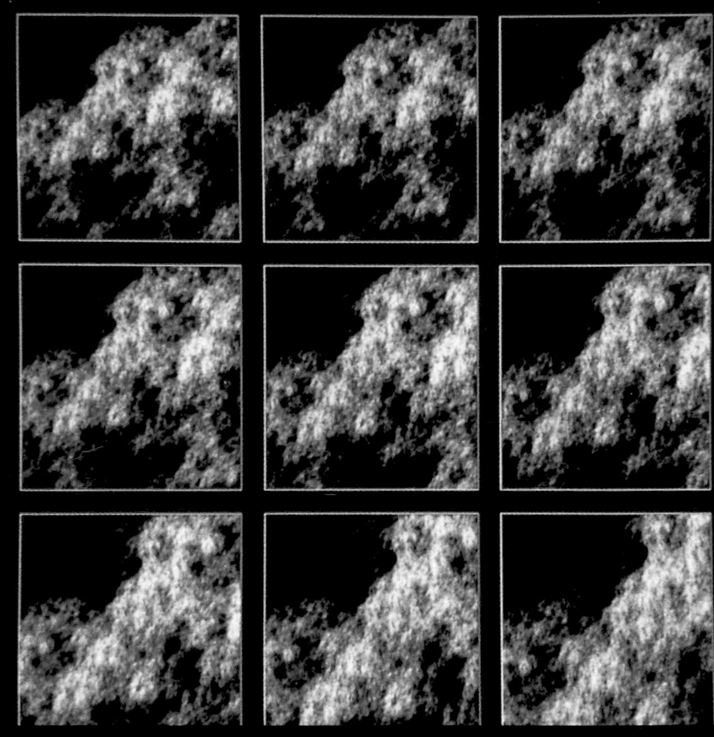

Plate 9.8.6
Monterey Coast.

Plate 9.8.7
Andes Girl.

Plate 9.8.8
Arctic Wolf.

Plate 9.8.9
Leaf and Sunflower, condensation sets used in *Color Plate* 9.8.10.

Plate 9.8.10
Sunflower Field.

Plate 9.8.11
Black Forest.

Plate 9.8.12
Zoom on Black Forest.

Plate 9.8.13
Black Forest in Winter.

Plate 9.8.14
Zoom on Black Forest in Winter.

satisfies Theorem 6.4.2, corresponding to the set of data

$$\{(0, F_0, H_0), (0.25, F_1, H_1), (0.5, F_2, H_2), (0.75, F_3, H_3), (1, F_4, H_4)\}.$$

It follows that the attractor of the IFS is the graph, G, of a continuous function f: $[0,1] \to \mathbb{R}^2$. What is the range of this function? It is

$$G_{yz} = \{(y, z) \in \mathbb{R}^2 : (x, y, z) \in G\},$$

namely the projection of G into the (y, z)-plane. It is straightforward to prove that G_{yz} is the attractor $A = ■$ of the IFS defined by the IFS code in Table 6.5.1. It follows that $f([0,1]) = ■$. So we have our space-filling curve!

We have something else very exciting as well. The attractor of the three-dimensional IFS is the graph of a function from $[0,1]$ to ■. The projections G_{xy} and G_{xz}, in the obvious notation, are graphs of hidden-variable fractal functions, while $G_{yz} = ■$. What does G look like from other points of view? Various views of the attractor are illustrated in Figures 6.5.3(a) and (b). We conclude that G is a curious complex three-dimensional object. It would be wonderful to have a three-dimensional model of G made out of very thin strong wire.

The following theorem summarizes what we have just learned

Theorem 1. *Let $A \subset \mathbb{R}^2$ be a nonempty pathwise-connected compact set, such that the following conditions hold. Let N be an integer greater than one. Let there be a hyperbolic IFS $\{\mathbb{R}^2; M_n,\ n = 1, 2, \ldots, N\}$ such that A is the attractor of the IFS. Let there be a set of distinct points $\{(F_i, G_i) \in A: i = 0, 1, 2, \ldots, N\}$ such that*

$$M_n(F_0, H_0) = (F_{n-1}, H_{n-1}) \quad \text{and} \quad M_n(F_N, H_N) = (F_n, H_n) \qquad \text{for } n = 1, 2, \ldots, N.$$

Then there is a continuous function f: $[0,1] \to \mathbb{R}^2$ such that $f([0,1]) = A$. One such function is the one whose graph is the attractor of the IFS

$$\left\{\mathbb{R}^3;\ w_n(x, y, z) = \left(\frac{1}{N}x + \frac{n-1}{N},\ M_n(y, z)\right), \quad n = 1, 2, \ldots, N\right\}.$$

Exercises & Examples

5.1. Let $\triangle$ denote the Sierpinski triangle with vertices at the points $(0,0)$, $(0,1)$ and $(1,0)$. Find an IFS of the form $\{\mathbb{R}^3; w_1, w_2, w_3\}$, where the maps are affine, such that the attractor of the IFS is the graph of a continuous function f: $[0,1] \to \mathbb{R}^2$ such that $f([0,1]) = \triangle$. Four projections of such an attractor are shown in Figure 6.5.4.

5.2. Find an IFS $\{\mathbb{R}^3; w_1, w_2, w_3, w_4\}$, where the transformations are affine, whose attractor is the graph of a continuous function f: $[0,1] \to \mathbb{R}^2$ such that $f([0,1]) = A$, where A is the set represented in Figure 6.5.5.

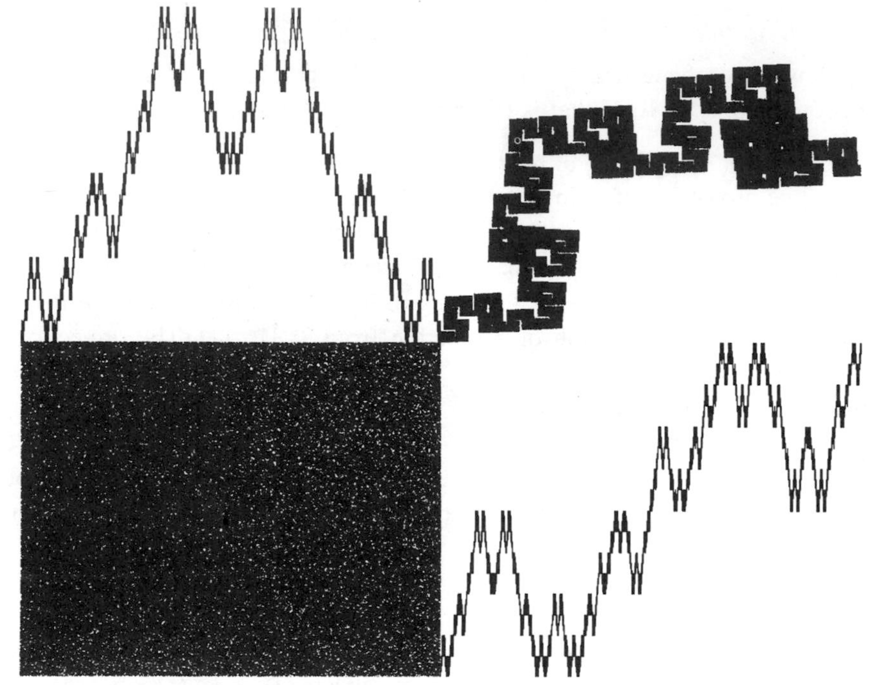

Figure 6.5.3(a)
Various views of the attractor of a certain IFS. From some points of view we see that it is the graph of a function. From one point of view it is clear that it is the graph of a space-filling curve!

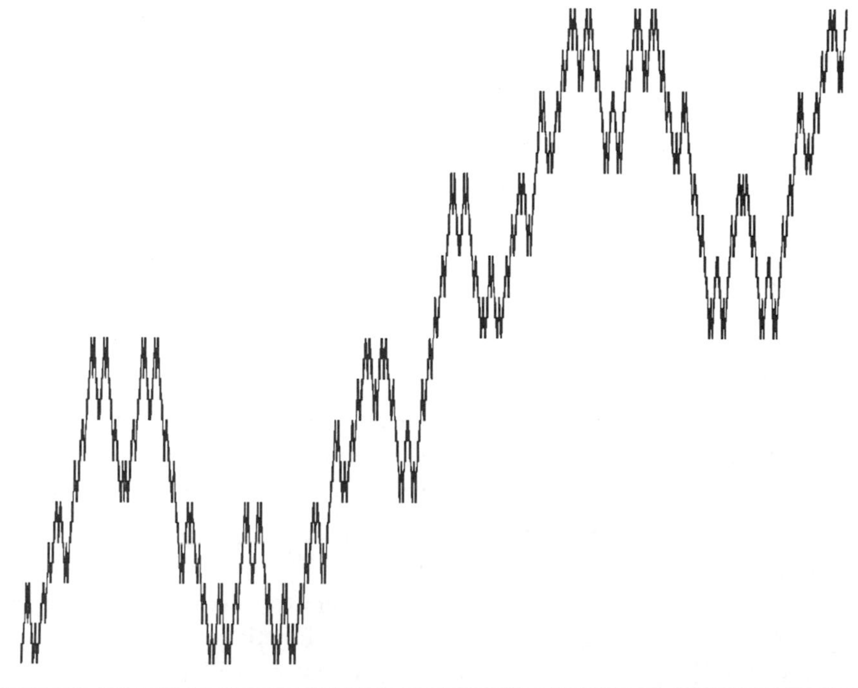

Figure 6.5.3(b)
Higher resolution view of the lower right panel in Figure 6.5.3(*a*).

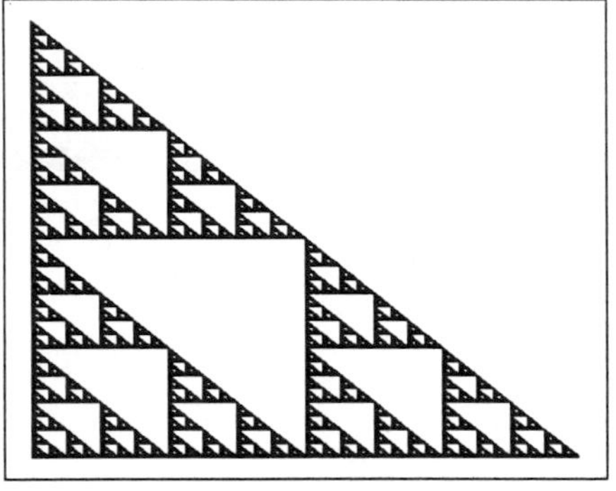

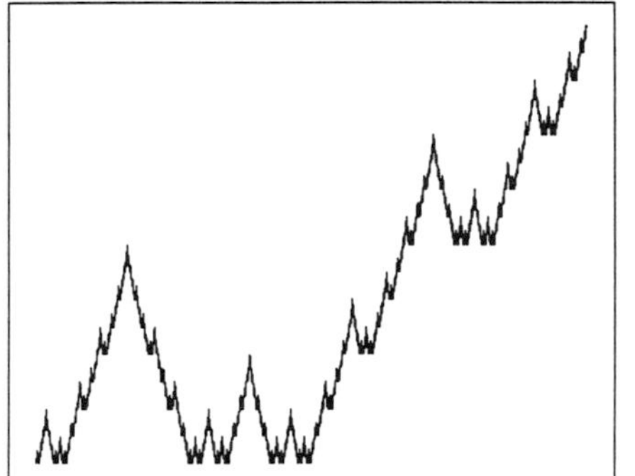
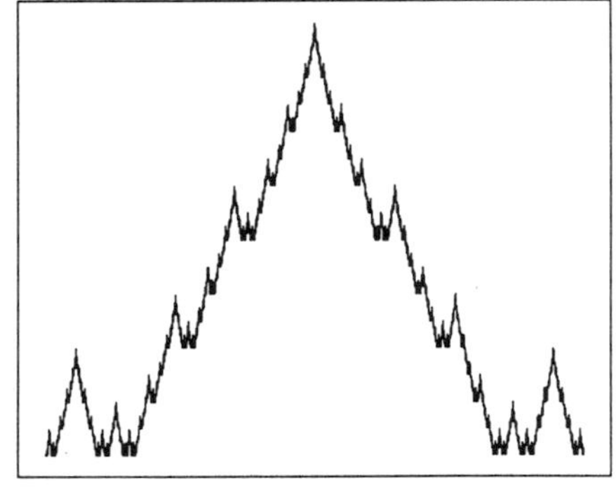

Figure 6.5.4
Four views of the attractor of an IFS. This attractor is the graph of a continuous function f: $[0,1] \to \mathbb{R}^2$ *such that* $f([0,1])$ *is a Sierpinski triangle. This function provides a "space-filling" curve, where the space is a fractal!*

Figure 6.5.5
Find an IFS $\{\mathbb{R}^3; w_1, w_2, w_3, w_4\}$, *where the transformations are affine, whose attractor is the graph of a continuous function* f: $[0,1] \to \mathbb{R}^2$ *such that* $f([0,1]) = A$, *where* A *is the set represented here.*

7 Julia Sets

7.1 THE ESCAPE TIME ALGORITHM FOR COMPUTING PICTURES OF IFS ATTRACTORS AND JULIA SETS

Let us consider the dynamical system $\{\mathbb{R}^2; f\}$ where $f\colon \mathbb{R}^2 \to \mathbb{R}^2$ is defined by

$$f(x, y) = \begin{cases} (2x, 2y - 1) & \text{if } y \geq 0.5, \\ (2x - 1, 2y) & \text{if } x \geq 0.5 \text{ and } y < 0.5, \\ (2x, 2y) & \text{otherwise.} \end{cases}$$

This dynamical system is related to the IFS

$$\{\mathbb{R}^2; w_1(x, y) = (0.5x, 0.5y + 0.5),$$
$$w_2(x, y) = (0.5x + 0.5, 0.5y), w_3(x, y) = (0.5x, 0.5y)\}.$$

The attractor of the IFS is a Sierpinski triangle with vertices at $(0, 0)$, $(0, 1)$, and $(1, 0)$. The relationship between the dynamical system $\{\mathbb{R}^2; f\}$ and the IFS $\{\mathbb{R}^2; w_1, w_2, w_3\}$ is that $\{\ ; f\}$ is a shift dynamical system associated with the IFS. (Shift dynamical systems are discussed in Chapter 4, Section 4.) One readily verifies that f restricted to $\triangle$ satisfies

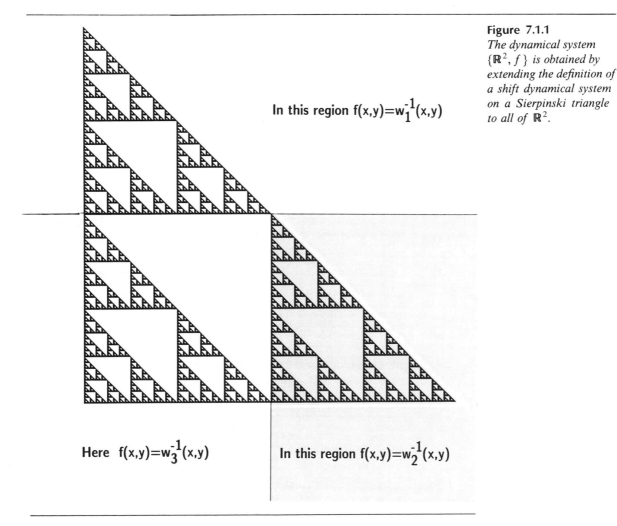

$$f(x,y) = \begin{cases} w_1^{-1}(x,y) & \text{if } (x,y) \in w_1(\triangle) \\ w_2^{-1}(x,y) & \text{if } (x,y) \in w_2(\triangle) \setminus \{(0.5,0.5)\}, \\ w_3^{-1}(x,y) & \text{if } (x,y) \in w_3(\triangle) \setminus \{(0,0.5),(0.5,0)\}. \end{cases}$$

In particular, f maps $\triangle$ onto itself. The dynamical system $\{\mathbb{R}^2; f\}$ is an extension of the shift dynamical system $\{\triangle; f\}$ to $\mathbb{R}^2$. The situation is illustrated in Figure 7.1.1.

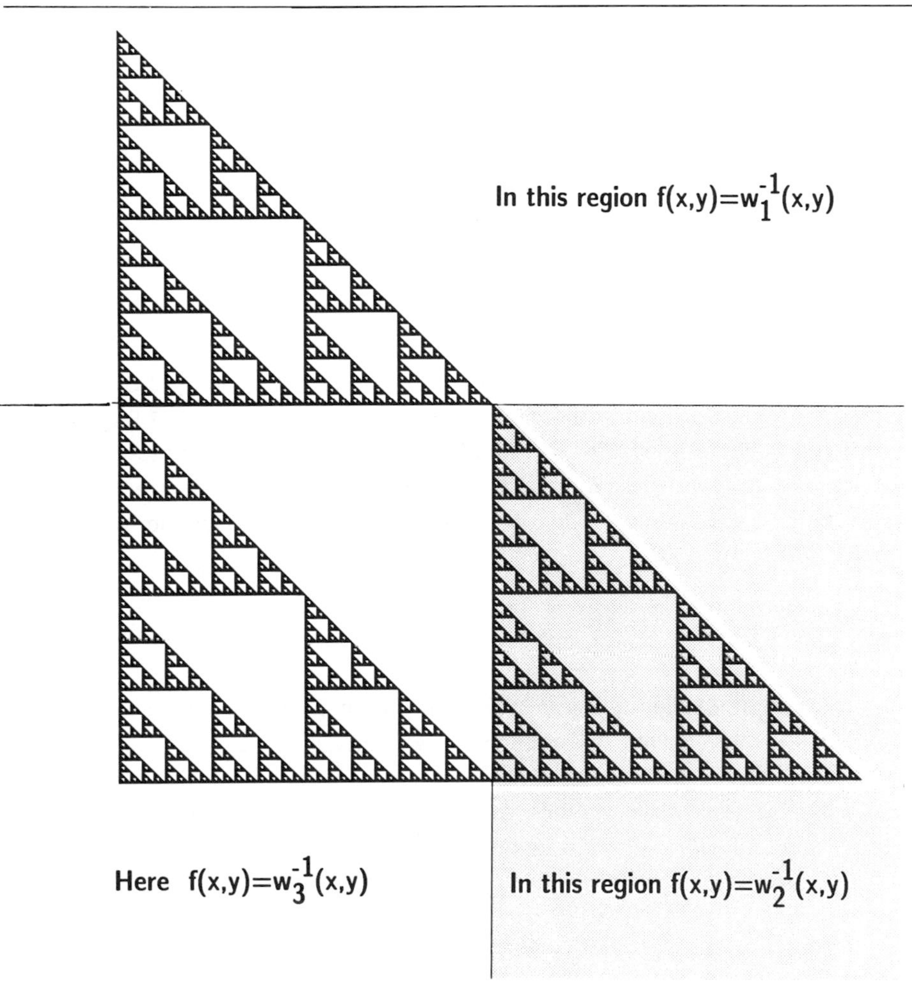

Figure 7.1.1 *The dynamical system $\{\mathbb{R}^2, f\}$ is obtained by extending the definition of a shift dynamical system on a Sierpinski triangle to all of $\mathbb{R}^2$.*

Let d denote the Euclidean metric on $\mathbb{R}^2$. The shift dynamical system $\{\mathbb{R}^2; f\}$ is "expanding:" for any pair of points x_1, x_2 lying in any one of the three domains associated with f, we have

$$d(f(x_1), f(x_2)) = 2d(x_1, x_2).$$

One can prove that the orbit $\{f^{\circ n}(x)\}_{n=0}^{\infty}$ diverges towards infinity if x does not belong to △. That is

$$d(O, f^{\circ n}(x)) \to \infty \text{ as } n \to \infty \qquad \text{for any point } x \in \mathbb{R}^2 \setminus △.$$

What happens if we compute numerically the orbit of a point $x \in$ △? Recall that the fractal dimension of △ is log(3)/log(2). This tells us that △ is "very small" compared to $\mathbb{R}^2$. Hence, although $f($△$) =$ △, errors in a computed orbit are likely to produce points which do not lie on △. This means that, in practice, most numerically computed orbits will diverge, regardless of whether or not the initial point lies on △. The Sierpinski triangle △ is an "unstable" invariant set for the transformation $f: \mathbb{R}^2 \to \mathbb{R}^2$. It is an attractive fixed point for the transformation $W: \mathscr{H}(\mathbb{R}^2) \to \mathscr{H}(\mathbb{R}^2)$, where $W = w_1 \cup w_2 \cup w_3$ is defined in the usual manner.

Intuitively, we expect that orbits of the dynamical system $\{\mathbb{R}^2; f\}$ which start close to △ should "take longer to diverge" than those which start far from △. How fast do different orbits diverge? Here we describe a numerical, computergraphical experiment to compare the number of iterations required for the orbits of different points to escape from a ball of large radius, centered at the origin. Let (a, b) and (c, d), respectively, denote the coordinates of the lower-left corner and the upper-right corner of a closed, filled rectangle $\mathscr{W} \subset \mathbb{R}^2$. Let M denote a positive integer, and define an array of points in $\mathscr{W}$ by

$$x_{p,q} = \left(a + p\frac{(c-a)}{M}, b + q\frac{(d-b)}{M}\right) \qquad \text{for } p, q = 0, 1, 2, \ldots, M.$$

In the experiment these points will be represented by pixels on a computer graphics display device. We compare the orbits $\{f^{\circ n}(x_{p,q}):\}_{n=0}^{\infty}$ for $p, q = 0, 1, 2, \ldots, M$.

Let R be a positive number, sufficiently large that the ball with center at the origin and radius R, contains both △ and $\mathscr{W}$. Define

$$\mathscr{V} = \{(x, y) \in \mathbb{R}: x^2 + y^2 > R\}.$$

A possible choice for the rectangle $\mathscr{W}$ and the set $\mathscr{V}$, in relation to △, is

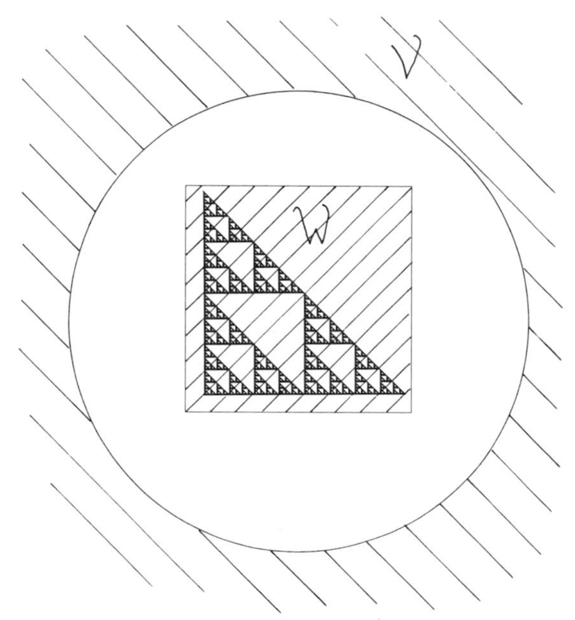

Figure 7.1.2
How long do orbits of points in $\mathscr{W}$ take to arrive in $\mathscr{V}$? We expect that the number of iterations required should tell us something about the structure of $\triangle$.

illustrated in Figure 7.1.2. In order that the comparison of orbits provides information about $\triangle$, one should choose $\mathscr{W}$ so that $\mathscr{W} \cap \triangle \neq \varnothing$.

Let *numits* denote a positive integer. The following Program computes a finite set of points $\{f^{\circ 1}(x_{p,q}), f^{\circ 2}(x_{p,q}), f^{\circ 3}(x_{p,q}), \ldots, f^{\circ n}(x_{p,q})\}$ belonging to the orbit of $x_{p,q} \in \mathscr{W}$, for each $p, q = 1, 2, \ldots, M$. The total number of points computed on an orbit is at most *numits*. If the set of computed points of the orbit of $x_{p,q}$ does not include a point in $\mathscr{V}$ when $n = numits$, then the computation passes to the next value of (p, q). Otherwise, the pixel corresponding to $x_{p,q}$ is rendered in a color indexed by the first integer n such that $f^{\circ n}(x_{p,q}) \in \mathscr{V}$, and then the computation passes to the next value of (p, q). This provides a computergraphical method for comparing how long the orbits of different points in $\mathscr{W}$ take to reach $\mathscr{V}$.

The following program is written in BASIC. It runs without modification on an IBM PC with Enhanced Graphics Adaptor and Turbobasic. On any line the words preceded by a ' are comments: they are not part of the program.

PROGRAM 7.1.1 (Example of the Escape Time Algorithm)

```
numits = 20: a = 0: b = 0: c = 1: d = 1: M = 100   'Define viewing
                                                    window, W, and numits.
R = 200   'Define the region V.
screen 9: cls   'Initialize graphics.
for p = 1 to M
```

```
for q = 1 to M
x = a + (c − a)*p/M: y = b + (d − b)*q/M   'Specify the initial
                                             point of an orbit, x(p, q).
for n = 1 to numits   'Compute at most numits points on the orbit of
                       x(p, q).
                                  'Evaluate f applied to the previous point on
if y > 0.5 then                    the orbit.
x = 2*x: y = 2*y − 1
else if x > 0.5 then
x = 2*x − 1: y = 2*y               THE FORMULA FOR THE FUNCTION f(x)
else
x = 2*x: y = 2*y
end if
150 if x*x + y*y > R then   'If the most recently computed point lies
                             in 𝒱 then . . .
160 pset(p, q), n: n = numits   '. . . render the pixel x(p, q) in color n, and
                                 go to the next (p, q).
170 end if
if instat then end   'Stop computing if any key is pressed!
next n: next q: next p
end
```

Color Plate 7.1.1 shows the result of running a version of Program 7.1.1 on a Masscomp 5600 workstation with Aurora graphics.

In Figure 7.1.3 we show the result of running a version of Program 7.1.1, but this time in black-and-white. A point is plotted in black if the number of iterations required to reach $\mathscr{V}$ is an odd integer, or if the orbit of the point does not reach $\mathscr{V}$ during the first *numits* iterations.

In Figure 7.1.4 we show the result of running a version of Program 7.1.1, with $(a, b) = (0, 0)$, $(c, d) = (5 \times 10^{-18}, 5 \times 10^{-18})$, and *numits* = 105. This viewing window is minute. See also Color Plate 7.1.2. *Now* you should be convinced that △ is not simplified by magnification.

The dynamical system $\{\mathbb{R}^2; f\}$ contains deep information about the "repelling" set △. Some of this information is revealed by means of the Escape Time Algorithm. The orbits of points which lie close to △ do indeed appear to take longer to escape from $\mathbb{R}^2 \setminus \mathscr{V}$ than those of points which lie further away from △.

Exercises & Examples

1.1. Let $\{\mathbb{R}^2, f\}$ denote the dynamical system defined at the start of this chapter, and let △ denote the associated Sierpinski triangle. Prove

Figure 7.1.3
Output from a modified version of Program 7.1.1. *A pixel is rendered in black if either the number of iterations required to reach* $\mathscr{V}$ *is an odd integer, or the orbit does not reach* $\mathscr{V}$ *during the first numits iterations.*

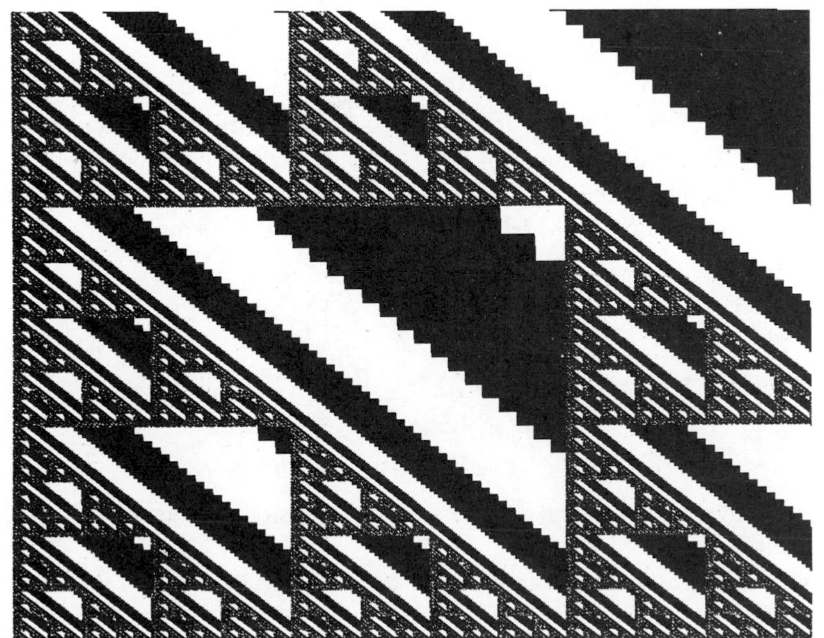

Figure 7.1.4
Here we show the result of running a version of Program 7.1.1, *with* $(a, b) = (0, 0)$, $(c, d) = (5 \times 10^{-18}, 5 \times 10^{-18})$, *and numits* = 65. *This viewing window is minute, yet the computation time was not significantly increased. If we did not know it before, we are now convinced that* *is not simplified by magnification.*

that the orbit $\{f^{\circ n}(x)\}_{n=0}^{\infty}$ diverges, for each $x \in \mathbb{R}^2 \backslash \blacktriangle$. That is, prove that $d(O, f^{\circ n}(x)) \to \infty$ as $n \to \infty$ for each $x \in \mathbb{R}^2 \backslash \blacktriangle$.

1.2. Rewrite Program 7.1.1 in a form suitable for your own computergraphical environment, then run it and obtain hardcopy of the output.

1.3. If the Escape Time Algorithm is applied to the dynamical system $\{\mathbb{R}^2;\ f(x, y) = (2x, 2y)\}$, what will be the general appearance of resulting colored regions?

1.4. By changing the window size in Program 7.1.1, obtain images of 'zooms' on the Sierpinski triangle. For example, use the following windows: $(0, 0)$–$(0.5, 0.5)$; $(0, 0)$–$(0.25, 0.25)$; $(0, 0)$–$(0.125, 0.125)$; .... How must the total number of iterations, *numits*, be adjusted as a function of window size in order that the quality of the images remains (approximately) uniform? Make a graph of the total number of iterations against the window size. Is there a possible relationship between the behaviour of *numits* as a function of window size, and the fractal dimension of the Sierpinski triangle? Make a hypothesis and test it experimentally.

Here we construct another example of a dynamical system whose orbits "try to escape" from the attractor of an IFS. This time we treat an IFS whose attractor has nonempty interior. Consider the hyperbolic IFS$\{\mathbb{R}^2;\ w_1, w_2\}$, where

$$w_1\begin{bmatrix} x \\ y \end{bmatrix} = \begin{bmatrix} 0 & -s^{-1} \\ s^{-1} & 0 \end{bmatrix}\begin{bmatrix} x \\ y \end{bmatrix} + \begin{bmatrix} 1 \\ 0 \end{bmatrix},$$

$$w_2\begin{bmatrix} x \\ y \end{bmatrix} = \begin{bmatrix} 0 & -s^{-1} \\ s^{-1} & 0 \end{bmatrix}\begin{bmatrix} x \\ y \end{bmatrix} - \begin{bmatrix} 1 \\ 0 \end{bmatrix}, \text{ and } s = \sqrt{2}.$$

The attractor of this IFS is a closed, filled rectangle, which we denote here by $\blacksquare$. This attractor is the union of two copies of itself, each scaled by a factor $1/\sqrt{2}$, rotated about the origin anticlockwise through $90°$, and then translated horizontally, one copy to the left and one to the right. The inverse transformations are

$$w_1^{-1}\begin{bmatrix} x \\ y \end{bmatrix} = \begin{bmatrix} 0 & s \\ -s & 0 \end{bmatrix}\begin{bmatrix} x \\ y \end{bmatrix} + \begin{bmatrix} 0 \\ s \end{bmatrix}, \quad \text{and} \quad w_2^{-1}\begin{bmatrix} x \\ y \end{bmatrix} = \begin{bmatrix} 0 & s \\ -s & 0 \end{bmatrix}\begin{bmatrix} x \\ y \end{bmatrix} - \begin{bmatrix} 0 \\ s \end{bmatrix}.$$

Define $f\colon \mathbb{R}^2 \to \mathbb{R}^2$ by

$$f(x, y) = \begin{cases} w_1^{-1}(x, y) & \text{if } x > 0 \\ w_2^{-1}(x, y) & \text{when } x \leq 0. \end{cases}$$

Then the dynamical system $\{\mathbb{R}^2;\ f\}$ is an extension of the shift dynamical system $\{\blacksquare;\ f\}$ to $\mathbb{R}^2$.

What happens when we apply the Escape Time Algorithm to this dynamical system? To see, one can replace the function $f(x)$ in Program 7.1.1 by

```
if x > 0 then
newx = s*y: newy = -s*x + s
else
newx = s*y: newy = -s*x - s
end if
x = newx: y = newy
```

THE FORMULA FOR THE FUNCTION f(x).

Results of running Program 7.1.1, thus modified, with the window $\mathscr{W}$ and the escape region $\mathscr{V}$ chosen appropriately, are shown in Figure 7.1.5 and in Color Plate 7.1.3.

It appears that the orbits of points in the interior ▬ do not escape. This is not surprising. The fractal dimension of the attractor of the IFS is the same as the fractal dimension of $\mathbb{R}^2$, so small computational errors are unlikely to knock the orbit off the invariant set. It also appears that the orbits of points which lie in $\mathbb{R}^2 \setminus$ ▬ reach $\mathscr{V}$ after fewer and fewer iterations, the further away from ▬ they start.

Again we see that the Escape Time Algorithm provides a means for the computation of the attractor of an IFS. Indeed, we have here the bare bones of

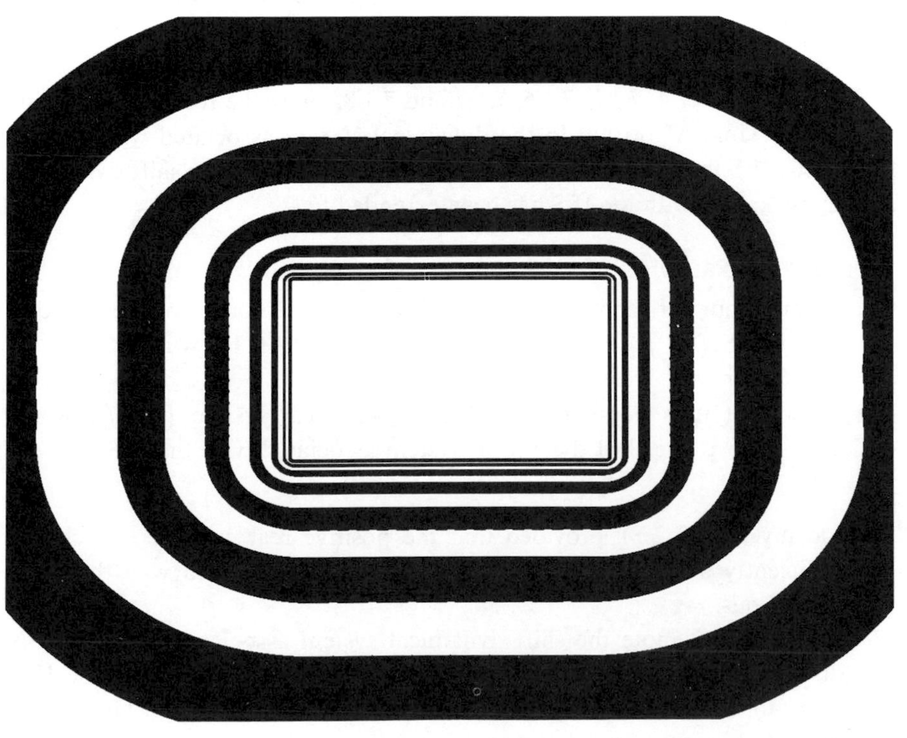

Figure 7.1.5 *An image of an IFS attractor computed using the Escape Time Algorithm. This time the attractor of the IFS is a filled rectangle and the computed orbits of points in* ▬ *seem never to escape.*

a new algorithm for computing images of the attractors of some hyperbolic IFS on $\mathbb{R}^2$. Here are the main steps: (a) Find a dynamical system $\{\mathbb{R}^2; f\}$ which is an extension of a shift dynamical system associated with the IFS, and which tends to transform points off the attractor of the IFS to new points which are further away from the attractor. (This is always possible if the IFS is totally disconnected. The tricky part is to find a formula for $f(x)$, one which can be input conveniently into a computer. In the case of affine transformations in $\mathbb{R}^2$, one can often define the extensions of the domains of the inverse transformations with the aid of straight lines.) (b) Apply the Escape Time Algorithm, with $\mathscr{V}$ and $\mathscr{W}$ chosen appropriately, but plot only those points whose numerical orbits require sufficiently many iterations before they reach $\mathscr{V}$.

For example, in Program 7.1.1 as it stands, one can replace the three lines 150, 160, and 170 by the two lines

```
150 if n = numits then pset(p, q), 1
160 if x*x + y*y > R then n = numits
```

and define *numits* = 10. If the value of *numits* is too high, then very few points will not escape from $\mathscr{W}$ and a poor image of 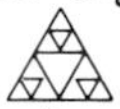 will result. If the value of *numits* is too low, then a coarse image of the  will be produced. An image of an IFS attractor computed using the Escape Time Algorithm, modified as described here, is shown in Figure 7.1.6.

Color Plates 7.1.4, 7.1.5, 7.1.6, 7.1.7 and 7.1.8, show the results of applying the Escape Time Algorithm to the dynamical system associated with various hyperbolic IFS in $\mathbb{R}^2$. In each case the maps are affine, and the shift dynamical system associated with the IFS has been extended to $\mathbb{R}^2$.

Exercises & Examples

1.5. Modify your version of Program 7.1.1 to compute images of the attractor of the IFS $\{\mathbb{C};\ w_1(z) = re^{i\theta}z - 1,\ w_2(z) = re^{i\theta}z + 1\}$, when $r = 1/\sqrt{2}$ and $\theta = \pi/2$.

1.6. Show that it is possible to define a dynamical system $\{\mathbb{C}; f\}$ which extends to $\mathbb{C}$ the shift dynamical system associated with the IFS

$$\{\mathbb{C};\ w_1(z) = re^{i\theta}z - 1,\ w_2(z) = re^{i\theta}z + 1\},$$

for any $\theta \in [0, 2\pi)$, provided that the positive real number r is chosen sufficiently small. Note that this can be done in such a way that f is continuous.

1.7. Let $\{A; f\}$ denote the shift dynamical system associated with a totally disconnected hyperbolic IFS in $\mathbb{R}^2$. A denotes the attractor of the IFS. Show that there are many ways to define a dynamical system $\{\mathbb{R}^2; g\}$ so that $f(x) = g(x)$ for all $x \in A$.

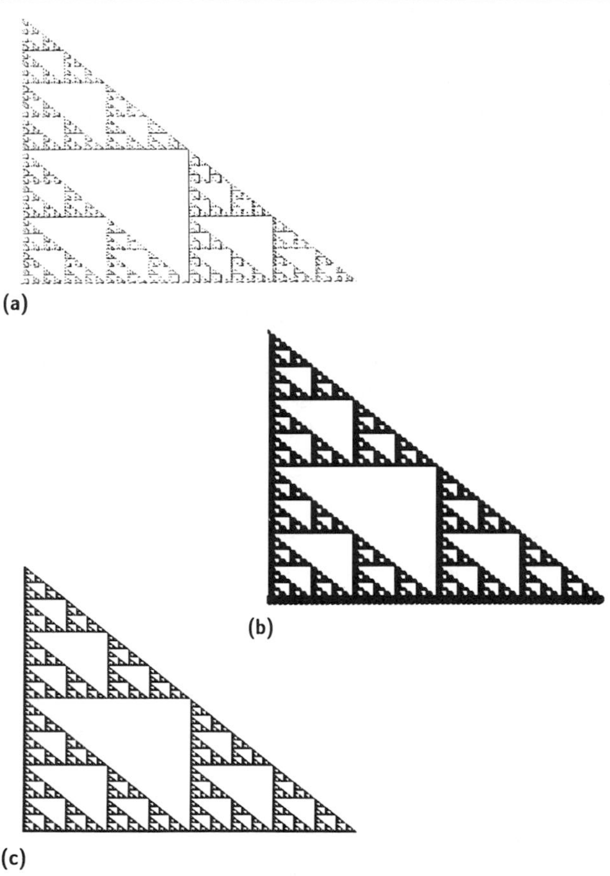

Figure 7.1.6
Images of an IFS attractor computed using the Escape Time Algorithm. Only points whose orbits have not escaped from $\mathbb{R} \setminus \mathscr{V}$ *after numits iterations are plotted. The value for numits must be chosen not too large, as in (a), and not too small, as in (b), but just right, as in (c).*

The Escape Time Algorithm can be applied, often with interesting results, to any dynamical system of the form $\{\mathbb{R}^2; f\}$, $\{\mathbb{C}; f\}$, or $\{\hat{\mathbb{C}}; f\}$. One needs only to specify a viewing window $\mathscr{W}$, and a region $\mathscr{V}$ to which orbits of points in $\mathscr{W}$ might escape. The result will be a "picture" of $\mathscr{W}$ wherein the pixel corresponding to the point z is colored according to the smallest value of the positive integer n such that $f^{\circ n}(z) \in \mathscr{V}$. A special color, such as black, may be reserved to represent points whose orbits do not reach $\mathscr{V}$ before $(numits + 1)$ iterations.

What would happen if the Escape Time Algorithm were applied to the dynamical system $f: \hat{\mathbb{C}} \to \hat{\mathbb{C}}$ defined by $f(z) = z^2$? This transformation can be expressed $f(x, y) = (x^2 - y^2, 2xy)$. From the discussion of the quadratic transformation in Chapter 3, Section 4, we know that the orbits of points in the complement of the unit disk $F = \{z \in \mathbb{C}: |z| \leq 1\}$ converge to the Point at Infinity. Orbits of points in the interior of F converge to the Origin. So if $\mathscr{W}$ is a rectangle which contains F and if the radius R, which defines $\mathscr{V}$, is sufficiently large, then we expect that the Escape Time Algorithm would yield pictures of F surrounded by concentric rings of different colors. The reader should verify this!

F is called the *filled Julia set* associated with the polynomial transformation $f(z) = z^2$. The boundary of F is called the *Julia set* of f, and we denote it by J. It consists of the circle of radius one, centered at the origin. One can think of J on the Riemann Sphere as being represented by the Equator on a globe. This Julia set separates those points whose orbits converge to the Point at Infinity from those whose orbits converge to the origin. Orbits of points on J itself cannot escape, either to infinity or to the origin. In fact $J \in \mathscr{H}(\hat{\mathbb{C}})$ and $f(J) = J = f^{-1}(J)$. It is an "unstable" fixed point for the transformation f: $\mathscr{H}(\hat{\mathbb{C}}) \to \mathscr{H}(\hat{\mathbb{C}})$.

Definition 1. Let f: $\hat{\mathbb{C}} \to \hat{\mathbb{C}}$ denote a polynomial of degree greater than one. Let F_f denote the set of points in $\mathbb{C}$ whose orbits do not converge to the Point at Infinity. That is

$$F_f = \left\{ z \in \mathbb{C} : \{ |f^{\circ n}(z)| \}_{n=0}^{\infty} \text{ is bounded} \right\}.$$

This set is called the *filled Julia set* associated with the polynomial f. The boundary of F_f is called the *Julia set* of the polynomial f, and it is denoted by J_f.

Theorem 1. *Let f: $\hat{\mathbb{C}} \to \hat{\mathbb{C}}$ denote a polynomial of degree greater than one. Let F_f denote the filled Julia set of f and let J_f denote the Julia set of f. Then F_f and J_f are nonempty compact subsets of $\mathbb{C}$; that is, $F_f \in \mathscr{H}(\mathbb{C})$ and $J_f \in \mathscr{H}(\mathbb{C})$. Moreover $f(J_f) = J_f = f^{-1}(J_f)$ and $f(F_f) = F_f = f^{-1}(F_f)$. The set $\mathscr{V}_\infty = \hat{\mathbb{C}} \setminus F_f$ is pathwise connected.*

Proof. We outline the proof for the one-parameter family of transformations f_λ: $\hat{\mathbb{C}} \to \hat{\mathbb{C}}$ defined by

$$f_\lambda(z) = z^2 - \lambda, \qquad \text{where } \lambda \in \mathbb{C} \text{ is the parameter.}$$

The general case is treated in [Blanchard 1984], [Brolin 1966], [Fatou 1919], [Julia 1918], for example. This outline proof is constructed to provide information about the relationship between the Theorem and the Escape Time Algorithm. Some of the ideas and notation used here are illustrated in Figure 7.1.7.

Let J_λ denote the Julia set for f_λ and let F_λ denote the filled Julia set for f_λ. Let d denote the Euclidean metric on $\mathbb{C}$ and let

$$R > 0.5 + \sqrt{0.25 + |\lambda|}.$$

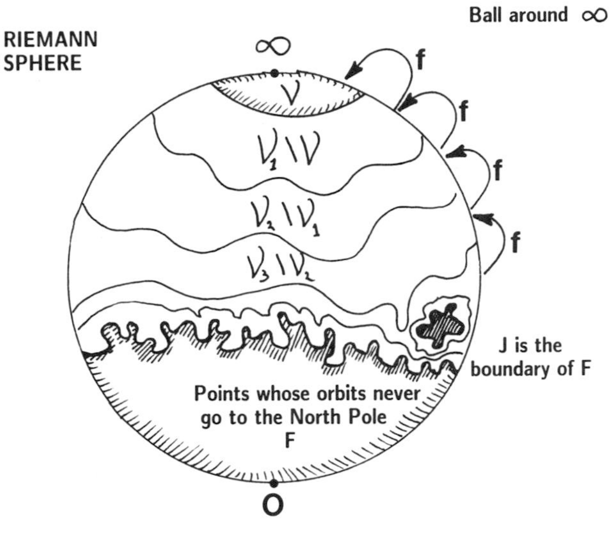

Figure 7.1.7 *Illustration showing what is going on in the proof of Theorem 7.1.1. This illustrates the increasing sequence of sets $\{\mathscr{V}_n\}$ which converges to the boundary of the basin of attraction $\mathscr{V}_\infty$ of the Point at Infinity. It also shows the decreasing sequence of sets, K_n, the complements of the latter, which converge to the filled Julia set F_f. In general the origin, O, need not belong to F_f.*

Then it is readily verified that

$$d(O, f(z)) > d(O, z) \qquad \text{for all } z \text{ such that } d(0, z) \geq R.$$

Define

$$\mathscr{V} = \{z \in \mathbb{C} : |z| > R\} \cup \{\infty\}.$$

Then it follows that

$$f(\mathscr{V}) \subset \mathscr{V}.$$

One can prove that the orbit $\{f^{\circ n}(z)\}$ converges to ∞ for all $z \in \mathscr{V}$. No bounded orbit intersects $\mathscr{V}$. It follows that

$$F_\lambda = \{z \in \hat{\mathbb{C}} : f^{\circ n}(z) \notin \mathscr{V} \quad \text{for each finite positive integer } n\}.$$

That is, F_λ is the same as the set of points whose orbits do not intersect $\mathscr{V}$.

Now consider the sequence of sets

$$\mathscr{V}_n = f^{\circ -n}(\mathscr{V}) \qquad \text{for } n = 0, 1, 2, \ldots.$$

For each non-negative integer n, $\mathscr{V}_n$ is an open connected subset of ($\hat{\mathbb{C}}$, Spherical). $\mathscr{V}_n$ is open because $\mathscr{V}$ is open and f is continuous. $\mathscr{V}_n$ is connected because of the geometry of the quadratic transformation, described in Chapter 3, Section 4: the inverse image of a path which joins the Point at Infinity to any other point on the sphere contains a path which contains the Point at Infinity.

Since $f(\mathscr{V}) \subset \mathscr{V}$, it follows that $\mathscr{V} \subset f^{-1}(\mathscr{V})$. This implies that

$$\mathscr{V} = \mathscr{V}_0 \subset \mathscr{V}_1 \subset \mathscr{V}_3 \subset \cdots \subset \mathscr{V}_n \subset \cdots \tag{7.1.1}$$

For each non-negative integer n,

$$\mathscr{V}_n = \{z \in \hat{\mathbb{C}}: \{z, f^{\circ 1}(z), f^{\circ 2}(z), f^{\circ 3}(z), \ldots, f^{\circ n}(z)\} \cap \mathscr{V} \neq \varnothing\}.$$

That is, $\mathscr{V}_n$ is the set of points whose orbits require at most n iterations to reach $\mathscr{V}$. Let

$$K_n = \hat{\mathbb{C}} \setminus \mathscr{V}_n \qquad \text{for } n = 0, 1, 2, 3, \ldots.$$

Then K_n is the set of points whose orbits do not intersect $\mathscr{V}$ during the first n iterations. That is

$$K_n = \{z \in \hat{\mathbb{C}}: \{z, f^{\circ 1}(z), f^{\circ 2}(z), f^{\circ 3}(z), \ldots, f^{\circ n}(z)\} \cap \mathscr{V} = \varnothing\}.$$

For each non-negative integer n, K_n is a nonempty compact subset of the metric space ($\hat{\mathbb{C}}$, Spherical). How do we know that K_n is nonempty? Because we can calculate that f possesses a fixed point $z_f \in \mathbb{C}$, by solving the equation

$$f(z_f) = z_f^2 - \lambda = z_f.$$

The orbit of z_f converges to z_f. Hence it cannot belong to $\mathscr{V}_n$ for any non-negative integer n. Hence $z_f \in K_n$ for each non-negative integer n.

Equation (7.1.1) implies that

$$K_0 \supset K_1 \supset K_2 \supset K_3 \supset \cdots \supset K_n \supset \cdots.$$

It follows that $\{K_n\}$ is a Cauchy sequence in $\mathscr{H}(\hat{\mathbb{C}})$. It follows that $\{K_n\}$ converges to a point in $\mathscr{H}(\hat{\mathbb{C}})$. The limit is the set of points whose orbits do not intersect $\mathscr{V}$. Hence

$$F_\lambda = \lim_{n \to \infty} K_n = \bigcap_{n=0}^{\infty} K_n$$

and we deduce that F_λ belongs to $\mathscr{H}(\hat{\mathbb{C}})$.

The equation

$$K_{n+1} = f^{\circ -1}(K_n) \qquad \text{for } n = 0, 1, 2, \ldots$$

now implies, as in the proof of Theorem 7.4.1, that

$$F_\lambda = f^{\circ -1}(F_\lambda).$$

Since f is an onto map, we obtain $f(F_\lambda) = F_\lambda$.

Let us now consider the boundary of F_λ, namely the Julia set J_λ for the dynamical system $\{\hat{\mathbb{C}}; f_\lambda\}$. Let $z \in$ interior (F_λ). Then the continuity of f implies $f^{-1}(z) \subset$ interior (F_λ). Hence $F_\lambda \supset f^{-1}(\partial F_\lambda) \supset \partial F_\lambda$. Now suppose that $z \in f^{-1}(\partial F_\lambda)$. Let O be any open ball which contains z. Since f is analytic, $f(O)$ is an open set, and it contains $f(z) \in \partial F_\lambda$. Hence $f(O)$ contains a point whose orbit converges to the Point at Infinity. It follows that $f^{-1}(\partial F_\lambda) \subset \partial F_\lambda$. We conclude that $f^{-1}(\partial F_\lambda) = \partial F_\lambda$ and in particular that $f(\partial F_\lambda) = \partial F_\lambda$. This completes the proof of the theorem.

We summarize some of what we discovered in the course of this proof. The filled Julia set F_λ is the limit of a decreasing sequence of compact sets. Its complement, which we denote by $\mathscr{V}_\infty$, is the limit of an increasing sequence $\{\mathscr{V}_n\}$ of open pathwise connected sets in $(\hat{\mathbb{C}}, \text{spherical})$. That is,

$$\mathscr{V}_\infty = \lim_{n\to\infty} \mathscr{V}_n = \bigcup_{n=0}^{\infty} \mathscr{V}_n.$$

The latter is called the *basin of attraction* of the Point at Infinity under the polynomial transformation f_λ. It is connected because each of the sets $\mathscr{V}_n$ is connected. We have

$$\hat{\mathbb{C}} = F_\lambda \cup \mathscr{V}_\infty.$$

$\mathscr{V}_\infty$ is open, connected, and nonempty. F_λ is compact and nonempty.

The Escape Time Algorithm provides us with a means for "seeing" the filled Julia sets F_λ, as well as the sequences of sets $\{\mathscr{V}_n\}$ and $\{K_n\}$ referred to in the Theorem. Let us look at what happens in the case $\lambda = 1.1$. Define $\mathscr{V}$ by choosing $R = 4$, and put $\mathscr{W} = \{(x, y)\colon -2 \le x \le 2, -2 \le y \le 2\}$. The function $f_{\lambda=1.1}\colon \mathbb{C} \to \mathbb{C}$ is given by the formula

$$f_{\lambda=1.1}(x, y) = (x^2 - y^2 - 1.1, 2xy) \qquad \text{for all } (x, y) \in \mathbb{C}.$$

An example of the result of running the Escape Time Algorithm, with $\mathscr{V}$, $\mathscr{W}$ and $f\colon \hat{\mathbb{C}} \to \hat{\mathbb{C}}$ thus defined, is shown in Figure 7.1.8. The black object represents the filled Julia set $F_{\lambda=1.1}$. The contours separate the regions $\mathscr{V}_{n+1} \setminus \mathscr{V}_n$, for some successive values of n. These contours also represent the

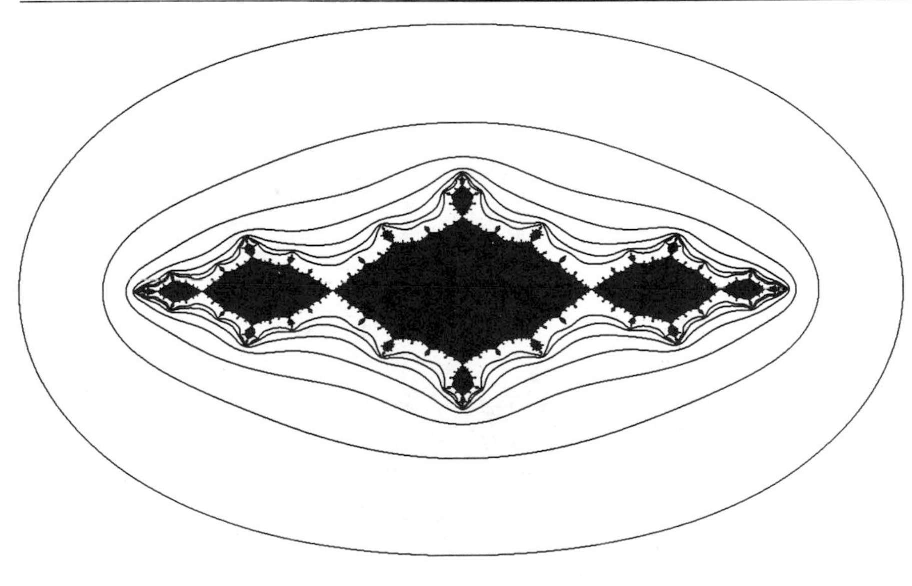

Figure 7.1.8 *The Escape Time Algorithm provides us with a means for "seeing" the filled Julia sets F_λ, as well as the sequences of sets $\{\mathscr{V}_n\}$ and $\{K_n\}$ referred to in Theorem* 7.1.1. *In this illustration,* $\lambda = 1.1$. *The black object represents the filled Julia set $F_{\lambda=1.1}$. The contours separate the regions $\mathscr{V}_{n+1} \setminus \mathscr{V}_n$, for some successive values of n. These contours also represent the boundaries of the regions K_n referred to in the proof of the theorem.*

boundaries of the regions K_n referred to in the proof of the theorem. We refer to them as escape time contours. Points in $\mathscr{V}_{n+1} \setminus \mathscr{V}_n$ have orbits which reach $\mathscr{V}$ in exactly $(n + 1)$ iterations. In Color Plate 7.1.8 we show another example of running the Escape Time Algorithm to produce an image of the same set. The regions $\mathscr{V}_{n+1} \setminus \mathscr{V}_n$ are represented by different colors.

Figure 7.1.9 shows a zoom on an interesting piece of $F_{\lambda=1.1}$, including parts of some escape time contours. This image was computed by choosing $\mathscr{W}$ to be a small rectangular subset of the window used in Figure 7.1.8.

Figures 7.1.10(a)–(e) shows pictures of the filled Julia sets F_λ for a set of real values of λ. These pictures also include a number of the escape time contours, to help indicate the location of F_λ. F_0 is a filled disk. As λ increases the set becomes more and more pinched together, until, when $\lambda = 2$, it is the closed interval $[-2, 2]$. For some values of $\lambda \in [0, 2]$, it appears that F_λ has no interior, and is "tree-like;" for other values it seems to possess a roomy interior. It also appears that F_λ is connected for all $\lambda \in [0, 2]$, and totally disconnected when $\lambda > 2$. In the latter case F_λ may be described as a "Cantor-like" set, or as a "dust." The transition between the totally disconnected set and the connected, bubbly set as the parameter λ is varied reminds

Figure 7.1.9
Zoom on an interesting piece of Figure 7.1.8.

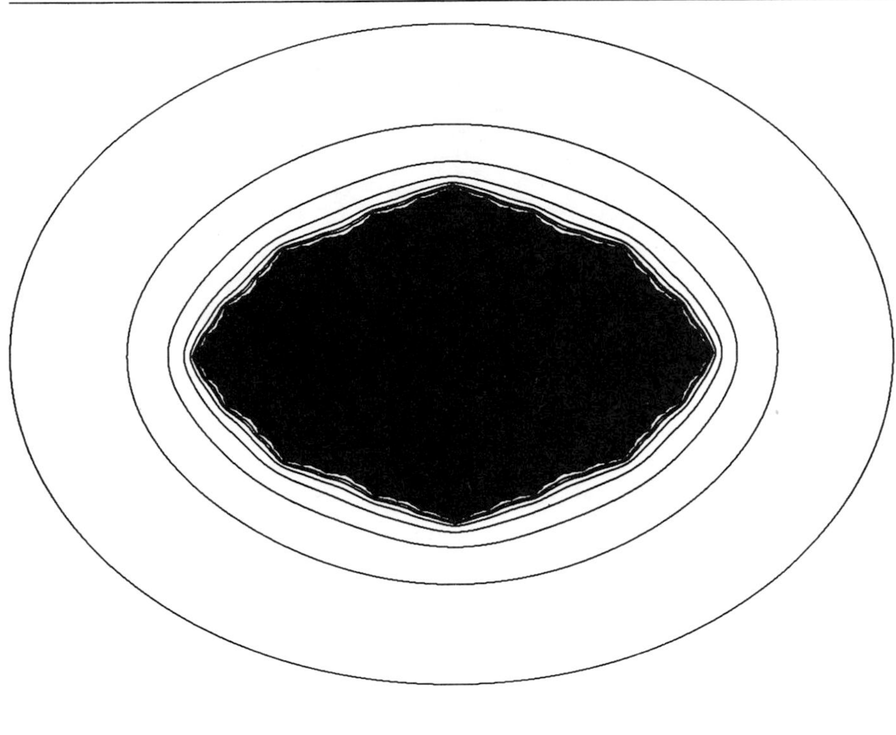

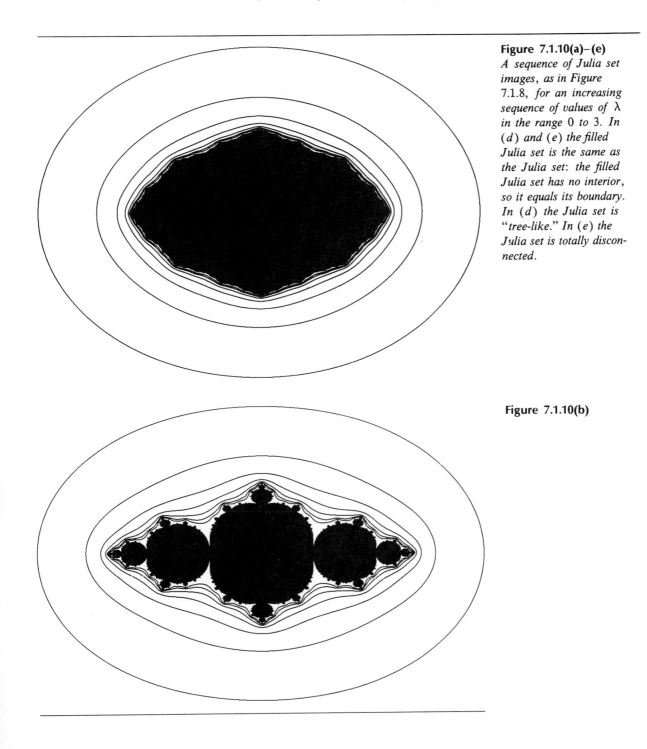

Figure 7.1.10(a)–(e) *A sequence of Julia set images, as in Figure* 7.1.8, *for an increasing sequence of values of* λ *in the range* 0 *to* 3. *In* (*d*) *and* (*e*) *the filled Julia set is the same as the Julia set: the filled Julia set has no interior, so it equals its boundary. In* (*d*) *the Julia set is "tree-like." In* (*e*) *the Julia set is totally disconnected.*

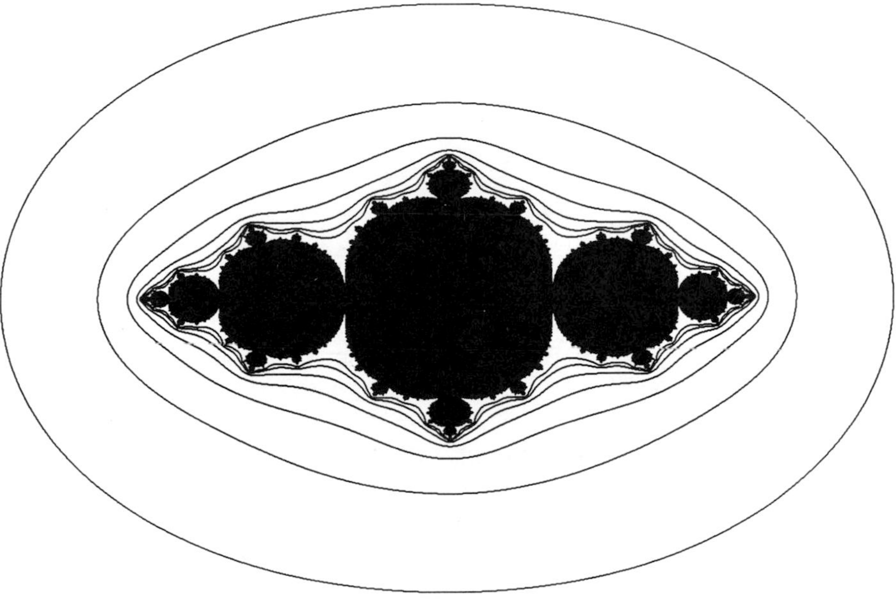

Figure 7.1.10(b)

Figure 7.1.10(c)

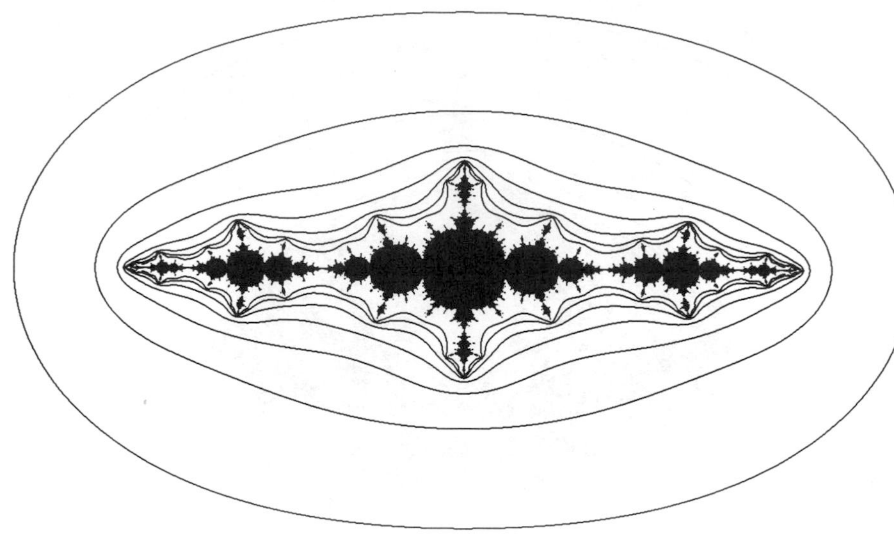

Figure 7.1.10(d)

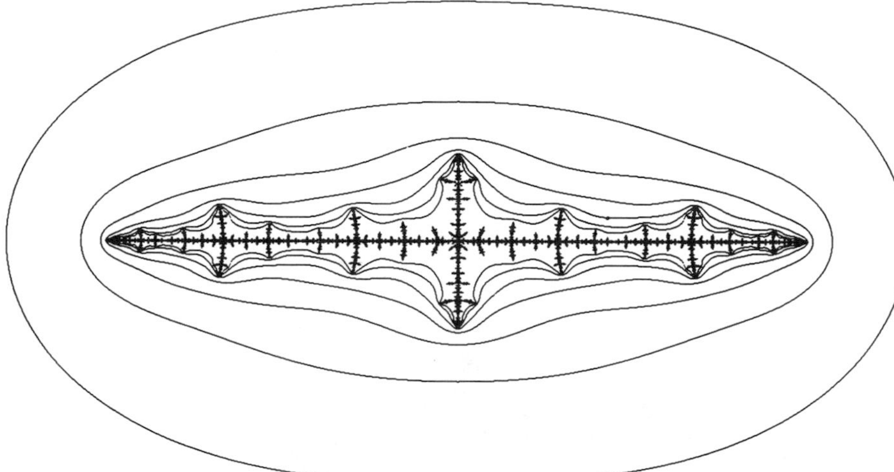

Figure 7.1.10(e)

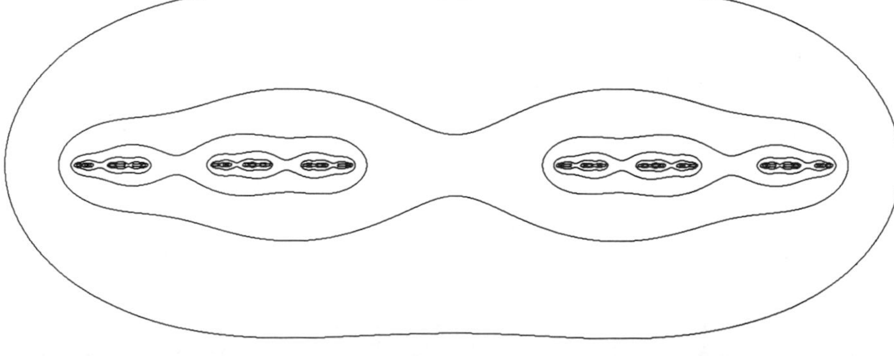

us of the transition between the Cantor Set and the Sierpinski triangle, discussed in connection with Figure 4.5.3.

Exercises & Examples

1.8. Modify your version of the Escape Time Algorithm to allow you to compute pictures of filled Julia sets for the family of quadratic polynomials $z^2 - \lambda$ for complex values of λ. Compute a picture of the filled Julia set for $\lambda = i$ and obtain hardcopy of the output.

1.9. Give the iteration formulas, and find a suitable value for R in terms of $|\lambda|$, so that the Escape Time Algorithm can be applied to the complex polynomial $z^3 - \lambda$.

1.10. Study web diagrams associated with $z^2 - \lambda$, for increasing values of $\lambda \in [0, 3]$. Speculate on the relation of these diagrams to the corresponding filled Julia sets.

1.11. Let $\lambda \in [0, 0.7] \cup [0.8, 1.2]$. Let $\mathscr{V}$ be an open ball of radius 0.00001 centered at the origin. Apply the Escape Time Algorithm to the dynamical system $\{\mathbb{C}, z^2 - \lambda\}$ with this choice of $\mathscr{V}$. Obtain computergraphical data in support of the hypothesis that, in this case, the algorithm yields approximate pictures of pieces of the closure of $\mathbb{C} \setminus F_\lambda$. Design an escape region $\mathscr{V}$ so that, for $\lambda \in [0, 0.7] \cup [0.8, 1.2]$, the Escape Time Algorithm yields approximate pictures of J_λ.

1.12. The Escape Time Algorithm introduces numerical errors in the computation of orbits. These errors should lead to inaccuracies in the computed pictures of Julia sets and IFS attractors. Consider the application to the filled Julia set for $z^2 - 1$. By means of computergraphical experiments, determine the importance of these errors in the images which you compute. One way to proceed is to choose successively smaller windows $\mathscr{W}$, which intersect the apparent boundary of the filled Julia set, and to seek the window size at which the quality of computed images seems to deteriorate. (You will need to increase the maximum number of iterations, M, as you zoom.) Can you give evidence to show that the apparently deteriorated images are not, in fact, correct?

1.13. Figure 7.1.11 was computed by applying the Escape Time Algorithm to the dynamical system $\{\mathbb{C};\ f(z) = z^4 - z - 0.78\}$. The viewing window is $\mathscr{W} = \{(x, y)\colon\ -1 \le x \le 1,\ -1 \le y \le 1\}$. Determine the escape region $\mathscr{V}$. Also, you might like to try magnifying one of the little faces in this image.

1.14. The images in Figure 7.1.12(a), (b), (c) and (d) represent the nontrivially distinct attractors of all IFS of the form $\{\blacksquare;\ w_1, w_2, w_3\}$ where the maps are similitudes of scaling factor one-half, and rotation angles in the set $\{0°, 90°, 180°, 270°\}$. The three translations $(0, 0)$, $(1, 0)$ and $(0, 1)$ are used. These IFS are all just-touching. For $i \ne j$ the set $w_i(A) \cap w_j(A)$, is contained in one of the two straight lines $x = 1$ or $y = 1$. Show that,

Figure 7.1.11 *This image was computed by applying the Escape Time Algorithm to the dynamical system* $\{\mathbb{C}; f(z) = z^4 - z - 0.78\}$. *The viewing window is* $\mathscr{W} = \{(x, y)\colon -1 \le x \le 1, -1 \le y \le 1\}$. *Can you determine the escape region* $\mathscr{V}$?

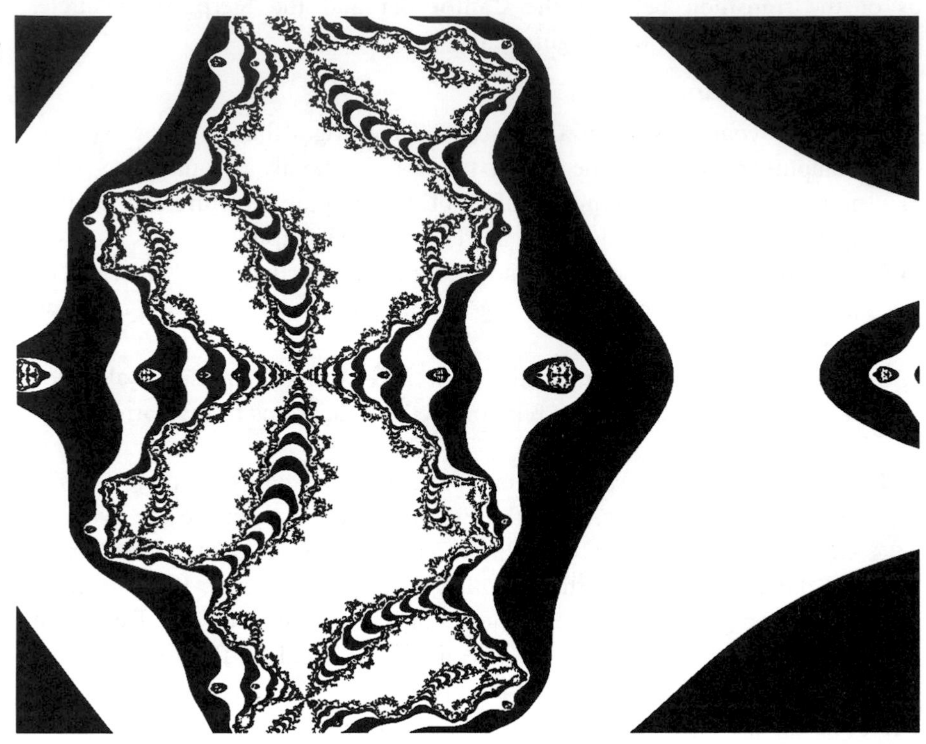

as a result, it is easy to compute these images using the Escape Time Algorithm.

Here are some observations about this "group" of images: Many of them contain straight lines. They all have the same fractal dimension. They all use approximately the same amount of ink. Many of them are connected. Make some more observations. Can you formalize and prove some of these observations?

1.15. Verify computationally that a "snowflake" curve is a basin boundary for the dynamical system $\{\mathbb{R}^2; f\}$ where, for all $(x, y) \in \mathbb{R}^2$:

$$f(x, y) = (0, -1) \text{ if } y < 0;\ f(x, y) = (3x, 3y)$$
$$\text{if } y \ge 0 \text{ and } x < -y/\sqrt{3} + 1;$$
$$f(x, y) = \big((9 - 3x - 3\sqrt{3}\,y)/2, (3\sqrt{3} - 3\sqrt{3}\,x + 3)/2$$
$$\text{if } y \ge 0 \text{ and } -y/\sqrt{3} + 1 \le x < 3/2;$$
$$f(x, y) = \big((3x - 3\sqrt{3}\,y)/2, \big(3\sqrt{3}\,x + 3y - 6\sqrt{3}\,\big)/2\big)/2\big),$$
$$\text{if } y \ge 0, \text{ and } 3/2 \le x < y/\sqrt{3} + 2;$$
$$f(x, y) = (9 - 3x, 3y), \text{ if } y \ge 0, \text{ and } x \ge y/\sqrt{3} + 2.$$

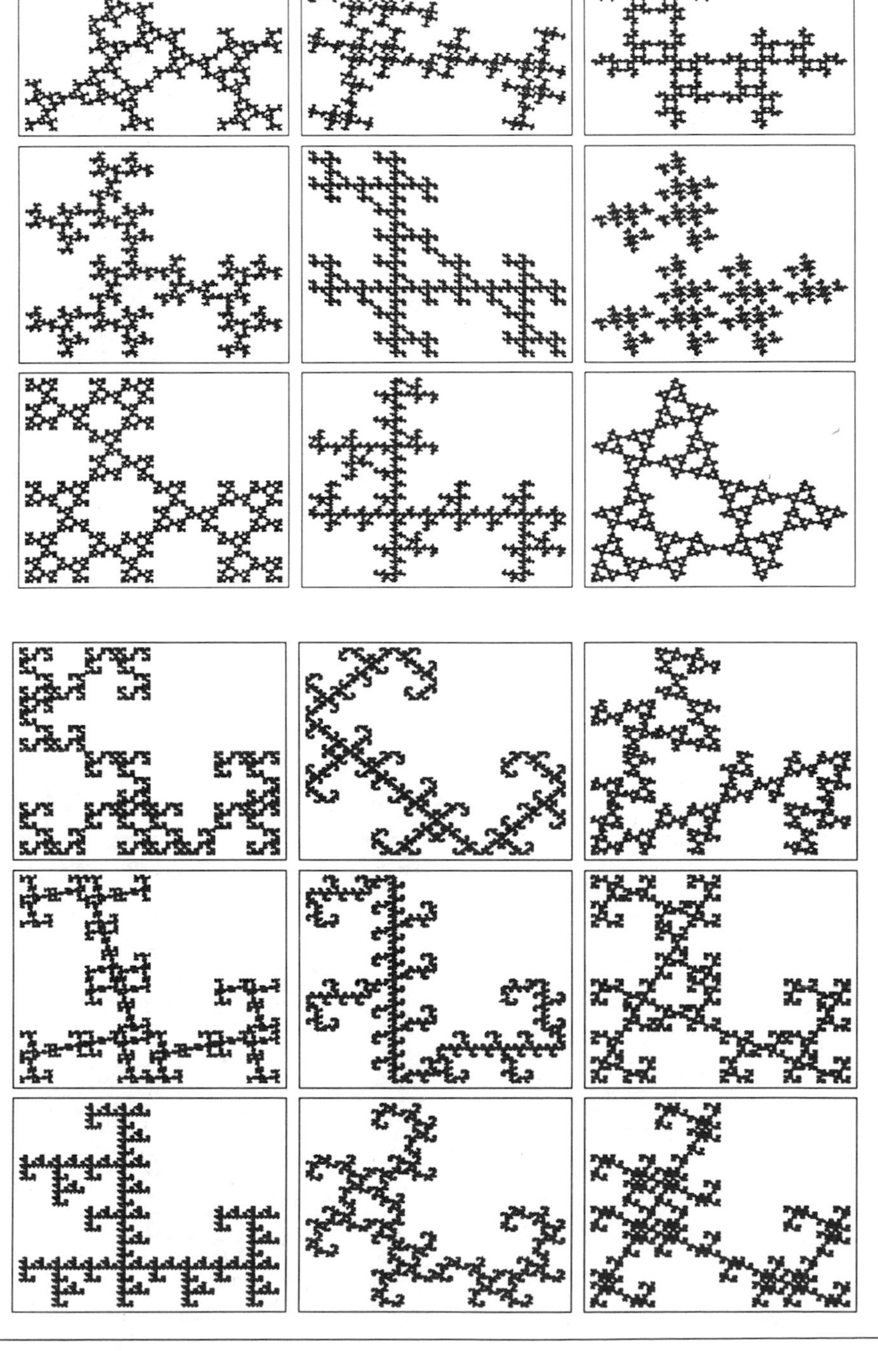

Figure 7.1.12(a) – (d) *The images in (a), (b), (c), and (d) represent the nontrivially distinct attractors of all IFS of the form* $\{\blacksquare;\ w_1, w_2, w_3\}$ *where the maps are similitudes of scaling factor one-half, and the rotation angles are in the set* $\{0°, 90°, 180°, 270°\}$. *The three translations* $(0, 0)$, $(1, 0)$ *and* $(0, 1)$ *are used. These IFS are all just-touching. For* $i \neq j$ *the set* $w_i(A) \cap w_j(A)$, *is contained in one of the two straight lines* $x = 1$ *or* $y = 1$. *Hence it is easy to compute images of these attractors using the Escape Time Algorithm.*

Figure 7.1.12(b)

Figure 7.1.12(c)

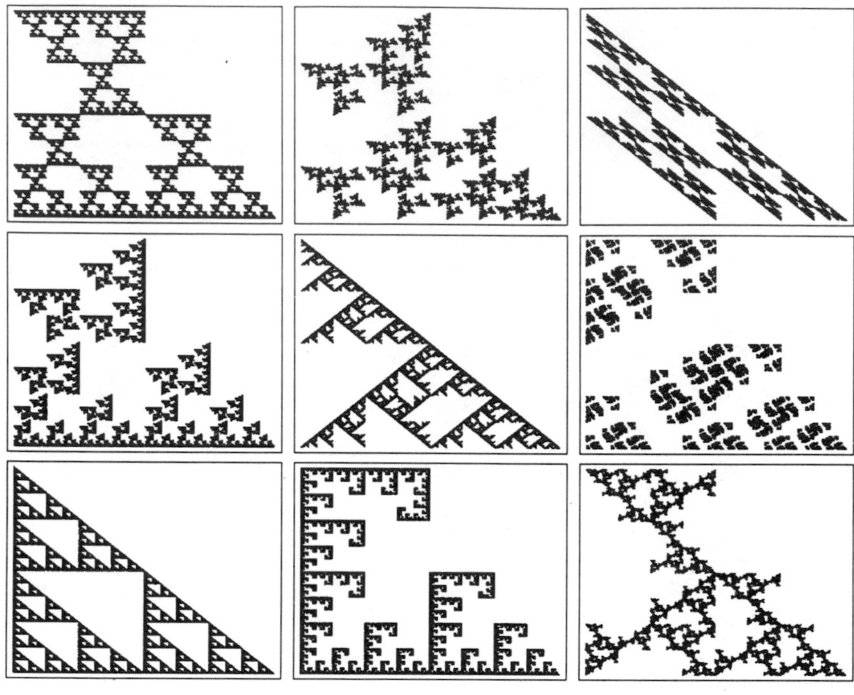

Figure 7.1.12(d)

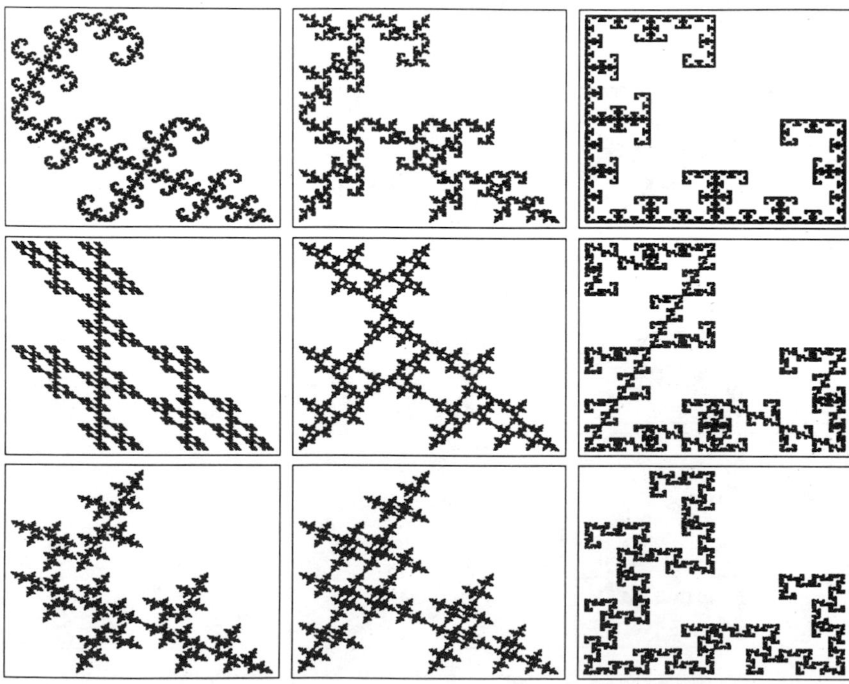

7.2 ITERATED FUNCTION SYSTEMS WHOSE ATTRACTORS ARE JULIA SETS

In Section 7.1 we learned how to define some IFS attractors and filled Julia sets with the aid of the Escape Time Algorithm applied to certain dynamical systems. In this section we explain how the Julia set of a quadratic transformation can be viewed as the attractor of a suitably defined IFS.

The Escape Time Algorithm compares how fast different points in $\mathscr{W}$ escape to $\mathscr{V}$, under the action of a dynamical system. Which set repells the orbits? From where do the escaping orbits originate? In the case of the dynamical systems considered at the start of Section 7.1, orbits were "escaping from" the attractor of the IFS.

Let $\lambda \in \mathbb{C}$ be fixed. Which set repells the orbits, in the case of the dynamical system $\{\hat{\mathbb{C}};\ f_\lambda(z) = z^2 - \lambda\}$? To find out let us consider the inverse of $f_\lambda(z)$. This is provided by a pair of functions, $f^{-1}(z) = \{+\sqrt{z+\lambda}, -\sqrt{z+\lambda}\}$ where, for example, the positive square root of a complex number is that complex root which lies on the non-negative real axis or in the upper half plane. Explicitly $\sqrt{z} = \sqrt{x_1 + ix_2} = (a(x_1, x_2), b(x_1, x_2))$ with

$$a(x_1, x_2) = \sqrt{\frac{\sqrt{x_1^2 + x_2^2} + x_1}{2}} \qquad \text{when } x_2 \geq 0,$$

$$a(x_1, x_2) = -\sqrt{\frac{\sqrt{x_1^2 + x_2^2} + x_1}{2}} \qquad \text{when } x_2 < 0, \quad \text{and}$$

$$b(x_1, x_2) = \sqrt{\frac{\sqrt{x_1^2 + x_2^2} - x_1}{2}}.$$

To find the "repelling" set, we must try to run the dynamical system backwards. This leads us to study the IFS

$$\{\hat{\mathbb{C}};\ w_1(z) = \sqrt{z+\lambda},\ w_2(z) = -\sqrt{z+\lambda}\}.$$

The natural idea is that this IFS has an attractor. This attractor is the set from which points try to flee, under the action of the dynamical system $\{\hat{\mathbb{C}};\ z^2 - \lambda\}$.

A few computergraphical experiments quickly suggest a wonderful idea: they suggest that the IFS indeed possesses an attractor, namely the Julia set $J_\lambda = \partial F_\lambda$ for $f_\lambda(z)$. Consider for example the case $\lambda = 1$. Figure 7.2.1(a) illustrates points in the window $\mathscr{W} = \{z = (x, y) \in \mathbb{C}: -2 \leq x \leq 2, -2 \leq y \leq 2\}$ whose orbits diverge. It was computed using the Escape Time Algorithm. Figure 7.2.1(b) shows the results of applying the Random Iteration Algorithm to the above IFS, with $\lambda = 1$ and the same screen coordinates, superimposed on (a). The boundary of the region $F_{\lambda=1}$ is outlined by points on the attractor of the IFS.

Figures 7.2.2(a)–(d) show the results of applying the Random Iteration Algorithm to the IFS $\{\hat{\mathbb{C}};\ w_1(z) = \sqrt{z+\lambda},\ w_2(z) = -\sqrt{z+\lambda}\}$ for various

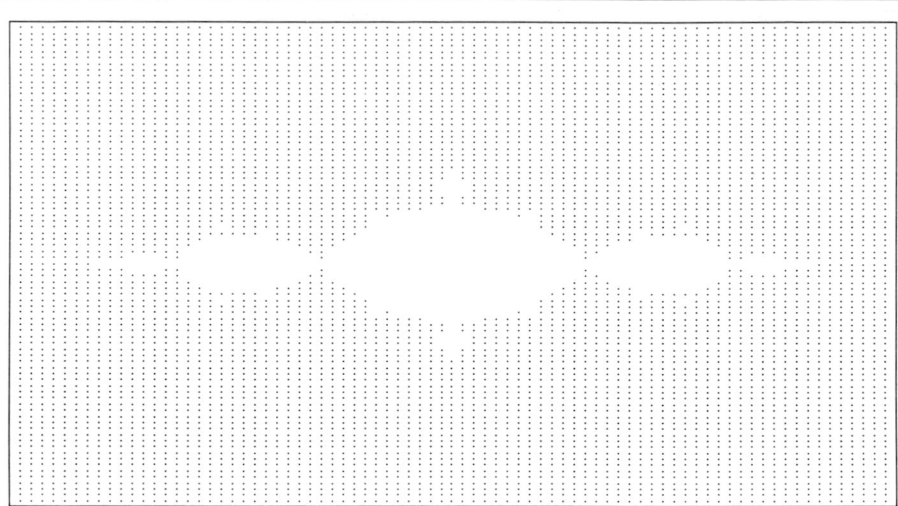

Figure 7.2.1(a), (b) *The attractor of the IFS* $\{\hat{\mathbb{C}};\ w_1(z) = \sqrt{z+1},\ w_2(z) = -\sqrt{z+1}\}$ *is the Julia set for the transformation* $f(z) = z^2 - 1$. *(a) illustrates points whose orbits "escape" when the Escape Time Algorithm is applied. (b) shows the results of applying the Random Iteration Algorithm to the IFS, superimposed on (a).*

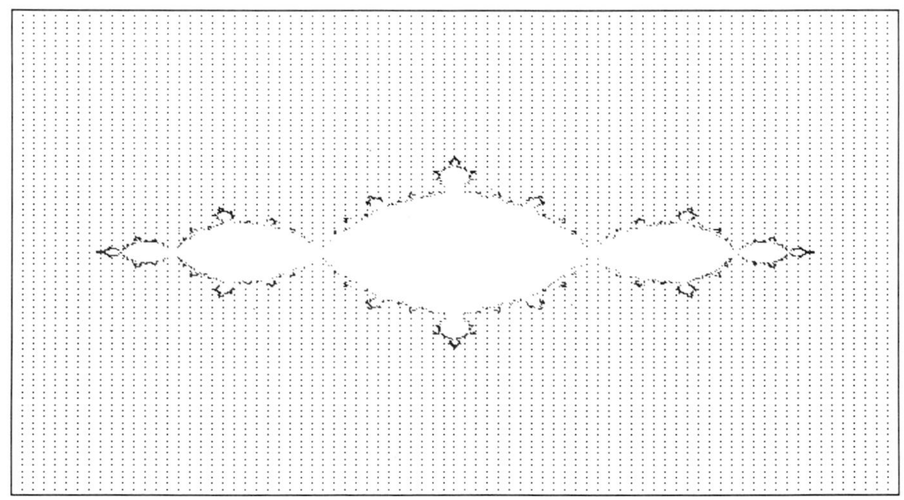

Figure 7.2.1(b)

$\lambda \in [0, 3]$. In all cases it appears that the IFS possesses an attractor, and this attractor is the Julia set J_λ.

Perhaps $\{\hat{\mathbb{C}};\ w_1(z) = \sqrt{z+\lambda},\ w_2(z) = -\sqrt{z+\lambda}\}$ is a hyperbolic IFS with J_λ as its attractor? No, it is not, because $\hat{\mathbb{C}} = w_1(\hat{\mathbb{C}}) \cup w_2(\hat{\mathbb{C}})$. The IFS is not associated with a unique fixed point in the space $\mathscr{H}(\hat{\mathbb{C}})$. In order to make the IFS have a unique attractor, we need to remove some pieces from $\hat{\mathbb{C}}$, to produce a smaller space on which the IFS acts.

Theorem 1. *Let* $\lambda \in \mathbb{C}$. *Suppose that the dynamical system* $\{\hat{\mathbb{C}};\ f(z) = z^2 - \lambda\}$ *possesses an attractive cycle* $\{z_1, z_2, z_3, \ldots, z_p\} \subset \mathbb{C}$. *Let* ϵ *be a very*

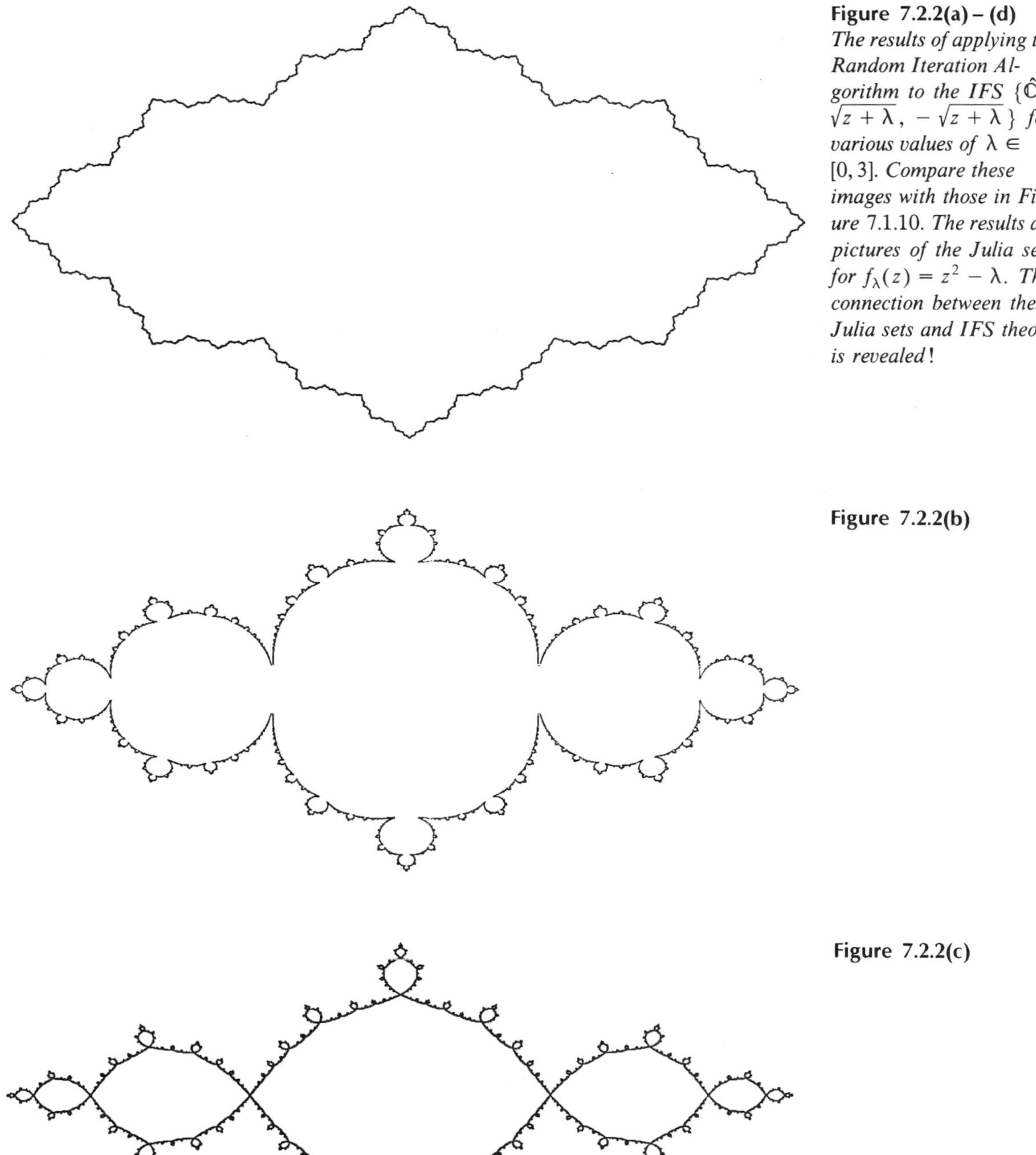

Figure 7.2.2(a) – (d) *The results of applying the Random Iteration Algorithm to the IFS* $\{\hat{\mathbb{C}}; \sqrt{z+\lambda}, -\sqrt{z+\lambda}\}$ *for various values of* $\lambda \in [0, 3]$. *Compare these images with those in Figure 7.1.10. The results are pictures of the Julia set for* $f_\lambda(z) = z^2 - \lambda$. *The connection between these Julia sets and IFS theory is revealed!*

Figure 7.2.2(b)

Figure 7.2.2(c)

Figure 7.2.2(d)

small positive number. Let $\mathbb{X}$ *denote the Riemann Sphere* $\hat{\mathbb{C}}$ *with* $(p+1)$ *open balls of radius* ϵ *removed. (The radius is measured using the spherical metric.) One ball is centered at each point of the cycle, and one ball is centered at the Point at Infinity, as illustrated in Figure 7.2.3. Define an IFS by* $\{\mathbb{X}; w_1(z) = \sqrt{z+\lambda}, w_2(z) = -\sqrt{z+\lambda}\}$. *Then the transformation* W *on* $\mathscr{H}(\mathbb{X})$, *defined by*

$$W(B) = w_1(B) \cup w_2(B) \qquad \text{for all } B \in \mathscr{H}(\mathbb{X}),$$

maps $\mathscr{H}(\mathbb{X})$ *into itself, continuously with respect to the Hausdorff metric on* $\mathscr{H}(\mathbb{X})$. *Moreover* $W\colon \mathscr{H}(\mathbb{X}) \to \mathscr{H}(\mathbb{X})$ *possesses a unique fixed point,* J_λ, *the Julia set for* $z^2 - \lambda$. *Also,*

$$\lim_{n\to\infty} W^{\circ n}(B) = J_\lambda \qquad \text{for all } B \in \mathscr{H}(\mathbb{X}).$$

These conclusions also hold if the orbit of the origin, $\{f^{\circ n}(O)\}$, *converges to the Point at Infinity, and* $\mathbb{X} = \hat{\mathbb{C}} \setminus B(\infty, \epsilon)$.

Sketch of Proof: The fact that W takes $\mathscr{H}(\mathbb{X})$ continuously into itself follows from Theorem 7.4.1. To apply Theorem 7.4.1, three conditions must be met. These conditions are (i), (ii) and (iii), stated next: f is analytic on $\hat{\mathbb{C}}$ so (i) it is continuous, *and* (ii) it maps open sets to open sets. The way in which $\mathbb{X}$ is constructed ensures that, for small enough ϵ, (iii) $f(\mathbb{X}) \supset \mathbb{X}$ and $W(\mathbb{X}) = f^{-1}(\mathbb{X}) \subset \mathbb{X}$.

To prove that W possesses a unique fixed point, we again make use of Theorem 7.4.1. Consider the limit $A \in \mathscr{H}(\mathbb{X})$ of the decreasing sequence of sets, $\{W^{\circ n}(\mathbb{X})\}$, namely,

$$A = \bigcap_{n=1}^{\infty} f^{\circ(-n)}(\mathbb{X}) = \lim_{n\to\infty} W^{\circ n}(\mathbb{X}).$$

This obeys $W(A) = A$. It follows from [Brolin 1965], Lemma 6.3, that $A = J_\lambda$, the Julia set. This completes the sketch of the proof.

Theorem 7.2.1 can be generalized to apply to polynomial transformations $f\colon \hat{\mathbb{C}} \to \hat{\mathbb{C}}$ of degree N greater than one. Here is a rough description: Let

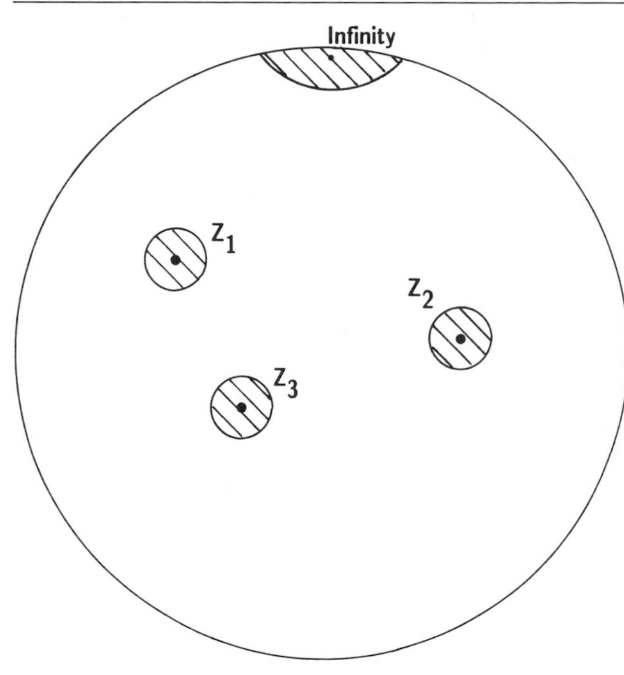

Figure 7.2.3 *The Riemann Sphere* $\hat{\mathbb{C}}$ *with a number of very small open balls of radius* ϵ *removed. One ball is centered at each of the points* $\{z_p \in \mathbb{C}\}$ *belonging to an attractive cycle of the transformation* $f_\lambda(z) = z^2 - \lambda$. *One ball is centered at the Point at Infinity.*

$f^{-1}(z) = \{w_1(z), w_2(z), \ldots, w_N(z)\}$ denote a definition of branches of the inverse of f. Then consider the IFS $\{\hat{\mathbb{C}};\ w_1(z), w_2(z), \ldots, w_N(z)\}$. This IFS is not hyperbolic: the "typical" situation is that the associated operator W: $\mathscr{H}(\hat{\mathbb{C}}) \to \mathscr{H}(\hat{\mathbb{C}})$ possesses a finite number of fixed points, all except one of which are "unstable." The one "stable" fixed point is J_f, and $W^{\circ n}(A) \to J_f$ for "most" $A \in \mathscr{H}(\hat{\mathbb{C}})$. In principle, J_f can be computed using the Random Iteration Algorithm.

Results like Theorem 7.2.1 are concerned with what are known as *hyperbolic* Julia sets. The Julia set of a rational transformation f: $\hat{\mathbb{C}} \to \hat{\mathbb{C}}$ is hyperbolic if, whenever $c \in \hat{\mathbb{C}}$ is a critical point of f, the orbit of c converges to an attractive cycle of f. The Julia set for $z^2 - 0.75$ is an example of a non-hyperbolic Julia set. We refer to [Peitgen 1986] as a good source of further information about Julia sets from the dynamical systems point of view.

Explicit formulas for the inverse maps, $\{w_n(z)\colon n = 1, 2, \ldots, N\}$, for a polynomial of degree N, are not generally available. So the Random Iteration Algorithm cannot usually be applied. Pictures of Julia sets and filled Julia sets are often computed with the aid of the Escape Time Algorithm. The case of quadratic transformations is somewhat special, because both algorithms can be used. The Random Iteration Algorithm can also be applied to compute

Julia sets of cubic and quartic polynomials, and of special polynomials of higher degree such as $z^n + \lambda$ where $n = 5, 6, 7, \ldots,$ and $\lambda \in \mathbb{C}$.

Exercises & Examples

2.1. Consider the dynamical system $\{\hat{\mathbb{C}};\ f(z) = z^2\}$. The origin, O, is an attractive cycle of period one: indeed $f(O) = O$ and $|f'(O)| = 0 < 1$. Notice that $\text{Lim}_{n\to\infty} f^{\circ n}(z) = O$ for all $z \in B(O, 0.99999999)$. Let $\mathring{B}(z, r)$ denote the open ball on $\hat{\mathbb{C}}$, with center at z and radius r. Theorem 7.2.1 tells us that the IFS

$$\{\mathbb{X} = \hat{\mathbb{C}} \setminus \{\mathring{B}(O, 0.0000001) \cup \mathring{B}(\infty, 0.0000001)\};\ w_1(z) = \sqrt{z},\ w_2(z) = -\sqrt{z}\}$$

possesses a unique attractor. The attractor is actually the circle of radius one centered at the origin. It can be computed by means of the Random Iteration Algorithm. Notice that if we extend the space $\mathbb{X}$ to include O, then $O \in \mathscr{H}(\mathbb{X})$ and $O = W(O) = w_1(O) \cup w_2(O)$. If we extend $\mathbb{X}$ to include $\mathring{B}(O, 0.0000001)$ then the filled Julia set F_0 belongs to $\mathscr{H}(\mathbb{X})$ and obeys $F_0 = W(F_0)$. If we take $\mathbb{X}$ to be all of $\hat{\mathbb{C}}$ then $\hat{\mathbb{C}} = W(\hat{\mathbb{C}})$. In other words, if the space on which the IFS acts is too large, then uniqueness of the "attractor" of the IFS is lost.

Can you find two more nonempty compact subsets of $\hat{\mathbb{C}}$ which are fixed points of W, in the case $\mathbb{X} = \hat{\mathbb{C}}$?

Establish that, for all $\lambda \in (-0.25, 0.75)$, the point $z_0 = 0.5 - \sqrt{0.25 + \lambda}$ is an attractive cycle of period one for $\{\hat{\mathbb{C}};\ z^2 - \lambda\}$. Deduce that the corresponding IFS, acting on a suitably chosen space $\mathbb{X}$, possesses a unique attractor.

2.2. Let $\lambda \in (0.75, 1.25)$. Consider the dynamical system $\{\hat{\mathbb{C}};\ f(z) = z^2 - \lambda\}$. Let $z_1, z_2 \in \mathbb{R}$, denote the two solutions of the equation $z^2 + z + (1 - \lambda) = 0$. Show $f(z_1) = z_2$, $f(z_2) = z_1$, $|(f^{\circ 2})'(z_1)| = |(f^{\circ 2})'(z_2)| < 1$ and hence that $\{z_1, z_2\}$ is an attractive cycle of period two. Deduce that the IFS

$$\{\hat{\mathbb{C}} \setminus \{\mathring{B}(z_1, \epsilon) \cup \mathring{B}(z_2, \epsilon) \cup \mathring{B}(\infty, \epsilon)\};\ +\sqrt{z + \lambda},\ -\sqrt{z + \lambda}\}$$

possesses a unique attractor when ϵ is sufficiently small.

2.3. The Julia set J_λ for the polynomial $z^2 - \lambda$ is a union of two "copies" of itself. Identify these two copies for various values of λ. Explain how, when $\lambda = 1$, the two inverse maps $w_1^{-1}(z)$ and $w_2^{-1}(z)$ rip the Julia set apart, and the set map $W = w_1 \cup w_2$ puts it back together again. Where is the rip? Describe the geometry of what is going on here.

2.4. Consider the one-parameter family of polynomials $f(z) = z^3 - \lambda$, where $\lambda \in \mathbb{C}$ is the parameter. Give explicit formulas for the real and imaginary parts of three inverse functions $w_1(z)$, $w_2(z)$, and $w_3(z)$ such that $f^{-1}(z) = \{w_1(z), w_2(z), w_3(z)\}$ for all $\lambda \in \mathbb{C}$. Compute images of the filled Julia set for $f(z)$ for $\lambda = 0.01$ and $\lambda = 1$. Compare these images with those obtained by applying the Random Iteration Algorithm to the IFS $\{\mathbb{C};\ w_1(z), w_2(z), w_3(z)\}$.

2.5. Consider the dynamical system $\{\hat{\mathbb{C}};\ f(z) = z^2 - \lambda\}$ for $\lambda > 2$. Show that $\{f^{\circ n}(O)\}$ converges to the Point at Infinity. Deduce that the IFS

$$\{\mathbb{X} = \hat{\mathbb{C}} \setminus B(\infty, \epsilon);\ +\sqrt{z+\lambda},\ -\sqrt{z+\lambda}\}$$

possesses a unique attractor $A(\lambda)$. $A(\lambda)$ is a generalized Cantor set. Compute some pictures of $A(3)$. Use the Collage Theorem to help find a pair of affine transformations $w_i\colon \mathbb{R} \to \mathbb{R}$, $i = 1, 2$, such that the attractor of the IFS $\{\mathbb{R};\ w_1, w_2\}$ is an approximation to $A(3)$. Define $\tilde{f}\colon \mathbb{R} \to \mathbb{R}$ by $\tilde{f}(x) = w_1^{-1}(x)$ when $x < 0$ and $\tilde{f}(x) = w_2^{-1}(x)$ when $x \geq 0$. Compare the graphs of the functions $f(x) = x^2 - 3$ and $\tilde{f}(x)$ for, say, $x \in [-4, 4]$. Compare one-dimensional "images" obtained by applying the Escape Time Algorithm in a similar manner to both $\{\mathbb{R};\ f\}$ and $\{\mathbb{R};\ \tilde{f}\}$.

One can sometimes obtain a *hyperbolic* IFS associated with a Julia set, if the domains and ranges of the inverse transformations are defined carefully. The following theorem provides such an example.

Theorem 2. *Let $\lambda \in [-0.249, 0.749]$, and let ϵ be a very small positive number. Let $a = 0.5 - \sqrt{0.25 + \lambda}$, an attractive fixed point of the dynamical system $\{\hat{\mathbb{C}};\ f(z) = z^2 - \lambda\}$. Let $\tilde{\mathbb{X}} = \hat{\mathbb{C}} \setminus \{\mathring{B}(a, \epsilon) \cup \mathring{B}(\infty, e) \cup (-\lambda, \infty)\}$. That is, $\tilde{\mathbb{X}}$ consists of the Riemann Sphere with a small open ball centered at a, a small open ball centered at ∞, and the open interval $(-\lambda, \infty)$ removed. (This space is not compact because the edges of the lips of the cut, from $-\lambda$ to ∞, are missing.) To each lip attach copies of the pieces of real interval $(-\lambda, \infty)$ which were removed, to provide a compact space $\mathbb{X}$, as illustrated in Figure 7.2.4. The distance $d(z_1, z_2)$ between a pair of points z_1 and $z_2 \in \mathbb{X}$ is the length (measured using the spherical metric) of the shortest path which lies in $\mathbb{X}$ and connects z_1 to z_2. (Paths in $\mathbb{X}$ cannot cross the cut, they have to go round it.) $(\mathbb{X}, d)$ is a compact metric space.*

Define $w_1\colon \mathbb{X} \to \mathbb{X}$ by $w_1(z) = \sqrt{z+\lambda}$, the root which lies in the "upper half plane." For z on the lower edge of the cut, $w_1(z)$ lies on the negative real axis, and on the upper edge if $w_1(z) > -\lambda$. Define $w_2\colon \mathbb{X} \to \mathbb{X}$ by $w_2(z) = -\sqrt{z+\lambda}$, the root which lies in the "lower half plane." For z on the upper edge of the cut, $w_2(z)$ lies on the negative real axis, and on the lower edge if $w_2(z) > -\lambda$. This makes w_1 and w_2 continuous on $\mathbb{X}$.

Then there is a metric on $\mathbb{X}$, equivalent to the metric d, such that the IFS $\{\mathbb{X};\ w_1, w_2\}$ is hyperbolic. The attractor is the Julia set J_λ for $z^2 - \lambda$, where the real point $0.5 + \sqrt{0.25 + \lambda}$ is repeated on both the upper and the lower edges of the cut.

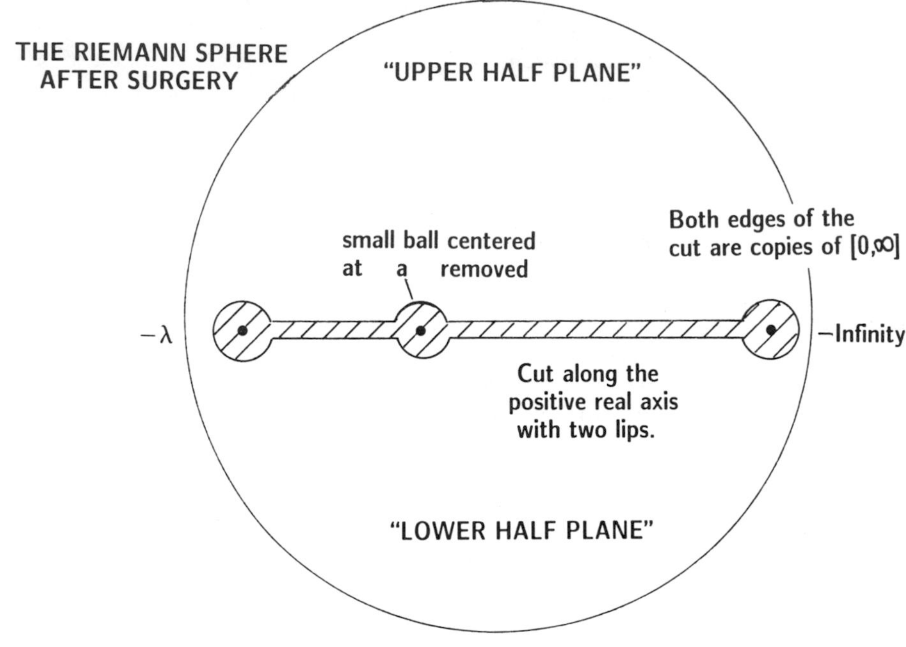

Figure 7.2.4 *Construction of a compact metric space $\mathbb{X}$ for the IFS $\{\mathbb{X}; \sqrt{z+\lambda}, -\sqrt{z+\lambda}\}$ with $\lambda \in (-0.25, 0.75)$, used in Theorem 7.2.2. Both sides of the "slit" from $-\lambda$ to infinity belong to the space. The distance between a pair of points on $\mathbb{X}$ is the length of the shortest path which connects the points without crossing the slit. The distance between points may be much greater than it looks.*

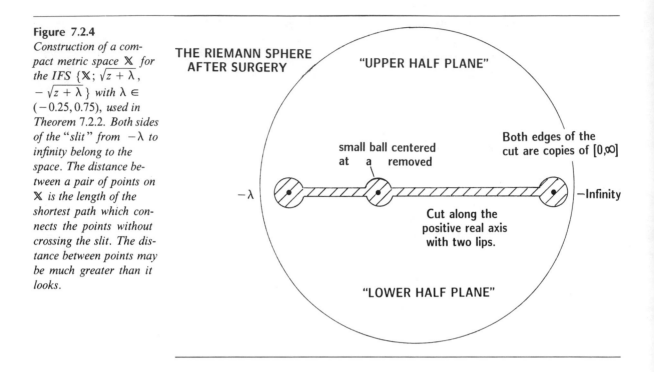

Outline of Proof. Let $e = (e_1, e_2, \ldots, e_n, \ldots) \in \Sigma$, the code space on the two symbols $\{1, 2\}$. Define a sequence of nonempty compact subsets of $\mathbb{X}$ by

$$\mathbb{X}_n(e) = w_{e_1} \circ w_{e_2} \circ w_{e_3} \circ w_{e_4} \circ w_{e_5} \circ w_{e_6} \circ \cdots \circ w_{e_n}(\mathbb{X}) \qquad \text{for } n = 1, 2, 3, \ldots .$$

It follows, using [Brolin 1965] Theorem 6.2 and Lemma 6.3, that the sequence $\{\mathbb{X}_n \in \mathscr{H}(\mathbb{X})\}$ converges to a singleton, say $\{\phi(e)\}$, where $\phi(e) \in J_\lambda$ and

$$\bigcup_{e \in \Sigma} \phi(e) = J_\lambda .$$

A beautiful theorem of Elton [Elton 1988] applies under just these conditions, and provides the conclusion of the theorem. This completes the outline of the proof.

In those situations where the IFS $\{\mathbb{X}; +\sqrt{z+\lambda}, -\sqrt{z+\lambda}\}$ is hyperbolic, one can use the associated code space to discuss both the Julia set and the associated shift dynamical system $\{J_\lambda; f(z) = z^2 - \lambda\}$. Here we give some of the flavor of such a discussion. More details can be found in [Barnsley 1984].

For the remainder of this section let $\lambda \in (-0.25, 0.75)$ and consider the IFS $\{\mathbb{X}; w_1(z) = \sqrt{z+\lambda}, w_2(z) = -\sqrt{z+\lambda}\}$, as defined in Theorem 7.2.2. Let Σ denote the code space on the two symbols $\{1, 2\}$, and let $\phi\colon \Sigma \to J_\lambda$

denote the associated code space map, introduced in Theorem 4.2.1. If $e = (e_1, e_2, \ldots, e_n, \ldots) \in \Sigma$, then

$$\phi(e) = \lim_{n\to\infty} w_{e_1} \circ w_{e_2} \circ w_{e_3} \circ w_{e_4} \circ w_{e_5} \circ w_{e_6} \circ \cdots \circ w_{e_n}(z).$$

Replace the symbol "1" by the symbol "+" and replace the symbol "2" by the symbol "−". Then the point $\phi(e)$ on the Julia set J_λ can be represented by the formula

$$\phi(e) = e_1\sqrt{\lambda e_2\sqrt{\lambda e_3\sqrt{\lambda e_4\sqrt{\lambda e_5\sqrt{\lambda e_6\sqrt{\lambda e_7\sqrt{\lambda e_8 \ldots e_n\sqrt{\lambda \ldots}}}}}}}}$$

where $e_i \in \{+, -\}$ for each positive integer i. The set J_λ itself can be represented by the collection of formulas

$$\pm\sqrt{\lambda \pm \sqrt{\lambda \pm \sqrt{\lambda \pm \sqrt{\lambda \pm \sqrt{\lambda \pm \sqrt{\lambda \pm \sqrt{\lambda \pm \cdots \pm \sqrt{\lambda \ldots}}}}}}}} \tag{7.2.1}$$

where all possible sequences of plus and minus signs are permitted. A particular sequence of signs, corresponding to a point in J_λ, is an address of the point.

In Figure 7.2.5 we show the Julia set for $z^2 - 0.7$, with the addresses of various points marked on it. Some points on $J_{0.7}$ have multiple addresses while others have single addresses. It appears that the IFS is just-touching.

The shift dynamical system associated with the IFS is $\{J_\lambda;\ f(z) = z^2 - \lambda\}$. Notice how the set of points represented by the formulas in equation (7.2.1) is mapped into itself by the function which "squares" a formula and subtracts λ. A point on a cycle of period two is represented by

$$+\sqrt{\lambda - \sqrt{\lambda + \sqrt{\lambda - \sqrt{\lambda + \sqrt{\lambda - \sqrt{\lambda + \sqrt{\lambda - \cdots + \sqrt{\lambda \ldots}}}}}}}}\ .$$

The other point on this cycle is obtained by squaring the formula and subtracting λ.

In Theorem 4.2.4, we learned that the set of periodic points of the shift dynamical system associated with a hyperbolic IFS is dense in the attractor of the IFS. Here this tells us that the set of periodic points of the dynamical system $\{J_\lambda;\ f(z) = z^2 - \lambda\}$ is dense in J_λ. In fact, a related idea was the starting point of Julia's original investigations. He considered dynamical systems of the form $\{\hat{\mathbb{C}};\ f(z)\}$, where $f(z)$ is analytic. He defined the (Julia) set to be the closure of the set of repulsive cycles of f.

Following Theorem 4.8.1 we explained the sense in which the shift dynamical system associated with a hyperbolic IFS is chaotic. In the present context we learn that the dynamical system $\{J_\lambda;\ z^2 - \lambda\}$ is chaotic.

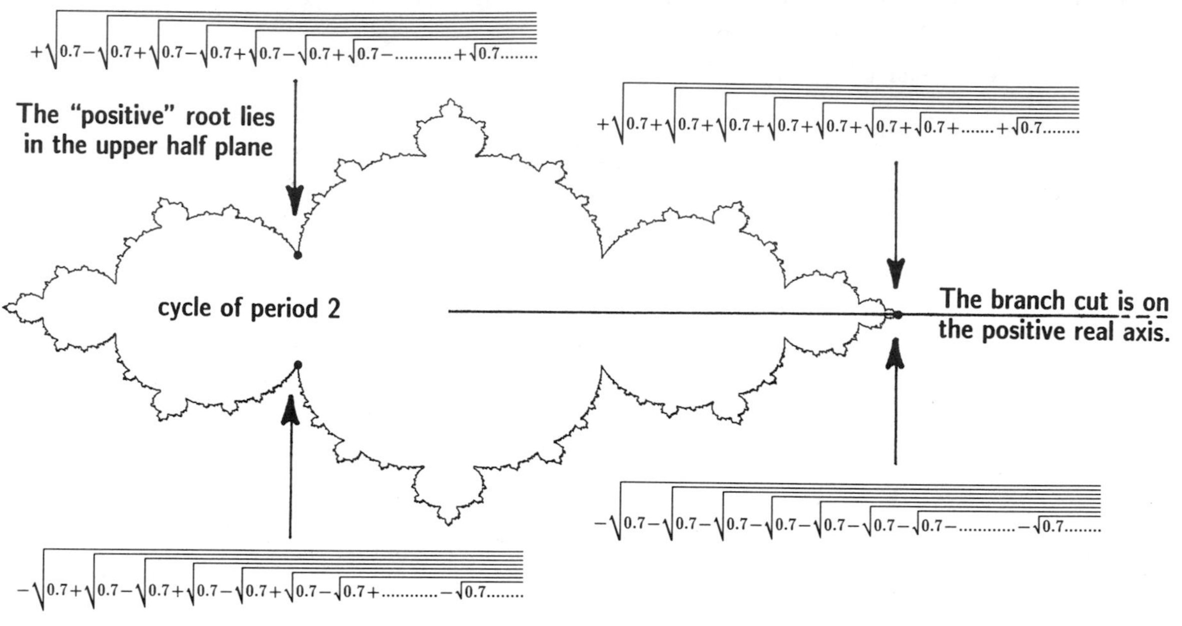

Figure 7.2.5
The Julia set for $z^2 - 0.7$, *labelled with various addresses. Chaotic dynamics takes place on the Julia set and orderly dynamics takes place off it. Boundaries of a fractal character often separate regions where the dynamical system behaves differently. The behaviour of the dynamical system on such a boundary may then be indecisive, and in some way chaotic.*

One can think of the dynamical system $\{\hat{\mathbb{C}};\ z^2 - \lambda\}$ as being the union of two dynamical systems, a chaotic one $\{J_\lambda;\ z^2 - \lambda\}$ and an orderly one $\{\hat{\mathbb{C}} \setminus J_\lambda;\ z^2 - \lambda\}$. The orbit of any point in the latter system converges to a fixed point of the transformation. The orbits of "most" points in the former system are wild. In practice they are usually so wild they cannot be constrained to remain on the repelling set J_λ. They escape, and thereafter behave in a rather predictable manner.

An example of chaotic dynamics on a Julia set is provided by the dynamical system $\{[0,1];\ f(x) = 4x(1 - x)\}$. The interval $[0,1]$ is exactly the Julia set for the transformation. This system is close to the "chaotic" one illustrated in Figure 4.3.5.

Exercises & Examples

2.6. The Julia set for $z^2 - 2$ is the interval $[-2,2]$. Show that the shift dynamical system associated with the IFS $\{[-2,2];\ +\sqrt{z+2},\ -\sqrt{z+2}\}$ is precisely the dynamical system $\{[-2,2];\ z^2 - 2\}$. Use a chain of square roots to locate a cycle of minimal period three.

2.7. Verify numerically that for various choices of $\pm$ on each square root, and for various complex numbers λ such that $|\lambda|$ is very small, the expression below evaluates approximately to a complex number which lies on the unit circle centered at the origin, if enough square roots are taken. ($+\sqrt{z}$ means the solution, w, of the equation $w^2 = z$, which lies either on the non-negative real axis or in the upper half plane.) Make a hand-waving explanation of why this is, in terms of Julia set theory.

$$\pm\sqrt{\lambda \pm \sqrt{\lambda \pm \sqrt{\lambda \pm \sqrt{\lambda \pm \sqrt{\lambda \pm \sqrt{\lambda \pm \sqrt{\lambda \pm \cdots \pm \sqrt{\lambda}}}}}}}}\ .$$

2.8. Design an IFS with condensation such that its attractor looks like an infinite nested chain of square root signs.

$$\sqrt{\sqrt{\sqrt{\sqrt{\sqrt{\cdots}}}}}$$

2.9. Figure 7.2.6 represents a sequence of sets $\{A_n\}$ which converges to the Julia set of $f(z) = z^2 - 1$. A_0 denotes the union of the two largest faces and $A_n = f^{\circ(-n)}(A_0)$. Identify the set A_2.

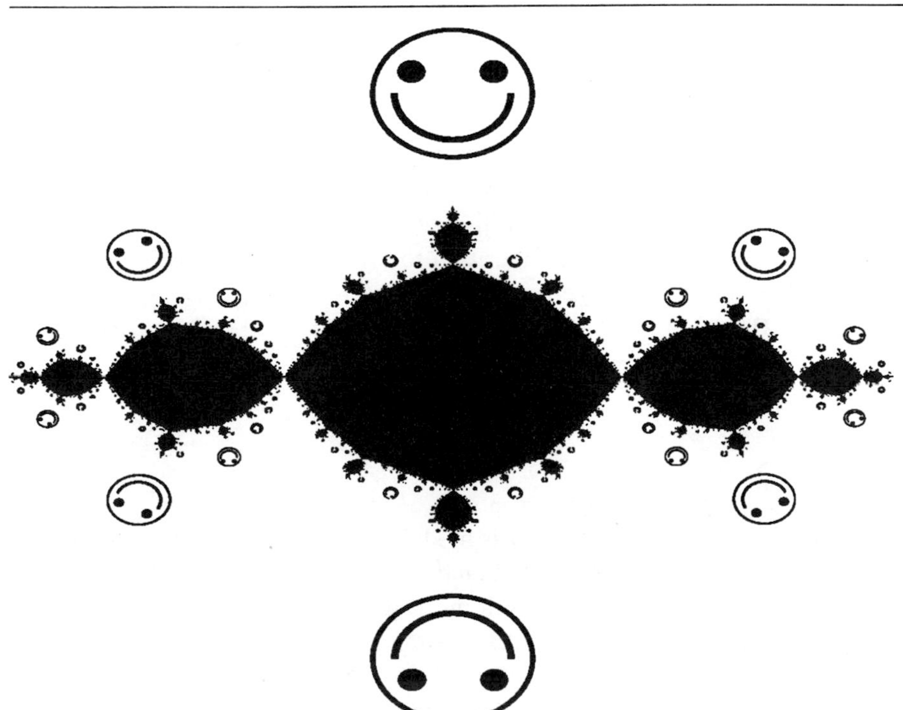

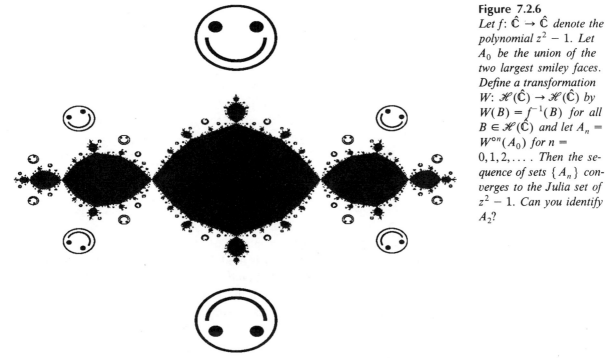

Figure 7.2.6
Let $f\colon \hat{\mathbb{C}} \to \hat{\mathbb{C}}$ denote the polynomial $z^2 - 1$. Let A_0 be the union of the two largest smiley faces. Define a transformation $W\colon \mathscr{H}(\hat{\mathbb{C}}) \to \mathscr{H}(\hat{\mathbb{C}})$ by $W(B) = f^{-1}(B)$ for all $B \in \mathscr{H}(\hat{\mathbb{C}})$ and let $A_n = W^{\circ n}(A_0)$ for $n = 0, 1, 2, \ldots$. Then the sequence of sets $\{A_n\}$ converges to the Julia set of $z^2 - 1$. Can you identify A_2?

7.3 THE APPLICATION OF JULIA SET THEORY TO NEWTON'S METHOD

We are familiar, since our first course in calculus, with Newton's method for computing solutions of the equation $F(x) = 0$. Or are we?

Consider the polynomial $F(z) = z^4 - 1$ for $z \in \mathbb{C}$. There are four distinct complex numbers, a_i $(i = 1, 2, 3, 4)$ such that $F(a_i) = 0$. These are called the roots, or the zeros, of the polynomial $F(z)$. Newton's method provides a means to compute them. Pretend that we do not know that $a_1 = 1$, $a_2 = -1$, $a_3 = i$, and $a_4 = -i$. Then Newton tells us to consider the dynamical system

$$\left\{\hat{\mathbb{C}};\ f(z) = z - \frac{F(z)}{F'(z)}\right\}.$$

We call $f(z)$ the *Newton transformation* associated with the function $F(z)$. The general expectation is that a typical orbit $\{f^{\circ n}(z_0)\}$, which starts from an initial "guess" $z_0 \in \mathbb{C}$, will converge to one of the roots of $F(z)$. In the present example, the Newton transformation is given by

$$f(z) = \frac{3z^4 + 1}{4z^3}.$$

We expect the orbit of z_0 to converge to one of the numbers a_1, a_2, a_3, or a_4. If we choose z_0 close enough to a_i then it is readily proved that

$$\lim_{n\to\infty} f^{\circ n}(z_0) = a_i, \qquad \text{for } i = 1, 2, 3, 4.$$

If, on the other hand, z_0 is far away from all of the a_i's, then what happens? Perhaps the orbit of z_0 converges to the root of $F(z)$ closest to z_0? Or perhaps the orbit does not settle down, but wanders, hopelessly, forever?

Let us make a computergraphical experiment to help answer these questions. We use the Escape Time Algorithm to produce a picture of those points $z_0 \in \hat{\mathbb{C}}$ whose orbits converge to a_1. Define $\mathscr{W} = \{(x, y) \in \mathbb{C}\colon -2 \le x \le 2, -2 \le y \le 2\}$ and $\mathscr{V} = \{z \in \mathbb{C}\colon |z - a_1| \le 0.0001\}$. The real and imaginary parts of $f(x + iy)$ are given by

$$f(x + iy) = \frac{(ce + df)}{(e^2 + f^2)} + i\frac{(de - cf)}{(e^2 + f^2)}$$

where $a = x^2 - y^2$, $b = 2xy$, $c = 3a^2 - 3b^2 + 1$, $d = 6ab$, $e = 4(xa - yb)$, and $f = 4(xb + ya)$. Program 7.1.1 is modified accordingly. Pixels corresponding to points in $\mathscr{W}$ whose orbits reach $\mathscr{V}$ in less than a fixed number of iterations are plotted. A picture resulting from such an experiment is shown in Figure 7.3.1(a). See also Figures 7.3.1(b) and (c).

Color Plate 7.3.1 shows the output from another such experiment. This time Mercator's projection is used to represent the Riemann sphere, and

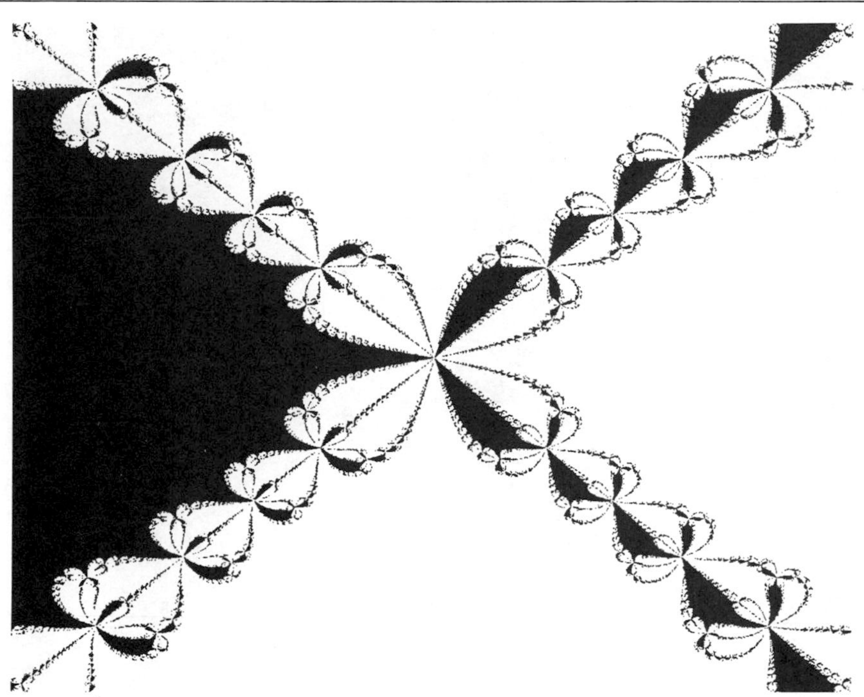

Figure 7.3.1(a)
The Escape Time Algorithm is applied to analyse Newton's method for finding the complex roots of the polynomial $z^4 - 1$. *The boundary of this region represents the Julia set for the rational function* $f(z) = (3z^4 + 1)/4z^3$. *The points plotted black are those points* $z = x + iy$ *with* $-2 \leq x \leq 2$, *and* $-2 \leq y \leq 2$, *whose orbits intersect* $\mathscr{V} = \{z \in \mathbb{C}: |z + 1| \leq 0.0001\}$ *in less than* 1000 *iterations.*

Figure 7.3.1(b)
The boundary of this region represents the Julia set for the rational function $f(z) = (3z^4 + 1)/4z^3$. *The two shades of grey, black, and white, correspond to the basins of attraction of the four attractive fixed points of* $f(z)$. *To which point in* $\mathbb{C}$ *do the orbits of points in the white region converge?*

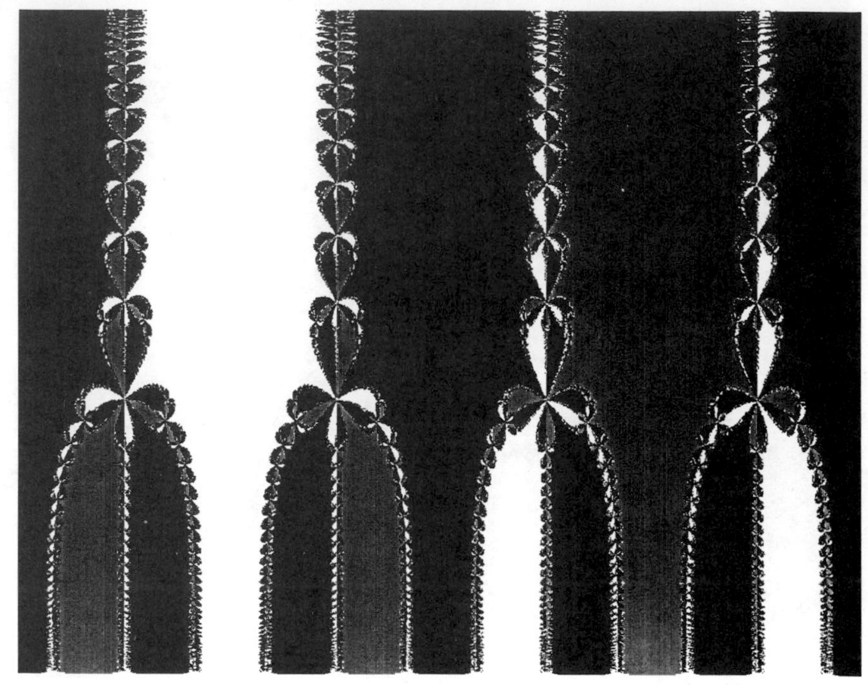

Figure 7.3.1(c) *Mercator's projection of the Riemann sphere showing the basins of attraction of the four attractive fixed points of the Newton transformation of* $z^4 - 1$. *The top of the rectangle corresponds to the Point at Infinity and the bottom of the box corresponds to the origin. The points* -1, $+i$, $+1$, *and* -1 *lie on the equator. The shading follows the same convention as in* (*a*) *and* (*b*). *Which point on* $\hat{\mathbb{C}}$ *is represented by the midpoint of this image?*

points whose orbits converge to the different points a_1, a_2, a_3, and a_4 are plotted in different colors.

The following definition is equivalent to Definition 7.1.1 in the case of polynomials, [Brolin 1966].

Definition 1. The *Julia set of a rational function* f: $\hat{\mathbb{C}} \to \hat{\mathbb{C}}$, of degree greater than one, is the closure of the set of repulsive periodic points of the dynamical system $\{\hat{\mathbb{C}}; f\}$.

For the rational function $f(z)$ considered above, one can prove that the Julia set J is the same as the set of points whose orbits do not converge to any one of the points a_1, a_2, a_3, a_4. In Figure 7.3.1, $J \cap \mathscr{W}$ is represented by the boundary between the black and the white regions. In Color Plate 7.3.1, $J \cap \mathscr{W}$ is the place where the four colors meet. The complement of the Julia set consists of four open sets, the *basins of attraction* of the four attractive fixed points of the Newton iteration scheme. In Color Plate 7.3.1 the red region represents part of the basin of attraction of a_1. The black regions in the color plate are caused by (a) rounding errors, and (b) the fact that only one hundred points on each orbit are tested, for convergence to one of the points a_1, a_2, a_3, a_4.

The Julia set J is the part of $\hat{\mathbb{C}}$ on which chaotic dynamics occurs. It can be characterized as the closure of the set of points whose orbits wander, hopelessly, forever. Orderly, slightly boring motion takes place on $\hat{\mathbb{C}} \setminus J$. J is the boundary of the blue region. It is the boundary of the red region. It is a bonafide fractal, yet nobody knows its fractal dimension.

There is a beautiful theorem of Sullivan, which can be illustrated using the "petals" in Color Plate 7.3.1. The complement of the Julia set is the union of a countable collection of connected open sets, which we call petals. If P is a petal then $f(P)$ is another petal. The Non-Wandering Domain Theorem [Sullivan 1982] says that no connected component of the complement of the Julia set wanders, hopelessly, forever. It always settles into a periodic orbit of petals. If P is a petal in the present example, then one can prove there is a positive integer S so that $f^{\circ S}(P) = f^{\circ(S+1)}(P) = f^{\circ(S+2)}(P) = f^{\circ(S+3)}(P) = \cdots$. The final petal $f^{\circ S}(P)$ is one of the connected components of the complement of the Julia set which contains one of the points a_1, a_2, a_3, a_4. Each petal is eventually periodic. The orbit of a petal ends up in a cycle of petals of period one.

How are we to think about this fabulous Julia set? IFS theory provides a simple point of view, as we show next. We begin by defining the inverse map associated with f. Let $z \in \hat{\mathbb{C}}$ be given and solve

$$z = \frac{3w^4 + 1}{4w^3},$$

to find w in terms of z. This leads to the quartic equation

$$3w^4 - 4zw^3 + 1 = 0.$$

This has four solutions, when we count solutions according to their multiplicities. We can organize these solutions to provide four functions; that is, we write $f^{-1}(z) = \{w_1(z), w_2(z), w_3(z), w_4(z)\}$. Then the Julia set is the "attractor" for the IFS $\{\hat{\mathbb{C}};\ w_i,\ i = 1, 2, 3, 4\}$. However, as in the case of quadratic transformations on $\hat{\mathbb{C}}$, this statement must be treated cautiously: for example, clearly this IFS admits more than one invariant set.

Theorem 1. *Let $f\colon \hat{\mathbb{C}} \to \hat{\mathbb{C}}$ be the Newton transformation associated with the polynomial $z^4 - 1$. Let $\epsilon > 0$ be very small. Let $\mathbb{X} = \hat{\mathbb{C}} \setminus \bigcup_{i=1}^{4} \mathring{B}(a_i, \epsilon)$ where $a_1 = 1$, $a_2 = -1$, $a_3 = i$, and $a_4 = -i$. As above, define $W\colon \mathscr{H}(\mathbb{X}) \to \mathscr{H}(\mathbb{X})$ by*

$$W(B) = \bigcup_{i=1}^{4} w_i(B) = f^{-1}(B) \qquad \text{for all } B \in \mathscr{H}(\mathbb{X}).$$

Then W is continuous, possesses a unique fixed point, J, the Julia set of f, and

$$\lim_{n\to\infty} W^{\circ n}(B) = J \qquad \text{for all } B \in \mathscr{H}(\mathbb{X}).$$

Sketch of Proof. This is essentially the same as the sketch of the proof of Theorem 7.2.1.

The Newton transformation associated with a polynomial may possess an attractive cycle of minimal period greater than one. This cycle may not be directly related to the roots of the polynomial. As an example consider the Newton transformation $f(z)$ associated with the polynomial

$$F(z) = z^3 + (\lambda - 1)z + 1.$$

$\lambda \in \mathbb{C}$ can be chosen so that $f(z)$ possesses an attractive cycle $\{b_1, b_2\}$ of minimal period 2. Figure 7.3.2 illustrates the basin of attraction of the cycle. The Escape Time Algorithm was used to obtain this image. Points whose orbits arrive within a distance 0.01 of the cycle, prior to one hundred iterations, are plotted in white. Accordingly, the escape region is $\mathscr{V} = B(b_1, 0.00001) \cup B(b_2, 0.00001)$. Notice the resemblance of the basin of at-

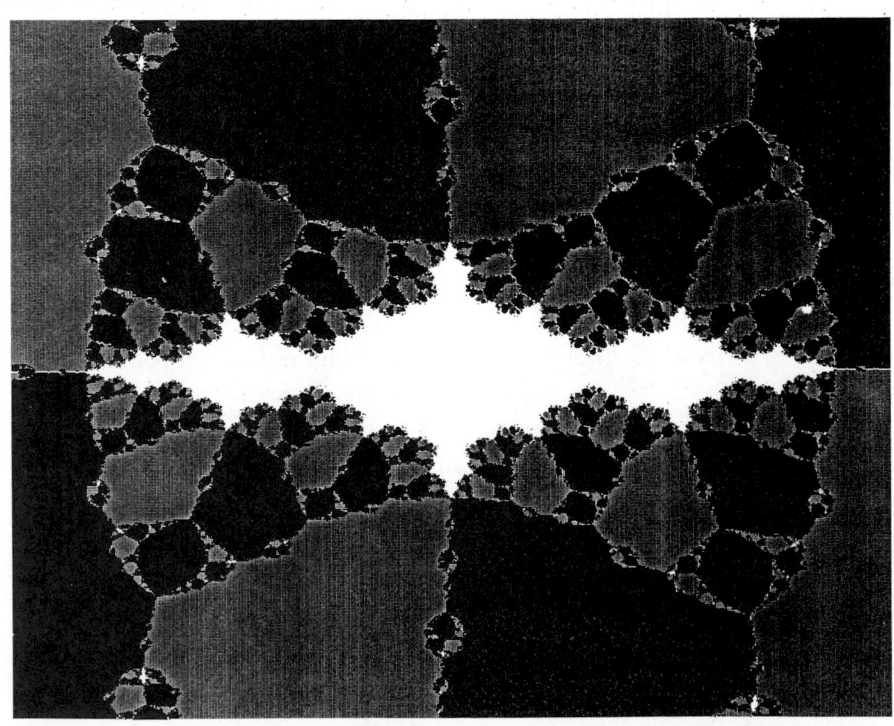

Figure 7.3.2 *The Escape Time Algorithm is applied to a Newton transformation $f(z)$ associated with a cubic polynomial. $f(z)$ possesses an attractive cycle $\{b_1, b_2\}$ of minimal period 2. The basin of attraction of the two-cycle is represented in white. Points whose orbits arrive within a distance of 0.01 of the cycle, prior to one hundred iterations, are plotted in white. Does the basin of attraction of the cycle look familiar?*

traction of $\{b_1, b_2\}$ to the filled Julia set for $z^2 - 1$. This similarity is not accidental. It can be explained using the theory of "polynomial-like" mappings [Douady 1986].

Some interesting computergraphical experiments involving Julia sets for Newton's method are described in [Curry 1983], [Peitgen 1986], and [Vrscay 1986].

Exercises & Examples

3.1. Verify that $z = 1$ is an attractive fixed point for the Newton transformation associated with $F(z) = z^4 - 1$.

3.2. The Newton transformation associated with the polynomial $F(z) = z^2 + 1$ is

$$f(z) = \frac{1}{2}\left(z - \frac{1}{z}\right).$$

Show that the corresponding IFS is $\{\hat{\mathbb{C}};\ w_1(z) = z + \sqrt{z^2+1},\ w_2(z) = z - \sqrt{z^2+1}\}$, where the square root is defined appropriately. Verify that $A = \mathbb{R} \cup \{\infty\}$ is an attractor of the IFS. Prove, or give evidence to show, that A is the Julia set for $f(z)$. (Hint: see exercise 3.7.) How could the space $\hat{\mathbb{C}}$ be modified, so that the IFS has a unique attractor? Notice that numerically computed orbits of points on A, under the dynamical system $\{\hat{\mathbb{C}};\ f\}$, can be constrained from escaping from A by keeping imaginary parts equal to zero. Verify numerically that the dynamics of $\{A;\ f\}$ are wild.

3.3. Find the Newton transformation $f(z)$ associated with the polynomial $F(z) = z^3 - 1$. Use the Escape Time Algorithm to obtain an image, analogous to Figure 7.3.1, which illustrates this Julia set. Discuss the dynamics of the "petals" in the image.

3.4. In this example we speculate on the application of fractal geometry to biological modelling. Let $F_\lambda(z) = (z - i\lambda)(z - 1)(z + 1)$, where λ is a real parameter, and let $f_\lambda(z)$ denote the associated Newton transformation. Let J_λ denote the Julia set for $f_\lambda(z)$. In Figure 7.3.3 we show images relating to J_λ, for an increasing sequence of values of λ. These images were computed by applying the Escape Time Algorithm to f_λ.

These images show complex blobs that are reminiscent of something small, biological, and organic. They make one think of the nuclei of cells; of collections of cells during the early stages of development of an embryo; of the process of cell division; and of protozoans. As we track the images we see that the blobs pass through one another. Somehow they do so while preserving their complex geometries. Their geometries seem to interact with one another. Such images suggest that fractal geometry can

Figure 7.3.3(a) – (h) *Julia sets associated with a one-parameter family of dynamical systems. Can such systems be used to model biological processes such as myosis?*

Figure 7.3.3(b)

Figure 7.3.3(c)

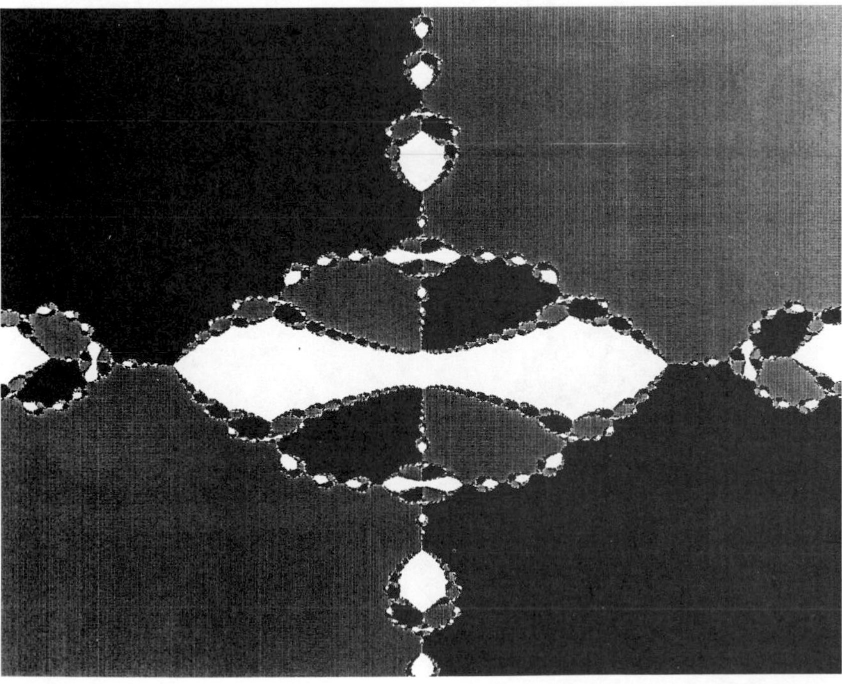

Figure 7.3.3(d)

Figure 7.3.3(e)

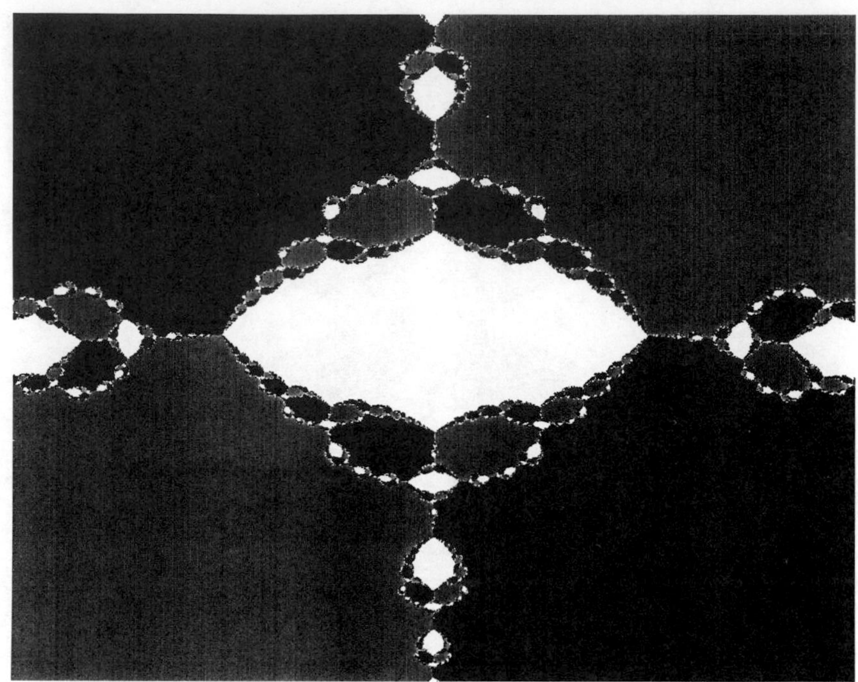

Figure 7.3.3(f)

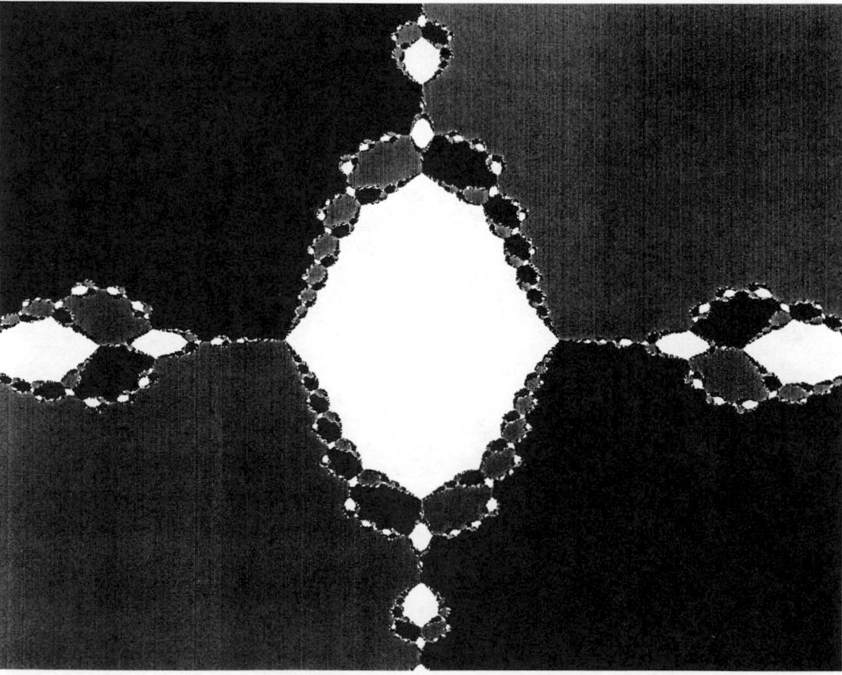

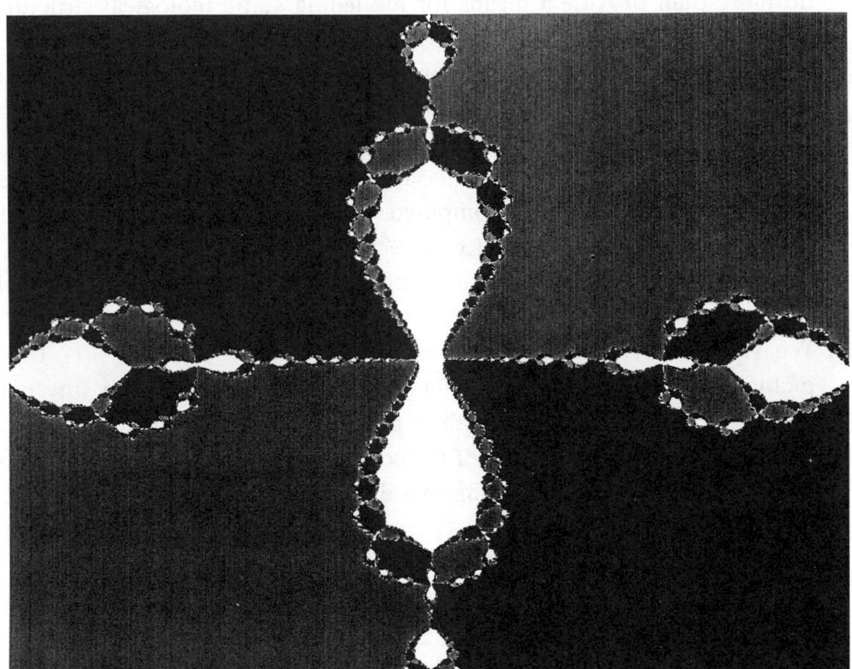

Figure 7.3.3(g)

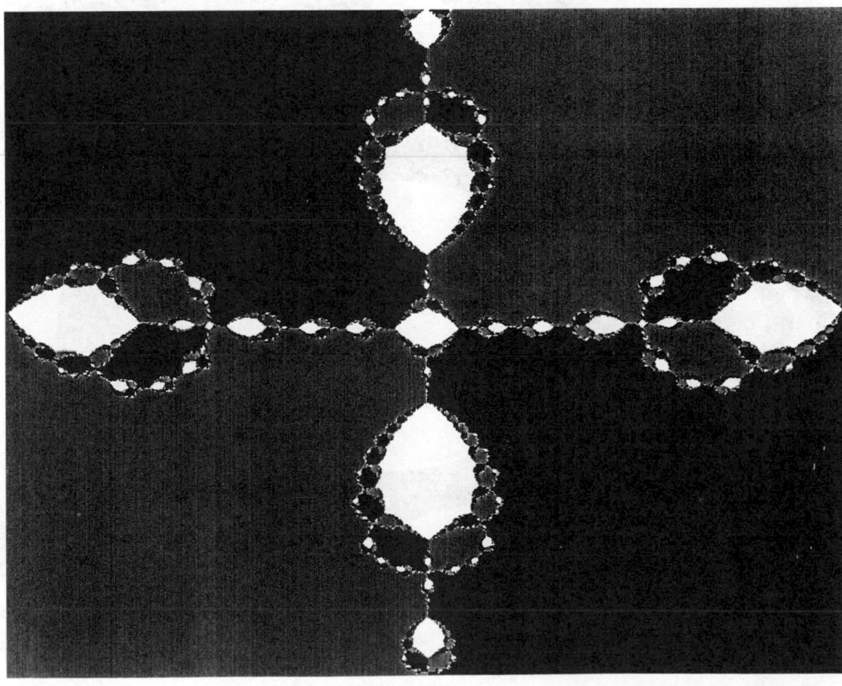

Figure 7.3.3(h)

do more than provide a means for modelling static biological structures, such as ferns: it appears feasible to construct deterministic fractal models, which describe the processes of physiological change which occur during the growth, metamorphosis, and movement of living organisms.

3.5. Find the Newton transformation $f(z)$ associated with the function $F(z) = e^z - 1$. What are the attractive fixed points of the dynamical system $\{\mathbb{C};\ f\}$? Figure 7.3.4 was computed using the Escape Time Algorithm applied to $f(z)$, with $\mathscr{W} = \{(x, y) \in \mathbb{R}^2\colon\ -2.5 \le x \le 2.5,\ -2.5 \le y \le 2.5\}$. Describe the main features of the image. Explain, roughly, the causes of some of these features.

3.6. What are the "petals", in the case of the Julia set for $z^2 - 1$? Use a picture of the Julia set for $z^2 - 1$ to illustrate the orbit of a tiny petal which is eventually periodic with minimal period two.

3.7. By making an explicit change of the coordinates, using a Möbius transformation, show that the following two dynamical systems are equivalent:

$$\left\{\hat{\mathbb{C}};\ f(z) = \frac{1}{2}\left(z - \frac{1}{z}\right)\right\} \quad \text{and} \quad \{\hat{\mathbb{C}};\ f(z) = z^2\}.$$

Figure 7.3.4 *This image was computed using the Escape Time Algorithm applied to the Newton transformation associated with* $f(z) = e^z - 1$. *The viewing window is* $\mathscr{W} = \{(x, y) \in \mathbb{R}^2\colon -2.5 \le x \le 2.5,\ -2.5 \le y \le 2.5\}$. *Can you work out what "escape region" was used?*

7.4 A RICH SOURCE OF FRACTALS: INVARIANT SETS OF CONTINUOUS OPEN MAPPINGS

Let f be a transformation which acts on a space $\mathbb{X}$. Recall that a set A is invariant under f if $f^{-1}(A) = A$. We are interested in invariant sets of f which belong to $\mathscr{H}(\mathbb{X})$. The following theorem is a theoretical tool which provides both the existence and a means for the computation of invariant sets. Most of the material in this chapter is based on it.

Theorem 1. *Let $(\mathbf{Y}, d)$ be a metric space. Let $\mathbb{X} \subset \mathbf{Y}$ be compact and non-empty. Let $f\colon \mathbb{X} \to \mathbf{Y}$ be continuous and such that $f(\mathbb{X}) \supset \mathbb{X}$. Then* (1) *a transformation*
$W\colon \mathscr{H}(\mathbb{X}) \to \mathscr{H}(\mathbb{X})$ is defined by

$$W(A) = f^{-1}(A) \qquad \text{for all } A \in \mathscr{H}(\mathbb{X}).$$

(2) *W possesses a fixed point $A \in \mathscr{H}(\mathbb{X})$, given by*

$$A = \bigcap_{n=0}^{\infty} f^{\circ(-n)}(\mathbb{X}) = \lim_{n\to\infty} W^{\circ n}(\mathbb{X}).$$

Suppose f obeys the additional condition that $f(\mathscr{O})$ is an open subset of the metric space $(f(\mathbb{X}), d)$ whenever $\mathscr{O} \subset \mathbb{X}$ is an open subset of the metric space $(\mathbb{X}, d)$. Then (3) *W is a continuous transformation from the metric space $(\mathscr{H}(\mathbb{X}), h(d))$ into itself.* (4) *If f has domain all of $\mathbf{Y}$ rather than just $\mathbb{X}$, but $f^{-1}(\mathbb{X}) \subset \mathbb{X}$ as well as $f(\mathbb{X}) \supset \mathbb{X}$, then the above applies and* (1), (2), *and* (3) *hold.* (*For example, see the proof of Theorem* 7.2.1).

Proof of (1) and (2). (The proof of (3) can be found in [Barnsley 1988c].)

(1) We begin by proving that W maps $\mathscr{H}(\mathbb{X})$ into $\mathscr{H}(\mathbb{X})$. Let $B \in \mathscr{H}(\mathbb{X})$. The condition $f(\mathbb{X}) \supset \mathbb{X}$ implies that $f^{-1}(B)$ is nonempty; $f^{-1}(B) \subset \mathbb{X}$ since f has domain $\mathbb{X}$. B is compact, so it is a closed set in the metric space $(\mathbb{X}, d)$. The continuity of f implies that $f^{-1}(B)$ is closed in the metric space $(\mathbb{X}, d)$. Since $\mathbb{X}$ is compact it follows that $f^{-1}(B)$ is compact. This completes the proof of (1).

(2) Since f has domain $\mathbb{X}$, it follows that $\mathbb{X} \supset f^{\circ(-1)}(\mathbb{X})$. Application of $f^{\circ(-n)}$ to both sides of the latter equation yields

$$\mathbb{X} \supset f^{\circ(-1)}(\mathbb{X}) \supset f^{\circ(-2)}(\mathbb{X}) \supset f^{\circ(-3)}(\mathbb{X}) \cdots \supset f^{\circ(-n)}(\mathbb{X}) \supset \cdots.$$

It follows that $\{f^{\circ(-n)}(\mathbb{X})\}$ is a Cauchy sequence in $\mathscr{H}(\mathbb{X})$, and it possesses a limit $A \in \mathscr{H}(\mathbb{X})$, given by $A = \bigcap_{n=0}^{\infty} f^{\circ(-n)}(\mathbb{X}) = \operatorname{Lim}_{n\to\infty} W^{\circ n}(\mathbb{X})$.

It remains to be proved that A is a fixed point of W. We need to show that $f^{\circ(-1)}(\bigcap_{n=0}^{\infty} A_n) = \bigcap_{n=0}^{\infty} A_n$ where $A_n = f^{\circ(-n)}(\mathbb{X})$ for $n = 1, 2, \ldots$. But for any sequence $\{B_n\}$ of sets whatever and any function f, $f^{-1}(\bigcap B_n) = \bigcap f^{-1}(B_n)$ as is easily checked, so we have here $f^{-1}(\bigcap_{n=0}^{\infty} A_n) = \bigcap_{n=0}^{\infty} f^{-1}(A_n) = \bigcap_{n=0}^{\infty} A_{n+1} = \bigcap_{n=1}^{\infty} A_n = \bigcap_{n=0}^{\infty} A_n$, since $A_0 \supset A_1 \supset A_2 \supset \ldots$. This completes the proof of (2).

The invariant set A referred to in Theorem 7.4.1 can be expressed

$$A = \{x \in \mathbb{X}: f^{\circ n}(x) \in \mathbb{X} \quad \text{for all } n = 1, 2, 3, \ldots\}.$$

That is, A is the set of points whose orbits do not escape from $\mathbb{X}$. Also, A is the complement of the set of points whose orbits do escape. If $\mathbb{X} \subset \mathbb{R}^2$ then pictures of A can be computed by using the Escape Time Algorithm.

The last statement in the last theorem expresses a desirable property for a transformation W on $\mathscr{H}(\mathbb{X})$. If W is not continuous, yet $A_0 \in \mathscr{H}(\mathbb{X})$ and $\{W^{\circ n}(A_0)\}$ converges to $A \in \mathscr{H}(\mathbb{X})$, one cannot conclude that $W(A) = A$. Without continuity of W, one should not trust the results of applying the Escape Time Algorithm. For example, slight numerical errors may mean that a computed sequence of sets $\{\tilde{A}_n \approx W^{\circ n}(\mathbb{X})\}$ is not decreasing. One may still wish to define $\tilde{A} = \operatorname{Lim} \tilde{A}_n$. Without continuity one cannot suppose that $f^{-1}(\tilde{A}) = \operatorname{Lim} W(\tilde{A}_n) = \tilde{A}$, even approximately.

Analytic transformations map open sets to open sets. Hence their inverses act continuously on the space $\mathscr{H}(\mathbb{X})$, where $\mathbb{X} \subset \hat{\mathbb{C}}$ is chosen appropriately. To help visualize this, look back at Figure 3.4.3. If the Sierpinski triangle ABO is deformed or moved, its inverse image $POQ \cup \tilde{P}O\tilde{Q}$ will move continuously with it.

The Hausdorff metric is the metric of perception: what we call a small change in the appearance of a picture is probably a small change in a Hausdorff distance. When one talks about continuous motion in the context of graphics, continuous growth in the context of botany, or continuous change in the context of a chemical system, the word "continuous" can often be replaced, pedantically, by "continuous in the Hausdorff metric." Theorem 7.4.1 suggests that one could use continuous *open* maps to model such systems.

Exercises & Examples

4.1. Let $\lambda \in [-1, 1]$. Define a transformation $f: \mathbb{R}^2 \to \mathbb{R}^2$ by

$$f(x, y) = \begin{cases} (x^2 - y^2 - 1, 2xy) & \text{when } x > 0, \\ (x^2 - y^2 - 1 + \lambda x, 2xy) & \text{when } x \leq 0. \end{cases}$$

Show f is continuous. Show that if $\mathbb{X}$ denotes a ball, centered at the

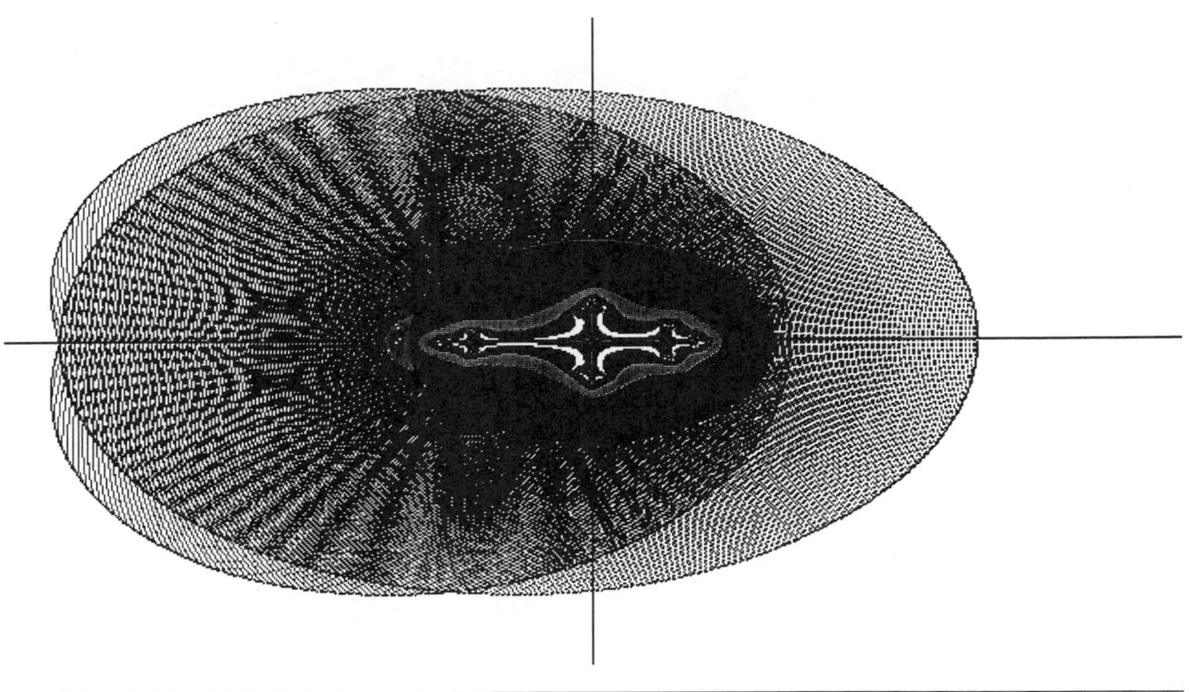

Figure 7.4.1
The result of applying the Escape Time Algorithm to the function f in example 7.4.1 with $\lambda = 1$. This function is continuous, and on such that $f(\mathbb{X}) \supset \mathbb{X}$, where $\mathbb{X}$ denotes the Fabergé egg in the middle. $f(\mathbb{X})$ is the region bounded by the outer curve. Different "escape times" of orbits of points in $\mathbb{X}$ are represented by different greytones.

origin, of sufficiently large radius, then $f(\mathbb{X}) \supset \mathbb{X}$ and $f^{-1}(\mathbb{X}) \subset \mathbb{X}$. Also verify that if $\lambda \in [-1, 0]$ then f maps open sets into open sets. Show that this is not the case for $\lambda = 1$. (Hint: look at what the map does to a very small disk centered at the origin.)

Figure 7.4.1 shows the result of applying the Escape Time Algorithm to f when $\lambda = 1$. The inner region, bounded by an ellipse, actually represents a disk $\mathbb{X}$ such that $f(\mathbb{X}) \supset \mathbb{X}$. Different scales have been used in the x and y directions. $f(\mathbb{X})$ is the region bounded by the outer curve. The image of a point which goes once around the inner ellipse is a point which goes twice around the origin, following the outer curve which looks like a folded figure eight. Different "escape times" of orbits of points in $\mathbb{X}$ are represented by different greytones. A magnified version of $\mathbb{X}$, painted by escape times, is shown in Figure 7.4.2. Roughly speaking, regions closest to the outside escape fastest. Points in the white region also escape. So where is the invariant set A? It is right in the middle. It appears to be a branching, connected, tree-like set, with no interior.

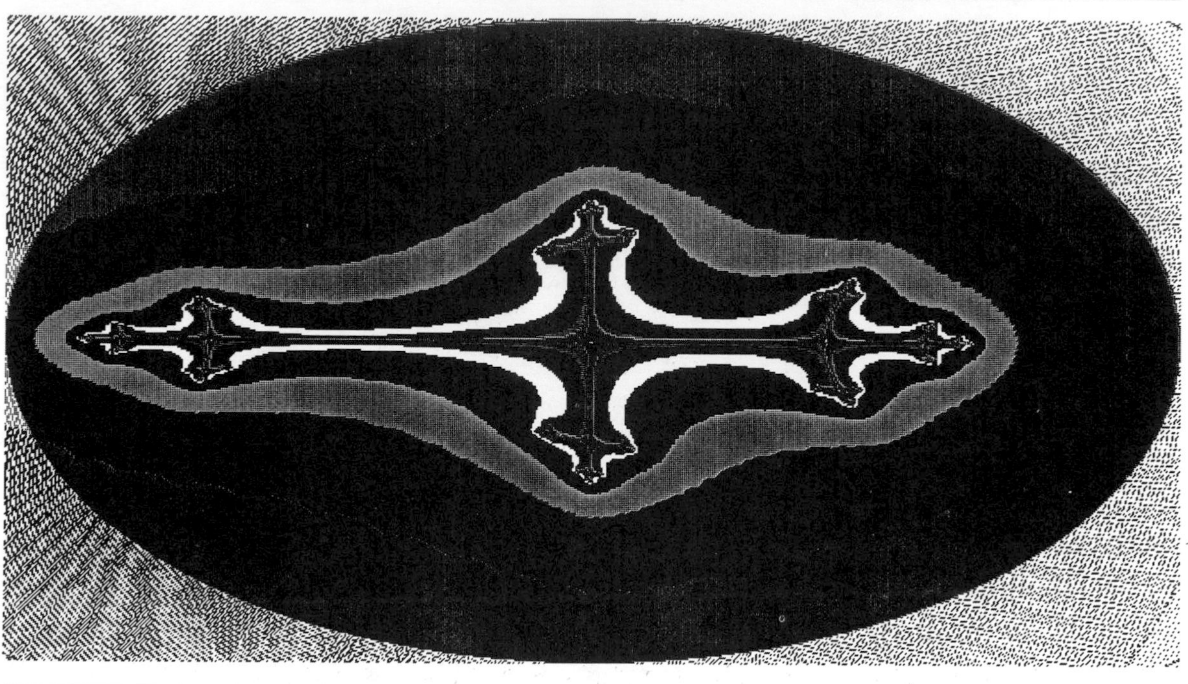

Figure 7.4.2
A magnified version of the region $\mathbb{X}$, *in Figure* 7.4.1. *Approximately, orbits of points closest to the outside escape fastest. Points in the white region also escape. The invariant set* A, *is in the middle. It appears to be a branching, connected, tree-like set. It is surrounded by layers, just as the center of a real tree is surrounded by layers of growth.*

Figure 7.4.3 shows the result of applying the Escape Time Algorithm to f when $\lambda = 0$. This time we see that the set invariant A, in the center, in white, is just the filled Julia set for $z^2 - 1$.

What happens if we choose $\lambda = -1$? This time we obtain the image shown in Figure 7.4.4. However, this time things may not be as simple as they appear to be. The inner "layers" which surround the apparent invariant set A are highly irregular and unstable. That is, points which are very close together seem to have orbits which have very different escape times.

4.2. Construct a function $f\colon \mathbb{X} \to \mathbb{R}^2$, where $\mathbb{X} \subset \mathbb{R}^2$, which obeys the conditions of Theorem 7.4.1. Use the Escape Time Algorithm to analyse the associated invariant set A described in the statement of the theorem. Your example should be interesting, and of a different character from those specifically described in this chapter.

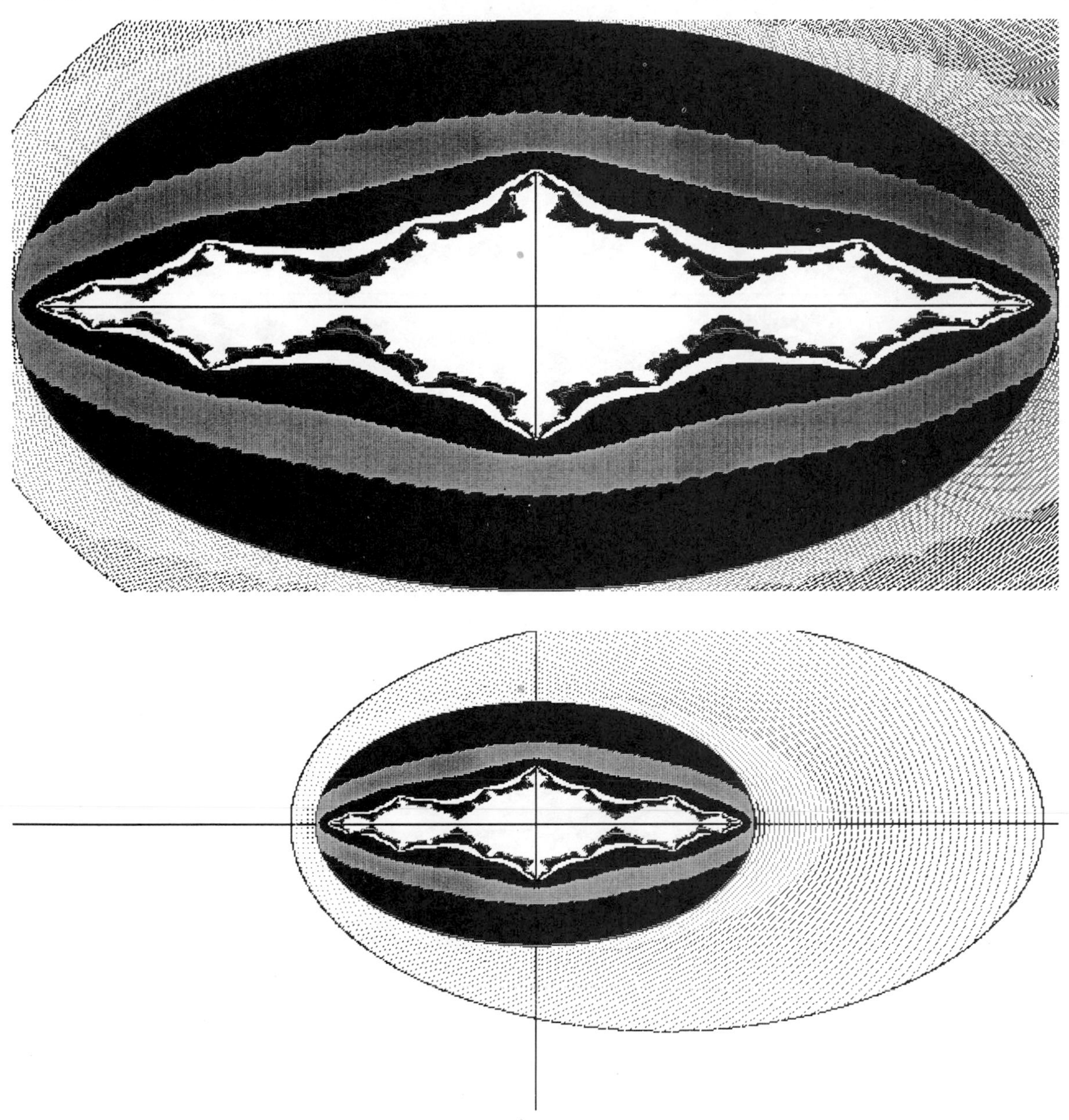

Figure 7.4.3
The result of applying the Escape Time Algorithm to f in example 7.4.1 with $\lambda = 0$. This time we see that the invariant set A, in the center, in white, is just the filled Julia set for $z^2 - 1$.

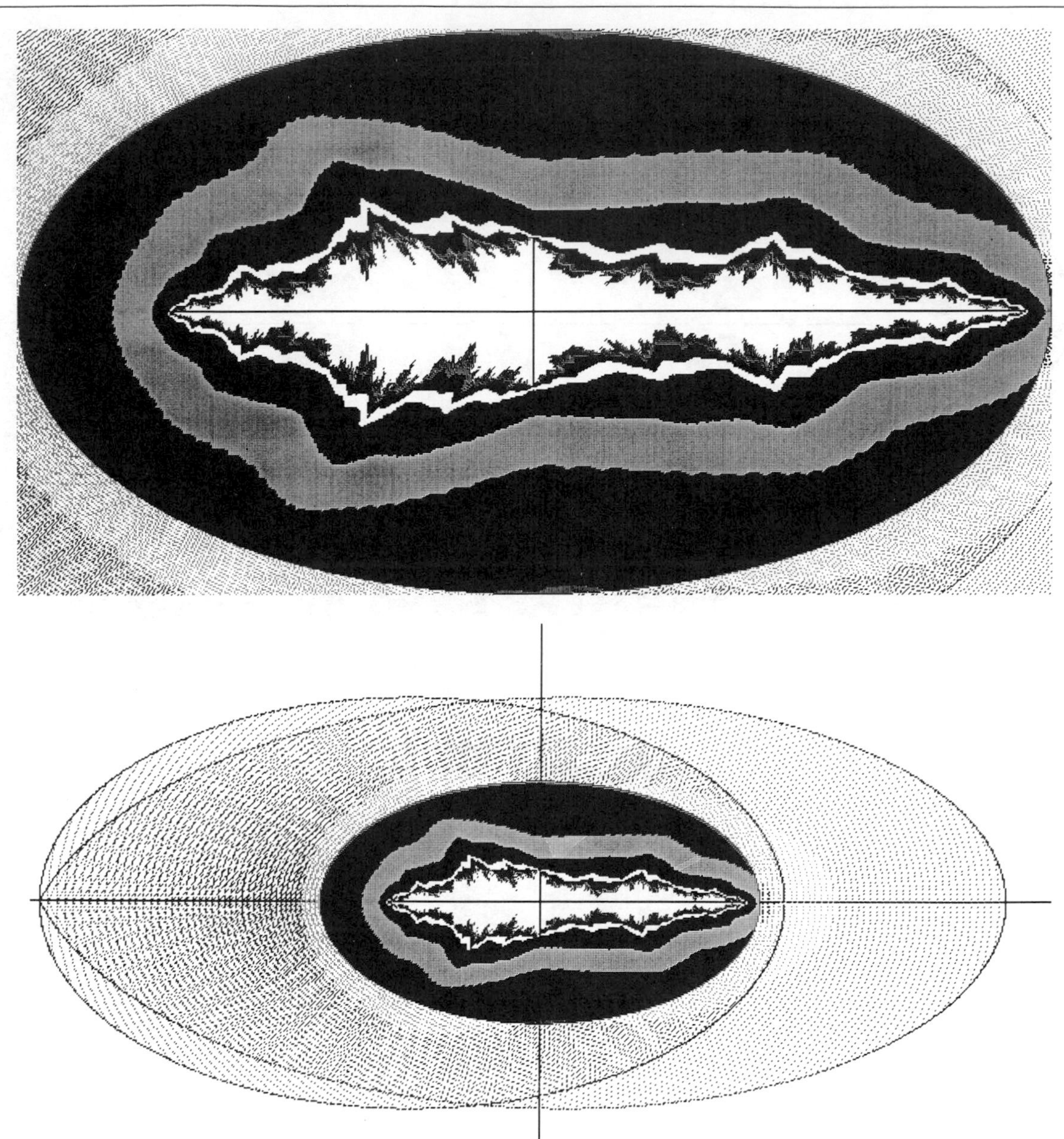

Figure 7.4.4
What happens if we choose $\lambda = -1$? *This time things may not be as simple as they appear to be. The inner "layers" which surround the apparent invariant set A are highly irregular and difficult to pin down numerically.*

Parameter Spaces and Mandelbrot Sets

8

8.1 THE IDEA OF A PARAMETER SPACE: A MAP OF FRACTALS

A map, with nothing marked on it, is practically useless. A map of a 1000×1000 square-mile region containing the British Isles is shown in Figure 8.1.1; it does not convey much information. However, as a concept it is quite exciting. Each location on the map corresponds to somewhere on the Earth. For example, the dot with coordinates (750, 227.3) represents the town of Maidstone. A point on the map may represent a certain grain of soil in a ploughed field, or a molecule of flotsam on the top of some foam on the surface of the sea. Nearby points in the map correspond to nearby points on the Earth. Connected sets with interiors correspond to physical regions.

How could a perfect map be made? Ideally, it should specify locations on the Earth's surface at a certain instant. The coordinates would be relative to some absolute coordinate system, perhaps determined by reference to the fixed stars. Moreover, the surface of the Earth would have to be defined precisely, up to the last molecule of water, soil, and plant matter; for this purpose one can imagine using a straight line from the center of the Earth as suggested in Figure 8.1.2. Of course, maps are not made like this, but the goal is the same:

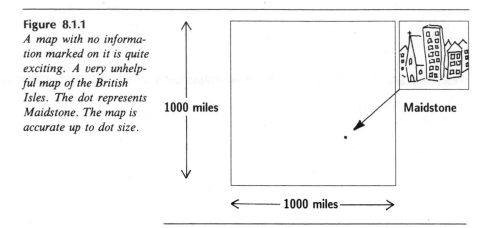

Figure 8.1.1
A map with no information marked on it is quite exciting. A very unhelpful map of the British Isles. The dot represents Maidstone. The map is accurate up to dot size.

to have an accurate correspondence between points on the physical surface of the Earth and the physical surface of the paper.

We must be careful how we interpret a map. Geographical maps are complicated by the real number system and the unphysical notion of infinite divisibility. Mathematically, the map is an abstract place. A point on the map cannot represent a certain physical atom in the real world, not just because of inaccuracies in the map, but because of the dual nature of matter: according to current theories one cannot know the exact location of an atom, at a given instant.

Fractal geometers avoid this problem by pretending that the surface of the Earth is an abstract place too; we imagine, once again, that matter is infinitely

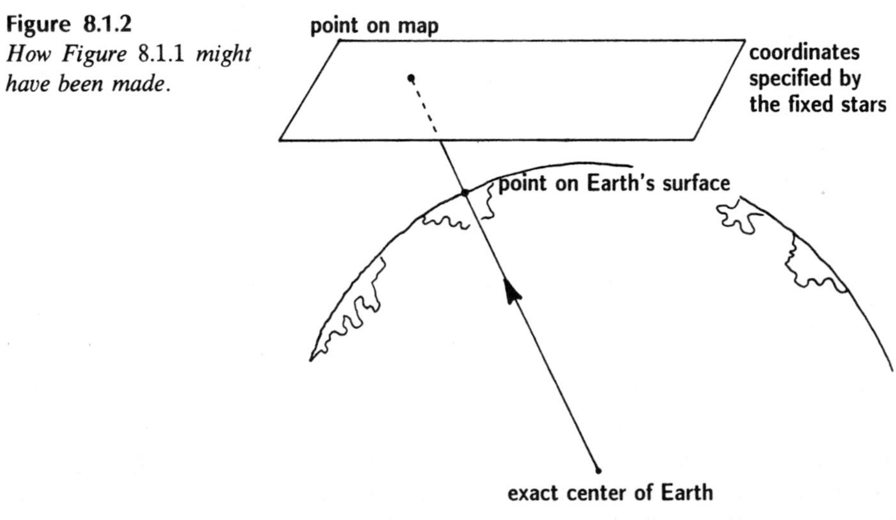

Figure 8.1.2
How Figure 8.1.1 *might have been made.*

divisible, and that we can address every point. In the same spirit, we presume that we can model trees and clouds, horizons, churning seas, and infinitely finely defined coastlines. Then, for example, we can define the Hausdorff-Besicovitch dimension of the coastline of the British Isles.

For a map to be useful it must have information marked on it, such as heights above sea-level, population densities, roads, vegetation, rainfall, types of underlying rock, ownership, names, incidence of volcanoes, malarial infestation, and so on. A good way of providing such information is with colors. For example, if we use blue for water and green for land then we can "see" the land on the map and we can understand some geometrical relationships. We can estimate overland distances between points, land areas of islands, the shortest sea passage from Llanellian Bay to Amylwch Harbour, the length of the coastline, etc. All this is achieved through the device of marking some colors on a blank map!

Let us consider the boundary of the shaded region in Figure 8.1.3. It is here that the map conveys extra information. In the interior of the shaded region we learn no more about the surface of the Earth than that "there is land there." However, on the boundary we learn not only that "there is land and sea there," but also, if the map is accurate enough, a feature which we will actually "see" on the surface, namely the local shape of the coastline.

The latter idea can be extended. If we include more colors on a geographical map, to provide more information about properties of the Earth's surface, we produce more boundaries on the map. These boundaries can give information about local geometry. For example, a map finely colored according to elevation reveals the shapes of the bases of the mountains, the paths of rivers, and—if we look closely enough—the outlines of buildings. Such a map, made abstract and perfect, placed in a metric space, would contain much detailed information about what, at each point, the local observer would see.

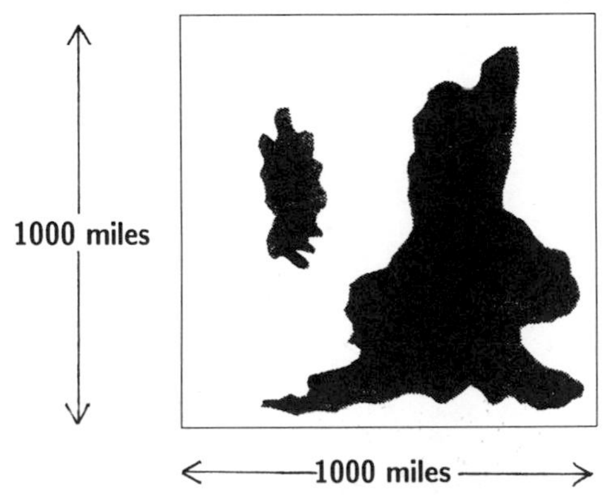

Figure 8.1.3
In this map points corresponding to land have been shaded. A fascinating entity, the coastline, is revealed.

Exercises & Examples

1.1. Study an atlas which contains maps colored according to diverse criteria, such as rainfall, population density, vegetation, and elevation. Discuss to what extent these maps provide information about the local geometry of the surface of the Earth.

We turn attention to making colored maps of parameterized families of fractals. We consider families of iterated function systems and families of dynamical systems which depend on two real parameters. The collection of possible parameter values defines a *parameter space* associated with the family. We use the notation P to denote a parameter space. Typically P is a subspace of ($\mathbb{R}^2$, Euclidean) such as ■, a closed ball, or $\mathbb{R}^2$.

An example of a parameter space is $P = \{(\lambda_1, \lambda_2) \in \mathbb{R}^2\colon |\lambda_1|, |\lambda_2| < 2^{-0.5}\}$. This is a parameter space for the family of hyperbolic IFS $\{\mathbb{C}; (\lambda_1 + i\lambda_2)z + 1, (\lambda_1 + i\lambda_2)z - 1\}$. Each point $\lambda = (\lambda_1, \lambda_2) \in P$ corresponds to an IFS. Each IFS possesses a unique attractor, say $A(\lambda)$. Hence each point of P corresponds to a single fractal. We can think of P as representing part of $\mathscr{H}(\mathbb{C})$, a space of fractals. A map of P, with a few points marked on it, is shown in Figure 8.1.4. Each point in P corresponds to a single

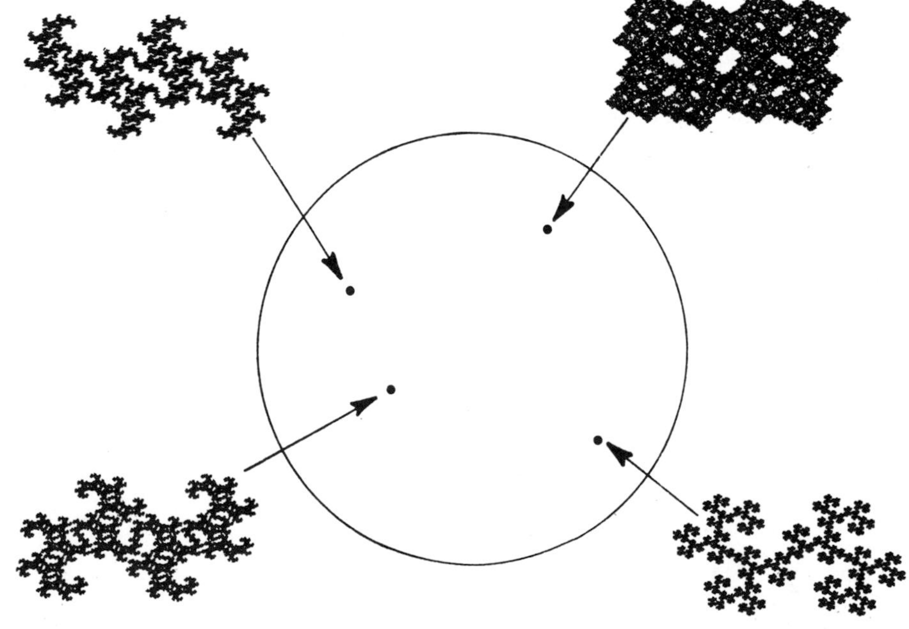

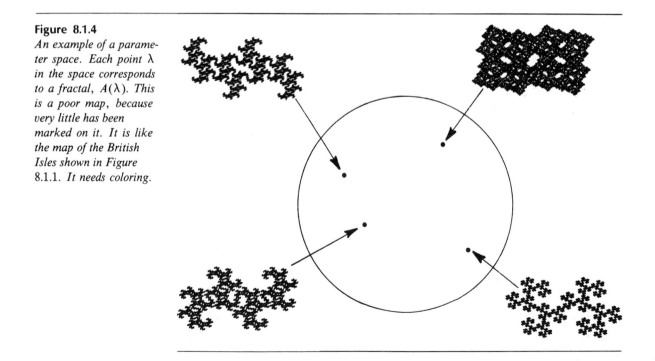

Figure 8.1.4 *An example of a parameter space. Each point λ in the space corresponds to a fractal, $A(\lambda)$. This is a poor map, because very little has been marked on it. It is like the map of the British Isles shown in Figure 8.1.1. It needs coloring.*

fractal. Nearby points in P correspond to nearby fractals, that is, points in $\mathscr{H}(\mathbb{C})$ whose Hausdorff distance apart is small.

Another example is $P = \mathbb{C}$, which provides a parameter space for the family of dynamical systems $\{\hat{\mathbb{C}}; f_\lambda(z) = z^2 - \lambda\}$. Each point in the parameter space corresponds to a different dynamical system. Each dynamical system is associated with a unique Julia set, $J(\lambda)$. The collection of fractals $\{J(\lambda): \lambda \in P\}$, associated with the parameter space, is vast and diverse.

Let $\mathbb{X}$ be a metric space such as $\mathbb{R}^2$ or $\hat{\mathbb{C}}$. Let P denote a parameter space corresponding to a family of fractals $\{A(\lambda) \in \mathscr{H}(\mathbb{X}): \lambda \in P\}$. Can we provide the explorer who wishes to investigate this collection of fractals, with a colored map? This map should give him information about the sets $A(\lambda)$, to be found at different points on P.

Suppose $P = \blacksquare$. To make a map, let us represent P by the pixels on the screen of a computer graphics monitor. The idea is to color the pixel λ according to some property of $A(\lambda)$. (We write λ = pixel = point in P, without repeatedly explaining that λ is a point in the small rectangle in P which corresponds to the pixel.) Suitable properties could relate to the connectivity of $A(\lambda)$, the fractal dimension of $A(\lambda)$, the escape time of a special point on $A(\lambda)$ under an associated dynamical system, the number of holes in $A(\lambda)$, or the presence of straight lines in $A(\lambda)$, for example.

If we make a good selection of the properties to associate with colors, the result will be a useful map containing various differently colored regions. This map will be a ready reference for the explorer of fractals. It will tell him something about what to expect as he travels about P. He might be surprised nonetheless.

The boundaries of the colored regions can provide additional geometrical information to the explorer, over and above the information which the map was originally designed to convey. It sometimes occurs that the local shapes of the boundaries in the map reflect the shapes of the corresponding fractals. There is a deep principle here, which we shall not pin down as a theorem, but which we will illustrate in a number of cases.

Exercises & Examples

1.2. Let $P = \{(\lambda_1, \lambda_2) \in \mathbb{R}^2: |\lambda_1|, |\lambda_2| \le 0.9\}$. The family of IFS $\{\mathbb{R}; \lambda_1 x, \lambda_2 x + 1 - \lambda_2, 0.5x + 0.5\}$ is hyperbolic, with contractivity factor $s = 0.9$, for all $\lambda \in P$. Use Theorem 3.11.1 to prove that the attractor depends continuously on λ.

1.3. The family of IFS $\{[0,1]; \lambda_1 x^2, \lambda_2 x + (1 - \lambda_2)\}$ is hyperbolic, with contractivity factor $s = 0.9$, for all λ in the parameter space $P = \{(\lambda_1, \lambda_2) \in \mathbb{R}^2: 0 \le \lambda_1 \le 0.9, 0 \le \lambda_2 \le 0.9\}$. The attractor depends continuously on λ.

1.4. An example of a parameter space is $(P, \text{Euclidean})$ where $P = \{(\lambda_1, \lambda_2) \in \mathbb{R}^2: |\lambda_1|, |\lambda_2| \leq 0.999\}$. This space can be used to represent the family of hyperbolic IFS $\{\mathbb{R}^2; w_1, w_2\}$, where

$$w_1\begin{pmatrix} x \\ y \end{pmatrix} = \begin{pmatrix} \lambda_1 & 0 \\ 0 & \lambda_2 \end{pmatrix}\begin{pmatrix} x \\ y \end{pmatrix} + \begin{pmatrix} 0 \\ 1 \end{pmatrix}; \quad \text{and} \quad w_2\begin{pmatrix} x \\ y \end{pmatrix} = \begin{pmatrix} 0.3 & -0.2 \\ 0.1 & 0.4 \end{pmatrix}\begin{pmatrix} x \\ y \end{pmatrix}.$$

8.2 MANDELBROT SETS FOR PAIRS OF TRANSFORMATIONS

Let $P \subset \mathbb{R}^2$ be a parameter space corresponding to a family of fractals. That is, we have a function $A: P \to \mathscr{H}(\mathbb{X})$, so that each point $\lambda \in P$ corresponds to a set $A(\lambda) \in \mathscr{H}(\mathbb{X})$. One way to make a map is to color the parameter space according to whether or not $A(\lambda)$ is connected.

Theorem 1. *Let $\{\mathbb{X}; w_1, w_2\}$ be a hyperbolic IFS with attractor A. Let w_1 and w_2 be one-to-one on A. If*

$$w_1(A) \cap w_2(A) = \varnothing$$

then A is totally disconnected. If

$$w_1(A) \cap w_2(A) \neq \varnothing$$

then A is connected.

Proof. Suppose that $w_1(A) \cap w_2(A) = \varnothing$. Let Σ denote the code space map associated with the IFS. By Theorem 4.2.2 the code space map $\phi: \Sigma \to A$ is invertible. ϕ is also a continuous transformation between two compact metric spaces. Hence, by Theorem 2.8.5, ϕ is a homeomorphism. Hence A is homeomorphic to code space, which is totally disconnected. (Recall that code space on two or more symbols is metrically equivalent to a classical Cantor set.) It follows that A is totally disconnected.

Suppose that $w_1(A) \cap w_2(A) \neq \varnothing$. Then there is at least one point $x \in w_1(A) \cap w_2(A)$. This point x has two addresses, say

$$x = \phi(\zeta) = \phi(\sigma) \qquad \text{where } \zeta_1 = 1 \quad \text{and} \quad \sigma_1 = 2.$$

Let us see what happens if we additionally suppose that "A is not connected." Then, since A is compact, we can find two nonempty compact sets E and F so that

$$A = E \cup F, \qquad E \cap F = \varnothing.$$

Using compactness, there is a positive real number δ so that

$$d(e, f) \geq \delta \qquad \text{for all } e \in E,\, f \in F.$$

Let π and ψ be a pair of codes which agree through the first K symbols, for some positive integer K. That is $\pi_i = \psi_i$ for $i = 1, 2, \ldots, K$. Then

$$d(\phi(\pi), \phi(\psi)) \leq s^K \text{diam}(A),$$

where $\text{diam}(A) = \text{Max}\{d(x, y)\text{: } x, y \in A\}$, and $s \in [0, 1)$ is a contractivity factor for the IFS. Suppose also that $\phi(\pi) \in E$ and $\phi(\psi) \in F$. Then

$$\delta \leq d(\phi(\pi), \phi(\psi)).$$

Combining the latter two inequalities we discover $\delta \leq s^K \text{diam}(A)$, which implies

$$K \leq \frac{\text{Log}(\delta/\text{diam}(A))}{\text{Log}(s)}.$$

We conclude that if $e \in E$ and $f \in F$ then the number of successive agreements between an address of e and an address of f cannot exceed the number on the right-hand side here. It follows that there is a maximum number, M, of initial agreements between the address of a point $e \in E$ and a point $f \in F$; and this maximum is achieved on some pair of points, say, e and f. Then we can find $\rho_i \in \{1, 2\}$ for $i = 1, 2, \ldots, M$ such that

$$\phi(\rho_1, \rho_2, \rho_3, \ldots, \rho_M, 1, \ldots) = e \in E$$

and

$$\phi(\rho_1, \rho_2, \rho_3, \ldots, \rho_M, 2, \ldots) = f \in F$$

Now consider the point $z \in A$ which has the two addresses

$$z = \phi(\rho_1, \rho_2, \rho_3, \ldots, \rho_M, 1, \zeta_2, \zeta_3, \zeta_4, \ldots) = \phi(\rho_1, \rho_2, \rho_3, \ldots, \rho_M, 2, \sigma_2, \sigma_3, \sigma_4, \ldots).$$

Suppose $z \in E$. Then its address agrees with that of $f \in F$ through $(M + 1)$ initial symbols. Hence $z \in F$. But then its address agrees with that of $e \in E$ through $(M + 1)$ initial symbols, which is not possible. We have a contradiction. Hence "A is *not* disconnected." It follows that A is connected. This completes the proof of the theorem.

Definition 1. Let $\{\mathbf{X}; w_1, w_2\}$ be a family of hyperbolic IFS which depends on a parameter $\lambda \in P \subset \mathbb{R}^2$. Let $A(\lambda)$ denote the attractor of the IFS. The set of points $\mathcal{M} \subset P$ defined by

$$\mathcal{M} = \{\lambda \in P\text{: } A(\lambda) \text{ is connected}\}$$

is called the *Mandelbrot set* for the family of IFS.

For the rest of this section we consider the family of IFS

$$\{\mathbb{C}; \lambda z - 1, \lambda z + 1\},$$

where the parameter space is

$$P = \{\lambda = (\lambda_1, \lambda_2) \in \mathbb{C}\text{: } \lambda_1^2 + \lambda_2^2 < 1\}.$$

Figure 8.2.1 shows a picture of the associated Mandelbrot Set, $\mathcal{M}$. This is a map for the collection of fractals associated with the IFS. It has been colored

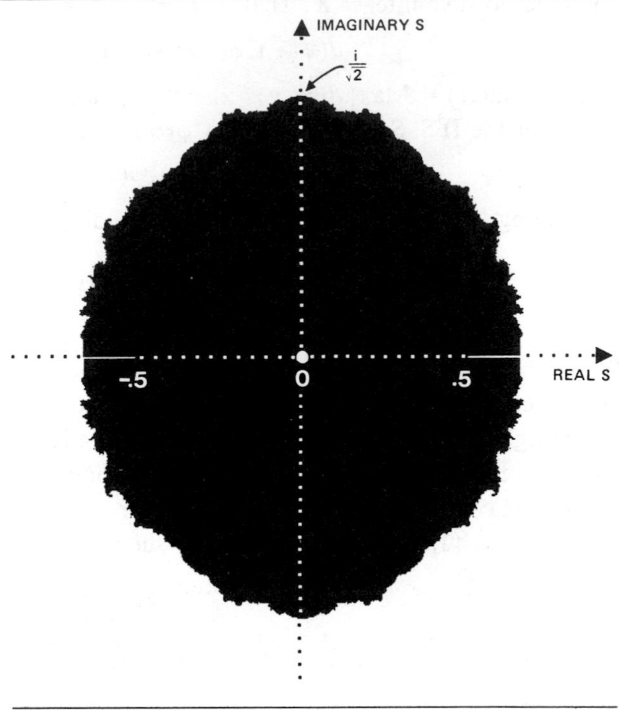

Figure 8.2.1
A map of the family of IFS $\{\mathbb{C};\ \lambda z - 1,\ \lambda z + 1\}$ *where the parameter space is* $P = \{\lambda = (\lambda_1, \lambda_2) \in \mathbb{C}: |\lambda| < 1\}$. *This picture of parameter space is obtained by "painting" black where the attractor of the IFS is disconnected and light where it is connected. The Mandelbrot set is the light region, the sea. It contains a dragon at* $\lambda = (0.5, 0.5)$.

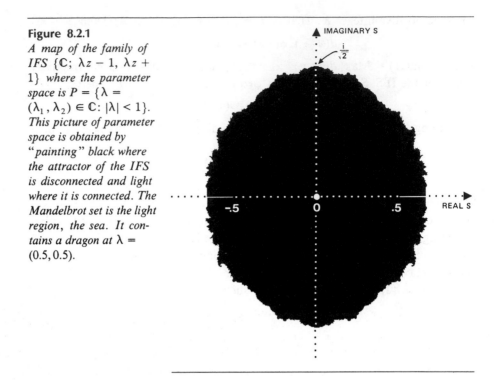

dark where the attractor is totally disconnected, and light where it is connected.

Here is an outline of an algorithm to compute images of the Mandelbrot Set $\mathcal{M}$ associated with the family $\{\mathbb{C};\ w_1(z) = \lambda z - 1,\ w_2(z) = \lambda z + 1\}$. It is based on Theorem 8.2.1.

Algorithm 1 (Example of Method for Making Pictures of the Mandelbrot Set of a Family of IFS)

(i) Choose a positive integer, L, corresponding to the amount of computation one is able to do. The greater the value of L, the more accurate the resulting map image will be.

(ii) Represent the parameter space $P = \{\lambda \in \mathbb{C}: |\lambda| < 1\}$ by an array of pixels. Carry out the following steps for each λ in the array.

(iii) Calculate a number R, so that the attractor is contained in a ball of radius R, centered at the origin; that is, choose $R > 0$ so that $A(\lambda) \subset B(0, R)$.

(iv) Compute the number

$$H = \mathrm{Min}\{\, d(x, y) : x \in w_1(W^{\circ L}(\{0\})),\ y \in w_2(W^{\circ L}(\{0\}))\},$$

where $W = w_1 \cup w_2$. If $H \leq 2|\lambda|^{L+1}R$ then the pixel λ is assumed to belong to $\mathcal{M}$, and is colored accordingly.

Step (iv) is based on the following observation. The attractor of the IFS is contained in the set $W^{\circ(L+1)}(B(0, R))$, which consists of 2^{L+1} balls of radius $|\lambda|^{L+1}R$. The centers of these balls lie in the union of the two sets $w_1(W^{\circ L}(0))$ and $w_2(W^{\circ L}(0))$. If H is greater than $2|\lambda|^{L+1}R$ then $A(\lambda)$ must be disconnected.

Figure 8.2.2 shows the "coastal region" of a quarter of the complement of the Mandelbrot set in Figure 8.2.1. It has been laid over a grid, in order to help you locate points where interesting fractals lie.

The boundary of $\mathcal{M}$ is complicated and intricate. Close-ups of the "coastline" near the places marked (a), (b), and (c) are shown in Figure 8.2.3. Figure 8.2.4 shows a zoom on the spiral peninsula in Figure 8.2.2.

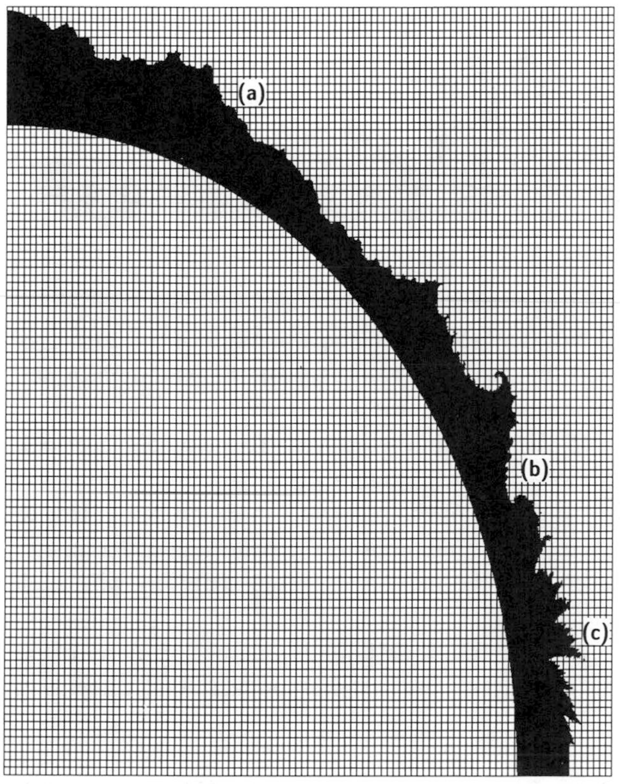

Figure 8.2.2 *This shows the coastal region of a quarter of the complement of the Mandelbrot set in Figure 8.2.1. It has been laid over a grid, in order to help you locate points where interesting fractals lie. Close-ups of the coast at (a), (b), and (c) are shown in Figure 8.2.3. The coordinates of the grid are* (0, 0)–(0.71, 0.71).

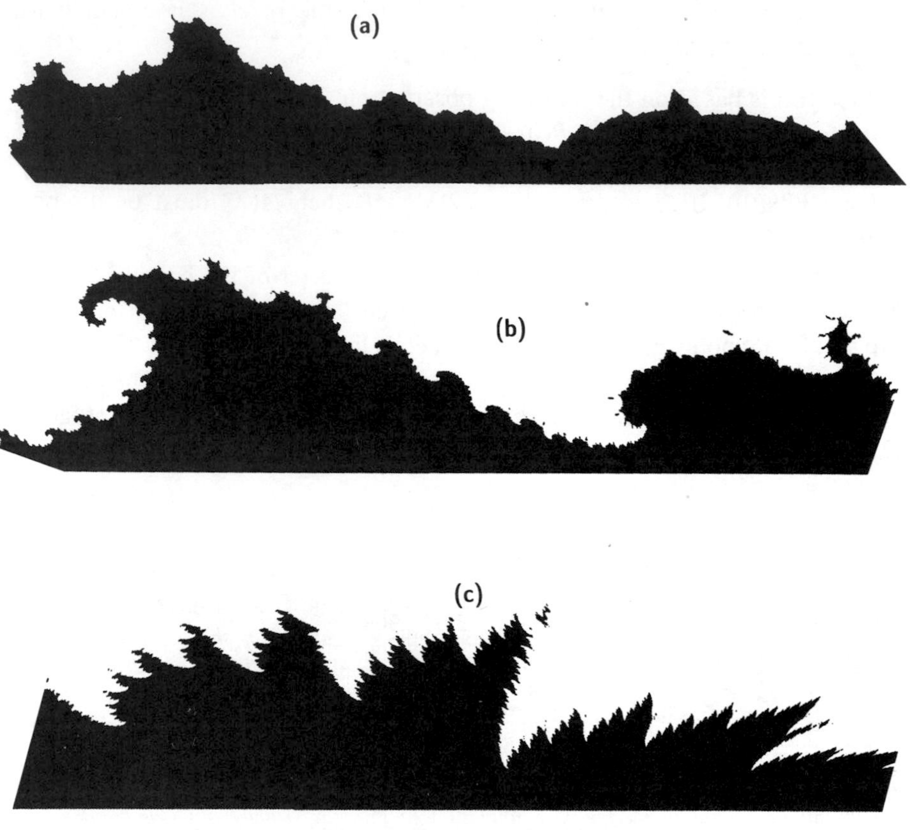

Figure 8.2.3 *Close-ups of the boundary of the Mandelbrot set at (a), (b), and (c). The diverse structures in this boundary echo the shapes of the attractors of the corresponding IFS.*

Now look at Figures 8.2.5 (a) and (b), which show pictures of the attractors $A(\lambda)$ for some points λ located near the boundary of the Mandelbrot set. There is a "family resemblance" between the places on the boundary from which the fractals come, and the fractals themselves. To help see this, look back at the close-ups on the coastline in Figure 8.2.3. Figure 8.2.6 shows the IFS attractor corresponding to the tip of the peninsula in Figure 8.2.4. Notice how it contains spirals, very much like the ones in the peninsula in parameter space. At the end of this chapter we make some comments on why such "family resemblances" occur.

The following theorem provides rigorous bounds on the locations of $\mathcal{M}$ and $\partial\mathcal{M}$. The proof is delightful, because it relies on a fractal dimension estimate.

Theorem 2. [Barnsley 1985c] *The attractor* $A(\lambda)$ *of the IFS* $\{\mathbb{C};\ \lambda z - 1, \lambda z + 1\}$ *is totally disconnected if* $|\lambda| < 0.5$ *and connected if* $1 > |\lambda| > 1/\sqrt{2}$.

Figure 8.2.4 *Close-up on the spiral peninsula on the edge of the Mandelbrot set in Figures* 8.2.1 *and* 8.2.2. *What information about the corresponding fractals does this boundary convey?*

The boundary of the associated Mandelbrot set is contained in the annulus $1/2 \leq |\lambda| \leq 1/\sqrt{2}$.

Proof. Let A denote the attractor of the IFS and let $D(A)$ denote its fractal dimension. The two maps in the IFS are similitudes of scaling factor $|\lambda|$. This means that Theorem 5.2.3 can be applied.

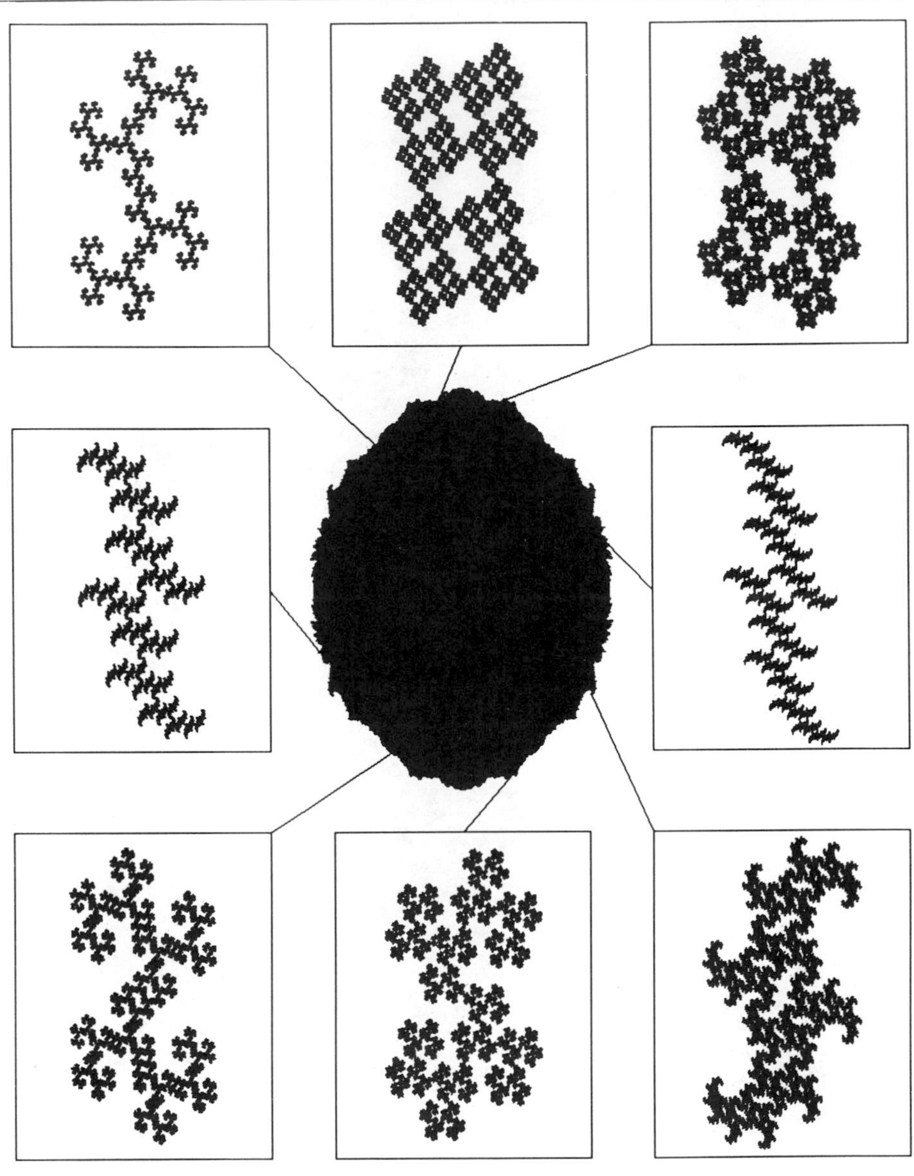

Figure 8.2.5(a) *Some of the fractals to be found at various points near the boundary of the Mandelbrot set associated with the parameterized family of IFS* $\{\mathbb{C};\ \lambda z - 1,\ \lambda z + 1\}$.

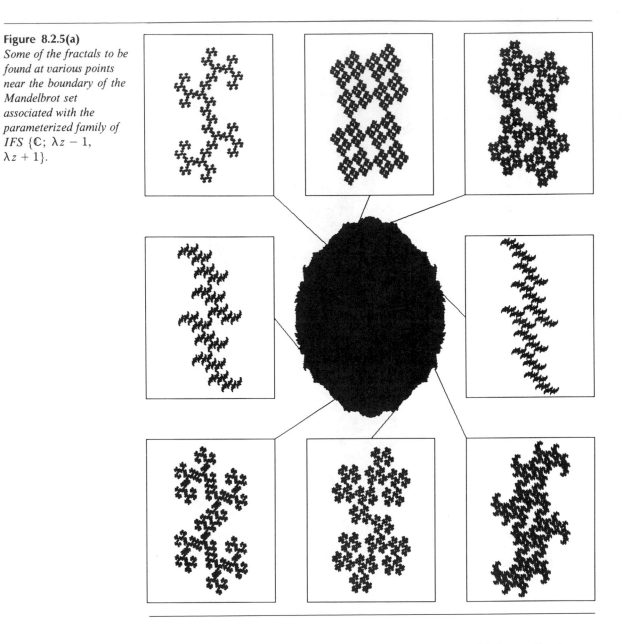

Suppose that A is totally disconnected. Then the IFS is totally disconnected and, by Theorem 5.2.3,

$$D(A) = \log(1/2)/\log(|\lambda|).$$

By Theorem 5.2.1, $D(A) \leq 2$. Hence

$$\log(1/2)/\log(|\lambda|) \leq 2.$$

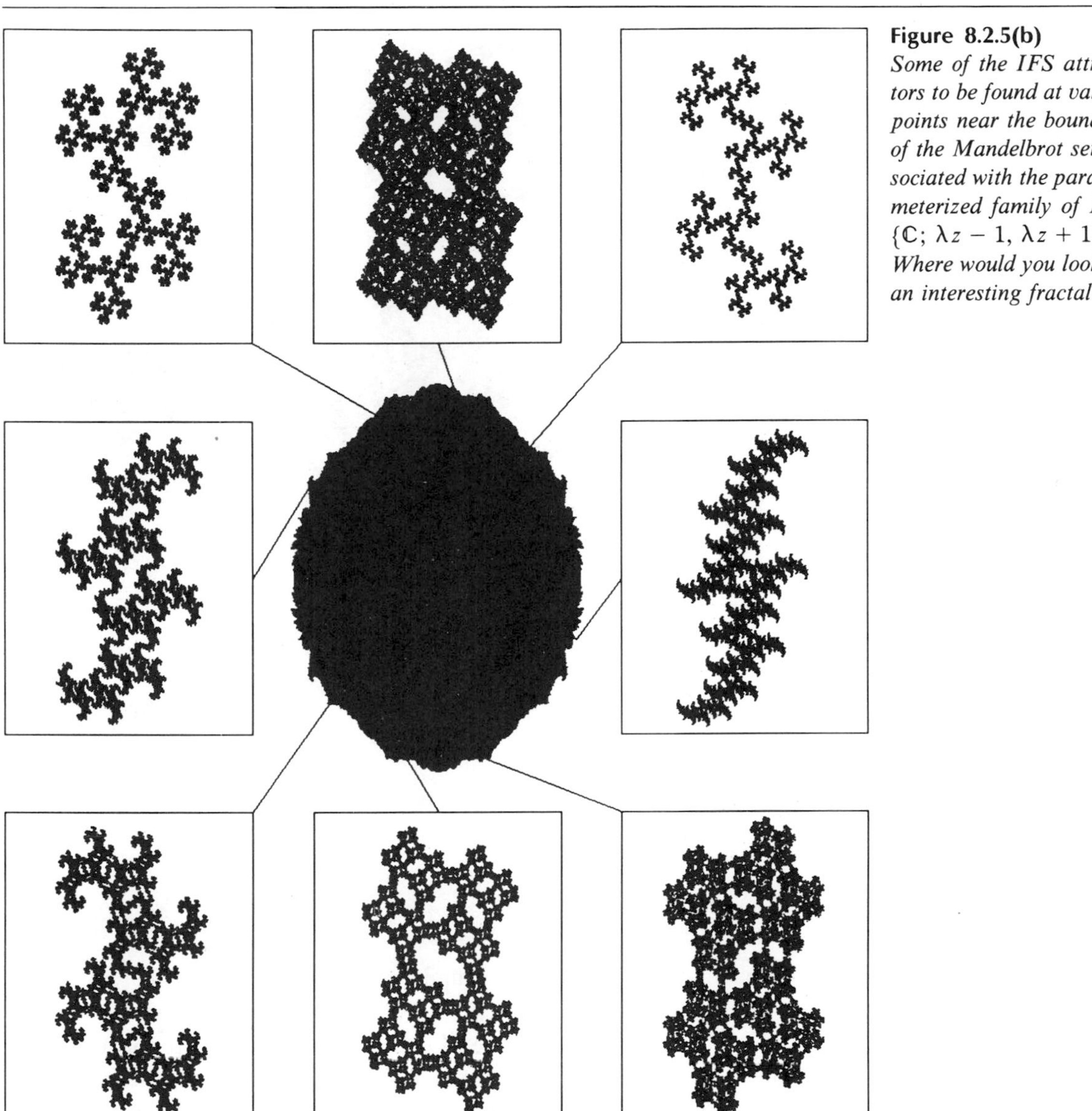

Figure 8.2.5(b) *Some of the IFS attractors to be found at various points near the boundary of the Mandelbrot set associated with the parameterized family of IFS* $\{\mathbb{C}; \lambda z - 1, \lambda z + 1\}$. *Where would you look for an interesting fractal?*

This implies that $|\lambda| \leq 1/\sqrt{2}$.

Suppose that A is connected. Then it contains a path which connects two distinct points. The fractal dimension of any path is greater than or equal to one. Hence $D(A) \geq 1$. However, by Theorem 5.2.3,

$$D(A) \leq \log(1/2)/\log(|\lambda|).$$

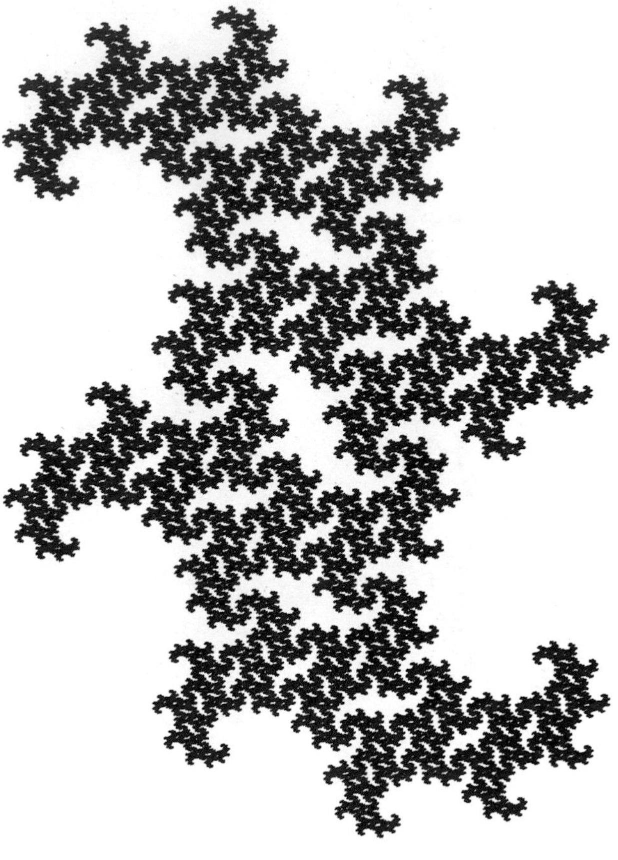

Figure 8.2.6
Attractor of the IFS $\{\mathbb{C}; \lambda z - 1, \lambda z + 1\}$ *corresponding to the value of* λ *at the tip of the spiral peninsula, shown in Figure* 8.2.4.

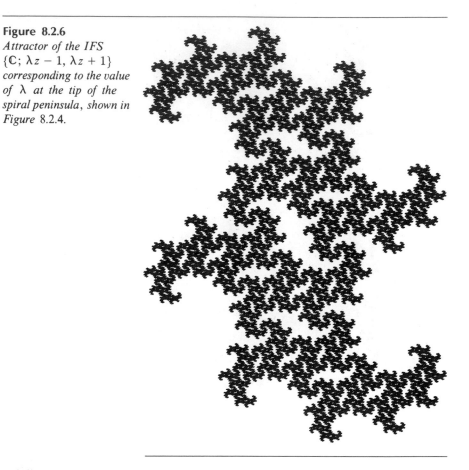

It follows that

$$1 \leq \log(1/2)/\log(|\lambda|).$$

This implies that $|\lambda| \geq 1/2$. This completes the proof of the Theorem.

A different point of view on the Mandelbrot set considered above is given by Figure 8.2.7. $\mathscr{M}$ has been turned inside-out by making the change of variable $\lambda' = \lambda^{-1}$. The inner white disk is no-man's land it does not belong to the parameter space. Also included are the two bounds provided by Theorem 8.2.2, namely the circle $|\lambda'| = 2$ and the circle $|\lambda'| = \sqrt{2}$. The fractal dimension decreases with increasing distance from the origin.

Exercises & Examples

2.1. Sketch the Mandelbrot set for the family of IFS $\{\mathbb{R}; \lambda_1 x + \lambda_2, \lambda_2 x + \lambda_1\}$ where the parameter space is $P = \{(\lambda_1, \lambda_2): |\lambda_1|, |\lambda_2| < 1\}$.

2.2. Let $\{\mathbb{X}; w_1, w_2\}$ be a family of hyperbolic IFS which depends on a

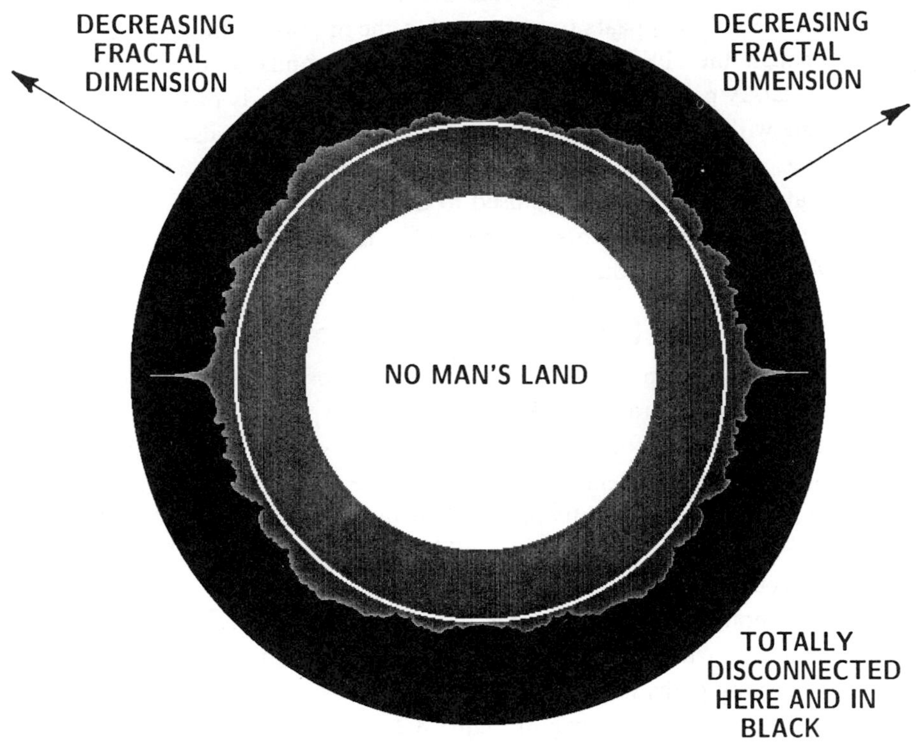

Figure 8.2.7 *Inside-out picture of the Mandelbrot set for* $\{\mathbb{C};\ \lambda z - 1, \lambda z + 1\}$. *It has been turned inside-out by making the change of variables* $\lambda' = \lambda^{-1}$. *The inner white disk is no-man's land; it does not belong to the parameter space. The figure also includes the two bounds provided by Theorem* 8.2.2, *namely the circle* $|\lambda'| = 2$ *and the circle* $|\lambda'| = \sqrt{2}$.

parameter $\lambda \in P \subset \mathbb{R}^2$. Let w_1 and w_2 depend continuously on λ for fixed $x \in \mathbb{X}$. Assume that the IFS has contractivity factor $s \in [0, 1)$ which is independent of $\lambda \in P$. Then by Theorem 3.11.1, the function $A\colon P \to \mathscr{H}(\mathbb{X})$ is continuous. Use this continuity to prove that the Mandelbrot set associated with the family of IFS is closed. It is suggested that you begin by showing that the set $S = \{B \in \mathscr{H}(\mathbb{X})\colon B \text{ is not connected}\}$ is an open subset of $\mathscr{H}(\mathbb{X})$.

2.3. Use Figure 8.2.2, to determine some values of λ which belong, approximately, to the boundary $\partial\mathscr{M}$ of the Mandelbrot set. Compute images of the corresponding attractors. Compare images corresponding to two points λ_1 and $\lambda_2 \in \partial\mathscr{M}$, with $|\lambda_1| < |\lambda_2|$. Explain why the picture of $A(\lambda_1)$ is more delicate than the picture of $A(\lambda_2)$. Also comment on similarities and differences between your images and the local geography of the parts of $\partial\mathscr{M}$ from which they come.

2.4. The pictures of the Mandelbrot set associated with the family of IFS $\{\mathbb{C};\ \lambda z - 1, \lambda z + 1\}$ suggest the conjecture that $\mathscr{M}$ is symmetric about the x-axis and about the origin. Prove the conjecture.

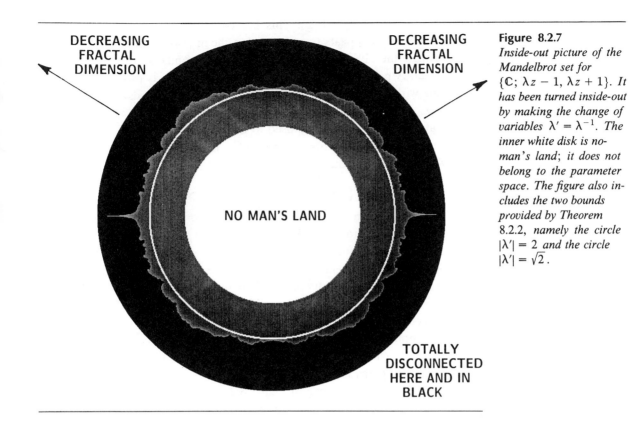

2.5. An interesting point in the parameter space for the family $\{\mathbb{C};\ \lambda z - 1,$

$\lambda z + 1\}$ is $\lambda = (1/2, 1/2)$. This lies on the circle $1/|\lambda| = |\lambda'| = \sqrt{2}$ in Figure 8.2.7. It appears to be located in the interior of the Mandelbrot set, although the IFS is just-touching. It corresponds to the Twin-Dragon Fractal. A picture of it is shown in Figure 8.2.8. It is possible to tile the plane with Twin-Dragons. Various other values of λ also correspond to tilings of the plane. See [Gilbert 1982]. Show that the attractor at the point $\lambda = (0, 1/\sqrt{2})$ can be used to tile the plane.

2.6. Notice the line segments on the real axis in Figure 8.2.7. In [Barnsley 1985c] it is proved that

$$\{\lambda \in \mathbb{C}: 0.5 \leq \lambda_1 \leq 0.53;\ \lambda_2 = 0\} \subset \mathcal{M},$$

but neighboring points in $\mathbb{C}$ are not in $\mathcal{M}$. For λ in such a line segment, what does the attractor look like? Are you surprised, in view of what you know about maps of coastlines?

2.7. Some of the most delicate attractors of the family $\{\mathbb{C};\ \lambda z - 1,\ \lambda z + 1\}$ are associated with points on $\partial\mathcal{M}$ where it touches the circle $1/|\lambda| = |\lambda'| = 2$. These have the lowest possible fractal dimension while still being connected. Let us call these attractors *tree-like* if $w_1(A) \cap w_2(A)$ is a single point. Argue (or, better yet, prove) that a tree-like attractor A contains no trapped holes; that is, A contains no nontrivial non-self-intersecting paths which start and finish at the same point. A picture of a tree-like attractor is shown in Figure 8.2.9.

2.8. Let $e = e_1 e_2 e_3 \ldots e_n \ldots$ be a point in the code space Σ of two symbols,

Figure 8.2.8
The Twin-Dragon Fractal. You can tile the plane with these sets. Although it is just-touching, it appears to lie in the interior of the Mandelbrot set.

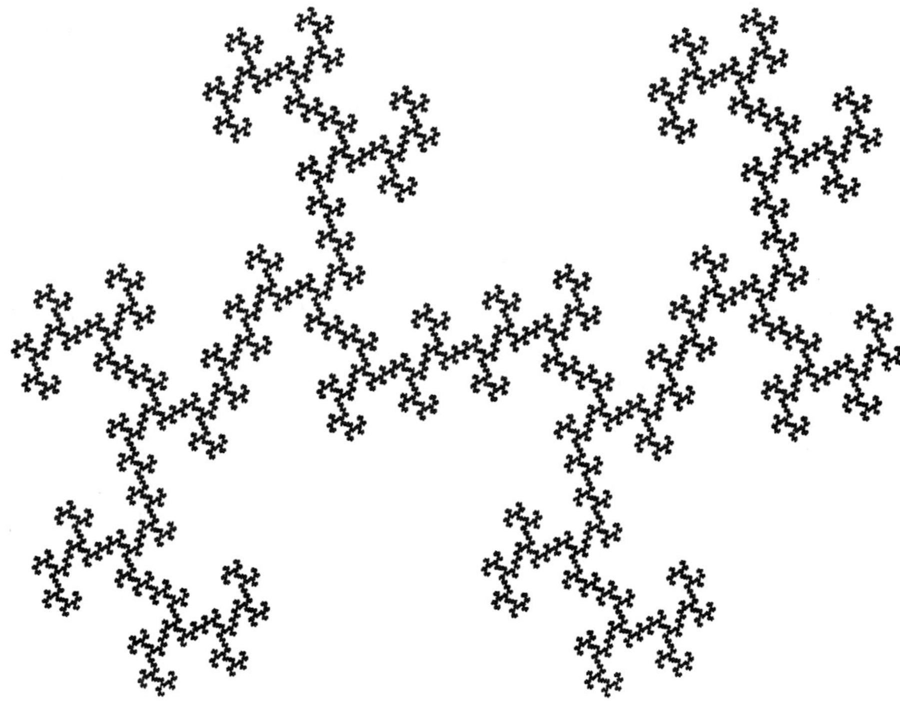

Figure 8.2.9
A "tree-like" attractor A from the family $\{\mathbb{C};\ w_1(z) = \lambda z - 1,\ w_2(z) = \lambda z + 1\}$. *The two sets* $w_1(A)$ *and* $w_2(A)$ *meet approximately at a single point.*

with $e_n \in \{+1, -1\}$ for all n. Let $\lambda \in \mathbb{C}$. Prove that the series

$$f(\lambda) = e_1 + e_2\lambda + e_3\lambda^2 + e_4\lambda^3 + e_5\lambda^4 + e_6\lambda^5 \cdots + e_n\lambda^n + \cdots$$

has radius of convergence one. What is the relationship between $f(\lambda)$ and the code space map $\phi\colon \Sigma \to A(\lambda)$ associated with the family of IFS $\{\mathbb{C};\ \lambda z - 1,\ \lambda z + 1\}$? Let $|\lambda| < 1$. Show that the attractor of the IFS is the set of all points which can be written in the form

$$\pm 1 \pm \lambda \pm \lambda^2 \pm \lambda^3 \pm \lambda^4 \pm \lambda^5 + \lambda^6 \pm \lambda^7 \pm \lambda^8 \pm \lambda^9 \pm \lambda^{10} \pm \lambda^{11} \pm \cdots$$

8.3 THE MANDELBROT SET FOR JULIA SETS

In this section we introduce a good method for making maps, such as might be found in an atlas, of families of dynamical systems. The method is based on the use of escape times and is discussed more generally in Section 8.4. Here we restrict attention to the family

$$\{\hat{\mathbb{C}};\ f_\lambda(z) = z^2 - \lambda\},$$

where the parameter space is $P = \mathbb{C}$. This family is of special importance because it provides a model for the onset of chaotic behaviour in physical and biological systems; see [May 1986], and [Feigenbaum 1979]. Moreover, it was the first family of dynamical systems for which a useful computergraphical map was constructed, by Mandelbrot. We concentrate on map making.

The Julia set $J(\lambda)$ associated with $f_\lambda(z)$ is symmetric about the origin, O. We know this because the filled Julia set, of which $J(\lambda)$ is the boundary, is the set of points whose orbits remain bounded. The orbit of $z \in \mathbb{C}$ remains bounded if and only if the orbit of $-z$ remains bounded.

For some values of $\lambda \in P$, O belongs to the filled Julia set, $F(\lambda)$, while for others it is quite far from $F(\lambda)$. This suggests that we try to color the parameter space according to the distance from O to $F(\lambda)$. How can we estimate this distance? An approximate method is to look at the "escape time" of the orbit of O. That is, we can color the parameter space according to the number of steps along the orbit of O that are required before it lands in a ball around the Point at Infinity, from where we know that all orbits diverge. The intuitive idea is that the longer an orbit of O takes to reach the ball, the closer O must be to $F(\lambda)$. Of course, if an orbit does not diverge then we know that $O \in F(\lambda)$.

Suppose that we want to make a map corresponding to a region $\mathscr{W} \subset P$. Here we choose

$$\mathscr{W} = \{\lambda = (\lambda_1, \lambda_2) \in \mathbb{C}: |\lambda_1|, |\lambda_2| \leq 2\}.$$

Let $R > 0$ and define

$$\mathscr{V}(R) = \{z \in \mathbb{C}: |z| > R\} \cup \{\infty\}.$$

Suppose

$$R > 0.5 + 0.25 + |\lambda|.$$

Then it is readily proved that the orbit $\{f_\lambda^{\circ n}(z)\}$ diverges if and only if it intersects $\mathscr{V}(R)$. So if we choose $R = 10$ we are sure that, for all $\lambda \in \mathscr{W}$, the orbit $\{f_\lambda^{\circ n}(O)\}$ diverges if and only if it intersects $\mathscr{V}(R)$. Let us see what happens if we color the pixels of $\mathscr{W}$ according to the number of iterations required to enter $\mathscr{V}(10)$.

The following program is written in BASIC. It runs without modification on an IBM PC with Enhanced Graphics Adaptor and Turbobasic. On any line the words preceded by ' are comments: they are not part of the program.

Program 8.3.1 (Example of algorithm for coloring parameter space according to an escape time.)

```
numits = 20: a = -2: b = - 2: c = 2: d = 2: M = 100   ' Define
                                     viewing window, W, and numits.
```

```
R = 10   ' Define the region 𝒱.
screen 9: cls   ' Initialize graphics.
for p = 1 to M
for q = 1 to M
k = a + (c - a)*p/M: l = b + (d - b)*q/M   ' Specify the value of
                                             lambda (k, l) ∈ P
x = 0: y = 0   ' Specify the initial point, O, on the orbit
for n = 1 to numits   ' Compute at most numits points on the orbit of O
newx = x*x - y*y - k
newy = 2*x*y - l
x = newx: y = newy
if x*x + y*y > R then   ' If the most recently computed point lies in 𝒱
                          then...
pset(p, q), n: n = numits   '... render the pixel (p, q) in color n, and
                              go to the next (p, q).
end if
if instat then end   ' Stop computing if any key is pressed!
next n: next q: next p
end
```

Color Plate 8.3.1 shows the result of running a version of Program 8.3.1 on a Masscomp 5600 workstation with Aurora graphics.

In Figure 8.3.1 we show the result of running a version of Program 8.3.1, but this time in halftones. The central white object corresponds to values of λ for which the computed orbit of O does not reach $\mathscr{V}$ during the first *numits* iterations. It represents the Mandelbrot set (defined below) for the dynamical system $\{\hat{\mathbb{C}};\ z^2 - \lambda\}$. The bands of colors (or white and shades of grey) surrounding the Mandelbrot set correspond to different numbers of iterations required before the orbit of O reaches $\mathscr{V}(10)$. The bands which are furthest away from the center represent orbits which reach O most rapidly. Approximately, the distance from O to $F(\lambda)$ increases with the distance from λ to the Mandelbrot set.

Definition 1. The *Mandelbrot set* for the family of dynamical systems $\{\hat{\mathbb{C}};\ z^2 - \lambda\}$ is

$$\mathscr{M} = \{\lambda \in P\colon J(\lambda) \text{ is connected}\}.$$

The relationship between escape times of orbits of O and the connectivity of $J(\lambda)$ is provided by the following theorem.

Theorem 1. *The Julia set for a member of the family of dynamical systems*

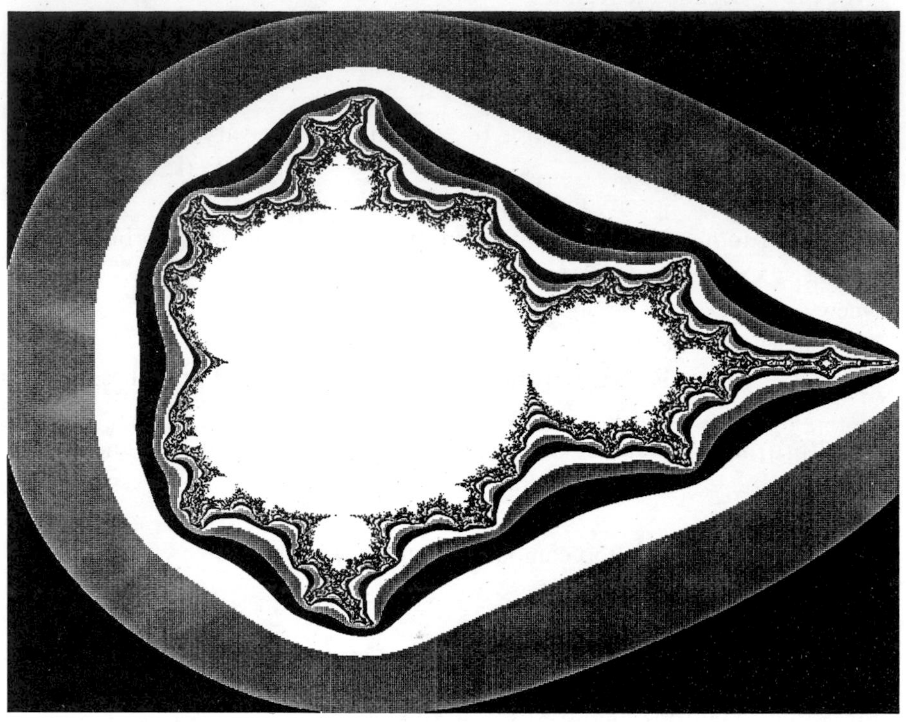

Figure 8.3.1 *The Mandelbrot set for $z^2 - \lambda$, computed by escape times.*

$\{\hat{\mathbb{C}};\ f_\lambda(z) = z^2 - \lambda\}$, $\lambda \in P = \mathbb{C}$ *is connected if and only if the orbit of the origin escapes to infinity; that is*

$$\mathcal{M} = \left\{ \lambda \in \mathbb{C} : |f_\lambda^{\circ n}(0)| \to \infty \text{ as } n \to \infty \right\}.$$

Proof. This theorem follows from [Brolin 1966], Theorem 11.2, which says that the Julia set of a polynomial, of degree greater than one, is connected if and only if all of the critical points lie in the basin of attraction of the Point at Infinity. $f_\lambda(z)$ possesses two critical points, O and ∞. Hence $J(\lambda)$ is connected if and only if $|f^{\circ n}_\lambda(0)| \to \infty$ as $n \to \infty$.

In this paragraph we discuss the relationship between the Mandelbrot set for the family of dynamical systems $\{\hat{\mathbb{C}};\ z^2 - \lambda\}$, and the corresponding family of IFS $\{\hat{\mathbb{C}};\ \sqrt{z + \lambda},\ -\sqrt{z + \lambda}\}$. We know that for various values of λ in $\mathbb{C}$ the IFS can be modified so that it is hyperbolic, with attractor $J(\lambda)$. For the purposes of this paragraph *let us pretend that the IFS is hyperbolic, with attractor* $J(\lambda)$, *for all* $\lambda \in \mathbb{C}$. Then Definition 8.2.1 would be equivalent to Definition 8.3.1. By Theorem 8.2.1, the attractor of the IFS would be connected if and only if $w_1(J(\lambda)) \cap w_2(J(\lambda)) \neq \varnothing$. But $w_1(\mathbb{C}) \cap w_2(\mathbb{C}) = \{0\}$. Then it would follow that the attractor of the IFS is connected if and only if $O \in J(\lambda)$. In other words: we discover the same criteria for connectivity of $J(\lambda)$ if we argue informally using the IFS point of view, as can be proved using Julia set theory. This completes the discussion.

We return to the theme of coastlines, and the possible resemblance between fractal sets corresponding to points on boundaries in parameter space and the local geometry of the boundaries. Figures 8.3.2 (a) and (b) show the Mandelbrot set for $z^2 - \lambda$, together with pictures of filled Julia sets corresponding to various points around the boundary. If one makes a very high resolution image of the boundary of the Mandelbrot set, at a value of λ corresponding to one of these Julia sets, one "usually" finds structures which resemble the Julia set. It is as though the boundary of the Mandelbrot set is made by stitching together microscopic copies of the Julia sets which it represents. An example of such a magnification of a piece of the boundary of $\mathcal{M}$, and a picture of a corresponding Julia set, are shown in Figures 8.3.3 and 8.3.4.

If you look closely at the pictures of the Mandelbrot set $\mathcal{M}$ considered in this section, you will see that there appear to be some parts of the set which are not connected to the main body. Pictures can be misleading.

Theorem 2. [Mandelbrot-Douady-Hubbard] *The Mandelbrot set for the family of dynamical systems* $\{\hat{\mathbb{C}};\ z^2 - \lambda\}$ *is connected.*

Proof. This can be found in [Douady 1982].

The Mandelbrot set for $z^2 - \lambda$ is related to the exciting subject of cascades of bifurcations, quantitative universality, chaos, and the work of Feigenbaum. To learn more you could consult [Feigenbaum 1979], [Douady 1982], [Barnsley 1984], [Devaney 1986], [Peitgen 1986], [Scia 1987].

Exercises & Examples

3.1. Rewrite Program 8.3.1 in a form suitable for your own computergraphical environment. Run your program and obtain hardcopy of the output. Adjust the window parameters a, b, c, and d, to allow you to make zooms on the boundary of the Mandelbrot set.

3.2. Figure 8.3.5 shows a picture of the Mandelbrot set for the family of dynamical systems $\{\hat{\mathbb{C}};\ z^2 - \lambda\}$ corresponding the coordinates $-0.5 \leq \lambda_1 \leq 1.5$, $-1.0 \leq \lambda_2 \leq 1.0$. It has been overlayed on a coordinate grid. The middle of the first bubble has not been plotted, to clarify the coordinate grid. Let $B_0, B_1, B_2, B_3, \ldots$ denote the sequence of bubbles

Figure 8.3.2(a) *Mandelbrot set for $z^2 - \lambda$, decorated with various Julia sets and filled Julia sets.*

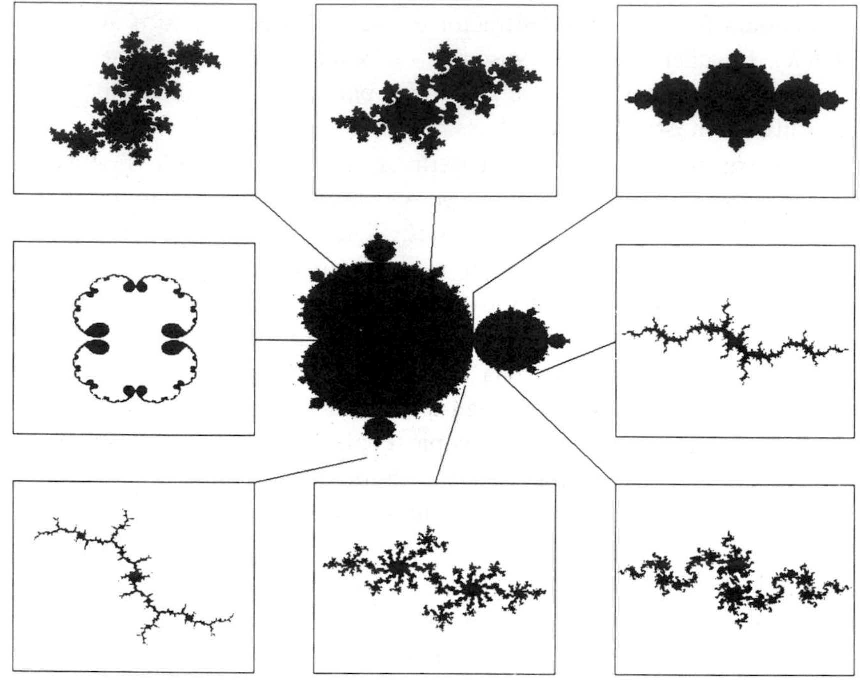

Figure 8.3.2(b) *Mandelbrot set for $z^2 - \lambda$, decorated with various Julia sets and filled Julia sets. These often resemble the place on the boundary from which they come, especially if one magnifies up enough.*

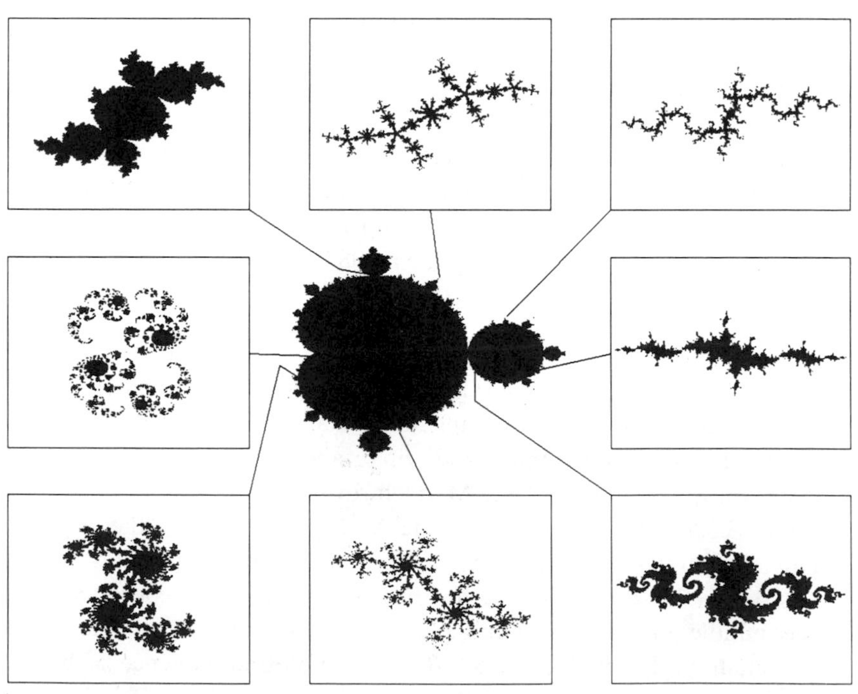

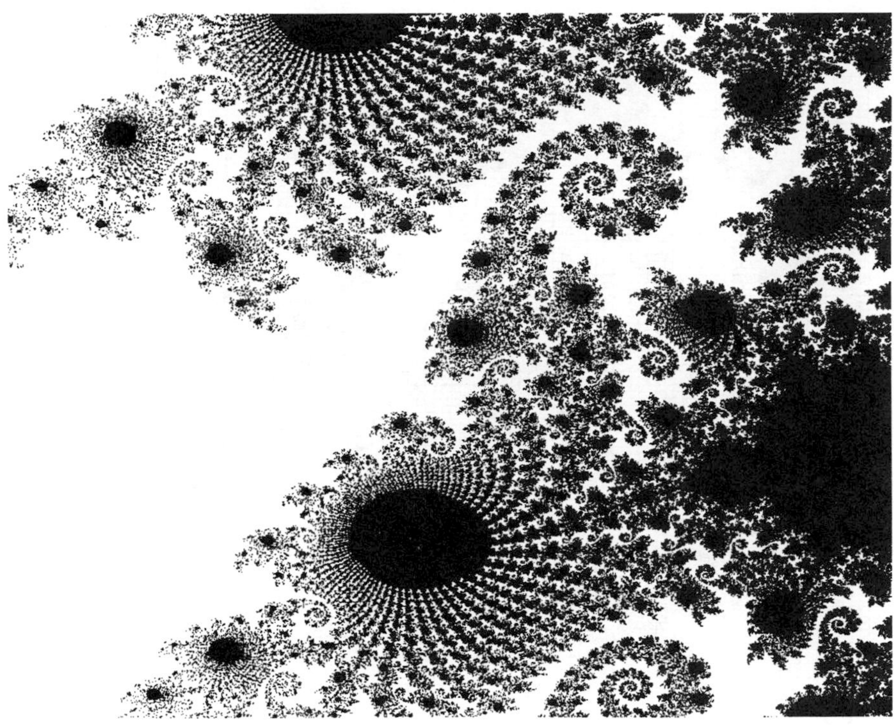

Figure 8.3.3
A zoom on a piece of the boundary of the Mandelbrot set for $z^2 - \lambda$.

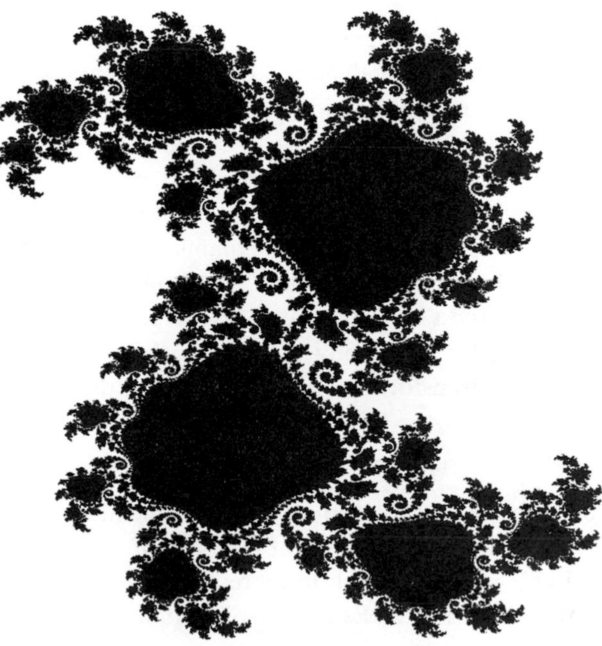

Figure 8.3.4
A filled Julia set corresponding to the piece of the coastline of the Mandelbrot set in Figure 8.3.3. *Notice the family resemblances.*

Figure 8.3.5
A picture of the Mandelbrot set for the family of dynamical systems $\{\hat{\mathbb{C}};\ z^2 - \lambda\}$. It has been overlayed on a coordinate grid. The middle of the first bubble has not been plotted, to clarify the coordinate grid.

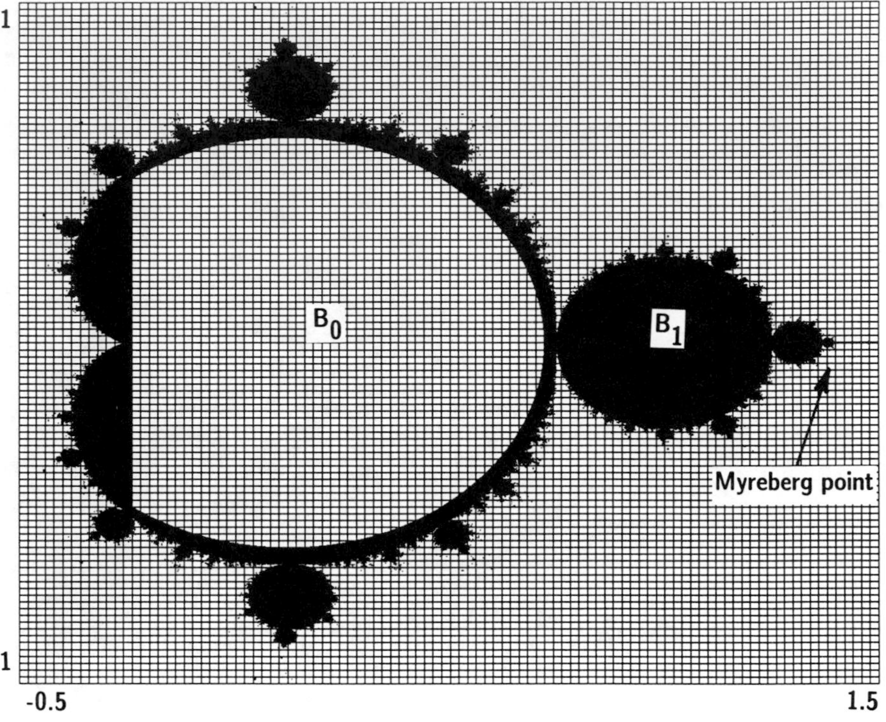

on the real axis, reading from left to right. Verify computationally that when λ lies in the interior of B_n the dynamical system possesses an attractive cycle, located in $\mathbb{C}$, of minimal period 2^n, for $n = 0, 1, 2,$ and 3.

3.3. The sequence of bubbles $\{B_n\}_{n=0}^{\infty}$ in exercise 3.2 converges to the *Myreberg point*, $\lambda = 1.40115\ldots$ The ratios of the widths of successive bubbles converges to the *Feigenbaum ratio* 4.66920. . . . Make a conjecture about what sort of "attractive cycle" the dynamical system $\{\hat{\mathbb{C}};\ z^2 - \lambda\}$ might possess at the Myreberg point. Test your conjecture numerically. You will find it easiest to restrict attention to real orbits.

3.4. Make a parameter space map for the family of dynamical systems $\{\hat{\mathbb{C}};\ f_\lambda(z)\}$ where f_λ is the Newton transformation associated with the family of polynomials

$$F(z) = z^3 + (\lambda - 1)z - \lambda, \qquad \lambda \in P = \mathbb{C}.$$

Notice that the polynomial has a root located at $z = 1$, independent of λ. Color your map according to the "escape time" of the orbit of O to a ball of small radius centered at $z = 1$. Use black to represent values of λ for which O does not converge to $z = 1$. Examine some Julia sets of f_λ

corresponding to points on the boundary of the black region. Are there resemblances between structures which occur in your map of parameter space, and some of the corresponding collection of Julia sets? (The correct answer to this question can be found in [Curry 1983].)

8.4 HOW TO MAKE MAPS OF FAMILIES OF FRACTALS USING ESCAPE TIMES

We begin by looking at the Mandelbrot set for a certain family of IFS. It is disappointing and we do not learn much. We then introduce a related family of dynamical systems and color the parameter space using escape times. The result is a map which is packed with information. We generalize the procedure to provide a method for making maps of other families of dynamical systems. We discover how certain boundaries in the resulting maps can yield information about the appearance of the fractals in the family. That is, we begin to learn to read the maps.

Figures 8.4.1 (a) and (b) show the Mandelbrot set $\mathcal{M}_1$ for the family of hyperbolic IFS

$$\{\mathbb{C};\, w_1(z) = \lambda z + 1,\, w_2(z) = \lambda^* z - 1\}, \qquad P = \{\lambda \in \mathbb{C}: |\lambda| < 1\}.$$

We use the notation $\lambda^* = (\lambda_1 + i\lambda_2)^* = (\lambda_1 - i\lambda_2)$, for the complex conjugate of λ. The two transformations are similitudes of scaling factor $|\lambda|$. At fixed λ, they rotate in opposite directions through the same angle. The figures also show attractors of the IFS corresponding to various points around the boundary of the Mandelbrot set. What a disappointing map this is! There are no secret bays, jutting peninsulas, nor ragged rocks in the coastline.

Theorem 1. [Hardin 1985] *The Mandelbrot* $\mathcal{M}_1$ *is connected. Its boundary is the union of a countable set of smooth curves, and is piecewise differentiable.*

Proof. This can be found in [Barnsley 1988d].

Let us try to obtain a better map of this family of attractors. In order to do so we begin by defining an extension of the associated shift dynamical system, for each $\lambda \in P\backslash\mathcal{M}_1$. Let $A(\lambda)$ denote the attractor of the IFS. One can prove that $A(\lambda)$ is symmetric about the y-axis. Hence $\lambda \in \mathcal{M}_1$ if and only if $A(\lambda)$ intersects the y-axis. Define $f_\lambda: \mathbb{C} \to \mathbb{C}$ by

$$f_\lambda(z) = \begin{cases} w_1^{-1}(z) & \text{if } \operatorname{Re} z \geq 0; \\ w_2^{-1}(z) & \text{if } \operatorname{Re} z < 0. \end{cases}$$

Then, when λ is such that $A(\lambda)$ is disconnected, $\{A(\lambda);\, f_\lambda\}$ is the shift dynamical system associated with the IFS; $\{\mathbb{C};\, f_\lambda\}$ is an extension of the shift

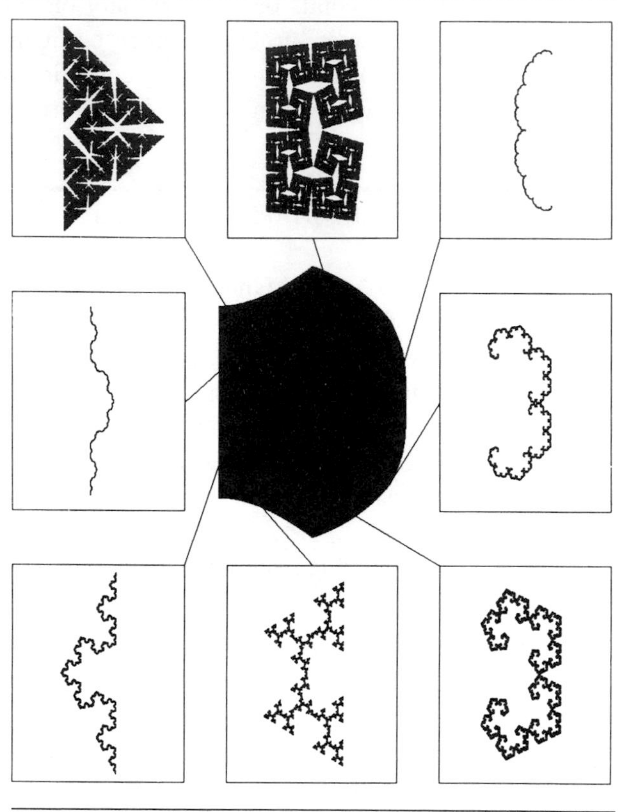

Figure 8.4.1(a) *The complement of the Mandelbrot set $\mathcal{M}_1$ associated with the family of IFS* $\{\mathbb{C};\ w_1(z) = \lambda z + 1,\ w_2(z) = \lambda^* z - 1\}$. *Points in the complement of the Mandelbrot set are colored black. The boundary of $\mathcal{M}_1$ is smooth and does not reveal much information about the family of fractals which it represents. The figure also shows attractors of the IFS corresponding to various points on the boundary of $\mathcal{M}_1$. What a disappointing map this is!*

dynamical system to all of $\mathbb{C}$; and $A(\lambda)$ is the "repelling set" of $\{\mathbb{C};\ f_\lambda\}$. This system can be used to compute images of $A(\lambda)$ in the just-touching and totally disconnected cases, using the Escape Time Algorithm, as discussed in Chapter 7, Section 1.

We make a map of the family of dynamical systems $\{\mathbb{R}^2;\ f_\lambda\}$, $\lambda \in P$. To do this we use the following algorithm, which was illustrated in Program 8.3.1. The algorithm applies to any family of dynamical systems $\{\mathbb{R}^2;\ f_\lambda\}$ which possesses a "repelling set" $A(\lambda)$, such that P is a two-dimensional parameter space with a nice classical shape, such as a square or a disk.

Algorithm 1 (Method for Coloring Parameter Space According to an Escape Time.)

(i) Choose a positive integer, *numits*, corresponding to the amount of computation one is able to do. Fix a point $Q \in \mathbb{R}^2$ such that $Q \in A(\lambda)$ for some, but not all, $\lambda \in P$.

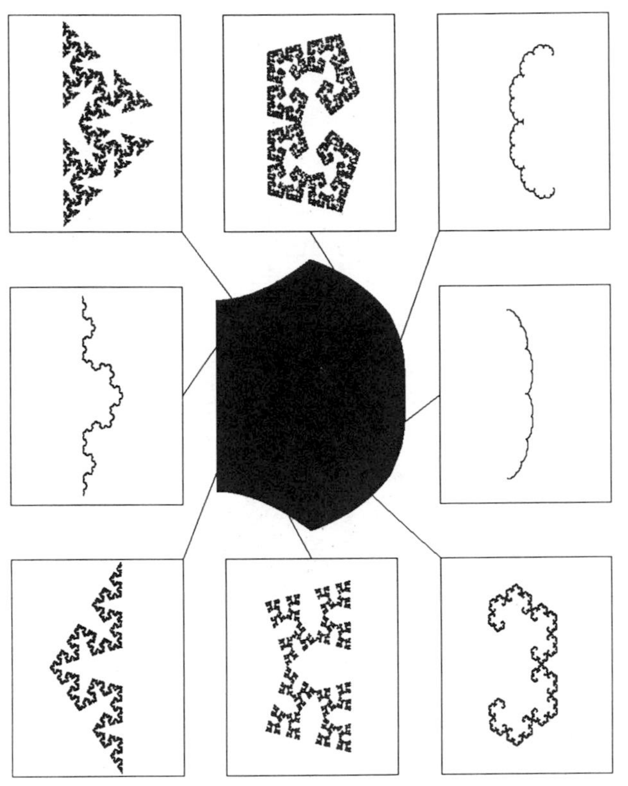

Figure 8.4.1(b) *The complement of the Mandelbrot set $\mathscr{M}_1$ associated with the family of IFS $\{\mathbb{C}; w_1(z) = \lambda z + 1, w_2(z) = \lambda^* z - 1\}$, together with some of the corresponding fractals. Notice how these have subsets of points which lie on straight lines, like the local structure of $\partial \mathscr{M}_1'$.*

(ii) Fix a ball $B \subset \mathbb{R}^2$ such that $A(\lambda) \subset B$ for all $\lambda \in P$. Define an escape region to be $\mathscr{V} = \mathbb{R}^2 \setminus B$.

(iii) Represent the parameter space P by an array of pixels. Carry out the following step for each λ in the array.

(iv) Compute $\{f_\lambda^{\circ n}(Q)\colon n = 0, 1, 2, 3, \ldots, numits\}$. Color the pixel λ according to the least value of n such that $f_\lambda^{\circ n}(Q) \in \mathscr{V}$. If the computed piece of the orbit does not intersect $\mathscr{V}$, color the pixel black.

The result of applying this algorithm to the dynamical system defined above, with $Q = O$, is illustrated in Figures 8.4.2, 8.4.3 (a)–(g), and Color Plates 8.4.1 and 8.4.2.

Figure 8.4.2 contains four different regions. The first is a neighborhood of O, surrounded by almost concentric bands of black, grey, and white. The location of this region is roughly the same as that of $P \setminus \mathscr{M}_1$, which corresponds to totally disconnected and just-touching attractors. The second region is the grainy area, which we refer to as the foggy coastline. Here, upon magnification, one finds complex geometrical structures. An example is illustrated in the sequence of zooms in Figures 8.4.3 (a)–(g). The structures

Figure 8.4.2
A map of the family of dynamical systems $\{\mathbb{C}; f_\lambda\}$, *where*

$$f_\lambda(z) = \begin{cases} (z-1)/\lambda & \text{if } \operatorname{Re} z \geq 0; \\ (z+1)/\lambda^* & \text{if } \operatorname{Re} z < 0. \end{cases}$$

The parameter space is $P = \{\lambda \in \mathbb{C}; 0 < \lambda_1 < 1, 0 < \lambda_2 \leq 0.75\}$. *The map is obtained by applying Algorithm* 8.4.1. *Pixels are shaded according to the "escape time" of a point* $O \in \mathbb{R}^2$. *The exciting places where the interesting fractals are to be found are not within the solid bands of black, grey, or white, but within the foggy coastline. This coastline is itself a fractal object, revealing infinite complexity under magnification. In it one finds approximate pictures of some of the connected and "almost connected" repelling sets of the dynamical system. Why are they there?*

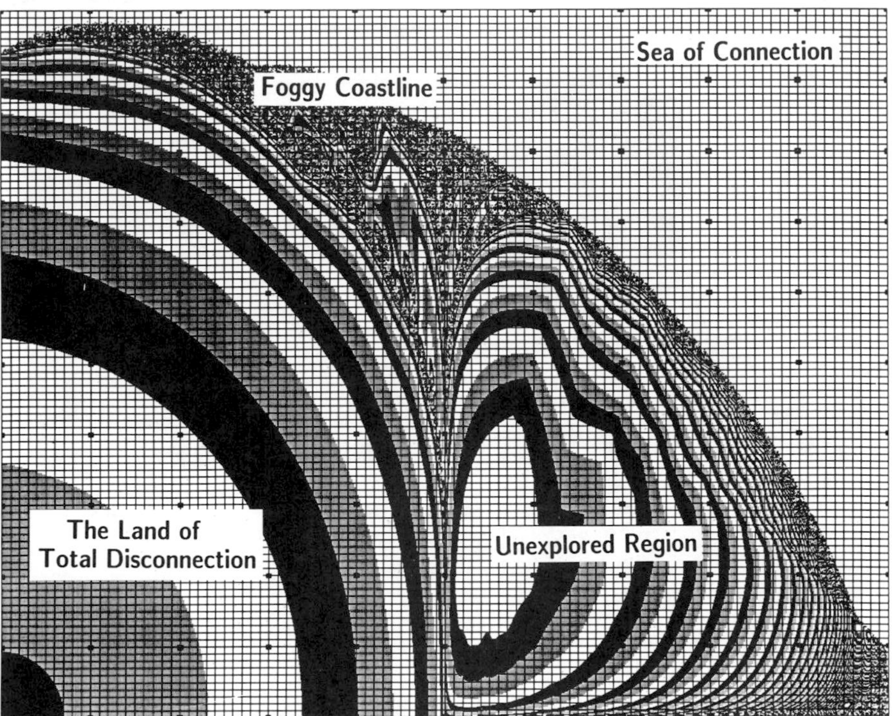

appear to be subtly different from one another. Early experiments show that if λ is chosen in the vicinity of one of these structures, then images of the "repelling set" of the dynamical system $\{\mathbb{R}^2; f_\lambda\}$, computed using the Escape Time Algorithm, contains similar structures. An example of such an image is shown in Figure 8.4.4. The third region, at the lower right in Figure 8.4.2, is made up of closed contours of black, grey, and white. Here the map conveys little information about the family of dynamical systems. To obtain information in this region one should examine the orbits of a point Q, different from O. The fourth region, the outer white area in Figure 8.4.2, corresponds to dynamical systems for which the orbit of O does not escape. It is likely that for λ in this region, the "repelling set" of the dynamical system possesses an interior.

Figures 8.4.3(a) – (g) *A sequence of zooms on a piece of the foggy coastline in Figure* 8.4.2. *The window coordinates of the highest power zoom are* $0.4123 \leq \lambda_1 \leq 0.4139$, $0.6208 \leq \lambda_2 \leq 0.6223$. *Can you find where each picture lies within the one that precedes it?*

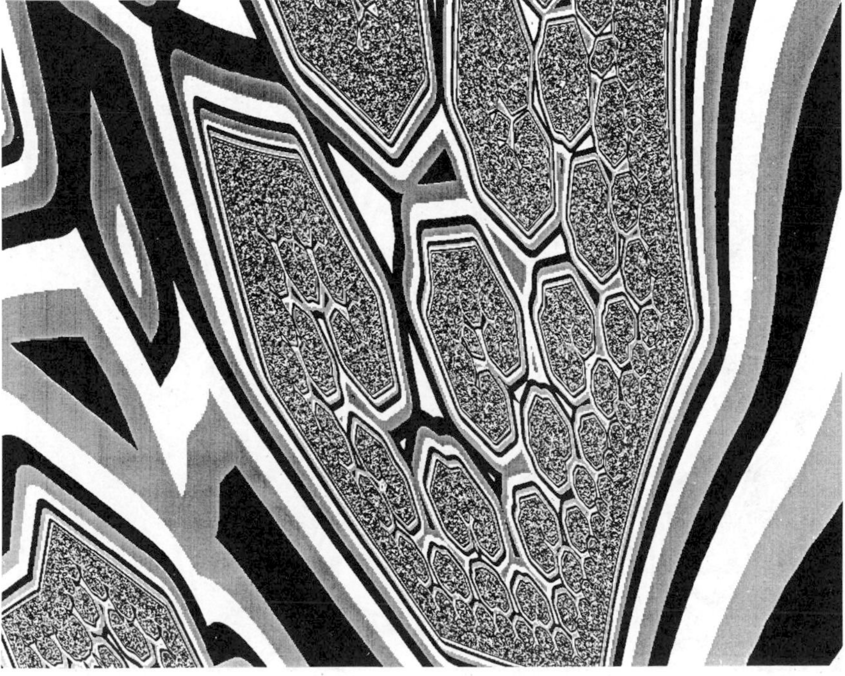

Figure 8.4.3(b)

Figure 8.4.3(c)

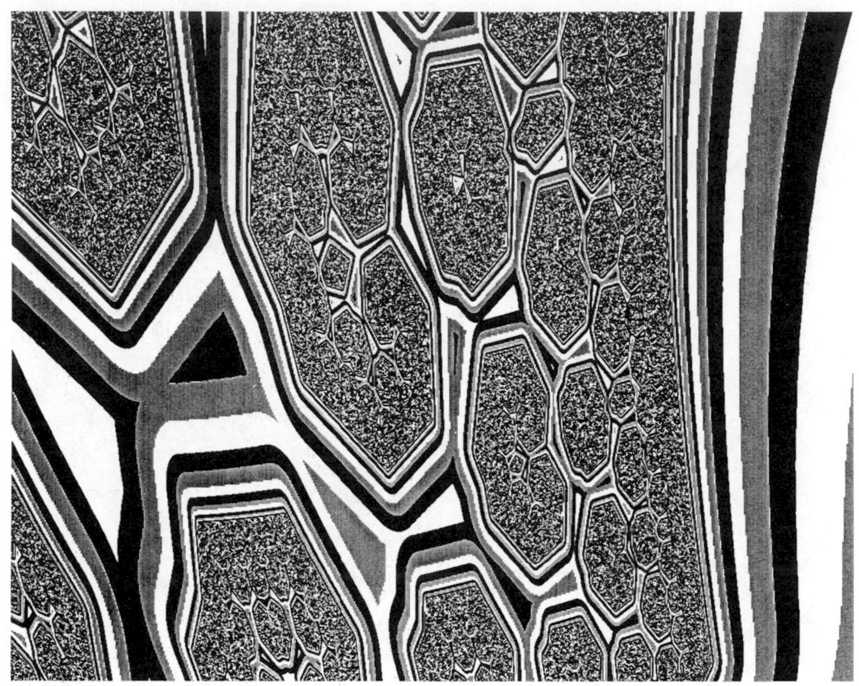

Figure 8.4.3(d)

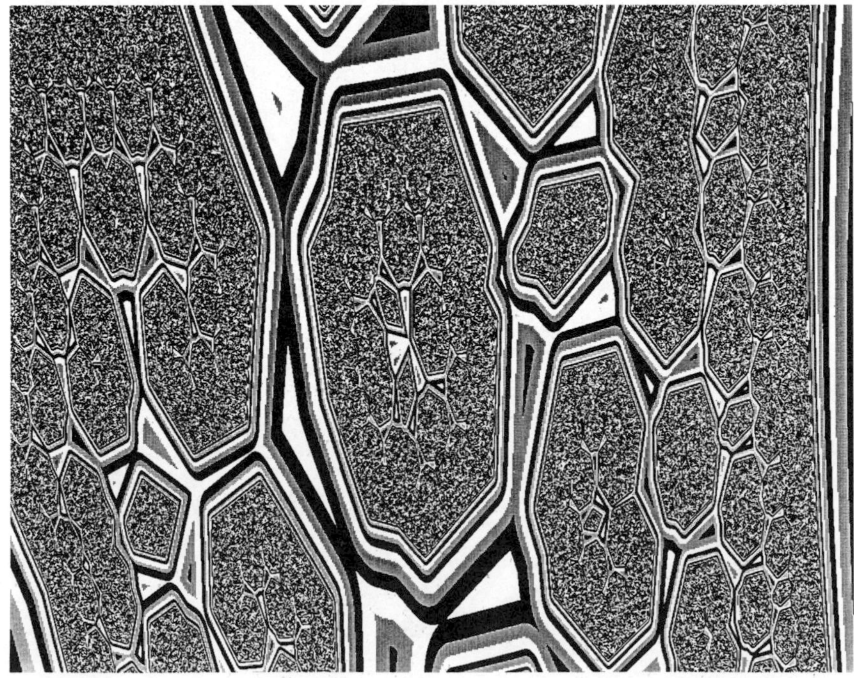

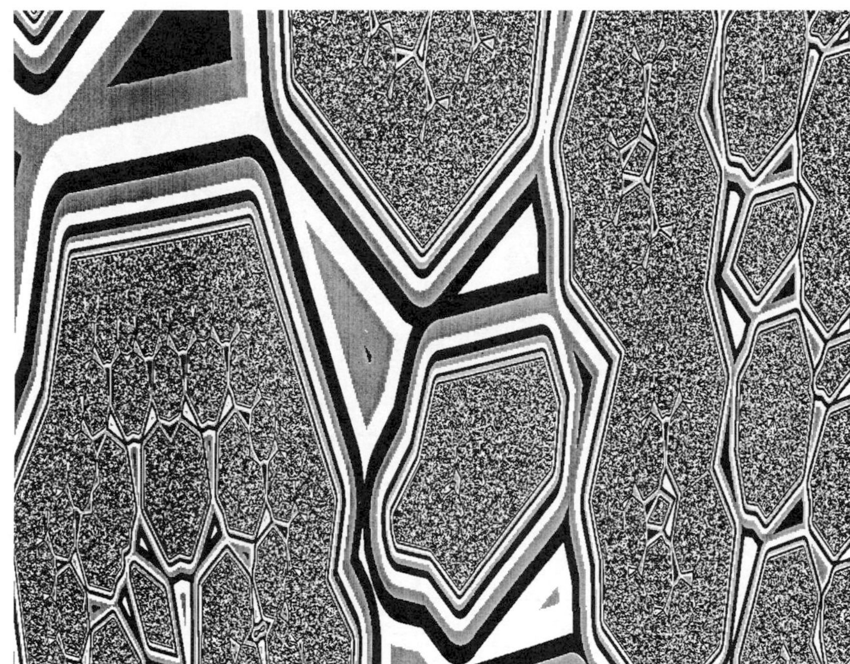

Figure 8.4.3(e)

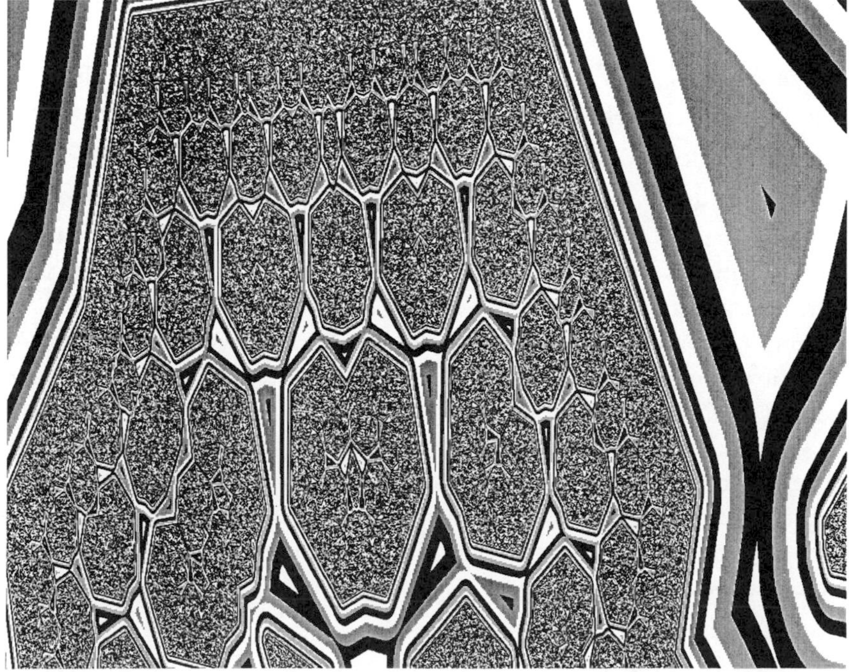

Figure 8.4.3(f)

Figure 8.4.3(g)

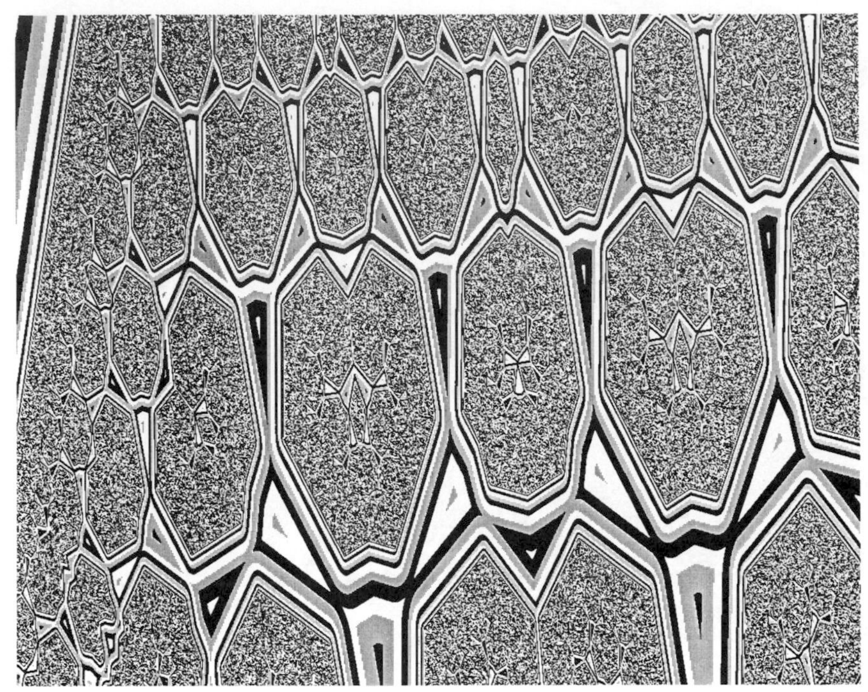

Our new maps, such as Figure 8.4.2, can provide information about the family of IFS

$$\{\mathbb{C};\ w_1(z) = \lambda z + 1,\ w_2(z) = \lambda^* z - 1\}, \qquad P = \{\lambda \in \mathbb{C}: |\lambda| < 1\}.$$

in the vicinity of the boundary of the Mandelbrot set. For $\lambda \in \partial\mathscr{M}_1$ the attractor of the IFS is the same as the repelling set of the dynamical system. For λ close to $\partial\mathscr{M}_1$ the attractor of the IFS "looks like" the repelling set of the dynamical system.

Figure 8.4.5 shows a transverse section through the anther of a lily. We include it because some of the structures in Figures 8.4.3 (a)–(g) are reminiscent of cells.

Algorithm 8.4.1 can be applied to families of dynamical systems of the type described in Theorem 7.4.1. For example, let $\{\mathbb{R}^2;\ f_\lambda\}$, where $\lambda \in P = \blacksquare \subset \mathbb{R}^2$, denote a family of dynamical systems. Let $\mathbb{X} \subset \mathbb{R}^2$ be compact. Let f_λ: $\mathbb{X} \to \mathbb{R}^2$ be continuous and such that $f(\mathbb{X}) \supset \mathbb{X}$. Then f_λ possesses an invariant set $A(\lambda) \in \mathscr{H}(\mathbb{X})$, given by

$$A(\lambda) = \bigcap_{n=0}^{\infty} f_\lambda^{0(-n)}(\mathbb{X}).$$

$A(\lambda)$ is the set of points whose orbits do not escape from $\mathbb{X}$. The set of points

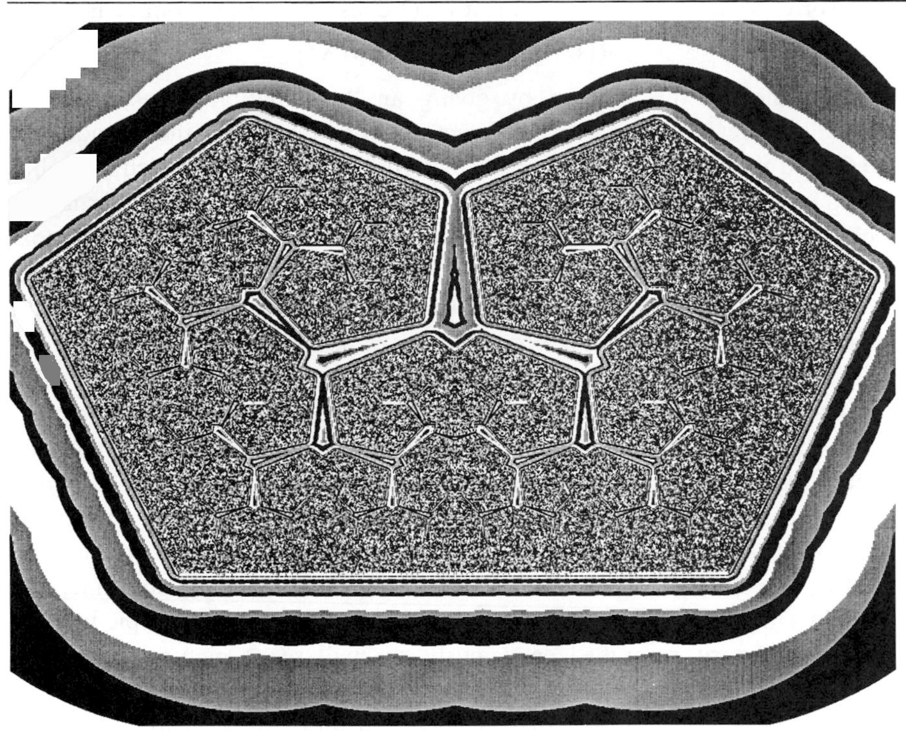

Figure 8.4.4
Image of the repelling set for one of the family of the dynamical systems whose parameter space was mapped in Figure 8.4.2. This image corresponds to a value of λ *which lies within the highest power zoom in Figure* 8.4.3. *Notice how the objects here resemble those in the corresponding position in the parameter space.*

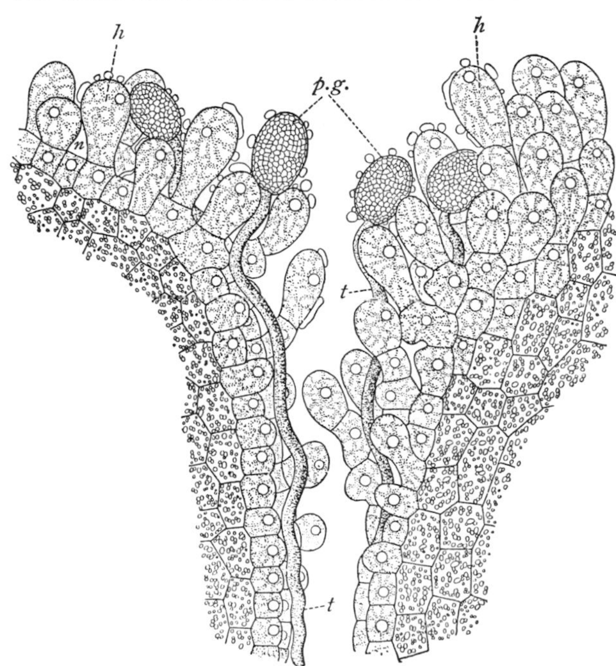

Figure 8.4.5
Longitudinal section through part of the stigma of a lily, showing germinating pollen-grains. h, papillae of stigma; p.g., pollen grains; t, pollen tubes. Highly magnified. (*After DodelPort*, [*Scott* 1917].)

in P corresponding to which the orbit of Q does not escape from $\mathbb{X}$ is

$$\mathcal{M}(Q) = \{\lambda \in P\colon Q \in A(\lambda)\}.$$

We conclude this chapter by giving an "explanation" of how family resemblances can happen between structures which occur on the boundary of $\mathcal{M}(Q)$ and the sets $A(\lambda)$. (1) Suppose that $A(\lambda)$ is a set in $\mathbb{R}^2$ which looks like a map of Great Britain, translated by λ. Then what does $\mathcal{M}(Q)$ look like? It looks like a map of Great Britain. (2) Suppose that $A(\lambda)$ is a set which looks like a map of Great Britain at time λ_1, translated by λ. We picture the set $A(\lambda)$ varying slowly, perhaps its boundary changing continuously in the Hausdorff metric as λ varies. Now $A(\lambda)$ looks like a deformed map of Great Britain. The local coves and inlets will be accurate representations of those coves at about the time λ_1 to which they correspond in the parameter space map. That is, the boundary of $\mathcal{M}(Q)$ will consist of neighboring bays and inlets at different times stitched together. It will be a map which is microscopically accurate (at some time) and globally inaccurate. (3) Now pretend in addition that the coastline of Great Britain is self-similar at each time λ_1. That is, imagine that little bays look like whole chunks of the coastline, at a given instant. Now what will $\mathcal{M}(Q)$ look like? At a given microscopic location on the boundary, magnified enormously, we will see a picture of a whole chunk of the coastline of Great Britain, at that instant. (4) Now imagine that for some values of λ, Great Britain, in the distant future, is totally disconnected, reduced to grains of isolated sand. It is unlikely that those values of λ belong to $\mathcal{M}(Q)$. As λ varies in a region of parameter space for which $A(\lambda)$ is totally disconnected, it is not probable that $Q \in A(\lambda)$. In these regions we would expect $\mathcal{M}(Q)$ to be totally disconnected.

The families of sets $\{A(\lambda) \in \mathbb{X}\colon \lambda \in P\}$ considered in this chapter broadly fit into the description in the preceding paragraph. Both P and $\mathbb{X}$ are two-dimensional. The sets $A(\lambda)$ are derived from transformations which behave locally like similitudes. For each $\lambda \in P$, $A(\lambda)$ is either connected or totally disconnected. Finally, the sets $A(\lambda)$ and their boundaries appear to depend continuously on λ.

Exercises & Examples

4.1. In the above section we applied Algorithm 8.4.1, with $Q = (0, 0)$, to compute a map of the family of dynamical systems

$$f_\lambda(z) = \begin{cases} (z-1)/\lambda & \text{if } \operatorname{Re} z \geq 0; \\ (z+1)/\lambda^* & \text{if } \operatorname{Re} z < 0. \end{cases}$$

The resulting map was shown in Figure 8.4.2. This map contains an unexplored region. Repeat the computation, but with (a) $Q = 0.5$, and (b) $Q = -0.5$, to obtain information about the Unexplored Region.

4.2. In this example we consider the family of dynamical systems $\{\mathbb{C}; f_\lambda\}$ where

$$f_\lambda(z) = \begin{cases} (z-1)/\lambda & \text{if } \lambda_2 x - \lambda_1 y \geq 0, \\ (z+1)/\lambda & \text{if } \lambda_2 x - \lambda_2 y < 0. \end{cases}$$

The parameter space is $\lambda \in P = \{\lambda \in \mathbb{C}: 0 < |\lambda| < 1\}$. This family is related to the family of IFS

$$\{\mathbb{C}; w_1(z) = \lambda z + 1, w_2(z) = \lambda z - 1\}.$$

Let $A(\lambda)$ denote the attractor of the IFS and let $\tilde{A}(\lambda)$ denote the "repelling set" associated with the dynamical system. Let

$$S = \{\lambda \in P: \text{the line } \lambda_2 x - \lambda_1 y = 0 \text{ separates the two sets } w_1(A(\lambda)) \text{ and } w_2(A(\lambda))\}.$$

If $\lambda \in S$ then $\{A(\lambda); f_\lambda\}$ is the shift dynamical system associated with the IFS, and $A(\lambda) = \tilde{A}(\lambda)$. Even when $\lambda \notin S$ we expect there to be similarities between $A(\lambda)$ and $\tilde{A}(\lambda)$.

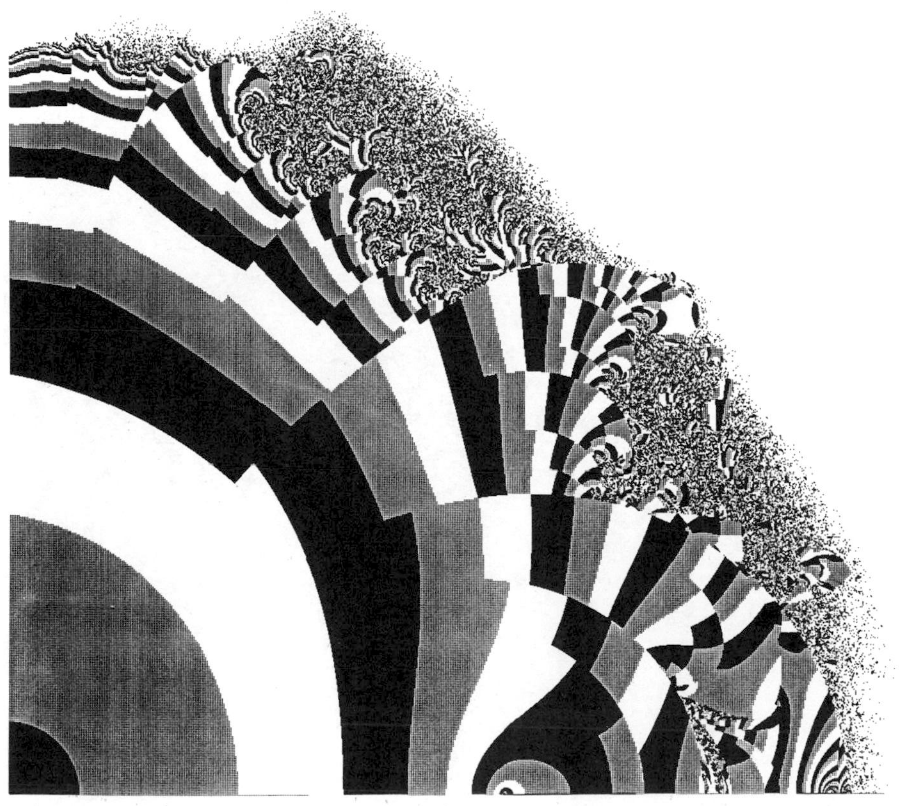

Figure 8.4.6 *A map of the family of dynamical systems described in Example* 8.4.2, *computed using Algorithm* 8.4.1. *The parameter space is* $P = \{\lambda \in \mathbb{C}: 0 < \lambda_1 < 1, 0 < \lambda_2 < 1\}$. *The grainy grey area is the interesting region. This is the "coastline;" it is itself a fractal object, revealing infinite complexity under magnification.*

In Figures 8.4.6, 8.4.7, 8.4.8, and Color Plates 8.4.3 and 8.4.4, we show some results of applying Algorithm 8.4.1 to the dynamical system $\{\mathbb{C}; f_\lambda\}$.

In Figure 8.4.6, the outer white region represents systems for which the orbit of the point O do not diverge, and probably corresponds to "repelling sets" with nonempty interiors. The inner region, defined by the patchwork of grey, black, and white sections, bounded by line segments, represents systems for which the orbit of O diverges and corresponds to totally disconnected "repelling sets." The grainy grey area is the interesting region. This is the "coastline;" it is itself a fractal object, revealing infinite complexity under magnification. Figures 8.4.7 and 8.4.8 show magnifications at two places on the coastline. The grainy areas which are revealed by magnification resemble pictures of the repelling set of the dynamical system at the corresponding values of λ.

4.3. This exercise refers to the family of dynamical systems $\{\mathbb{C}; z^2 - \lambda\}$. Use Algorithm 8.4.1 with $-0.25 \leq \lambda_1 \leq 2$, $-1 \leq \lambda_2 \leq 1$, and $Q = (0.5, 0.5)$, to make a picture of the "Mandelbrot set" $\mathcal{M}(0.5, 0.5)$. An example of such a set, for a different choice of Q, is shown in Figure 8.4.9.

Figure 8.4.7
Zoom on a small piece of the foggy area in Figure 8.4.5. In it one finds grainy areas which resemble the repelling sets of the corresponding dynamical systems. At what value of λ does one find them? At the value of λ in the map where the picture you are interested in occurs.

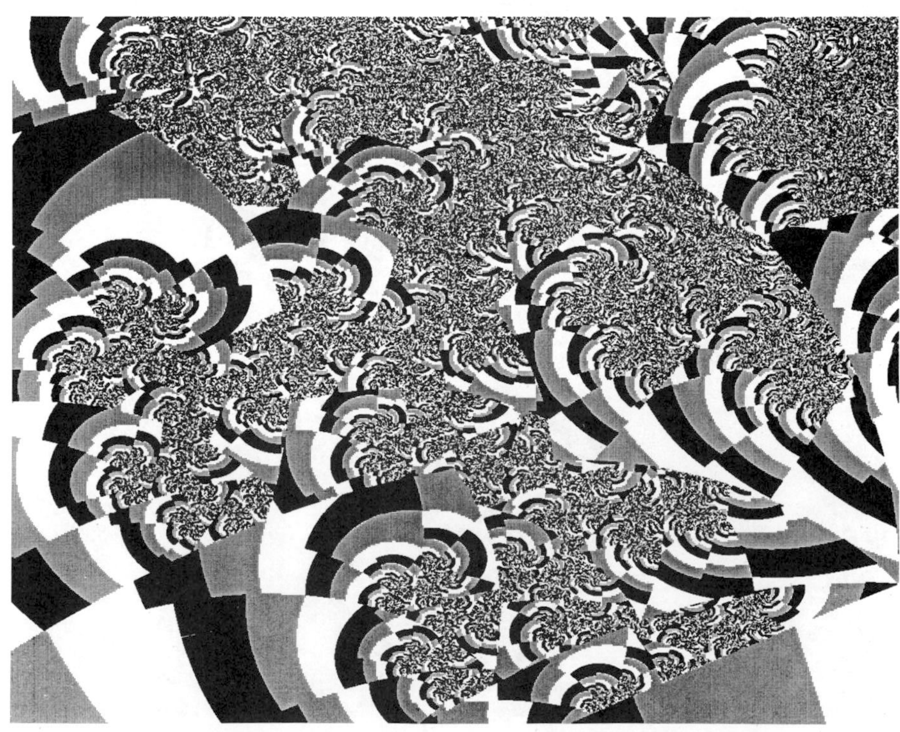

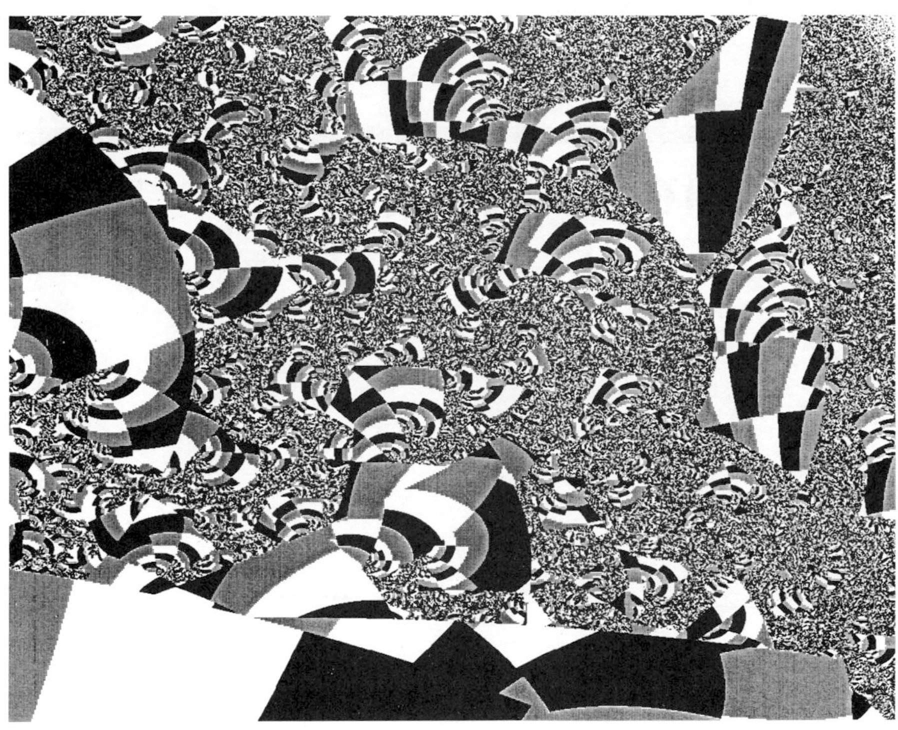

Figure 8.4.8
Zoom on a small piece of the foggy area in Figure 8.4.5. *The grainy areas in this picture here have different shapes from those in Figure* 8.4.7.

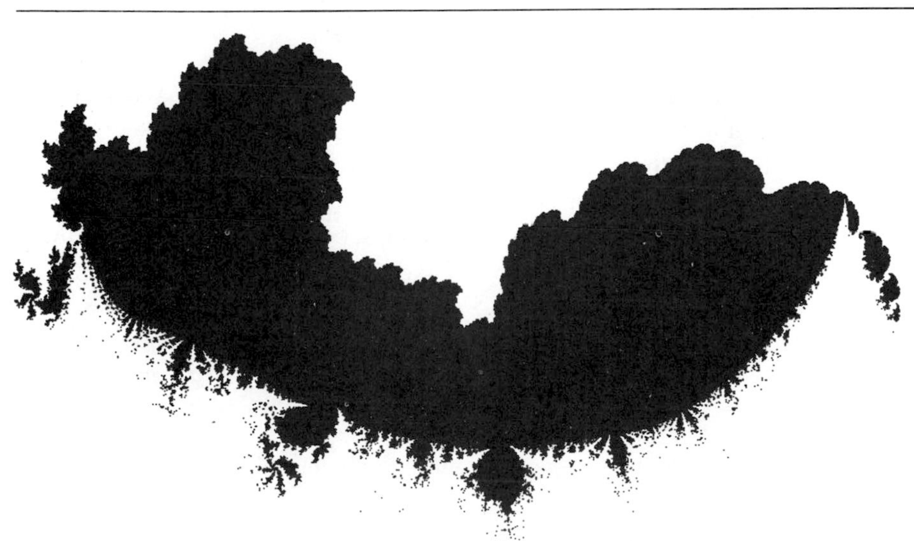

Figure 8.4.9
A "Mandelbrot set" $\mathcal{M}(z_0)$ associated with the family of dynamical systems $\{\mathbb{C};\ z^2 - \lambda\}$. This was computed using escape times of orbits of a point $z = z_0$ different from the critical point, $z = 0$.

9 Measures on Fractals

9.1 INTRODUCTION TO INVARIANT MEASURES ON FRACTALS

In this section we give an intuitive introduction to measures. We focus on measures which arise from iterated function systems in $\mathbb{R}^2$.

In Chapter 3, Section 8, we introduced the Random Iteration Algorithm. This algorithm is a means for computing the attractor of a hyperbolic IFS in $\mathbb{R}^2$. In order to run the algorithm one needs a set of probabilities, in addition to the IFS.

Definition 1. An iterated function system *with probabilities* consists of an IFS $\{\mathbb{X}; w_1, w_2, \ldots, w_N\}$ together with a set of numbers $\{p_1, p_2, \ldots, p_N\}$, such that

$$p_1 + p_2 + p_3 + \cdots + p_N = 1 \quad \text{and} \quad p_i > 0 \text{ for } i = 1, 2, \ldots, N.$$

The probability p_i is associated with the transformation w_i. The nomenclature "IFS with probabilities" is used for "iterated function system with probabilities." The full notation for such an IFS is $\{\mathbb{X};\ w_1, w_2, \ldots, w_N;\ p_1, p_2, \ldots, p_N\}$. Explicit reference to the probabilities may be suppressed.

An example of an IFS with probabilities is

$$\{\mathbb{C};\ w_1(z) = 0.5z,\ w_2(z) = 0.5z + 0.5,\ w_3(z) = 0.5z + 0.5i,$$
$$w_4(z) = 0.5z + 0.5 + (0.5)i;\ 0.1, 0.2, 0.3, 0.4\}.$$

It can be represented by the IFS code in Table 9.1.1. The attractor is the filled square ■, with corners at (0, 0), (1, 0), (1, 1), and (0, 1).

Here is how the Random Iteration Algorithm proceeds in the present case. An initial point, $z_0 \in \mathbb{C}$ is chosen. One of the transformations is selected "at random" from the set $\{w_1, w_2, w_3, w_4\}$. The probability that w_i is selected is p_i, for $i = 1, 2, 3, 4$. The selected transformation is applied to z_0 to produce a new point $z_1 \in \mathbb{C}$. Again a transformation is selected, in the same manner, independently of the previous choice, and applied to z_1 to produce a new point z_2. The process is repeated a number of times, resulting in a finite sequence of points $\{z_n\colon n = 1, 2, \ldots, numits\}$, where *numits* is a positive integer. For simplicity, we assume that $z_0 \in$ ■. Then, since w_i(■) ⊂ ■, for $i = 1, 2, 3, 4$, the "orbit" $\{z_n\colon n = 1, 2, \ldots, numits\}$ lies in ■.

Consider what happens when we apply the algorithm to the IFS code in Table 9.1.1. If the number of iterations is sufficiently large, a picture of ■ will be the result. That is, every pixel corresponding to ■ is visited by the "orbit" $\{z_n\colon n = 1, 2, \ldots, numits\}$. The rate at which a picture of ■ is produced depends on the probabilities. If *numits* = 10000, then we expect that, because the images of ■ are just-touching,

the number of computed points in w_1(■) ≈ 1000,

the number of computed points in w_2(■) ≈ 2000,

the number of computed points in w_3(■) ≈ 3000,

the number of computed points in w_4(■) ≈ 4000.

These estimates are supported by Figure 9.1.1, which shows the result of running a modified version of Program 3.8.2, with the IFS code in Table 9.1.1, and *numits* = 10000.

Table 9.1.1
IFS code for a measure on ■.

w	a	b	c	d	e	f	p
1	0.5	0	0	0.5	1	1	0.1
2	0.5	0	0	0.5	50	1	0.2
3	0.5	0	0	0.5	1	50	0.3
4	0.5	0	0	0.5	50	50	0.4

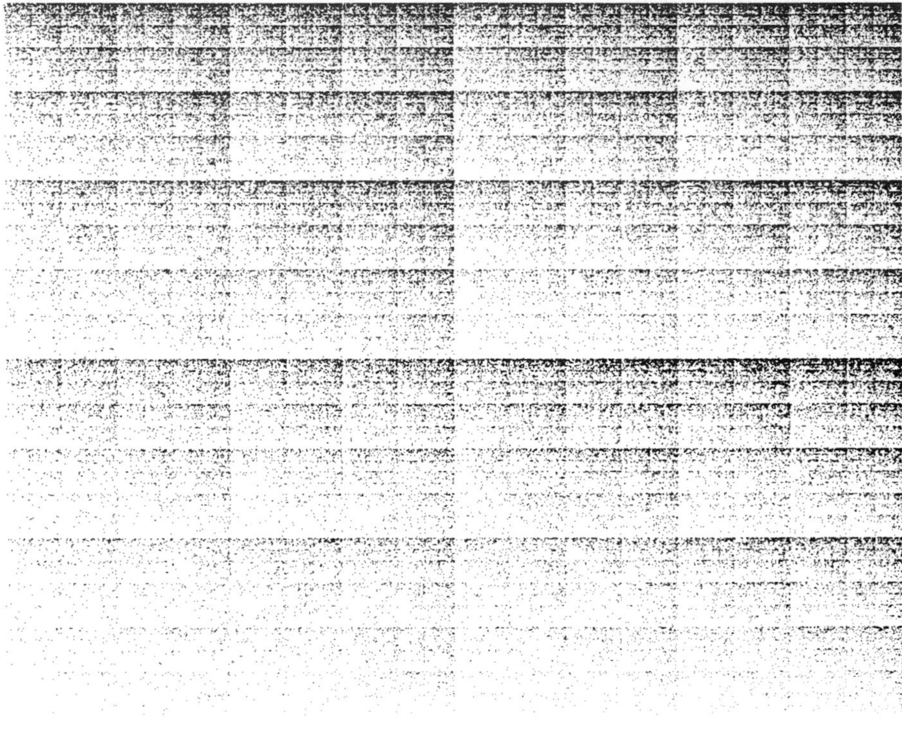

Figure 9.1.1
The Random Iteration Algorithm, Program 3.8.1, *is applied to the IFS code in Table* 9.1.1, *with numits* = 100,000. *Verify that the number of points which lie in* $w_i(\blacksquare)$ *is approximately* $(numits)\,p_i$, *for* $i = 1, 2, 3, 4$.

In Figures 9.1.2 (a)–(c) we show the result of running a modified version of Program 3.8.2, for the IFS code in Table 9.1.1, with various choices for the probabilities. In each case we have halted the program after a relatively small number of iterations, to stop the image becoming "saturated." The results are diverse textures. In each case the attractor of the IFS is the same set, ■. However, the points produced by the Random Iteration Algorithm "rain down" on ■ with different frequencies at different places. Places where the "rainfall" is highest appear "darker" or "more dense" than those places where the "rainfall" is lower. In the end all places on the attractor get wet.

The pictures in Figure 9.1.2 (a)–(c) suggests a wonderful idea. They suggest that, associated with an IFS with probabilities, there is a unique "density" on the attractor of the IFS. The Random Iteration Algorithm gives one a glimpse of this "density," but one loses sight of it as the number of iterations is increased. This is true, and much more as well! As we will see, the "density" is so beautiful that we need a new mathematical concept to describe it. The concept is that of a *measure*. Measures can be used to describe intricate distributions of "mass" on metric spaces. They are introduced formally further

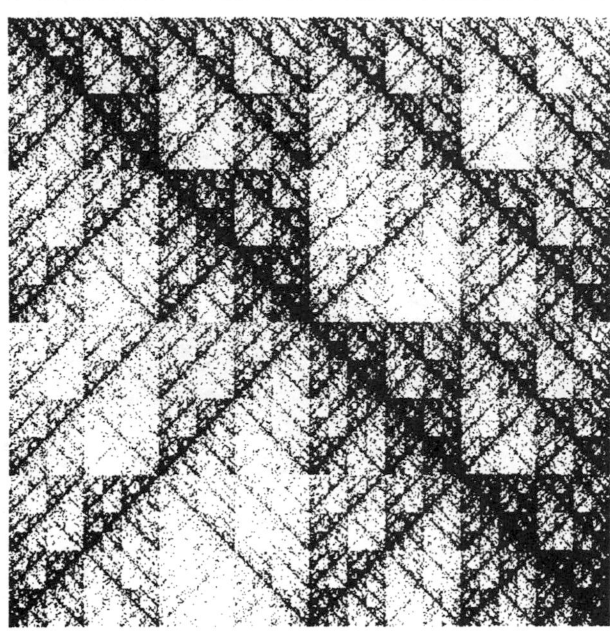

Figure 9.1.2(a) – (c) *The Random Iteration Algorithm is applied to the IFS code in Table* 9.1.1, *but with various different sets of probabilities. The result is that points rain down on the attractor of the IFS at different rates at different places. What we are seeing are the faint traces of wonderful mathematical entities called* measures. *These are the true fractals. Their supports, the attractors of IFS, are merely sets upon which measures live.*

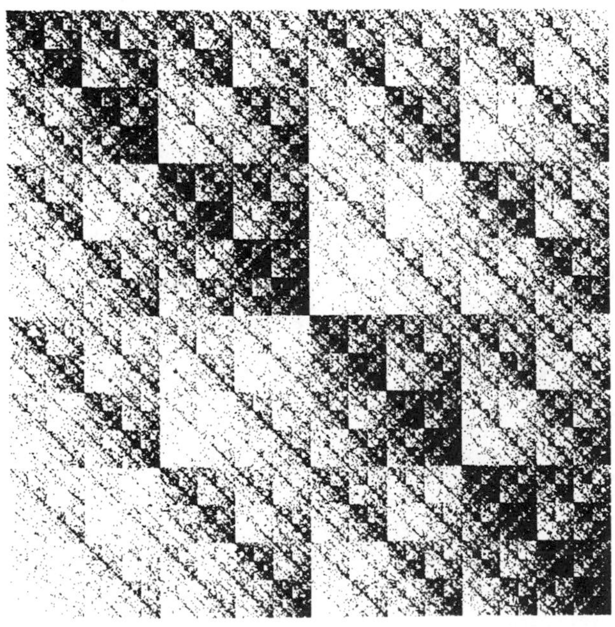

Figure 9.1.2(b)

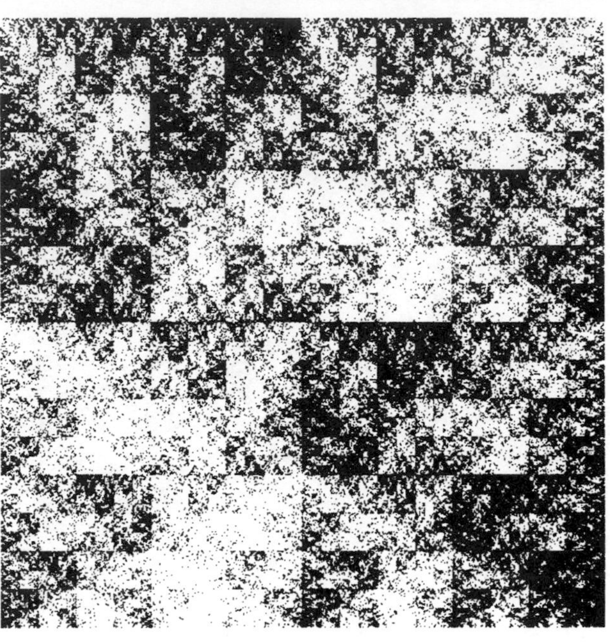

Figure 9.1.2(c)

on in this chapter. The present section provides an intuitive understanding of what measures are, and of how an interesting class of measures arises from IFS's with probabilities.

As a second example, consider the IFS with probabilities

$$\{\mathbb{C};\ w_1(z) = 0.5z,\ w_2(z) = 0.5z + 48,\ w_3(z) = 0.5z + 24 + 48i;\ 0.25, 0.25, 0.5\}.$$

The attractor is a Sierpinski triangle, . The probability associated with w_3 is twice that associated with either w_1 or w_2. In Figure 9.1.3 we show the result of applying the Random Iteration Algorithm, with these probabilities, to compute 10000 points belonging to . There appear to be different "densities" at different places on . For example, $w_3($ $)$ appears to have more "mass" than either $w_1($ $)$ or $w_2($ $)$.

In Figure 9.1.4 we show the result of applying the Random Iteration Algorithm to another IFS with probabilities, for three different sets of probabilities. The IFS is $\{\mathbb{R}^2;\ w_1, w_2, w_3, w_4\}$, where w_i is an affine transformation for $i = 1, 2, 3, 4$. The attractor of the IFS is a leaf-like subset of $\mathbb{R}^2$. In each case we see a different pattern of "mass" on the attractor of the IFS. It appears that each "density" is itself a fractal object.

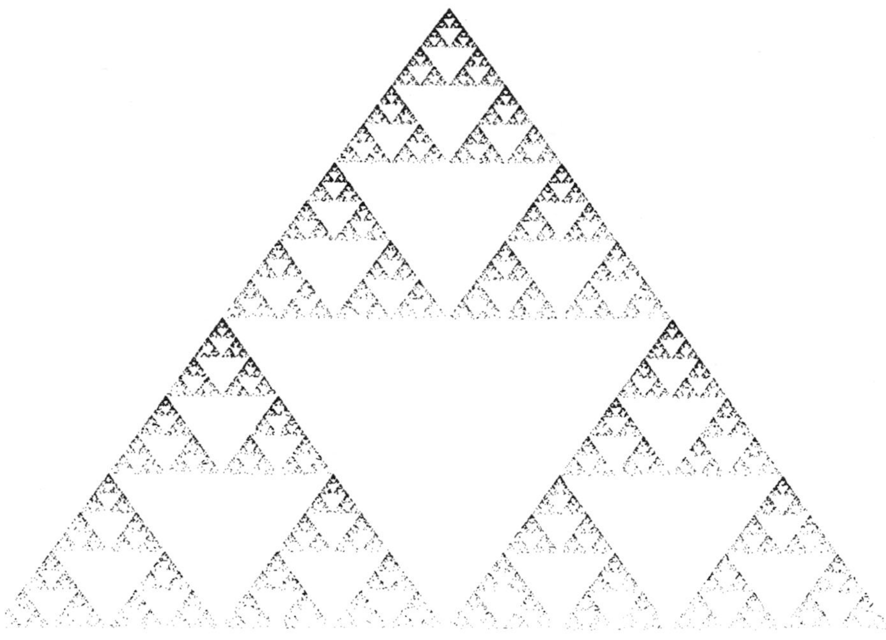

Figure 9.1.3 *The Random Iteration Algorithm is used to compute an image of the Sierpinski triangle* ▲. *The probability associated with w_3 is twice that associated with w_1 or w_2. One thousand points have been computed. The result is that w_3*(▲) *appears denser than w_1*(▲) *or w_2*(▲). *This appearance is lost when the number of iterations is increased. We are led to the idea of a "mass" or measure which is supported on the fractal.*

Exercises & Examples

1.1. Carry out the following numerical experiment. Apply the Random Iteration Algorithm to the IFS code in Table 9.1.1, for *numits* = 1000, 2000, 3000, ... In each case record the number, $\mathcal{N}$, of computed points which land in $B = \{(x, y) \in \mathbb{R}^2: (x-1)^2 + (y-1)^2 \leq 1\}$, and make a table of your results. Verify that the ratio $\mathcal{N}/numits$ appears to approach a constant.

1.2. Repeat the computergraphical experiment which produced Figure 9.1.1. Verify that you obtain "similar looking" output to that shown in Figure 9.1.1, even though you (probably) use a different random number sequence.

1.3. The Random Iteration Algorithm is used to compute 100,000 points belonging to ■, using the IFS code in Table 9.1.1. How many of these points, do you expect, would belong to $w_1 \circ w_3$(■)? Why?

Let $(\mathbb{X}, d)$ be a complete metric space. Let $\{\mathbb{X}; w_1, \ldots, w_N; p_1, \ldots, p_N\}$ be an IFS with probabilities. Let A denote the attractor of the IFS. Then there exists a thing called the *invariant measure* of the IFS, which we denote here by

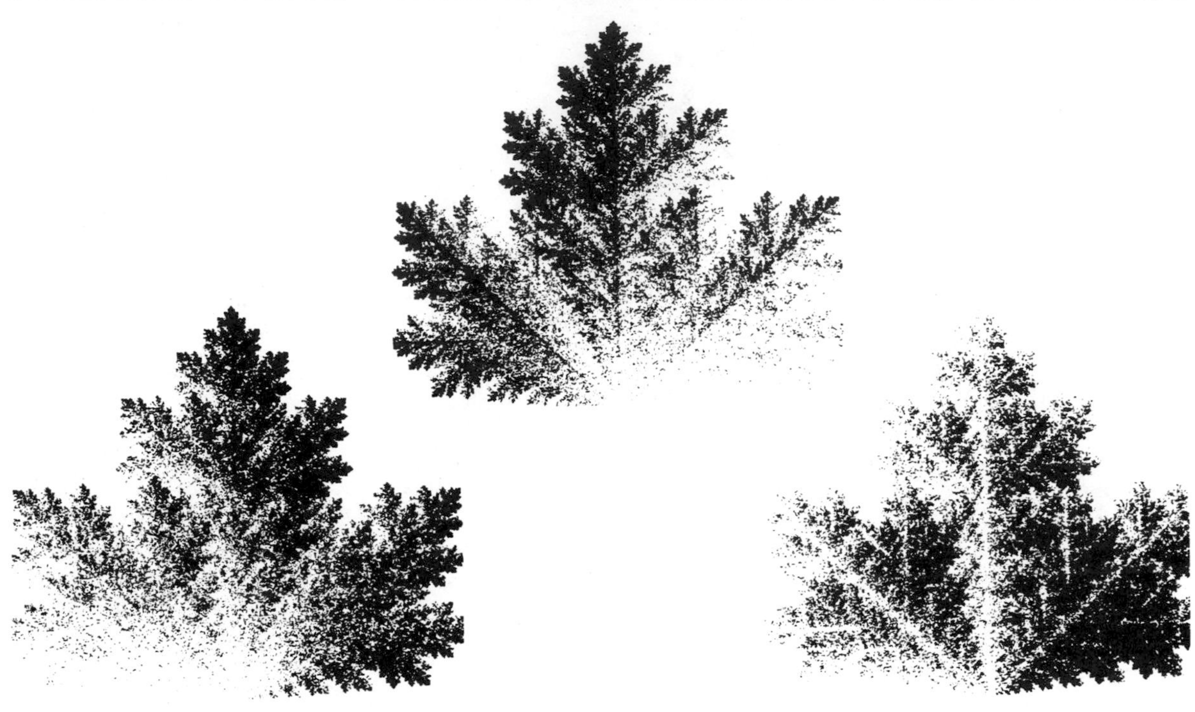

Figure 9.1.4
The Random Iteration Algorithm is used to compute an image of a leaf. Different sets of probabilities lead to different distributions of "mass" on the leaf.

μ. μ assigns "mass" to many subsets of $\mathbb{X}$. For example $\mu(A) = 1$ and $\mu(\varnothing) = 0$. That is, the "mass" of the attractor is one unit, and the "mass" of the empty set is zero. Also $\mu(\mathbb{X}) = 1$, which says that the whole space has the same "mass" as the attractor of the IFS; the "mass" is located on the attractor.

Not all subsets of $\mathbb{X}$ have a "mass" assigned to them. The subsets of $\mathbb{X}$ which do have a "mass" are called the *Borel subsets* of $\mathbb{X}$, denoted by $\mathscr{B}(\mathbb{X})$. The Borel subsets of $\mathbb{X}$ include the compact nonempty subsets of $\mathbb{X}$, so that $\mathscr{H}(\mathbb{X}) \subset \mathscr{B}(\mathbb{X})$. Also, if $\mathscr{O}$ is an open subset of $\mathbb{X}$, then $\mathscr{O} \in \mathscr{B}(\mathbb{X})$. So there are plenty of sets which have "mass."

Let B denote a closed ball in $\mathbb{X}$. Here is how to calculate the "mass" of the ball, $\mu(B)$. Apply the Random Iteration Algorithm to the IFS with probabilities, to produce a sequence of points $\{z_n\}_{n=0}^{\infty}$. Let

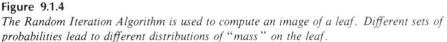

$$\mathscr{N}(B, n) = \text{number of points in } \{z_0, z_1, z_2, z_3, \ldots, z_n\} \cap B, \qquad \text{for } n = 0, 1, 2, \ldots .$$

Then, almost always,

$$\mu(B) = \lim_{n\to\infty} \left\{ \frac{\mathcal{N}(B, n)}{(n+1)} \right\}.$$

That is, the "mass" of the ball B is the proportion of points, produced by the Random Iteration Algorithm, which land in B. (To be precise, we also have to require that the "mass" of the boundary of B is zero; see Corollary 9.7.1)

By now you should be bursting with questions. How do we know that this formula "almost always" gives the same answer? What are Borel sets? Why don't all sets have "mass?" Welcome to measure theory!

As an example, we evaluate the measure of some subsets of $\mathbb{C}$, for the IFS with probabilities

$$\{\mathbb{C};\ w_1(z) = 0.5z,\ w_2(z) = 0.5z + (0.5)i,\ w_3(z) = 0.5z + 0.5;\ 0.33, 0.33, 0.34\}.$$

The attractor is a Sierpinski triangle with vertices at 0, i, and 1. We compute the measures of the following sets:

$$B_1 = \{z \in \mathbb{C}: |z| \le 0.5\}$$

$$B_2 = \{z \in \mathbb{C}: |z - (0.5 + 0.5i)| \le 0.2\}$$

$$B_3 = \{z \in \mathbb{C}: |z - (0.5 + 0.5i)| \le 0.5\}$$

$$B_4 = \{z \in \mathbb{C}: |z - (2 + i)| \le \sqrt{2}\}.$$

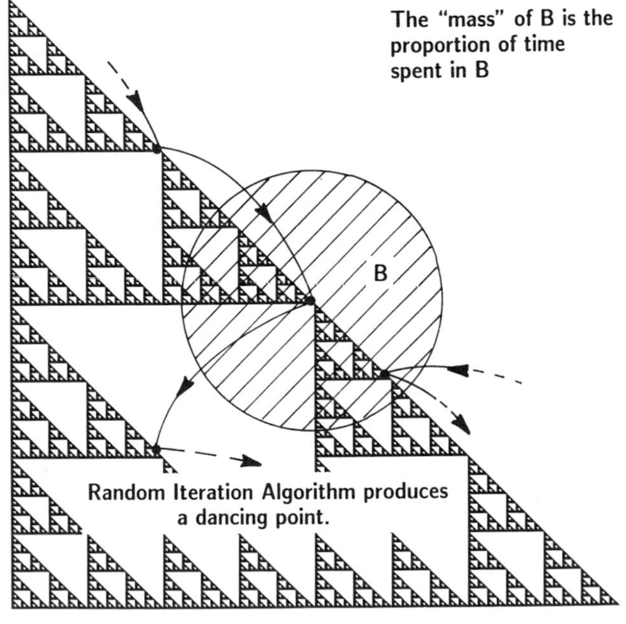
The "mass" of B is the proportion of time spent in B
B
Random Iteration Algorithm produces a dancing point.

Figure 9.1.5 *Diagram of the Random Iteration Algorithm running, and a dancing point coming and going from the ball B. The "mass" or measure of the ball is $\mu(B)$. It is equal to the proportion of points which land in B.*

Table 9.1.2
The measures of some subsets of are computed by random iteration.

n	$\mathcal{N}(B_1, n)/n$	$\mathcal{N}(B_2, n)/n$	$\mathcal{N}(B_3, n)/n$	$\mathcal{N}(B_4, n)/n$
5,000	0.3313	0.1036	0.6385	0.0004
10,000	0.3314	0.1050	0.6500	0.0002
15,000	0.3323	0.1041	0.6512	0.0001
20,000	0.3330	0.1030	0.6525	0.0000
50,000	0.3326	0.1041	0.6527	0.0000
100,000	0.3325	0.1054	0.6497	0.0000
	$\mu(B_1) \approx 0.33$	$\mu(B_2) \approx 0.10$	$\mu(B_3) \approx 0.65$	$\mu(B_4) \approx 0.00$

The results are presented in Table 9.1.2.
Figure 9.1.5 illustrates the ideas introduced here.

Exercises & Examples

1.4. Explain why $\mu(B_4) \approx 0$ in Table 9.1.2.

1.5. What value, approximately, would have been obtained for $\mu(B_1)$ in Table 9.1.2, if the probabilities on the three maps had been $p_1 = 0.275$, $p_2 = 0.125$, and $p_3 = 0.5$?

1.6. Why, do you think, is the phrase "almost always" written in connection with the formula for $\mu(B)$, given above?

9.2 FIELDS AND SIGMA-FIELDS

Definition 1. Let $\mathbb{X}$ be a space. Let $\mathscr{F}$ denote a nonempty class of subsets of a space $\mathbb{X}$, such that

(i) $A, B \in \mathscr{F} \Rightarrow A \cup B \in \mathscr{F}$;
(ii) $A \in \mathscr{F} \Rightarrow \mathbb{X} \setminus A \in \mathscr{F}$.

Then $\mathscr{F}$ is called a *field*. (In example 9.2.13 you are asked to prove that $\mathbb{X} \in \mathscr{F}$.)

Theorem 1. *Let $\mathbb{X}$ be a space. Let $\mathscr{G}$ be a nonempty set of subsets of $\mathbb{X}$. Let $\mathscr{F}$ be the set of subsets of $\mathbb{X}$ which can be built up from finitely many sets in $\mathscr{G}$ using the operations of union, intersection, and complementation with respect to $\mathbb{X}$. Then $\mathscr{F}$ is a field.*

Proof. Elements of $\mathscr{F}$ consist of sets such as

$$\mathbb{X} \setminus (((\mathbb{X} \setminus (G_1 \cup G_2)) \cap G_3) \cup (G_5 \cap G_6)),$$

where $G_1, G_2, G_3, G_3, \ldots$ denote elements of $\mathscr{G}$. That is, $\mathscr{F}$ is made of all those sets which can be expressed using finite chains of parentheses, "$\backslash$", "$\cup$", "$\cap$", elements of $\mathscr{G}$, and $\mathbb{X}$. (In fact, using de Morgan's laws one can prove that it is not necessary to use the intersection operation.) If we form the union of any two such expressions, we obtain another one. Similarly, if we form the complement of such an expression with respect to $\mathbb{X}$, we obtain another such expression. So conditions (i) and (ii) in Definition 9.2.1 are satisfied. This completes the proof.

***Definition* 2.** The field referred to in Theorem 9.2.1 is called the field *generated* by $\mathscr{G}$.

Exercises & Examples

2.1. Let $\mathbb{X}$ be a space and let $A \subset \mathbb{X}$. Then $\mathscr{F} = \{\mathbb{X}, A, \mathbb{X} \setminus A, \varnothing\}$ is a field.

2.2. Let $\mathbb{X}$ be the set of all leaves on a certain tree, and $\mathscr{F}$, the set of all subsets of $\mathbb{X}$. Then $\mathscr{F}$ is a field. Let A denote the set of all the leaves on the lowest branch of the tree. Then $A \in \mathscr{F}$. Prove that $\mathscr{F}$ is generated by the leaves.

2.3. Let $\mathbb{X} = [0,1] \subset \mathbb{R}$. Let $\mathscr{G}$ denote the set of all of subintervals (open, closed, half-open) of $[0,1]$. Let $\mathscr{F}$ denote the field generated by $\mathscr{G}$. Examples of members of $\mathscr{F}$ are $[0.5, 0.6) \cup (0.7, 0.81)$; $[0,1]$; $[1,1]$; and $(\frac{1}{2}, 1) \cup (\frac{1}{4}, \frac{1}{3}) \cup \cdots \cup (\frac{1}{100}, \frac{1}{99})$. Show that

$$\bigcup_{n=1}^{\infty} \left(\frac{1}{(n+1)}, \frac{1}{n} \right) = \left(\frac{1}{2}, 1 \right) \cup \left(\frac{1}{3}, \frac{1}{2} \right) \cup \left(\frac{1}{4}, \frac{1}{3} \right) \cup \cdots$$

is a subset of $\mathbb{X}$ but it is not a member of $\mathscr{F}$.

2.4. Let $\mathbb{X} = \blacksquare \subset \mathbb{R}^2$. Let $\mathscr{G}$ denote the set of closed rectangles contained in $\mathbb{X}$, whose sides are parallel to the coordinate axes and whose corners have rational coordinates. Let $\mathscr{F}$ denote the field generated by $\mathscr{G}$. An example of an element of $\mathscr{F}$ is

$$((\blacksquare \setminus ((\blacksquare \setminus R_1) \cup R_2) \cap R_3) \cup (R_4 \cap (\blacksquare \setminus R_5))$$

where R_1, R_2, R_3, R_4, and R_5 are rectangles in $\mathscr{G}$. Let $S \in \mathscr{F}$. Prove that the area of S is a rational number. Deduce that $\mathscr{F}$ does not contain the ball $B(O,1) = \{(x,y) \in \blacksquare: x^2 + y^2 \leq 1\}$.

2.5. Let $\mathbb{X}$ denote the set of pixels corresponding to a certain computer graphics display device. The set of all monochrome images which can be produced on this device forms a field. Figure 9.2.1 shows an example of a small field of subsets of $\mathbb{X}$. It is generated by the pair of images, G_1 and G_2, in the middle of the first row, together with the set $\mathbb{X}$. $\mathbb{X}$ is represented by the black rectangle. The empty set is represented by a blank screen. Find formulas for all of the images in Figure 9.2.1, in terms of G_1, G_2, and $\mathbb{X}$.

Figure 9.2.1
A field whose elements are sets of pixels. Can you find two elements of the field which generate the field?

2.6. Let Σ denote the code space on two symbols 1 and 2. Let $n \in \{1, 2, 3, \ldots\}$ and $e_i \in \{1, 2\}$ for $i = 1, 2, \ldots, n$. Let

$$C(e_1, e_2, \ldots, e_n) = \{x \in \Sigma: x_i = e_i \text{ for } i = 1, 2, \ldots, n\}.$$

Any subset of Σ which can be written in this form is called a *cylinder subset* of Σ. Let $\mathscr{F}$ denote the field generated by the cylinder subsets of Σ. Find a subset of Σ which is not in $\mathscr{F}$.

2.7. Let $\mathbb{X}$ be a space. Let $\mathscr{F}$ denote the set of all subsets of $\mathbb{X}$. The customary notation for this field is $\mathscr{F} = 2^{\mathbb{X}}$. Show that $\mathscr{F}$ is a field.

Definition 3. Let $\mathscr{F}$ be a field such that

$$A_i \in \mathscr{F} \text{ for } i \in 1, 2, 3, \ldots \Rightarrow \bigcup_{i=1}^{\infty} A_i \in \mathscr{F}.$$

Then $\mathscr{F}$ is called a *σ-field* (sigma-field). Given any field, there always is a minimal, or smallest, σ-field which contains it.

Theorem 2. *Let $\mathbb{X}$ be a space and let $\mathscr{G}$ be a set of subsets of $\mathbb{X}$. Let $\{\mathscr{F}_\alpha: \alpha \in I\}$ denote the set of all σ-fields on $\mathbb{X}$ which contain $\mathscr{G}$. Then $\mathscr{F} = \cap_\alpha \mathscr{F}_\alpha$ is a σ-field containing $\mathscr{G}$.*

Proof. Note that there is at least one σ-field which contains $\mathcal{G}$, namely $2^{\mathbb{X}}$, the field consisting of all subsets of $\mathbb{X}$. We have to show that $\cap_{\alpha}\mathcal{F}_{\alpha}$ is a σ-field if each $\mathcal{F}_{\alpha}$ is a σ-field which contains $\mathcal{G}$. Suppose that $A_i \in \cap_{\alpha}\mathcal{F}_{\alpha}$; then, for each α, A_i is an element of the σ-field $\mathcal{F}_{\alpha}$ and so $\cup_{i=1}^{\infty} A_i \in \mathcal{F}_{\alpha}$. Suppose $A \in \cap_{\alpha}\mathcal{F}_{\alpha}$; then, for each α, $A \in \mathcal{F}_{\alpha}$ and so $\mathbb{X} \setminus A \in \mathcal{F}_{\alpha}$. Hence $\mathbb{X} \setminus A \in \cap_{\alpha}\mathcal{F}_{\alpha}$. This completes the proof.

Definition 4. Let $\mathcal{G}$ be a set of subsets of a space $\mathbb{X}$. The minimal σ-field which contains $\mathcal{G}$, defined in Theorem 9.2.2, is called the σ-field *generated* by $\mathcal{G}$.

Definition 5. Let $(\mathbb{X}, d)$ be a metric space. Let $\mathcal{B}$ denote the σ-field generated by the open subsets of $\mathbb{X}$. $\mathcal{B}$ is called the *Borel field* associated with the metric space. An element of $\mathcal{B}$ is called a *Borel subset* of $\mathbb{X}$.

The following theorem gives the flavor of ways in which the Borel field can be generated.

Theorem 3. *Let $(\mathbb{X}, d)$ be a compact metric space. Then the associated Borel field $\mathcal{B}$ is generated by a countable set of balls.*

Proof. We prove a more general result first. Let $\mathcal{G} = \{b_n \subset \mathbb{X}: n = 1, 2, 3, \ldots;\ b_n \text{ open}\}$ be a *countable base* for the open subsets of $\mathbb{X}$. That is, every open set in $\mathbb{X}$ can be written as a union of sets in $\mathcal{G}$. Then $\mathcal{B}$ is generated by $\mathcal{G}$. To see this, let $\tilde{\mathcal{B}}$ denote the σ-field generated by $\mathcal{G}$. Then $\tilde{\mathcal{B}} \subset \mathcal{B}$ because $\mathcal{G}$ is contained in the set of open subsets of $\mathbb{X}$. On the other hand $\mathcal{B} \subset \tilde{\mathcal{B}}$ because $\tilde{\mathcal{B}}$ contains all the generators of $\mathcal{B}$. Hence $\mathcal{B} = \tilde{\mathcal{B}}$.

It remains to construct a countable base for the open subsets of $\mathbb{X}$ using balls. For $R > 0$ let

$$B(x, R) = \{y \in \mathbb{X}: d(x, y) < R\}.$$

Let n be a positive integer. Then $\mathbb{X} = \cup_{x \in \mathbb{X}} B(x, 1/n)$. Hence $\{B(x, 1/n): x \in \mathbb{X}\}$ is an open covering of $\mathbb{X}$. Since $\mathbb{X}$ is compact, it contains a finite subcovering $\{B(x_m^{(n)}, 1/n): m = 1, 2, \ldots, M(n)\}$ for some integer $M(n)$. We claim that

$$\mathcal{D} = \left\{B\left(x_m^{(n)}, \frac{1}{n}\right): m = 1, 2, \ldots, M(n);\ n = 1, 2, 3, \ldots\right\}$$

is a countable base for the open subsets of $\mathbb{X}$. For let $\mathcal{O}$ be an open subset of $\mathbb{X}$, and let $x \in \mathcal{O}$. Then there is an open ball, of radius $R > 0$, such that $B(x, R) \subset \mathcal{O}$. Let n be large enough that $1/n < R/2$. Then there is $m \in \{1, 2, \ldots, M(n)\}$ so that x is in the ball $B(x_m^{(n)}, 1/n)$, and this ball is contained in $\mathcal{O}$. Each x in $\mathcal{O}$ is contained in such a ball, belonging to $\mathcal{D}$. Hence $\mathcal{D}$ is indeed a countable base for the open subsets of $\mathbb{X}$. This completes the proof.

Exercises & Examples

2.8. This example takes place in the metric space ([0, 1], Euclidean). Let $\mathscr{B}$ denote the σ-field generated by the real intervals which are contained in [0, 1]. Then $\mathscr{B}$ is the Borel field associated with the metric space ([0, 1], Euclidean metric). Show that $\mathscr{B}$ is the same as the σ-field generated by the field in exercise 9.2.3.

2.9. Let $\mathscr{B}$ denote the σ-field generated by the field in exercise 9.2.4. Then $\mathscr{B}$ contains the ball $B(O, 1)$. Similarly, it contains all balls in $\blacksquare \subset R^2$. Show that $\mathscr{B}$ is the Borel field associated with ($\blacksquare$, Manhattan).

2.10. Let Σ denote the code space on the two symbols $\{0, 1\}$. Show that the Borel field associated with (Σ, code space metric) is generated by the cylinder subsets of Σ, defined in exercise 9.2.6.

2.11. Let △ $\subset R^2$ denote a Sierpinski triangle. Let $\mathscr{G}$ denote the set of connected components of $R^2 \setminus$ △. Let $\mathscr{F}$ denote the σ-field generated by $\mathscr{G}$. Show that $\mathscr{F}$ is contained in, but not equal to, the Borel field associated with ($\mathbb{R}^2$, Euclidean).

2.12. Let $\mathbb{X}$ be a space and let $\mathscr{G}$ be a set of subsets of $\mathbb{X}$. Let $\mathscr{F}_1$ be the field generated by $\mathscr{G}$, let $\mathscr{F}_2$ be the σ-field generated by $\mathscr{G}$, and let $\mathscr{F}_3$ be the σ-field generated by $\mathscr{F}_1$. Prove that $\mathscr{F}_3 = \mathscr{F}_2$.

2.13. Let $\mathscr{F}$ be a field of subsets of a space $\mathbb{X}$. Prove that $\mathbb{X} \in \mathscr{F}$.

9.3 MEASURES

A measure is defined on a field. Each member of the field is assigned a non-negative real number, which tells us its "mass."

Definition 1. A *measure* μ, on a field $\mathscr{F}$, is a real non-negative function μ: $\mathscr{F} \to [0, \infty) \subset \mathbb{R}$, such that whenever $A_i \in \mathscr{F}$ for $i = 1, 2, 3, \ldots$, with $A_i \cap A_j = \varnothing$ for $i \neq j$ and $\cup_{i=1}^{\infty} A_i \in \mathscr{F}$, we have

$$\mu\left(\bigcup_{i=1}^{\infty} A_i\right) = \sum_{i=1}^{\infty} \mu(A_i).$$

(In other texts, a measure as defined here is usually referred to as a finite measure.)

Definition 2. Let $(\mathbb{X}, d)$ be a metric space. Let $\mathscr{B}$ denote the Borel subsets of $\mathbb{X}$. Let μ be a measure on $\mathscr{B}$. Then μ is called a *Borel measure*.

Some basic properties of measures are summarized below.

Theorem 1. *Let $\mathscr{F}$ be a field and let μ: $\mathscr{F} \to \mathbb{R}$ be a measure. Then*

(i) *If $B \supset A$ then $\mu(B) = \mu(B \setminus A) + \mu(B)$, for $A, B \in \mathscr{F}$;*

(ii) *If* $B \supset A$ *then* $\mu(B) \geq \mu(A)$;

(iii) $\mu(\varnothing) = 0$;

(iv) *If* $A_i \in \mathscr{F}$ *for* $i = 1, 2, 3, \ldots$ *and* $\cup_{i=1}^{\infty} A_i \in \mathscr{F}$ *then* $\mu(\cup_{i=1}^{\infty} A_i) \leq \sum_{i=1}^{\infty} \mu(A_i)$;

(v) *If* $\{A_i \in \mathscr{F}\}$ *obeys* $A_1 \subset A_2 \subset A_3 \subset \ldots$, *and if* $\cup_{i=1}^{\infty} A_i \in \mathscr{F}$, *then* $\mu(A_i) \to \mu(\cup_{i=1}^{\infty} A_i)$.

(vi) *If* $\{A_i \in \mathscr{F}\}$ *obeys* $A_1 \supset A_2 \supset A_3 \supset \ldots$, *and if* $\cap_{i=1}^{\infty} A_i \in \mathscr{F}$, *then* $\mu(A_i) \to \mu(\cap_{i=1}^{\infty} A_i)$.

Proof. [Rudin 1966] Theorem 1.19, p. 17. These are fun to prove for yourself!

We are concerned with measures on compact subsets of metric spaces such as ($\mathbb{R}^2$, Euclidean). The natural underlying σ-field is the Borel field, generated by the open subsets of the metric space. The following theorem allows us to work with the restriction of the measure to any field which generates the σ-field.

Theorem 2. (Carathéodory) *Let* μ *denote a measure on a field* $\mathscr{F}$. *Let* $\hat{\mathscr{F}}$ *denote the* σ*-field generated by* $\mathscr{F}$. *Then there exists a unique measure* $\hat{\mu}$ *on* $\hat{\mathscr{F}}$ *such that* $\mu(A) = \hat{\mu}(A)$ *for all* $A \in \mathscr{F}$.

Sketch of proof. The proof can be found in most books on measure theory, see [Eisen 1969] Theorem 5, Chapter 6, p. 180, for example. First μ is used to define an "outer measure" μ^0 on the set of subsets of $\mathbb{X}$. μ^0 is defined by

$$\mu^0(A) = \inf\left\{ \sum_{n=1}^{\infty} \mu(A_n) \colon A \subset \bigcup_{n=1}^{\infty} A_n,\ A_n \in \mathscr{F}\ \forall n \in \mathbb{Z}^+ \right\}.$$

μ^0 is not usually a measure. However, one can show that the class $\mathscr{F}^0$ of subsets A of $\mathbb{X}$ such that—this was Caratheodory's smart idea—

$$\mu^0(E) = \mu^0(A \cap E) + \mu^0((\mathbb{X} \setminus A) \cap E) \qquad \text{for all } E \in 2^{\mathbb{X}},$$

is a σ-field which contains $\mathscr{F}$. One can also show that μ^0 is a measure on $\mathscr{F}^0$. Note that $\mathscr{F}^0 \supset \hat{\mathscr{F}}$. $\hat{\mu}$ is defined by restricting μ^0 to $\hat{\mathscr{F}}$. Finally one shows that this extension of μ to $\hat{\mathscr{F}}$ is unique. This completes the sketch.

In the above sketch we have discovered how to evaluate the extended measure $\hat{\mu}$ in terms of its values on the original field.

Theorem 3. *Let a measure* μ *on a field* $\mathscr{F}$ *be extended to a measure* $\hat{\mu}$ *on the minimal* σ*-field* $\hat{\mathscr{F}}$ *which contains* $\mathscr{F}$. *Then, for all* $A \in \hat{\mathscr{F}}$,

$$\hat{\mu}(A) = \inf\left\{ \sum_{n=1}^{\infty} \mu(A_n) \colon A \subset \bigcup_{n=1}^{\infty} A_n,\ A_n \in \mathscr{F}\ \forall n = 1, 2, \ldots \right\}.$$

Exercises & Examples

3.1. Consider the field $\mathscr{F} = \{\mathbb{X}, A, \mathbb{X} \setminus A, \varnothing\}$, where $A \neq \mathbb{X}$ and $A \neq \varnothing$. A measure $\mu: \mathscr{F} \to \mathbb{R}$ is defined by $\mu(\mathbb{X}) = 7.2$, $\mu(A) = 3.5$, $\mu(\mathbb{X} \setminus A) = 3.7$, and $\mu(\varnothing) = 0$. $\mathscr{F}$ is also a σ-field. The extension of the measure promised by Caratheodory's theorem is just the measure itself.

3.2. Let $\mathscr{F}$ be the field made of sets of leaves on a certain tree, at a certain instant in time, and let $\mu(A)$ be the number of aphids on all the leaves in $A \in \mathscr{F}$. Then μ is a measure on a finite σ-field.

3.3. Let $\mathbb{X} = [0,1] \subset \mathbb{R}$. Let $\mathscr{F}$ be the field generated by the set of subintervals of $[0,1]$. Let $a, b \in [0,1]$ and define $\mu((a,b)) = \mu([a,b]) = b - a$, for $a \leq b$; and more generally let

$$\mu(\text{element of } \mathscr{F}) = \text{sum of lengths of disjoint subintervals which comprise the element.}$$

Show that μ is a measure on $\mathscr{F}$. The σ-field $\hat{\mathscr{F}}$ generated by $\mathscr{F}$ is the Borel field for ([0,1]), Euclidean). Show that $S = \{x \in [0,1]: x \text{ is a rational number}\}$ belongs to $\hat{\mathscr{F}}$ but not to $\mathscr{F}$. Evaluate $\hat{\mu}(S)$, where $\mu\hat{}$ is the extension of μ to $\hat{\mathscr{F}}$.

3.4. Let $\mathbb{X} = \Sigma$, the code space on the two symbols 1 and 2. Let $\mathscr{F}$ denote the field generated by the cylinder subsets of Σ, as defined in exercise 9.2.6. Let $0 \leq p_1 \leq 1$ and $p_2 = 1 - p_1$. Define

$$\mu(C(e_1, e_2, \ldots, e_n)) = p_{e_1} p_{e_2} \cdots p_{e_n},$$

for each cylinder subset $C(e_1, e_2, \ldots, e_n)$ of Σ. Show how μ can be defined on the other elements of $\mathscr{F}$ in such a way as to provide a measure on $\mathscr{F}$. Evaluate

$$\mu(\{x \in \Sigma: x_7 = 1\}), \qquad \text{and } \mu(\Sigma).$$

Extend $\mathscr{F}$ to the field $\hat{\mathscr{F}}$ generated by $\mathscr{F}$, and correspondingly extend μ to $\hat{\mu}$. Show that

$$S = \{x \in \Sigma: x_{\text{odd}} = 1\} \in \hat{\mathscr{F}}$$

and evaluate $\hat{\mu}(S)$.

3.5. This example takes place in the metric space $\{[0,1]; \text{Euclidean}\}$. Consider the IFS with probabilities

$$\{[0,1]; w_1(x) = \tfrac{1}{3}x; w_2(x) = \tfrac{1}{3}x + \tfrac{2}{3}; p_1, p_2\}.$$

Let $\mathscr{F}$ denote the field generated by the set of intervals which can be expressed in the form

$$w_{e_1} \circ w_{e_2} \circ \cdots \circ w_{e_n}([0,1]),$$

where $n \in \{1, 2, \ldots\}$ and $e_i \in \{1, 2\}$ for each $i = 1, 2, \ldots, n$. Let $0 \leq p_1$

≤ 1 and $p_2 = 1 - p_1$. Show that one can define a measure on $\mathscr{F}$ so that, for every such interval,

$$\mu\left(w_{e_1} \circ w_{e_2} \circ \cdots \circ w_{e_n}([0,1])\right) = p_{e_1} p_{e_2} \cdots p_{e_m}.$$

Let A denote the attractor of the IFS. Evaluate $\mu(A)$, $\mu(\mathbb{X} \setminus A)$, and $\mu[1/3, 2/3])$.

3.6. What happens in exercise 9.3.5 if the IFS is replaced by

$$\{[0,1];\, w_1(x) = \tfrac{1}{2}x,\, w_2(x) = \tfrac{1}{2}x + \tfrac{1}{2};\, p_1, p_2\}?$$

For what value of p_1 is the extension of the measure to the σ-field generated by $\mathscr{F}$ the same as the Borel measure defined in exercise 9.3.3?

Definition 3. Let $(\mathbb{X}, d)$ be a metric space, and let μ be a Borel measure. Then the *support* of μ is the set of points $x \in \mathbb{X}$ such that $\mu(B(x, \epsilon)) > 0$ for all $\epsilon > 0$, where $B(x, \epsilon) = \{y \in \mathbb{X}: d(y, x) < \epsilon\}$.

The support of a measure is the set on which the measure lives. The following is an easy exercise.

Theorem 4. *Let $(\mathbb{X}, d)$ be a metric space, and let μ be a Borel measure. Then the support of μ is closed.*

Exercises & Examples

3.7. Let $(\mathbb{X}, d)$ be a compact metric space and let μ be a Borel measure on $\mathbb{X}$ such that $\mu(\mathbb{X}) \neq 0$. Show that the support of μ belongs to $\mathscr{H}(\mathbb{X})$, the space of nonempty compact subsets of $\mathbb{X}$.

3.8. Prove the following. "Let μ be a measure on a σ-field $\mathscr{F}$, and let $\tilde{\mathscr{F}}$ be the class of all sets of the form $A \cup B$ where $A \in \mathscr{F}$ and B is a subset of a set of measure zero. Then $\tilde{\mathscr{F}}$ is a σ-field and the function $\bar{\mu}$: $\tilde{\mathscr{F}} \to \mathbb{R}$ defined by $\bar{\mu}(A \cup B) = \bar{\mu}(A)$ is a measure." The measure $\bar{\mu}$ referred to here is called the *completion* of μ. The completion of the measure in Exercise 3.3 is called the *Lebesgue* measure on [0, 1].

9.4 INTEGRATION

In the next section we will introduce a remarkable compact metric space. It is a space whose points are measures! In order to define the metric on this space, we need to be able to integrate continuous real-valued functions with respect to measures.

Can one integrate a continuous function defined on a fractal? How does one evaluate the "average" temperature of the coastline of Sweden? Here we learn how to integrate functions with respect to measures. Let $(\mathbb{X}, d)$ be a compact metric space. Let μ be a Borel measure on $\mathbb{X}$. Let $f\colon \mathbb{X} \to \mathbb{R}$ be a continuous function. We will explain the meaning of integrals such as

$$\int_{\mathbb{X}} f(x)\, d\mu(x).$$

Definition 1. We reserve the notation χ_A for the *characteristic function* of a set $A \subset \mathbb{X}$. It is defined by

$$\chi_A(x) = \begin{cases} 1 & \text{for } x \in A, \\ 0 & \text{for } x \in \mathbb{X} \setminus A. \end{cases}$$

A function $f\colon \mathbb{X} \to \mathbb{R}$ is called *simple* if it can be written in the form

$$f(x) = \sum_{i=1}^{N} y_i \chi_{I_i}(x),$$

where N is a positive integer, $I_i \in \mathscr{B}$ and $y_i \in \mathbb{R}$ for $i = 1, 2, \ldots, N$, $\cup_{i=1}^{N} I_i = \mathbb{X}$, and $I_i \cap I_j = \varnothing$ for $i \neq j$.

The graphs of several simple functions, associated with different spaces, are shown in Figures 9.4.1 (a) and (b).

Definition 2. The *integral* (with respect to μ) of the simple function f in Definition 9.4.1, is

$$\int_{\mathbb{X}} f(x)\, d\mu(x) = \int_{\mathbb{X}} f\, d\mu = \sum_{i=1}^{N} y_i \mu(I_i).$$

This does not depend on how f is represented as a simple function.

Exercises & Examples

4.1. Let $f\colon [0, 1] \to \mathbb{R}$ be a piecewise constant function, with finitely many discontinuities. Show that f is a simple function. Let μ denote the Borel measure on $[0, 1]$ such that $\mu(I) =$ length of I, when I is a subinterval of $[0, 1]$. Show that

$$\int_0^1 f(x)\, dx = \int_{[0,1]} f(x)\, d\mu(x),$$

where the left-hand side denotes the area under the graph of f.

4.2. This example takes place in the metric space (■, Euclidean). Let $\mathscr{G}$ denote the set of rectangular subsets of ■. Let $\mathscr{F}$ denote the field generated by $\mathscr{G}$. Show that there is a unique measure μ on $\mathscr{F}$ such that $\mu(A) =$ area of A, for all $A \in \mathscr{G}$. Notice that the σ-field generated by $\mathscr{F}$ is precisely the Borel field $\mathscr{B}$ associated with (■, Euclidean). Let $\hat{\mu}$ denote the extension of μ to $\mathscr{B}$. Let △ denote a Sierpinski triangle

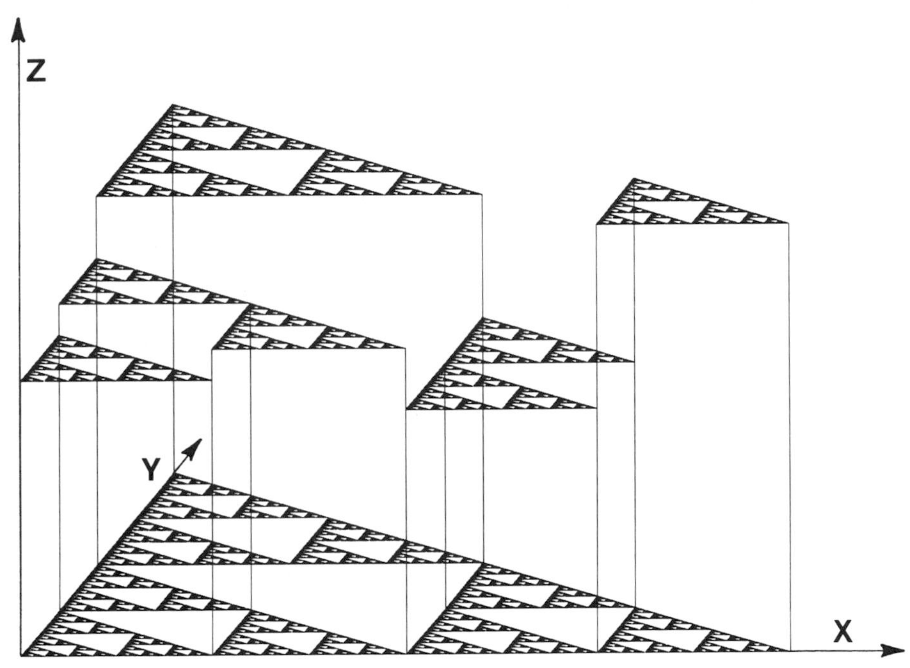

Figure 9.4.1(a)
The graph of a simple function on a Sierpinski triangle. The domain is a Sierpinski triangle in the (x, y)-plane. The function values are represented by the z-coordinates.

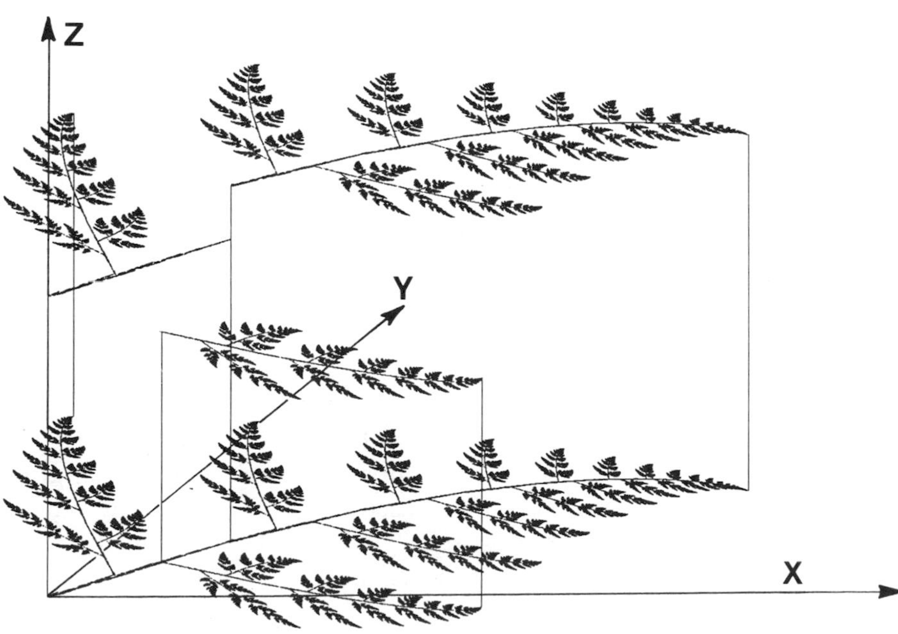

Figure 9.4.1(b)
The graph of a simple function whose domain is a fractal fern. If, instead, the function values were represented by colors, a painted fern would replace the graph.

contained in $\blacksquare$. Show that $\triangle \in \mathscr{B}$, and

$$\int_{\blacksquare} \chi_{\triangle}\, d\hat{\mu} = \hat{\mu}\left(\triangle \right) = 0.$$

4.3. This example concerns the IFS with probabilities

$$\{ C;\ w_1(z) = 0.5z,\ w_2(z) = 0.5z + (0.5)i,\ w_3(z) = 0.5z + 0.5;$$
$$p_1 = 0.2,\ p_2 = 0.3,\ p_3 = 0.5\}.$$

Let $\triangle$ denote the attractor of the IFS. Let $\mathscr{B}$ denote the Borel subsets of ($\triangle$, Euclidean). Let μ denote the unique measure on $\mathscr{B}$ such that

$$\mu\left(\triangle \right) = 1$$
$$\mu\left(w_i\left(\triangle \right)\right) = p_i \qquad \text{for } i \in \{1, 2, 3\};$$
$$\mu\left(w_i \circ w_j\left(\triangle \right)\right) = p_i p_j \qquad \text{for } i, j \in \{1, 2, 3\};$$
$$\cdot$$
$$\mu\left(w_i \circ w_j \ldots \circ w_k\left(\triangle \right)\right) = p_i p_j \ldots p_k \qquad \text{for } i, j, \ldots, k \in \{1, 2, 3\}.$$

Define a simple function on $\triangle$ by

$$f(x + iy) = \begin{cases} 1 \text{ for } x + iy \in \triangle \text{ and } 1/3 \le x \le 1, \\ -1 \text{ for } x + iy \in \triangle \text{ and } 0 \le x \le 1/3. \end{cases}$$

Calculate $\int_{\triangle} f(z)\, d\mu(z)$, accurate to two decimal places.

Based on the ideas of Section 1 of this chapter, can you guess a method for calculating the integral which makes use of the Random Iteration Algorithm? Try it!

4.4. Show that if $\alpha, \beta \in \mathbb{R}$ and f, g are simple functions then $\alpha f + \beta g$ is a simple function, and

$$\alpha \int_{\mathbf{X}} f\, d\mu + \beta \int_{\mathbf{X}} g\, d\mu = \int_{\mathbf{X}} (\alpha f + \beta g)\, d\mu.$$

4.5. Black ink is printed to make this page. Let $\blacksquare \subset \mathbb{R}^2$ be a model for the page, and represent the ink by means of a Borel measure μ, so that $\mu(A)$ is the mass of ink associated with the set $A \subset \blacksquare$. Let $\mathscr{A} \in \mathscr{F}$ denote the smallest Borel set which contains all of the letters "a" on the page. Assume that the total mass of ink on the page is one unit. Estimate $\int_{\blacksquare} \chi_{\mathscr{A}}\, d\mu$.

4.6. Let Σ denote code space on two symbols $\{1, 2\}$. Let $\mathscr{B}$ denote the Borel field associated with (Σ, code space metric). Consider the IFS $\{\Sigma; w_1(x) = 1x, w_2(x) = 2x; p_1 = 0.4, p_2 = 0.6\}$, where "$1x$" means the string "$1x_1x_2x_3\ldots$" and "$2x$" means the string "$2x_1x_2x_3\ldots$". The attractor of the IFS is Σ. Let μ denote the unique measure on $\mathcal{B}$ such that

$$\mu(w_i \circ w_j \cdots \circ w_k(\Sigma)) = p_i p_j \cdots p_k \qquad \text{for } i, j, \ldots, k \in \{1,2\}.$$

Define sets A and B in $\mathcal{B}$ by

$$A = \{x \in \mathcal{B}: x_1 = 1\} \quad \text{and} \quad B = \{x \in \mathcal{B}: x_2 = 2\}.$$

Define $f: \Sigma \to \mathbb{R}$ by

$$f(x) = \chi_A(x) + (2.3)\chi_B(x) \qquad \text{for all } x \in \Sigma.$$

Evaluate the integral

$$\int_\Sigma f(x)\, d\mu(x).$$

Definition 3. Let $(\mathbb{X}, d)$ be a compact metric space, and let $\mathcal{B}$ denote the associated Borel field. Let μ be a Borel measure. A *partition* of $\mathbb{X}$ is a finite set of nonempty Borel sets, $\{A_i \in \mathcal{B}: i = 1, 2, \ldots, M\}$, such that $\mathbb{X} = \cup_{i=1}^M A_i$, and $A_i \cap A_j = \varnothing$ for $i \neq j$. The *diameter* of the partition is $\text{Max}\{\text{Sup}\{d(x, y): x, y \in A_i\}: i = 1, 2, \ldots, M\}$.

Theorem 1. *Let $(\mathbb{X}, d)$ be a compact metric space. Let $\mathcal{B}$ denote the associated Borel field. Let μ be a Borel measure on $\mathbb{X}$. Let $f: \mathbb{X} \to \mathbb{R}$ be continuous.* (i) *Let n be a positive integer. Then there exists a partition $\mathcal{B}_n = \{A_{n,m} \in \mathcal{B}: m = 1, 2, \ldots, M(n)\}$ of diameter $\leq 1/n$.* (ii) *Let $x_{n,m} \in A_{n,m}$ for $m = 1, 2, 3, \ldots$ and define a sequence of simple functions by*

$$f_n(x) = \sum_{m=1}^{M(n)} f(x_{n,m})\chi_{A_{n,m}}(x) \qquad \text{for } n = 1, 2, 3, \ldots$$

Then $\{f_n\}$ converges uniformly to $f(x)$. (iii) *The sequence $\{\int_{\mathbb{X}} f_n\, d\mu\}$ converges.* (iv) *The value of the limit is independent of the particular sequence of partitions, and of the choices of $x_{n,m} \in A_{n,m}$.*

Sketch of Proof. (i) Since $\mathbb{X}$ is compact, it is possible to cover $\mathbb{X}$ by a finite set of closed balls of diameter $1/n$, say $b_{n,1}, b_{n,2}, \ldots, b_{n,M(n)}$. We can assume that each ball contains a point which is in none of the other balls. Then define $A_{n,1} = b_{n,1}$, and $A_{n,j} = b_{n,j} \setminus \cup_{k=1}^{j-1} A_{n,k}$, for $j = 2, 3, \ldots, M(n)$. Then $\mathcal{B}_n = \{A_{n,m} \in \mathcal{B}: m = 1, 2, \ldots, M(n)\}$ is a partition of $\mathbb{X}$ of diameter $\leq 1/n$.

(ii) Let $\epsilon > 0$. f is continuous on a compact space, so it is uniformly continuous. It follows that there exists an integer $N(\epsilon)$ so that if $x, y \in \mathbb{X}$ and $d(x, y) \leq 1/N(\epsilon)$ then $|f(x) - f(y)| \leq \epsilon$. It follows that $|f(x) - f_n(x)| \leq \epsilon$ when $n \geq N(\epsilon)$.

(iii) It is readily proved that $\{\int_{\mathbb{X}} f_n \, d\mu\}$ is a Cauchy sequence. Indeed, for all $n, m \geq N(\epsilon)$ we have

$$\left|\int_{\mathbb{X}} f_n \, d\mu - \int_{\mathbb{X}} f_m \, d\mu\right| \leq \int_{\mathbb{X}} |(f_n - f_m)| \, d\mu \leq 2\epsilon\mu(\mathbb{X}).$$

It follows that the sequence converges.

(iv) Let $\{\tilde{f}_n\}$ be a sequence of simple functions, constructed as above. Then there is an integer $\tilde{N}(\epsilon)$ such that $|f(x) - \tilde{f}_n(x)| \leq \epsilon$ when $n \geq \tilde{N}(\epsilon)$. It follows that, for all $n \geq \text{Max}\{N(\epsilon), \tilde{N}(\epsilon)\}$,

$$\left|\int_{\mathbb{X}} f_n \, d\mu - \int_{\mathbb{X}} \tilde{f}_n \, d\mu\right| \leq \int_{\mathbb{X}} |(f_n - \tilde{f}_n)| \, d\mu \leq 2\epsilon\mu(\mathbb{X}).$$

This completes the sketch of the proof.

Definition 4. The limit in Theorem 9.4.1 is called the *integral* of f (with respect to μ). It is denoted by

$$\lim_{n\to\infty} \int_{\mathbb{X}} f_n \, d\mu = \int_{\mathbb{X}} f \, d\mu.$$

Exercises & Examples

4.7. Let $(\mathbb{X}, d)$ be a metric space. Let $a \in \mathbb{X}$. Define a Borel measure δ_a by $\delta_a(B) = 1$ if $a \in B$ and $\delta_a(B) = 0$ if $a \notin B$, for all Borel sets $B \subset \mathbb{X}$. This measure is referred to as "a delta function" and "a point mass at a." Let $f\colon \mathbb{X} \to \mathbb{R}$ be continuous. Show that

$$\int_{\mathbb{X}} f(x) \, d\delta_a(x) = f(a).$$

4.8. This example takes place in the metric space ($\blacksquare$, Euclidean). Let μ be the measure defined in exercise 9.4.2, and define $f\colon \blacksquare \to \mathbb{R}$ by $f(x, y) = x^2 + 2xy + 3$. Evaluate

$$\int_{\blacksquare} f \, d\mu.$$

4.9. Make an approximate evaluation of the integral $\int_{\triangle} x^2 \, d\mu(x)$ where μ and $\triangle$ are as defined in Exercise 9.4.3.

4.10. Let $\mathbb{X}$ denote the set of pixels corresponding to a certain computer graphics display device. Define a metric d on $\mathbb{X}$ so that $(\mathbb{X}, d)$ is a compact metric space. Give an example of a Borel subset of $\mathbb{X}$ and of a nontrivial Borel measure on $\mathbb{X}$. Show that any function $f\colon \mathbb{X} \to \mathbb{R}$ is continuous. Give a specific example of such a function, and evaluate $\int_{\mathbb{X}} f \, d\mu$.

9.5 THE COMPACT METRIC SPACE $(\mathscr{P}(\mathbb{X}), d)$

We introduce the most exciting metric space in the book. It is the space where fractals really live.

Definition 1. Let $(\mathbb{X}, d)$ be a compact metric space. Let μ be a Borel measure on $\mathbb{X}$. If $\mu(\mathbb{X}) = 1$ then μ is said to be *normalized.*

Definition 2. Let $(\mathbb{X}, d)$ be a compact metric space. Let $\mathscr{P}(\mathbb{X})$ denote the set of *normalized Borel measures* on $\mathbb{X}$. The *Hutchinson metric* d_H on $\mathscr{P}(\mathbb{X})$ is defined by

$$d_H(\mu, \nu)$$

$$= \text{Sup}\left\{ \int_{\mathbb{X}} f\,d\mu - \int_{\mathbb{X}} f\,d\nu \text{: } f\text{: } \mathbb{X} \to \mathbb{R}, f \text{ continuous}, |f(x) - f(y)| \le d(x, y) \forall x,\right.$$

for all $\mu, \nu \in \mathscr{P}(\mathbb{X})$.

Theorem 1. *Let $(\mathbb{X}, d)$ be a compact metric space. Let $\mathscr{P}(\mathbb{X})$ denote the set of normalized Borel measures on $\mathbb{X}$ and let d_H denote the Hutchinson metric. Then $(\mathscr{P}(\mathbb{X}), d_H)$ is a compact metric space.*

Sketch of proof. A direct proof, using the tools in this book, is cumbersome. It is straightfoward to verify that d_H is a metric. It is most efficient to use the concept of the "weak topology" on $\mathscr{P}(\mathbb{X})$ to prove compactness. One shows that this topology is the same as the one induced by the Hutchinson metric, and then applies Alaoglu's Theorem. See [Hutchinson 1981] and [Dunford 1966].

Exercises & Examples

5.1. Let K be a positive integer. Let $X = \{(i, j)\text{: } i, j = 1, 2, \ldots, K\}$. Define a metric on $\mathbb{X}$ by $d((i_1, j_1), (i_2, j_2)) = |i_1 - i_2| + |j_1 - j_2|$. Then $(\mathbb{X}, d)$ is a compact metric space. Let $\mu \in \mathscr{P}(\mathbb{X})$ be such that $\mu((i, j)) = (i + j)/(K^3 + K^2)$ and let $\nu \in \mathscr{P}(\mathbb{X})$ be such that $\nu(i, j) = 1/K^2$, for all $i, j \in \{1, 2, \ldots, K\}$. Calculate $d_H(\mu, \nu)$.

5.2. Prove Theorem 9.4.6 for the space $\mathbb{X}$ in exercise 9.5.1. (See also Exercise 9.4.10.)

5.3. Let $(\mathbb{X}, d)$ be a compact metric space. Let $\mu \in \mathscr{P}(\mathbb{X})$. Prove that the support of μ belongs to $\mathscr{H}(\mathbb{X})$.

9.6 A CONTRACTION MAPPING ON $\mathscr{P}(\mathbb{X})$

Let $(\mathbb{X}, d)$ denote a compact metric space. Let $\mathscr{B}$ denote the Borel subsets of $\mathbb{X}$. Let $w: \mathbb{X} \to \mathbb{X}$ be continuous. Then one can prove that $w^{-1}: \mathscr{B} \to \mathscr{B}$. It follows that if ν is a normalized Borel measure on $\mathbb{X}$, then so is $\nu \circ w^{-1}$. In turn, this implies that the function defined next indeed takes $\mathscr{P}(\mathbb{X})$ into itself.

Definition 1. Let $(\mathbb{X}, d)$ be a compact metric space and let $\mathscr{P}(\mathbb{X})$ denote the space of normalized Borel measures on $\mathbb{X}$. Let $\{\mathbb{X}; w_1, w_2, \ldots, w_N; p_1, p_2, \ldots, p_N\}$ be a hyperbolic IFS with probabilities. The *Markov operator* associated with the IFS is the function $M: \mathscr{P}(\mathbb{X}) \to \mathscr{P}(\mathbb{X})$ defined by

$$M(\nu) = p_1 \nu \circ w_1^{-1} + p_2 \nu \circ w_2^{-1} + \cdots + p_N \nu \circ w_N^{-1}$$

for all $\nu \in \mathscr{P}(\mathbb{X})$.

Lemma 1. *Let M denote the Markov operator associated with a hyperbolic IFS, as in Definition 9.6.1. Let $f: \mathbb{X} \to \mathbb{R}$ be either a simple function or a continuous function. Let $\nu \in \mathscr{P}(\mathbb{X})$. Then*

$$\int_{\mathbb{X}} f \, d(M(\nu)) = \sum_{i=1}^{N} p_i \int_{\mathbb{X}} f \circ w_i \, d\nu.$$

Proof. Suppose that $f: \mathbb{X} \to \mathbb{R}$ is continuous. By Theorem 9.5.1 we can find a sequence of simple functions $\{f_n\}$ which converges uniformly to f. Let n be a positive integer. It is readily verified that

$$\int_{\mathbb{X}} f_n \, d(M(\nu)) = \sum_{i=1}^{N} p_i \int_{\mathbb{X}} f_n \, d\nu \circ w_i^{-1} = \sum_{i=1}^{N} p_i \int_{w_i(\mathbb{X})} f_n \, d\nu \circ w_i^{-1} = \sum_{i=1}^{N} p_i \int_{\mathbb{X}} f_n \circ w_i \, d\nu.$$

The sequence $\{\int f_n \, d(M(\nu))\}$ converges to $\int f \, d(M(\nu))$.

For each $i \in \{1, 2, \ldots, N\}$ and each positive integer n, $f_n \circ w_i$ is a simple function. The sequence $\{f_n \circ w_i\}_{n=1}^{\infty}$ converges uniformly to $f \circ w_i$. It follows that $\{\int f_n \circ w_i \, d\nu\}_{n=1}^{\infty}$ converges to $\int f \circ w_i \, d\nu$. It follows that $\{\sum_{i=1}^{N} p_i \int f_n \circ w_i \, d\nu\}_{n=1}^{\infty}$ converges to $\sum_{i=1}^{N} p_i \int f \circ w_i \, d\nu$. This completes the proof.

Theorem 1. *Let $(\mathbb{X}, d)$ be a compact metric space. Let $\{\mathbb{X}; w_1, w_2, \ldots, w_N; p_1, p_2, \ldots, p_N\}$ be a hyperbolic IFS with probabilities. Let $s \in (0,1)$ be a contractivity factor for the IFS. Let $M: \mathscr{P}(\mathbb{X}) \to \mathscr{P}(\mathbb{X})$ be the associated Markov operator. Then M is a contraction mapping, with contractivity factor s, with respect to the Hutchinson metric on $\mathscr{P}(\mathbb{X})$. That is,*

$$d_H(M(\nu), M(\mu)) \le s \, d_H(\nu, \mu) \qquad \text{for all } \nu, \mu \in \mathscr{P}(\mathbb{X}).$$

In particular, there is a unique measure $\mu \in \mathscr{P}(\mathbb{X})$ such that

$$M\mu = \mu.$$

Proof. Let L denote the set of continuous functions $f\colon \mathbb{X} \to \mathbb{R}$ such that $|f(x) - f(y)| \le d(x, y)\ \forall x, y \in \mathbb{X}$. Then

$$d_H(M(\nu), M(\mu)) = \operatorname{Sup}\left\{\int f d(M(\mu)) - \int f d(M(\nu))\colon f \in L\right\}$$

$$= \operatorname{Sup}\left\{\int \sum_{i=1}^{N} p_i f \circ w_i \, d\mu - \int \sum_{i=1}^{N} p_i f \circ w_i \, d\nu\colon f \in L\right\}.$$

Let $\tilde{f} = s^{-1}\sum_{i=1}^{N} p_i f \circ w_i$. Then $\tilde{f} \in L$. Let $\tilde{L} = \{\tilde{f} \in L\colon \tilde{f} = s^{-1}\sum_{i=1}^{N} p_i f \circ w_i$, some $f \in L\}$. Then we can write

$$d_H(M(\nu), M(\mu)) = \operatorname{Sup}\left\{s\int \tilde{f} d\mu - s\int \tilde{f} d\nu\colon \tilde{f} \in \tilde{L}\right\}.$$

Since $\tilde{L} \subset L$, it follows that

$$d_H(M(\nu), M(\mu)) \le s\, d_H(\nu, \mu).$$

This completes the proof.

Definition 2. Let μ denote the fixed point of the Markov operator, promised by Theorem 1. μ is called the *invariant measure* of the IFS with probabilities.

We have arrived at our goal! This invariant measure is the object which we discussed informally in Section 1 of this chapter. *Now* we know what fractals are.

Exercises & Examples

6.1. Verify that the Markov operator associated with a hyperbolic IFS on a compact metric space indeed maps the space into itself.

6.2. This example uses the notation in the proof of Theorem 9.6.1. Let $f \in L$ and let $\tilde{f} = s^{-1}\sum_{i=1}^{N} p_i f \circ w_i$. Prove that $\tilde{f} \in L$.

6.3. Consider the hyperbolic IFS $\{\blacksquare \subset \mathbb{R}^2;\ w_1, w_2, w_3, w_4;\ p_1, p_2, p_3, p_4\}$ corresponding to the collage in Figure 9.6.1(a). Let M be the associated Markov operator. Let $\mu_0 \in \mathscr{P}(\mathbb{X})$, so that $\mu_0(\blacksquare) = 1$. For example, μ_0 could be the uniform measure, for which $\mu_0(S)$ is the area of $S \in \mathscr{P}(\blacksquare)$. We look at the sequence of measures $\{\mu_n = M^{\circ n}(\mu_0)\}$. The measure $\mu_1 = M(\mu_0)$ is such that $\mu(w_i(\blacksquare)) = p_i$ for $i = 1, 2, 3, 4$, as illustrated in Figure 9.6.1(b). It follows that $\mu_2 = M^{\circ 2}(\mu_0)$ obeys $\mu(w_i \circ w_j(\blacksquare)) = p_i p_j$ for $i, j = 1, 2, 3, 4$, as illustrated in Figure 9.6.1(c). We quickly get the idea. When the Markov operator is applied, the "mass" in a cell $\blacksquare_{ij\ldots k} = w_i \circ w_j \circ \cdots \circ w_k(\blacksquare)$ is redistributed among the four smaller cells $w_1(\blacksquare_{ij\ldots k})$, $w_2(\blacksquare_{ij\ldots k})$, $w_3(\blacksquare_{ij\ldots k})$, and $w_4(\blacksquare_{ij\ldots k})$. Also, mass from other cells is mapped into subcells of $\blacksquare_{ij\ldots k}$ in such a way that the total mass of $\blacksquare_{ij\ldots k}$ remains the same as before the Markov operator

Figure 9.6.1(a), (b) *A collage for an IFS of four maps. The attractor of the IFS is ■, and the probability of the map w_i is p_i for $i = 1, 2, 3, 4$. Let M denote the associated Markov operator. Let $\mu_0 \in \mathscr{P}(\mathbb{X})$ so that $\mu_0(■) = 1$. Then $\mu_1 = M(\mu_0)$ is a measure such that $\mu(w_i(■)) = p_i$ for $i = 1, 2, 3, 4$, as illustrated in (b). The measure $\mu_2 = M^{\circ 2}(\mu_0)$ is such that $\mu(w_i \circ w_j(■)) = p_i p_j$ for $i, j = 1, 2, 3, 4$, as illustrated in (c). See also Figures 9.6.2 and 9.6.3.*

(a)

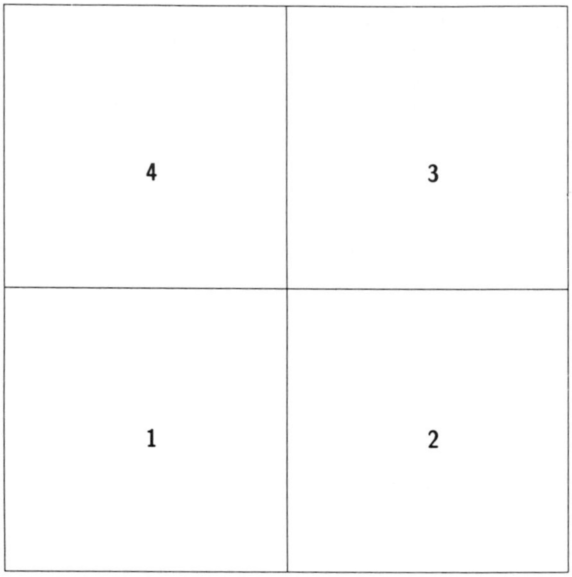

4	3
1	2

(b)

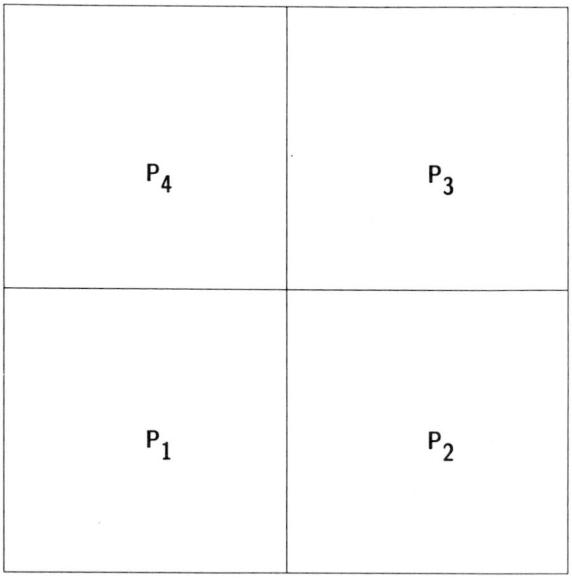

P_4	P_3
P_1	P_2

(c)

P_4P_4	P_4P_3	P_3P_4	P_3P_3
P_4P_1	P_4P_2	P_3P_1	P_3P_2
P_1P_4	P_1P_3	P_2P_4	P_2P_3
P_1P_1	P_1P_2	P_2P_1	P_2P_2

Figure 9.6.1(c)

was applied. In this manner the distribution of "mass" is defined on finer and finer scales as the Markov operator is repeatedly applied. What a wonderful idea. We have also illustrated this idea in Figures 9.6.2 and 9.6.3.

6.4. Apply the Random Iteration Algorithm to an IFS of the form considered in exercise 9.6.3. Choose the probabilities p_1, p_2, p_3, and p_4 so as to obtain a "picture" of the invariant measure which would occur at the end of the sequence which commences in Figures 9.6.3 (a), (b), (c), and (d).

6.5. Consider the IFS $\{[0,1] \subset \mathbb{R};\ w_1(x) = (0.5)x,\ w_2(x) = (0.7)x + 0.3;\ p_1 = 0.45,\ p_2 = 0.55\}$. The attractor of the IFS is $[0,1]$. Let M denote the associated Markov operator. Let $\mu_0 \in \mathscr{P}([0,1])$ be the uniform measure on $[0,1]$. The successive iterates $M(\mu_0)$, $M^{\circ 2}(\mu_0)$, $M^{\circ 3}(\mu_0)$ and $M^{\circ 4}(\mu_0)$ are represented in Figures 9.6.4 (a), (b), (c), and (d). Each measure is represented by a collection of rectangles whose bases are contained in the interval $[0,1]$. The area of a rectangle equals the measure of the base of the rectangle. Although the sequence of measures converges $\{M^{\circ n}(\mu_0)\}$ in the metric space $\{\mathscr{P}([0,1], d_H\}$ some of the rectangles would become infinitely tall as n tends to infinity.

6.6. Make a sequence of figures, analogous to Figures 9.6.4(a)–(d), to repre-

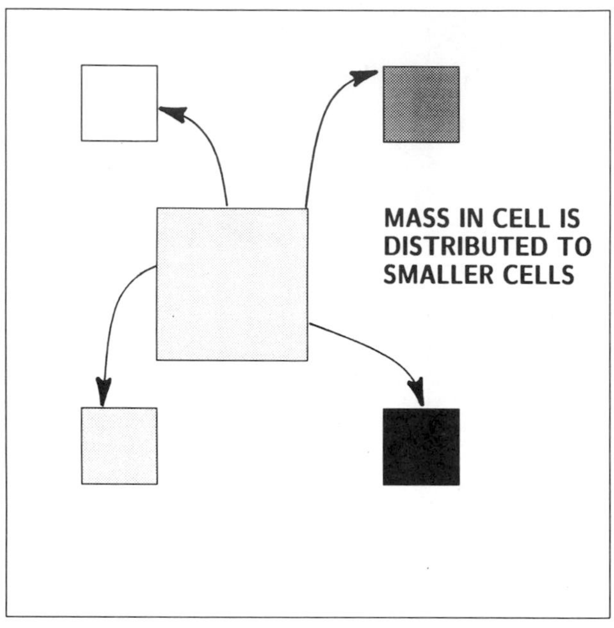

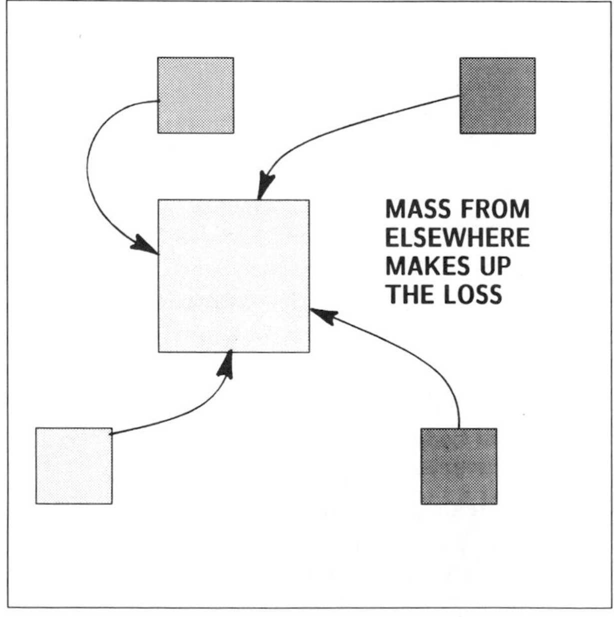

Figure 9.6.2(a), (b) *This illustrates the action of the Markov operator on one of the sequence of measures* $\{M^{\circ n}(\mu_0)\}$, *where* $\mu_0(\blacksquare) = 1$. *When the Markov operator is applied, the "mass" in a cell* $\blacksquare_{ij\ldots k} = w_i \circ w_j \circ \cdots \circ w_k(\blacksquare)$ *is redistributed among the four cells* $w_1(\blacksquare_{ij\ldots k})$, $w_2(\blacksquare_{ij\ldots k})$, $w_3(\blacksquare_{ij\ldots k})$, *and* $w_4(\blacksquare_{ij\ldots k})$. *Also, mass from other cells is mapped into subcells of* $\blacksquare_{ij\ldots k}$ *in such a way that the total mass of* $\blacksquare_{ij\ldots k}$ *remains the same as before the Markov operator was applied. In this manner the distribution of "mass" is defined on finer and finer scales as the Markov operator is repeatedly applied.*

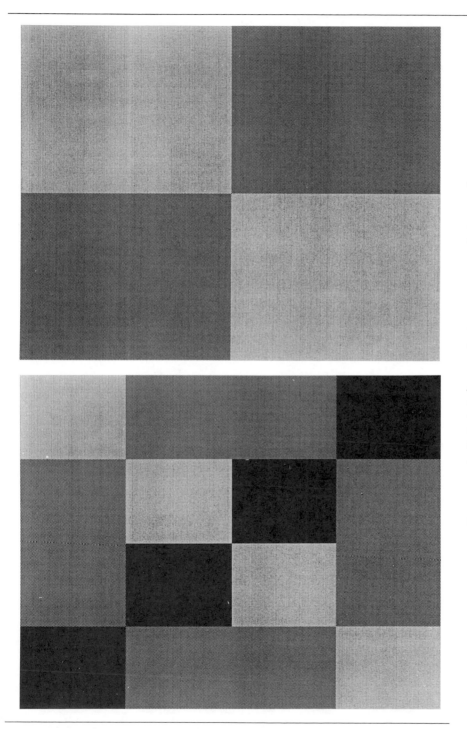

Figure 9.6.3(a)–(d) *This sequence of figures represents successive measures produced by iterative application of a Markov operator of the type considered in Figures* 9.6.1 *and* 9.6.2. *The result of one application of the operator to the uniform measure on* ■ *is represented in* (*a*). *Figures* (*b*), (*c*), *and* (*d*) *show the results of further successive applications of the Markov operator. The measures are represented in such a way as to keep the total number of dots constant. The measure of a set corresponds approximately to the number of dots which it contains. This represents the first few of a sequence of measures which converges in the metric space* ($\mathscr{P}$(■), d_H) *to the invariant measure of the IFS.*

Figure 9.6.3(b)

Figure 9.6.3(c)

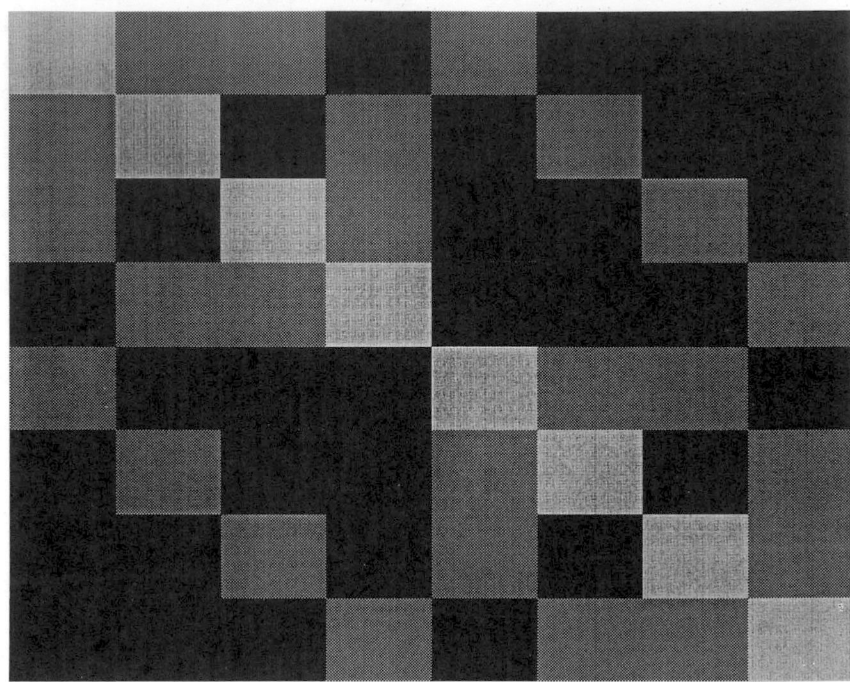

Figure 9.6.3(d)

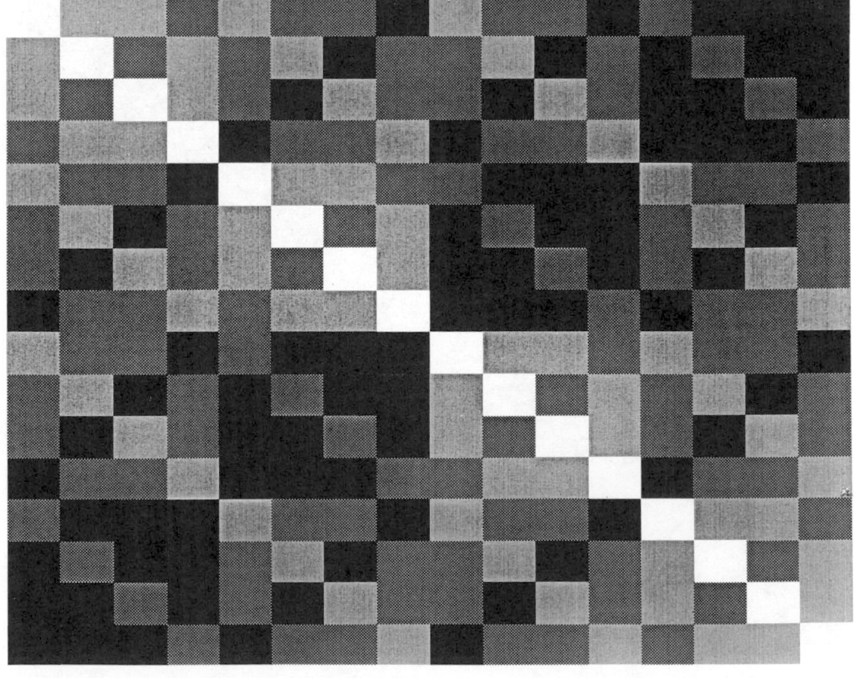

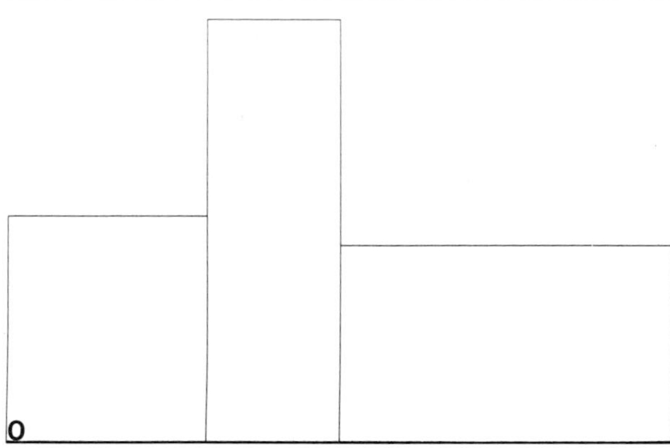

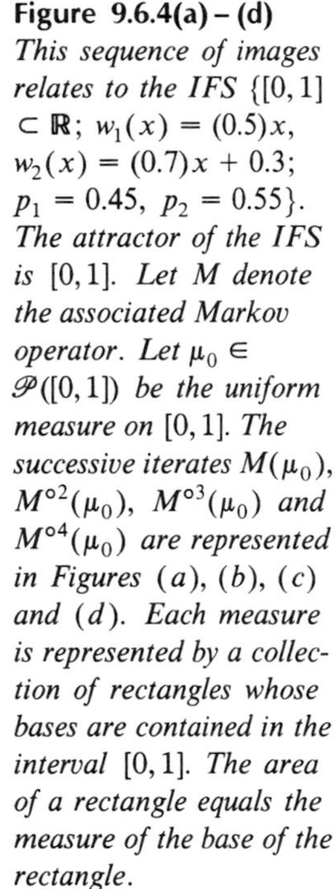

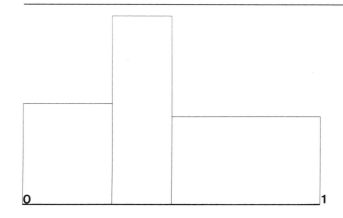

Figure 9.6.4(a) – (d) *This sequence of images relates to the IFS* $\{[0,1] \subset \mathbb{R};\ w_1(x) = (0.5)x,\ w_2(x) = (0.7)x + 0.3;\ p_1 = 0.45,\ p_2 = 0.55\}$. *The attractor of the IFS is* $[0,1]$. *Let* M *denote the associated Markov operator. Let* $\mu_0 \in \mathscr{P}([0,1])$ *be the uniform measure on* $[0,1]$. *The successive iterates* $M(\mu_0)$, $M^{\circ 2}(\mu_0)$, $M^{\circ 3}(\mu_0)$ *and* $M^{\circ 4}(\mu_0)$ *are represented in Figures* (a), (b), (c) *and* (d). *Each measure is represented by a collection of rectangles whose bases are contained in the interval* $[0,1]$. *The area of a rectangle equals the measure of the base of the rectangle.*

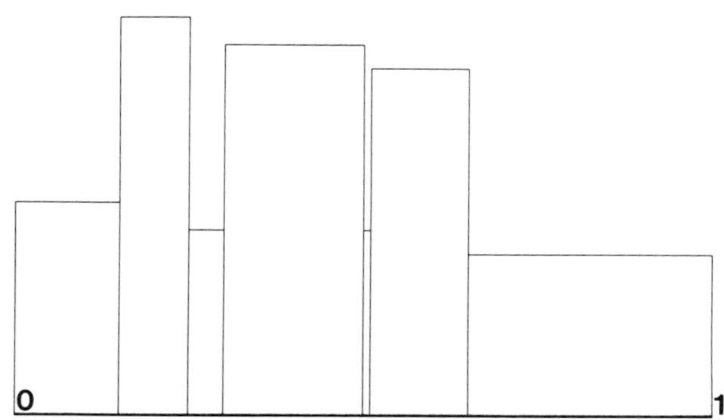

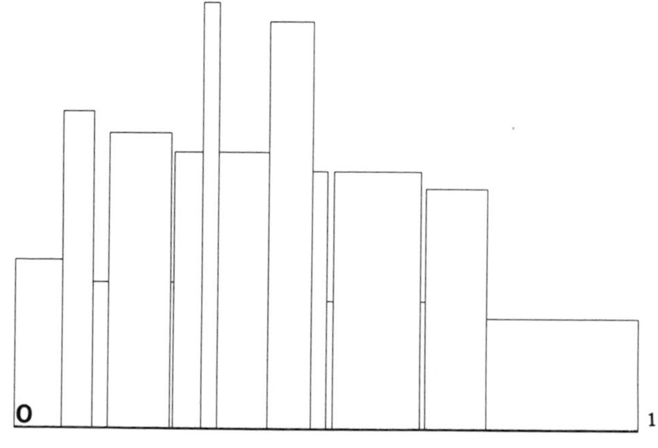

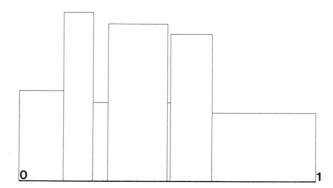

Figure 9.6.4(d)

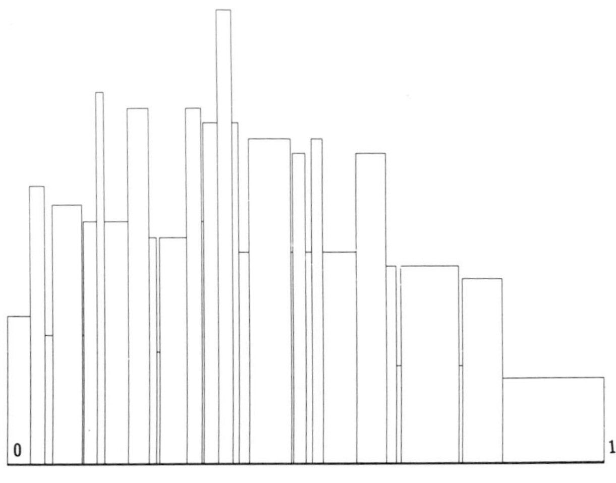

sent the Markov operator applied to the uniform measure μ_0, for each of the following IFS's with probabilities:

(i) $\{[0,1] \subset \mathbb{R};\ w_1(x) = (0.5)x,\ w_2(x) = (0.5)x + 0.5;\ p_1 = 0.5,$ $p_2 = 0.5\}$;

(ii) $\{[0,1] \subset \mathbb{R};\ w_1(x) = (0.5)x,\ w_2(x) = (0.5)x + 0.5;$ $p_1 = 0.99,\ p_2 = 0.01\}$;

(iii) $\{[0,1] \subset \mathbb{R};\ w_1(x) = (0.9)x,\ w_2(x) = (0.9)x + 0.1;$ $p_1 = 0.45,\ p_2 = 0.55\}$.

In each case describe the associated invariant measure.

6.7. Let $\mathbb{X} = \{A, B, C\}$ denote a space which consists of three points. Let $\mathcal{B}$ denote the σ-field which consists of all subsets of $\mathbb{X}$. Consider the IFS with probabilities $\{\mathbb{X};\ w_1, w_2;\ p_1 = 0.6,\ p_2 = 0.4\}$ where $w_1\colon \mathbb{X} \to \mathbb{X}$ is defined by $w_1(A) = B$, $w_1(B) = B$, $w_1(C) = B$, and $w_2\colon \mathbb{X} \to \mathbb{X}$ is defined by $w_2(A) = C$, $w_2(B) = A$, and $w_2(C) = C$. Let $\mathcal{P}(\mathbb{X})$ denote the set of normalized measures on $\mathcal{B}$. Let $\mu_0 \in \mathcal{P}(\mathbb{X})$ be defined by $\mu_0(A) = \mu_0(B) = \mu_0(C) = \frac{1}{3}$. Let M denote the Markov operator associated with the IFS, and let $\mu_n = M^{\circ n}(\mu_0)$ for $n = 1, 2, 3, \ldots$. Determine real numbers $a, b, c, d, e, f, g, h, i$ such that for each n

$$\begin{bmatrix} \mu_n(A) \\ \mu_n(B) \\ \mu_n(C) \end{bmatrix} = \begin{bmatrix} a & b & c \\ d & e & f \\ g & h & i \end{bmatrix} \begin{bmatrix} \mu_{n-1}(A) \\ \mu_{n-1}(B) \\ \mu_{n-1}(C) \end{bmatrix}.$$

Let $\tilde{M}$ denote the 3×3 matrix here. Explain how $\tilde{M}$ is related to M, and show that the invariant measure of the IFS can be described in terms of an eigenvector of $\tilde{M}$.

6.8. Let $\{\mathbb{X};\ w_1, w_2, \ldots, w_N;\ p_1, p_2, \ldots, p_N\}$ be a hyperbolic IFS with probabilities. Let μ denote the associated invariant measure. Let A denote the attractor of the IFS. Let $\mu_0 \in \mathscr{P}(\mathbb{X})$ be such that $\mu_0(A) = 1$. By considering the sequence of measures $\{\mu_n = M^{\circ n}(\mu_0)\}$, prove that

$$\mu\big(w_i \circ w_j \circ \cdots \circ w_k(A)\big) \geq p_i p_j \cdots p_k, \qquad \text{for all } i, j, \ldots, K \in \{1, 2, \ldots, N\}.$$

Show that if the IFS is totally disconnected then the equality sign holds.

Theorem 2. *Let $(\mathbb{X}, d)$ be a compact metric space. Let $\{\mathbb{X};\ w_1, w_2, \ldots, w_N;\ p_1, p_2, \ldots, p_N\}$ be a hyperbolic IFS with probabilities. Let μ be the associated invariant measure. Then the support of μ is the attractor of the IFS $\{\mathbb{X};\ w_1, w_2, \ldots, w_N\}$.*

Proof. Let B denote the support of μ. Then B is a nonempty compact subset of $\mathbb{X}$. Let A denote the attractor of the IFS. Then $\{A;\ w_1, w_2, \ldots, w_N;\ p_1, p_2, \ldots, p_N\}$ is a hyperbolic IFS. Let ν denote the invariant measure of the latter. Then ν is also an invariant measure for the original IFS. So, since μ is unique, $\nu = \mu$. It follows that $B \subset A$.

Let $a \in A$. Let $\mathcal{O}$ be an open set which contains a. We will use the notation of Theorem 4.2.1. Let Σ denote the code space associated with the IFS and let $\sigma \in \Sigma$ denote an address of a. It follows from Theorem 4.2.1 that $\text{Lim}_{n \to \infty}\phi(\sigma, n, A) = a$, where the convergence is in the Hausdorff metric. It follows that there is a positive integer n so that $\phi(\sigma, n, A) \subset \mathcal{O}$. But $\mu(\phi(\sigma, n, A)) \geq p_{\sigma_1} p_{\sigma_2} \cdots p_{\sigma_n} > 0$. It follows that $\mu(\mathcal{O}) > 0$. It follows that a is in the support of μ. It follows that $a \in B$. It follows that $A \subset B$. This completes the proof.

Theorem 3. (The Collage Theorem for Measures) *Let $\{\mathbb{X};\ w_1, w_2, \ldots, w_N;\ p_1, p_2, \ldots, p_N\}$ be a hyperbolic IFS with probabilities. Let μ be the associated invariant measure. Let $s \in (0, 1)$ be a contractivity factor for the IFS. Let $M\colon \mathscr{P}(\mathbb{X}) \to \mathscr{P}(\mathbb{X})$ be the associated Markov operator. Let $\nu \in \mathscr{P}(\mathbb{X})$. Then*

$$d_H(\nu, \mu) \leq \frac{d_H(\nu, M(\nu))}{(1 - s)}.$$

Proof. This is a corollary of Theorem 9.6.1.

We conclude this section with a description of the application of Theorem 9.6.3 to an inverse problem. The problem is to find an IFS with probabilities whose invariant measure, when represented by a set of dots, looks like a given texture.

A measure supported on a subset of $\mathbb{R}^2$ such as ■ can be represented by a lot of black dots on a piece of white paper. Figures 9.1.2 and 9.1.4 provide examples. The dots may be granules of carbon attached to the paper by means of a laser printer. The number of dots inside any circle of radius $\frac{1}{2}$ inch, say, should be approximately proportional to the measure of the corresponding ball in $\mathbb{R}^2$. A greytone image in a newspaper is made of small dots and can be thought of as representing a measure.

Let two such images, each consisting of the same number of points, be given. Then we expect that the degree to which they look alike corresponds to the Hutchinson distance between the corresponding measures.

Let such an image, L, be given. We imagine that it corresponds to a measure ν. Theorem 9.6.3 can be used to help find a hyperbolic IFS with probabilities whose invariant measure, represented with dots, approximates the given image. Let N be a positive integer. Let $w_i: \mathbb{R}^2 \to \mathbb{R}^2$ be an affine transformation, for $i = 1, 2, \ldots, N$. Let $\{\mathbb{R}^2; w_1, w_2, \ldots, w_N; p_1, p_2, \ldots, p_N\}$ denote the sought-after IFS. Let M denote the associated Markov operator.

Let $p_i \,\&\, L$ mean the set of dots L after the "density of dots" has been decreased by a factor p_i. For example, $0.5 \,\&\, L$ means L after "every second dot" in L has been removed. The action of the Markov operator on ν is represented by $\bigcup_{i=1}^{N} w_i(p_i \,\&\, L)$. This set consists of approximately the same number of dots as L. Then we seek contractive affine transformations and probabilities such that

$$\bigcup_{i=1}^{N} w_i(p_i \,\&\, L) \approx L. \tag{9.6.1}$$

That is, the coefficients which define the affine transformations and the probabilities must be adjusted so that the left-hand side "looks like" the original image.

Suppose we have found an IFS with probabilities so that (9.6.1) is true. Then generate an image $\tilde{L}$ of the invariant measure of the IFS, containing the same number of points as L. We expect that

$$\tilde{L} \approx L. \tag{9.6.2}$$

If the maps are sufficiently contractive, then the meaning of " $\approx$ " should be the same in both (9.6.1) and (9.6.2). These ideas are illustrated in Figure 9.6.5.

Exercises & Examples

6.9. Use the Collage Theorem for Measures to help find an IFS with probabilities for each of the images in Figure 9.6.6.

6.10. Let $\{\mathbb{X}; w_1, w_2, \ldots, w_N; p_1, p_2, \ldots, p_N\}$ be a hyperbolic IFS. Let μ denote the invariant measure. Let A denote the attractor. Let Σ denote the associated code space on the N symbols $\{1, 2, \ldots, N\}$. Let $T_i: \Sigma \to \Sigma$ be defined by $T_i(\sigma) = i\sigma$, for all $\sigma \in \Sigma$, for $i = 1, 2, 3, 4$. Let

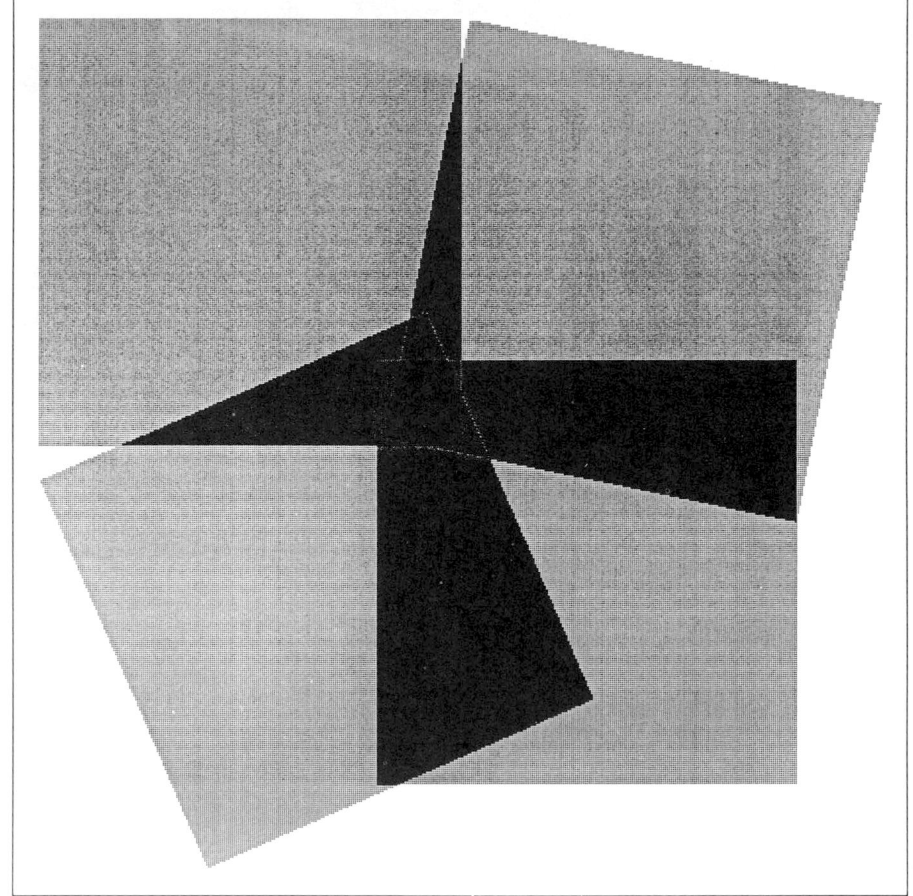

Figure 9.6.5
This illustration relates to the Collage Theorem for Measures. The shades of gray "add up" in the overlapping regions.

ρ denote the invariant measure for the hyperbolic IFS $\{\Sigma; T_1, T_2, T_3, T_4; p_1, p_2, p_3, p_4\}$. Let $\phi: \Sigma \to A$ denote the continuous map between code space and the attractor of the IFS introduced in Theorem 4.2.1. Prove that $\rho(\phi^{-1}(B)) = \mu(B)$ for all Borel subsets B of $\mathbb{X}$.

6.11. Figure 9.6.7 depicts the invariant measure for the IFS

$$\{[0,1] \subset \mathbb{R};\ w_1(x) = a_1 x,\ w_2(x) = a_2 x + e_2;\ p_1, p_2\},$$

where a_1, a_2 and e_2 are real constants such that the attractor is contained in $[0,1]$. The measure of a Borel subset of $[0,1]$ is approximately the amount of black which lies "vertically" above it. Find a_1, a_2, and e_2.

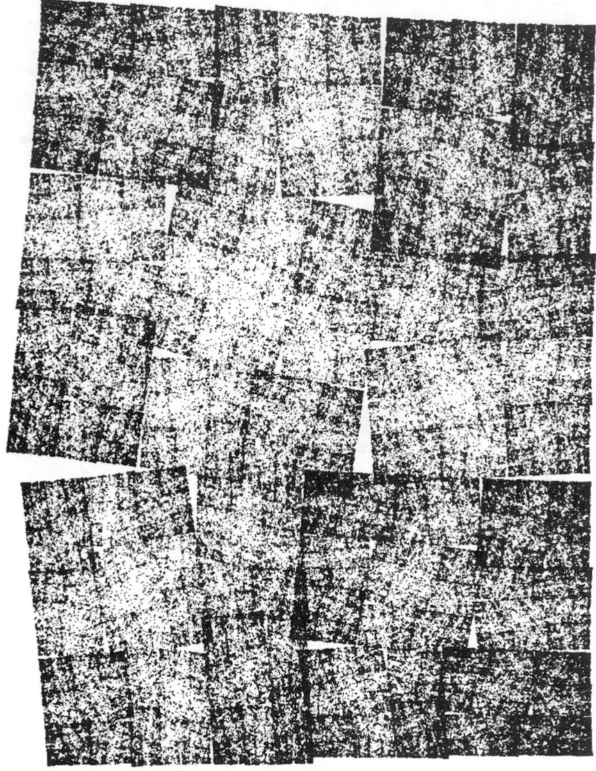

Figure 9.6.6(a)
Can you find the IFS and probabilities corresponding to this texture?

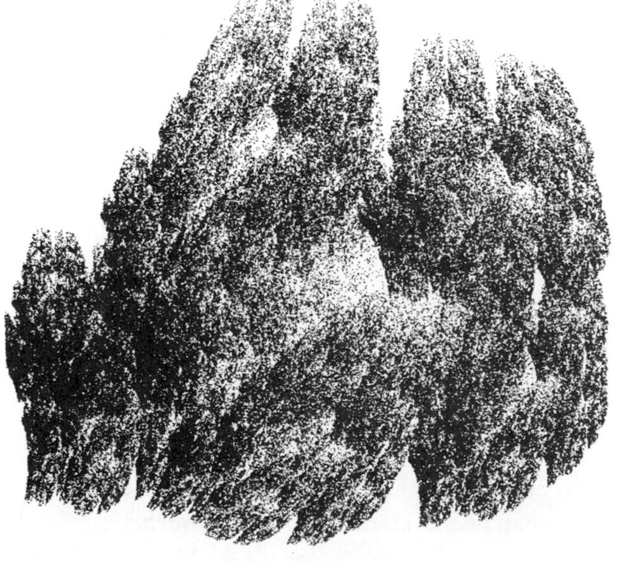

Figure 9.6.6(b)
Determine the IFS and probabilities for this cloud texture.

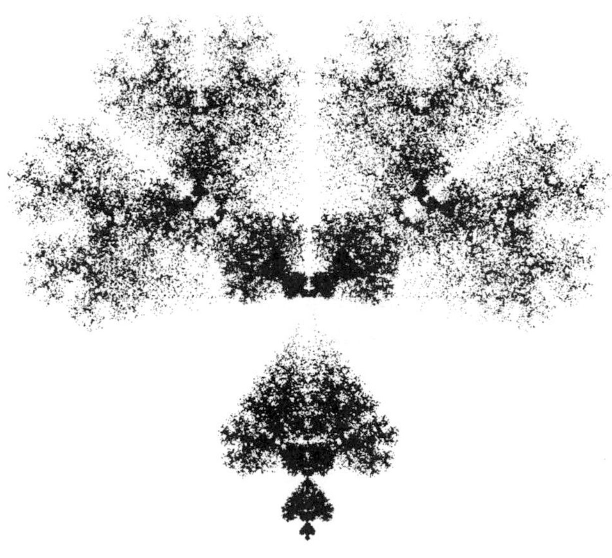

Figure 9.6.6(c)
Find the four affine maps and probabilities for this texture.

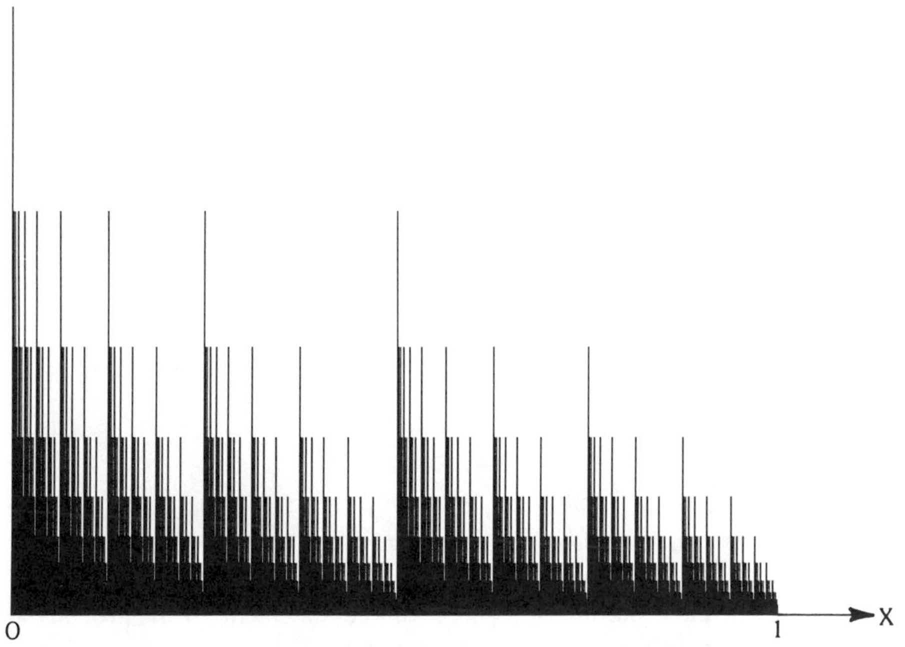

Figure 9.6.7
This figure depicts the invariant measure for the IFS: $\{[0,1] \subset \mathbb{R}; w_1(x) = a_1x, w_2(x) = a_2x + e_2; p_1, p_2\}$, where a_1, a_2, and e_2 are real constants such that the attractor is contained in $[0,1]$. The measure of a Borel subset of $[0,1]$ is approximately the amount of black which lies "vertically" above it. Can you find a_1, a_2, and e_2?

9.7 ELTON'S THEOREM

Both the following theorem and its corollary claim that certain events occur "with probability one." Although this has a very precise technical meaning, it is fine to interpret it in the same way as you would interpret the statement "There is 100% chance of rain tomorrow." After the statements we mention the mathematical framework which is used for dealing with probabilistic statements. To go further, we recommend reading parts of [Eisen 1969].

The theorem below is actually true when the p_i's are functions of x, the w_i's are only contraction mappings "on the average" and the space is "locally" compact.

Theorem 1. *Let $(\mathbb{X}, d)$ be a compact metric space. Let $\{\mathbb{X}; w_1, w_2, \ldots, w_N; p_1, p_2, \ldots, p_N\}$ be a hyperbolic IFS with probabilities. Let $(\mathbb{X}, d)$ be a compact metric space. Let $\{x_n\}_{n=0}^{\infty}$ denote an orbit of the IFS produced by the Random Iteration Algorithm, starting at x_0. That is,*

$$x_n = w_{\sigma_n} \circ w_{\sigma_{n-1}} \circ \cdots \circ w_{\sigma_1}(x_0),$$

where the maps are chosen independently according to the probabilities $p_1, p_2, \ldots, p_N$, for $n = 1, 2, 3, \ldots$. Let μ be the unique invariant measure for the IFS. Then, with probability one (that is, for all code sequences $\sigma_1, \sigma_2, \ldots$ except for a set of sequences having probability zero),

$$\lim_{n \to \infty} \frac{1}{n+1} \sum_{k=0}^{n} f(x_k) = \int_{\mathbb{X}} f(x)\, d\mu(x).$$

for all continuous functions $f: \mathbb{X} \to \mathbb{R}$ and all x_0.

Proof. See [Elton 1986].

Corollary 1. *Let B be a Borel subset of $\mathbb{X}$ and let μ(boundary of B) $= 0$. Let*

$$\mathcal{N}(B, n) = \text{number of points in } \{x_0, x_1, x_2, x_3, \ldots, x_n\} \cap B, \quad \text{for } n = 0, 1, 2, \ldots.$$

Then, with probability one,

$$\mu(B) = \lim_{n \to \infty} \left\{ \frac{\mathcal{N}(B, n)}{(n+1)} \right\},$$

for all starting points x_0. That is, the "mass" of B is the proportion of iteration steps, when running the Random Iteration Algorithm, which produce points in B.

We explain more deeply the context of the statement "with probability one." Let Σ denote the code space on the N symbols $\{1, 2, \ldots, N\}$. Let ρ

denote the unique Borel measure on Σ such that

$$\rho(C(\sigma_1, \sigma_2, \ldots, \sigma_m)) = p_{\sigma_1} p_{\sigma_2} \cdots p_{\sigma_m}$$

for each positive integer m and all $\sigma_1, \sigma_2, \ldots, \sigma_m \in \{1, 2, \ldots, N\}$, where

$$C(\sigma_1, \sigma_2, \ldots, \sigma_m) = \{\omega \in \Sigma: \omega_1 = \sigma_1, \omega_2 = \sigma_2, \ldots, \omega_m = \sigma_m\}.$$

Then $\rho \in \mathscr{P}(\Sigma)$. This measure provides a means for assigning probabilities to sets of possible outcomes of applying the Random Iteration Algorithm. Let us see how this works.

When the Random Iteration Algorithm is applied, an infinite sequence of symbols $\omega_1, \omega_2, \omega_3, \ldots$, namely a code $\omega = \omega_1\omega_2\omega_3 \ldots \in \Sigma$, is generated. Provided that we keep $x_0 \in \mathbb{X}$ fixed, we can describe the probabilities of orbits $\{x_n\}$ in terms of the probabilities of codes ω. So we examine how probabilities are associated to sets of codes.

The Random Iteration Algorithm is applied, and produces a code $\omega \in \Sigma$. What is the probability that $\omega_1 = 1$? Clearly it is $p_1 = \rho(C(1))$. What is the probability that $\omega_1 = \sigma_1$, $\omega_2 = \sigma_2, \ldots,$ and $\omega_n = \sigma_n$? Because the symbols are chosen independently, it is $\rho(C(\sigma_1, \sigma_2, \ldots, \sigma_m)) = p_{\sigma_1} p_{\sigma_2} \cdots p_{\sigma_m}$. Let B denote a Borel subset of Σ. What is the probability that the Random Iteration Algorithm produces a code $\sigma \in B$? It is at least intuitively reasonable that it is $\rho(B)$. This can be formalized; see for example [Eisen 1969]. The measure ρ provides a means of describing the probabilities of outcomes of the Random Iteration Algorithm.

Here is a heavy way of stating the central part of Theorem 1. "... Let $B \subset \Sigma$ denote the set codes $\sigma \in \Sigma$ such that

$$\lim_{n \to \infty} \frac{1}{n+1} \sum_{k=0}^{n} f(x_k) = \int_{\mathbb{X}} f(x)\, d\mu(x),$$

for all $x_0 \in \mathbb{X}$ and all continuous functions $f: \mathbb{X} \to \mathbb{R}$, where

$$x_n = w_{\sigma_n} \circ w_{\sigma_{n-1}} \circ \cdots \circ w_{\sigma_1}(x_0).$$

Then B is a Borel subset of Σ and $\rho(B) = 1$."

A similar equivalent restatement of the Corollary can be made.

Exercises & Examples

7.1. This example concerns the IFS $\{[0,1]; \frac{1}{2}x, \frac{1}{2}x + \frac{1}{2}; 0.5, 0.5\}$. Show that the invariant measure μ is such that $\mu([x, x+\delta]) = \delta$ when $[x, x+\delta]$ is a subinterval of $[0,1]$. Deduce that if $f: [0,1] \to \mathbb{R}$ is a continuous function then

$$\int_0^1 f(x)\, dx = \int_{[0,1]} f\, d\mu.$$

Let $f(x) = 1 + x^2$. Compute approximations to the latter integral with the aid of Elton's theorem and the Random Iteration Algorithm. Compare your results with the exact value $\frac{4}{3}$.

7.2. This example concerns the IFS $\{\blacksquare \subset \mathbb{R}^2;\ w_1, w_2, w_3, w_4; 0.25, 0.25, 0.25, 0.25\}$ corresponding to the collage in Figure 9.6.1(a). Let μ denote the invariant measure. Argue that μ is the uniform measure which assigns "measure" $dx\,dy$ to an infinitesimal rectangular cell of side lengths dx and dy. Use Elton's theorem and the Random Iteration Algorithm to evaluate approximations to

$$\int_{\blacksquare}(x^2 + 2xy + 3y^2)\,dx\,dy.$$

Compare your approximations with the exact value.

7.3. This example concerns the IFS

$$\left\{\triangle \subset \mathbb{R}^2;\ w_1(x, y) = \left(\tfrac{1}{2}x, \tfrac{1}{2}y\right),\ w_2(x, y) = \left(\tfrac{1}{2}x, \tfrac{1}{2}y + \tfrac{1}{2}\right),\right.$$

$$\left. w_3(x, y) = \left(\tfrac{1}{2}x + \tfrac{1}{2}, \tfrac{1}{2}y\right);\ \tfrac{1}{3}, \tfrac{1}{3}, \tfrac{1}{3}\right\}.$$

where $\triangle$ is the attractor of the IFS, our old friend. Let μ denote the invariant measure of the IFS. Argue that μ provides a good concept of a "uniform" measure on $\triangle$. Use Elton's theorem and the Random Iteration Algorithm to compute approximations to

$$\int_{\triangle}(x^2 + 2xy + 3y^2)\,dx\,dy.$$

9.8 APPLICATION TO COMPUTER GRAPHICS

We begin by illustrating how a color image of the invariant measure of an IFS with probabilities, can be produced. The idea is very simple. We start from an IFS such as

$$\{\mathbb{C}; 0.5z + 24 + 24i, 0.5z + 24i, 0.5z; 0.25, 0.25, 0.5\}.$$

A viewing window and a corresponding array of pixels P_{ij} is specified. The Random Iteration Algorithm is applied to the IFS, to produce an orbit $\{z_n : n = 0, 1, \ldots, \mathit{numits}\}$, where *numits* is the number of iterations. For each (i, j) the number of points $\mathcal{N}(P_{ij})$, which lie in the pixel P_{ij}, are counted. The pixel P_{ij} is assigned the value $\mathcal{N}(P_{ij})/\mathit{numits}$. By Elton's theorem, if *numits* is large, this value should be a good approximation to the measure of the pixel. The pixels are plotted on the screen in colors which are determined from their measures.

The following program implements this procedure. It is written in BASIC. It runs without modification on an IBM PC with Enhanced Graphics Adaptor and Turbobasic.

Program 9.8.1 (Uses the Random Iteration Algorithm to Make a "Picture" of the Invariant Measure Associated with an IFS with Probabilities.)

```
screen 9: cls   'Initialize graphics.
dim s(51, 51)   'Allocate array of pixels.
a(1) = 0.5: b(1) = 0: c(1) = 0: d(1) = 0.5: e(1) = 24: f(1) = 24   'IFS
                                   code for a Sierpinski triangle.
a(2) = 0.5: b(2) = 0: c(2) = 0: d(2) = 0.5: e(2) = 0: f(2) = 24
a(3) = 0.5: b(3) = 0: c(3) = 0: d(3) = 0.5: e(3) = 0: f(3) = 0
p(1) = 0.25: p(2) = 0.25: p(3) = 0.5   'Probabilities for the IFS; they must
                                              add to one!
mag = 1   'Magnification factor.
numits = 5000   'Increase the number of iterations as you magnify.
factor = 100   'Scales pixel values to color values.
numcols = 8   'This is the number of colors you are able to use.
for n = 1 to numits   'Random iteration begins!
r = rnd: k = 1   'Pick a number in [0, 1] at random.
if r > p(1) then k = 2
if r > p(1) + p(2) then k = 3
newx = a[k]*x + b[k]*y + e[k]   'Map k is picked with probability p(k).
newy = c[k]*x + d[k]*y + f[k]
x = newx: y = newy
i = int(mag*x): j = int(mag*y)   'Scale by magnification factor.
if (((i < 50) and (i >= 0)) and ((0 =< j) and (j < 50))) then   'If the
                                                scaled value is...
s(i,j) = s(i,j) + 1   '...in the array add one to pixel (i,j).
end if
pset(i,j)   'Plot the point.
if instat then end   'Stop if a key is pressed.
next
for i = 0 to 49   'Normalize values in pixel array, and plot...
for j = 0 to 49   '...in colors corresponding to the normalized...
col = s(i,j)*numcols*factor*mag*mag/numits   '...values of the
                                                numbers s(i,j).
pset(i,j), col   'Plot the pixel (i,j) in the color determined by...
next j   '...its measure.
next i
end
```

The program allows the user to zoom in on a piece of the rendered measure by altering the value of the magnification parameter *mag*. The result of running an adaptation of this program on a Masscomp workstation, and then printing the contents of the graphics screen, is shown in Figure 9.8.1.

Rendered invariant measures for IFS's acting in $\mathbb{R}^2$ are also shown in Figures 9.8.2 (a) and (b).

By carrying out some simple computergraphical experiments, using a program such as the one above, we discover that "pictures" of invariant measures of IFS's possess a number of properties. (i) Once the viewing window and color assignments have been fixed, the image produced is stable with respect to the number of iterations, provided that the number of iterations is sufficiently large. (ii) Images vary consistently with respect to translation and rotation of the viewing window, and with respect to changes in resolution. In particular, they vary consistently when they are magnified. (iii) The images depend continuously on the IFS code, including the probabilities. Property (i) ensures that the images are well-defined. The properties in

Figure 9.8.1
The result of running a modified version of Program 9.8.1 *and then printing the contents of the graphics screen in greytones. A rendered picture of a measure is the result.*

Figure 9.8.2(a), (b) *Rendered invariant measures for IFS's of two maps.*

(ii) are also true for views of the real world seen through the viewfinder of a camera. Property (iii) means that images can be controlled interactively. These properties suggest that IFS theory is applicable to computer graphics.

We should, if we have done our measure theory homework, understand the reasons for (i) and (ii). They are consequences of corresponding properties Borel of measures on $\mathbb{R}^2$. Property (iii) follows from a theorem by Withers [Withers 1987].

Exercises & Examples

8.1. Rewrite Program 9.8.1 in a form suitable for your own computer environment. Adjust *numits* and *factor* to ensure that a stable image results. Then make experiments to verify that the conditions (i)–(iii) above are verified. For example, to test the consistency of images with respect to changes in resolution, you should try *mag* = 0.5, 1, and 1.5.

Figure 9.8.2*b*

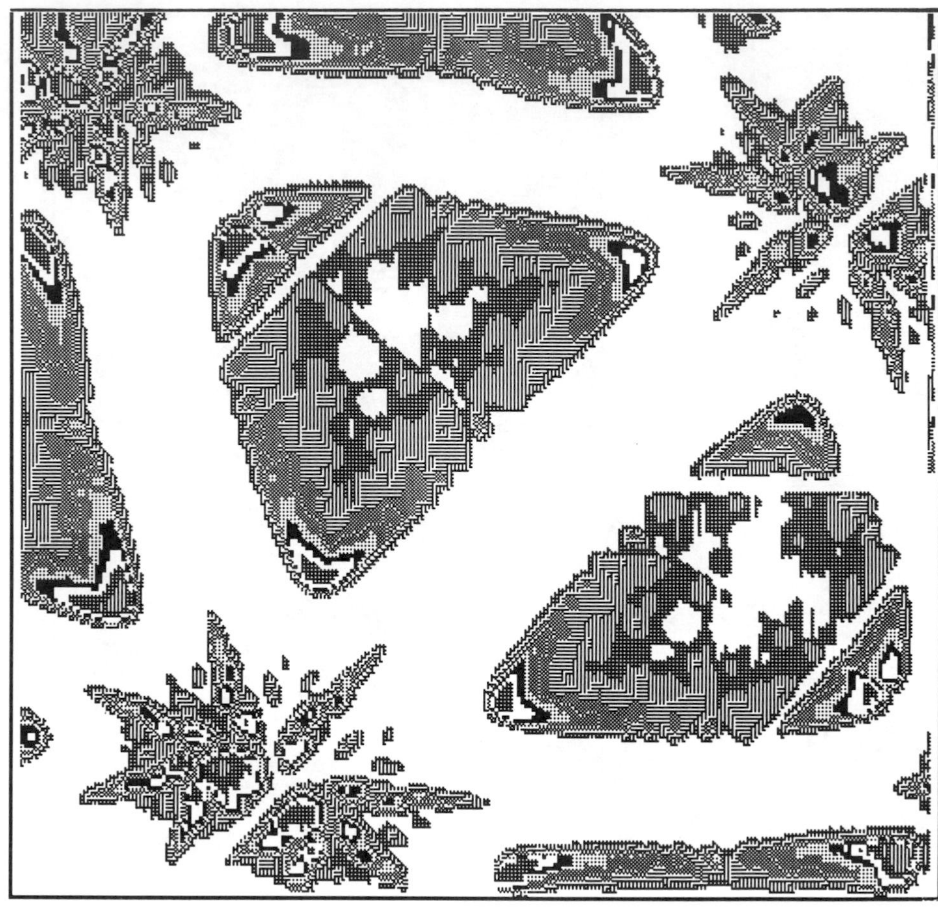

Unless you have a very powerful system, do not make extreme adjustments. For example, do not choose *mag* too small, otherwise you will need a very large value for *numits*.

Applications of fractal geometry to computer graphics have been investigated by a number of authors including [Mandelbrot 1982], [Kawaguchi 1982], [Oppenheimer 1986], [Fournier *et al.* 1982], [Smith 1984], [Miller 1986], and [Amburn *et al.* 1986]. In all cases the focus has been on the modelling of natural objects and scenes. Both deterministic and random geometries have been used. The application of IFS theory to computer graphics was first reviewed in [Demko 1985]. It provides a single framework which can reach an unlimited range of images. It is distinguished from other fractal approaches because it is the only one which uses measure theory.

The modeling of natural scenes is an important area of computer graphics. Photographs of natural scenes contain redundant information in the form of subtle patterns and variations. There are two characteristic features. (i) The presence of complex geometrical structure and distributions of color and brightness at many scales: Natural boundaries and textures are not smoothed out under magnification; they preserve some degree of geometrical complexity. (ii) Natural scenes are organized in hierarchical structures. For example, a forest is made of trees; a tree is a collection of boughs and limbs along a trunk; on each branch there are clusters of leaves; and a single leaf is filled with veins and covered with fine hairs. It appears often in a natural scene that a recognizable entity is built up from numerous near repetitions of some smaller structure. These two observations can be integrated into systems for modeling images using IFS theory.

Exercises & Examples

8.2. Examine a good quality color photograph of a natural scene, such as can be found in a *Sierra Club* calendar, or an issue of *National Geographic*. Discuss the extent to which (i) and (ii) are true for that photograph. Be specific.

In [Barnsley 1988a] it is reported that IFS theory can be used efficiently to model photographs of clouds, mountains, ferns, a field of sunflowers, a forest, seascapes and landscapes, a hat, the face of a girl, and a glaring arctic wolf.

There are two parts to making any computer graphics image. They are geometrical modeling and rendering. Consider an architect making a computergraphical house: first she defines the dimensions of the floor, the roof, the windows, the shapes of the gables, and so on, to produce the geometrical model. Traditionally this is specified in terms of polygons, circles, and other classical geometrical objects which can be conveniently input to the computer. This model is not a picture. To make a picture, the model must be projected into two-dimensions from a certain point of view and distance, discretized so that it can be represented with pixels, and finally rendered in colors on a display device.

Here we describe briefly the software system designed by the author, Alan Sloan, and Laurie Reuter, which was used to produce the color images which accompany this section. More details can be found in [Reuter 1987] and [Barnsley 1988a]. The system consists of two subsystems known as *Collage* and *Seurat*. *Collage* is used for geometrical modeling, while *Seurat* is used for rendering.

Collage and *Seurat* process IFS structures of the form

$$\left\{\mathbb{R}^2 ; w_1, w_2, \ldots, w_N ; p_1, p_2, \ldots, p_N : n = 1, 2, \ldots, N\right\},$$

where the maps are affine transformations in $\mathbb{R}^2$. An IFS is represented by a file which consists of an IFS code, where each coefficient is written with a fixed number of bits. Let μ denote the invariant measure of such an IFS and let A denote the attractor. The pair (A, μ) is referred to as an *underlying model*. The attractor A carries the geometry, while μ carries the rendering information. One can think of the IFS code, or equivalently (A, μ), as being analogous to the plans of an architect. It corresponds to many different pictures.

Collage is a geometrical modeling system which is used to determine the coefficients of the affine transformations $w_1, w_2, \ldots, w_N$. It is based on the Collage Theorem. *Seurat* is a software system for rendering images starting from an IFS code. An image is produced once a viewing window, color table, and resolution have been specified. This is achieved using the Random Iteration Algorithm. Its mathematical basis is Elton's Theorem. *Seurat* is also used in an interactive mode to determine the probabilities and color values.

The input to *Collage* is a target image, which we denote here by T. For example, T may be a polygonal approximation to a leaf. We suppose that $T \subset \blacksquare = \{(x, y) \in \mathbb{R}^2 \colon\ 0 \le x \le 1, 0 \le y \le 1\}$, and that the screen of the computer display device corresponds to $\blacksquare$. T is rendered on the graphics workstation monitor. An affine transformation

$$w_1(x, y) = \begin{pmatrix} a_1 & b_1 \\ c_1 & d_1 \end{pmatrix} \begin{pmatrix} x \\ y \end{pmatrix} + \begin{pmatrix} e_1 \\ f_1 \end{pmatrix} = A_1 x + t_1$$

is introduced, with coefficients initialized at $a_1 = d_1 = 0.25$, $b_1 = c_1 = e_1 = f_1$. The image $w_1(T)$ is displayed on the monitor in a different color from T. $w_1(T)$ is a quarter-sized copy of T, centered closer to the point $(0, 0)$. The user now interactively adjusts the coefficients with a mouse or some other interaction technique, so that the image $w_1(T)$ is variously translated, rotated, and sheared on the screen. The goal of the user is to transform $w_1(T)$ so that it lies over part of T. It is important that the dimensions of $w_1(T)$ are smaller than those of T, to ensure that w_1 is a contraction. Once $w_1(T)$ is suitably positioned, it is fixed, and a new subcopy of the target, $w_2(T)$, is introduced. w_2 is adjusted until $w_2(T)$ covers a subset of those pixels in T which are not in $w_1(T)$. Overlap between $w_1(T)$ and $w_2(T)$ is allowed, but in general it should be made as small as possible, for efficiency. New maps are added and adjusted until $\bigcup_{j=1}^{N} w_j(T)$ is a good approximation to T. The output from *Collage* is the IFS resulting code. The probability p_j is chosen proportional to $|a_j d_j - b_j c_j|$ if this number is non-zero, and equal to a small positive number if the determinant of A_j equals zero.

The input to *Seurat* is one or more IFS codes generated by *Collage*. The viewing window and the number of iterations are specified by the user. The measures of the pixels are computed. The resulting numbers are multiplied by the inverse of the maximum value so that all of them lie in [0, 1]. Colors are assigned to numbers in [0, 1] using a color assignment function. The default is a greyscale where the intensity is proportional to the number, such that 0 corresponds to black and 1 corresponds to brightest white. The coloring and texture of the image can be controlled through the probabilities and the color assignment function. Although one does not explicitly use it, Theorem 9.6.3 lies in the background and can help in the adjustment of the probabilities.

Color Plate 9.8.1 shows some smoking chimneys in a landscape. The IFS codes for the elements of this image we obtained using *Collage*. Different color assignment functions are associated to different elements in the image. The image was rendered using *Seurat*.

The consistency of images with respect to changes in resolution is illustrated in Color Plate 9.8.2, which shows a zoom on one of the smokestacks in Color Plate 9.8.1. The number of iterations must be increased with magnification to keep the number of points landing within the viewing window constant. This requirement ensures the consistency of the textures in an image throughout the magnification process.

Color Plates 9.8.3 and 9.8.4 show various renderings of leaves produced by *Seurat*.

Color Plate 9.8.5 shows a sequence of frames taken from an IFS encoded movie entitled *A Cloud Study* [Barnsley 1987a]. The smooth transition from frame to frame is a consequence of the continuous dependence on parameters of the invariant measure of the IFS for the cloud.

Color Plates 9.8.6, 9.8.7, and 9.8.8, were encoded from color photographs. Segmentation according to color was performed on the originals to define textured pieces. IFS codes for these components were obtained using *Collage*. The IFS database contained less than 180 maps for the Monterey seascape, and less than 160 maps for the Andes Indian girl.

The two primitives, a leaf and a flower, in Color Plate 9.8.9, were used as condensation sets in the picture *Sunflower Field*, Color Plate 9.8.10. Here we see the hierarchical structure: the leaf is itself the attractor of an IFS; and the flower is an overlay of four IFS attractors. The leaf is a condensation set for the IFS which generates all of the leaves. The flower is a condensation set to an IFS which generates many flowers, converging to the horizon. In the pictures *Sunflower Field* and *Black Forest*, shown in Color Plates 9.8.11, 9.8.12, 9.8.13, and 9.8.14, the primitives were displayed from back to front. The databases for *Sunflower Field* and *Black Forest* contain less than 100 and 120 maps, respectively. Notice the shadows behind the little trees in the background in Color Plate 9.8.12. The winter forest pictures were obtained by adjusting the color assignment function. The important point is that once the adjustment has been made, the image and the zoom are consistent.

Exercises & Examples

8.3. Use the Collage Theorem to help you find an IFS code for a leaf. Adjust your version of Program 9.8.1 to allow you to render images of associated invariant measures. Assign colors in the range from red through orange to green. Adjust the probabilities. Obtain a spectacular color picture of the leaf showing the veins. Make a color slide of the output. To photograph a picture on the screen of a computergraphics monitor, use a telephoto lens. Mount the camera on a tripod, and take the photograph in a darkened room, on Ektachrome 64 ASA color slide film, 0.1 sec exposure, f-stop 5.6. For possible publication, submit the color slide, together with a letter of copyright assignment, to Michael Barnsley, School of Mathematics, Georgia Institute of Technology, Atlanta, Georgia, 30332 USA. Include a self-addressed envelope.

8.4. Obtain a very powerful computer with good graphics. Find the heirarchical IFS codes for the *Sunflower Field*. Replace the sunflowers by roses. Fly into your picture, to explore forever that scent-filled horizon. You are on your own.

References

AMBURN, P., GRANT, E., AND WHITTED, T., "Managing Geometric Complexity with Enhanced Procedural Methods," *Computer Graphics 20*(4): (August 1986).

AONO, M., AND KUNII, T. L., "Botanical Tree Image Generation," *IEEE Computer Graphics and Applications*, *4*(5): 10–33 (May 1984).

BARNSLEY, M. F., GERONIMO, J. S., AND HARRINGTON, A. N., "Geometry and Combinatorics of Julia Sets for Real Quadratic Maps," *Journal of Statistical Physics 37*:51–92 (1984).

BARNSLEY, M. F. AND DEMKO, S., "Iterated Function Systems and the Global Construction of Fractals," *The Proceedings of the Royal Society of London A399*: 243–275 (1985).

BARNSLEY, M. F., ERVIN, V., HARDIN, D., AND LANCASTER, J., "Solution of an Inverse Problem for Fractals and Other Sets," *Proceedings of the National Academy of Science*, *83*:1975–1977 (April 1985).

BARNSLEY, M. F., AND HARRINGTON, A. N., "A Mandelbrot Set for Pairs of Linear Maps," *Physica 15D*: 421–432 (1985).

BARNSLEY, M. F., "Fractal Functions and Interpolation," *Constructive Approximation*, *2*:303–329 (1986).

BARNSLEY, M. F., JACQUIN, A., REUTER, L., AND SLOAN, A. D., "Harnessing Chaos for Image Synthesis." *Computer Graphics*, SIGGRAPH 1988 Conference Proceedings, 1988a.

BARNSLEY, M. F., AND SLOAN A. D., "A Better Way to Compress Images," *Byte Magazine*: 215–223, January 1988b.

BARNSLEY, M. F., CAIN, G., KASRIEL, R. K., "The Escape Time Algorithm." Preprint, School of Mathematics, Georgia Institute of Technology, 1988c.

BARNSLEY, M. F., HARDIN, D. P., "A Mandelbrot Set Whose Boundary is Piecewise Smooth," to appear in the *Transactions of the American Mathematical Society*, 1988d.

BARNSLEY, M. F., AND ELTON, J., "A New Class of Markov Processes for Image Encoding," *Journal of Applied Probability*, *20*:14–32 (1988e).

BARNSLEY, M. F., ELTON, J., HARDIN, D., AND MASSOPUST, P., "Hidden Variable Fractal Interpolation Functions," Georgia Institute of Technology Preprint, July 1986, to appear in *SIAM Journal of Analysis* (1988f).

BARNSLEY, M. F., JACQUIN, A., REUTER, L., AND SLOAN, A. D., "A Cloud Study." A videotape produced by the Computergraphical Mathematics Laboratory at Georgia Institute of Technology.

BILLINGSLEY, P., *Ergodic Theory and Information*, John Wiley and Sons, New York, 1965.

BEDFORD, T. J., "Dimension and Dynamics for Fractal Recurrent Sets," *Journal of the London Mathematical Society 2*(33):89–100 (1986).

BLANCHARD, P., "Complex Analytic Dynamics On The Riemann Sphere," *Bulletin of the American Mathematical Society 11*:88–144 (1984).

BROLIN, H., "Invariant Sets Under Iteration of Rational Functions," *Arkiv För Matematik*, Band 6 nr 6: 103 to 144.

BROWN, J. R., *Ergodic Theory and Topological Dynamics*, Academic Press, New York, 1976.

CURRY, J., GARNETT, L., SULLIVAN, D., "On the Iteration of Rational Functions: Computer Experiments with Newton's Method," *Communications In Mathematical Physics 91*:267–277 (1983).

DEMKO, S., HODGES, L., AND NAYLOR, B., "Construction of Fractal Objects with Iterated Function Systems," *Computer Graphics 19*(3):271–278 (July 1985). SIGGRAPH 1985 Proceedings.

DEVANEY, R., *An Introduction to Chaotic Dynamical Systems*, Addison-Wesley, New York, 1986.

DEWDNEY, A. K., "Beauty and Profundity: The Mandelbrot Set and a Flock of its Cousins called Julia," *Scientific American 257*:140–146 (November 1987).

DIACONIS, P. M., AND SHAHSHAHANI, M., "Products of Random Matrices and Computer Image Generation," *Contemporary Mathamatics*, *50*:173–182 (1986).

DOUADY, A., AND HUBBARD J., *Comptes Rendus* (*Paris*) *294*:123–126 (1982).

DOUADY, A., AND HUBBARD, J. H., "On the Dynamics of Polynomial-like Mappings," *Annales Scientifiques de l'Ecole Normale Supérieur*, 4e Série, *18*:287–343 (1985).

DUNFORD, N, AND SCHWARTZ, J. T. *Linear Operators Part I*: *General Theory*, Third edition, John Wiley and Sons, New York, 1966.

EISEN, M., *Introduction to Mathematical Probability Theory*, Prentice Hall, Englewood Cliffs, 1969.

ELTON, J., "An Ergodic Theorem for Iterated Maps," *Journal of Ergodic Theory and Dynamical Systems*, 7:481–488 (1987).

ELTON, J., "A Simultaneously Contractive Remetrization Theorem for Iterated Function Systems," Georgia Institute of Technology Preprint, 1988.

FATOU, P., "Sur les Equations Fonctionelles," *Bulletin Societé Math. France 47*:161–271 (1919); *48*:33–94, 208–314 (1920).

FEDERER, H., *Geometric Measure Theory*, Springer-Verlag, New York, 1969.

FEIGENBAUM, M. J., "The Universal Metric Properties of Nonlinear Transformations," *Journal of Statistical Physics 21*:669–706 (1979).

FOURNIER, A., FUSSELL, D., AND CARPENTER, L., "Computer Rendering of Stochastic Models," *Communications of the AMC 25*(6) (June 1982).

GILBERT, W. J., "Fractal Geometry Derived from Complex Bases," *The Mathematical Intelligencer 4*:78–86 (1982).

GLEICK, J., *Chaos*: *Making a New Science*, Viking Press, New York, 1987.

HALMOS, P. R., *Measure Theory*, Springer-Verlag, New York, 1974.

HARDIN, D. P., "Hyperbolic Iterated Function Systems and Applications." Ph.D. Thesis, Georgia Institute of Technology, 1985.

HATA, M., "On the Structure of Self-Similar Sets," *Japan Journal of Applied Mathematics 2*(2):381–414 (Dec. 1985).

HUTCHINSON, J., "Fractals and Self-similarity," *Indiana University Journal of Mathematics* 30:713–747 (1981).

JULIA, G., "Memoire sur l'Itération des Fonctions Rationelles," *Journal de Mathematiques Pures et Appliques 4*:47–245 (1918).

KASRIEL, R. H., *Undergraduate Topology*, Saunders, Philadelphia, 1971.

KAWAGUCHI, Y., "A Morphological Study of the Form of Nature," *Computer Graphics 16*(3), (July 1982). SIGGRAPH 1982 Proceedings.

MANDELBROT, B., *The Fractal Geometry of Nature*, W. H. Freeman and Co., San Francisco, 1982.

MASSOPUST, P., Ph.D. Thesis, Georgia Institute of Technology, 1986.

MAY, R. B., "Simple mathematical models with very complicated dynamics," *Nature 261*:459–467 (1976).

MENDELSON, B., *Introduction to Topology*, Blackie & Son Limited, London, 1963.

MILLER, G. S. P., "The Definition and Rendering of Terrain Maps," *Computer Graphics 20*(4), (August 1986). SIGGRAPH 1986 Proceedings.

OPPENHEIMER, P. E., "Real Time Design and Animation of Fractal Plants and Trees," *Computer Graphics 20*(4), (August 1986).

PEITGEN, H.-O., RICHTER, P. H., *The Beauty of Fractals*, Springer-Verlag, Berlin, New York, 1986.

RUDIN, W., *Principles of Mathematical Analysis*, 2nd Edition, McGraw-Hill, New York, 1964.

RUDIN, W., *Real and Complex Analysis*, McGraw-Hill, New York, 1966.

REUTER, L., "Rendering and Magnification of Fractals Using Iterated Function Systems," Ph.D. Thesis, Georgia Institute of Technology, December 1987.

SCIA 1987, see DEWDNEY 1987.

SCOTT, D. H., *An Introduction to Structural Botany*: *Part I*, *Flowering Plants*, A.& C. Black, Ltd., London, 1917.

SINAI, YA. G., *Introduction to Ergodic Theory*, Princeton University Press, Princeton, 1976.

SMITH, A. R., "Plants, Fractals, and Formal Languages," *Computer Graphics 18*(3):1–10 (July 1984). SIGGRAPH 1984 Proceedings.

SULLIVAN, D., "Quasi Conformal Homeomorphisms and Dynamics," I, II, and III. Preprints, Institut des Hautes Etudes Scientifiques, Bures-sur-Yvette, France, 1982.

VRSCAY, E. R., "Julia Sets and Mandelbrot-Like Sets Associated with Higher Order Schröder Iteration Functions: A Computer Assisted Study," *Mathematics of Computation 46*:151–169 (1986).

WITHERS, W. D., "Calculating Derivatives with Respect to Parameters of Average Values in Iterated Function Systems," *Physica 28D*:206–214 (1987).

LASOTA, A., AND YORKE, J. A., "On the Existence of Invariant Measures for Piecewise Monotonic Transformations," *Transactions of the American Mathematical Society 186*:481–488 (1973).

Additional Background References

T. Bedford, "Dimension and Dynamics for Fractal Recurrent Sets," *Journal of the London Mathematical Society (2)33*:89–100 (1986).

P. Collet, and J.-P. Eckmann, *Iterated Maps on the Interval as Dynamical Systems*, Birkäuser, Boston, 1980.

F. M. Dekking, "Recurrent Sets," *Advances in Mathematics 44*:78–104 (1982).

J. Dugundji, *Topology*, Allyn and Bacon, Boston, 1966, p. 253.

W. Feller, *An Introduction to Probability Theory and its Applications*, Wiley, London, 1957.

D. P. Hardin, and P. Massopust, "Dynamical Systems Arising from Iterated Function Systems," *Communications in Mathematical Physics 105*:455–460 (1986).

U. Krengel, *Ergodic Theorems*, de Gruyter, New York, 1985.

T. Li and J. A. Yorke, "Period Three Implies Chaos," *American Mathematics Monthly 82*:985–992 (1975).

E. Seneta, *Non-negative Matrices*, Wiley, New York, 1973.

Index

∀, definition, 11
 , definition, 10
 , definition, 10
∅ , definition, 10
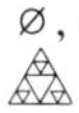 , definition, 10
■ , definition, 10
$|A|$, definition, 58
$A \subset X$, definition, 10

address, 3, 118–125, 129, 146, 151, 277, 302, 303, 365
  definition, 129
  illustration, 119, 120, 121, 122, 123, 124, 126
  periodic, *see* periodic address
affine transformation, 2, 43, 44, 45–46, 50, 52–53, 54, 56, 57, 71, 73, 74, 75, 76, 78, 97, 98, 102, 110, 117, 121, 123, 134, 141, 146, 187, 214, 221, 228, 231, 236, 239, 240, 244, 245, 256, 275, 338, 366, 378
  definition, 50
  illustration, 46, 50, 52, 53, 54, 56, 58, 73, 100, 101, 102, 103, 105, 107, 108, 233, Plates 7.1.5–7.1.7
  see also transformation, affine
algorithm, *see* fractals; specific algorithms for computing
analytic transformation, 62–68, 75, 292
  definition, 66
Andes Girl, 379
  illustration, Plate 9.8.7
Arctic Wolf, illustration, Plate 9.8.8
attractive cycle, 148, 270, 274, 320
  definition, 137
attractive fixed point, 137, 142, 148, 250
  definition, 136
  illustration, 138, 149
attractive periodic point
  definition, 137
attractor, 83–84, 85, 86, 87, 91, 92, 94, 96, 97, 98, 101, 103, 105, 110, 113, 118–119,

Credits for Figures and Color Plates

Dr. John Herndon collaborated with the author on the computation of many of the black and white figures.

The Orchard Subset of $\mathbb{R}^2$, Figure 3.2.8, was computed by Henry Strickland.

Figure 6.2.2 was produced by Peter Massopust.

All of the color images were produced in the Computergraphical Mathematics Laboratory at Georgia Institute of Technology. The following list gives the authors of each color image. The italics indicate who did most of the work.

Plate	**Author**
2.6.1	Michael Barnsley, *Alan Sloan*
3.10.1	*Michael Barnsley*
3.11.1	Michael Barnsley, *Arnaud Jacquin*, *Laurie Reuter*, Alan Sloan
7.1.1	Michael Barnsley, *John Herndon*
7.1.2	Michael Barnsley, *John Herndon*
7.1.3	Michael Barnsley, *John Herndon*
7.1.4	Michael Barnsley, *John Herndon*

7.1.5	Michael Barnsley, *John Herndon*
7.1.6	Michael Barnsley, *John Herndon*
7.1.7	Michael Barnsley, *John Herndon*
7.1.8	Michael Barnsley, *John Herndon*
7.1.9	Michael Barnsley, *John Herndon*
7.3.1	Michael Barnsley, *John Herndon*
8.3.1	Michael Barnsley, *John Herndon*
8.4.1	Michael Barnsley, *John Herndon*
8.4.2	Michael Barnsley, *John Herndon*
8.4.3	Michael Barnsley, *John Herndon*
8.4.4	Michael Barnsley, *John Herndon*
9.8.1	*Michael Barnsley*, *Laurie Reuter*, Alan Sloan
9.8.2	*Michael Barnsley*, *Laurie Reuter*, Alan Sloan
9.8.3	*Michael Barnsley*
9.8.4	*Michael Barnsley*
9.8.5	Michael Barnsley, *Arnaud Jacquin*, *Laurie Reuter*, Alan Sloan
9.8.6	Michael Barnsley, *Laurie Reuter*, Alan Sloan
9.8.7	Michael Barnsley, *Arnaud Jacquin*, *François Malassenet*, Laurie Reuter, Alan Sloan
9.8.8	Michael Barnsley, *Arnaud Jacquin*, *Laurie Reuter*, Alan Sloan
9.8.9	Michael Barnsley, *Arnaud Jacquin*, *Laurie Reuter*, Alan Sloan
9.8.10	Michael Barnsley, *Arnaud Jacquin*, *Laurie Reuter*, Alan Sloan
9.8.11	Michael Barnsley, *Arnaud Jacquin*, *François Malassenet*, Laurie Reuter, Alan Sloan
9.8.12	Michael Barnsley, *Arnaud Jacquin*, *François Malassenet*, Laurie Reuter, Alan Sloan
9.8.13	Michael Barnsley, *Arnaud Jacquin*, *François Malassenet*, Laurie Reuter, Alan Sloan
9.8.14	Michael Barnsley, *Arnaud Jacquin*, *François Malassenet*, Laurie Reuter, Alan Sloan

All of the color images were computed on a Masscomp 5600 with Aurora Graphics. Photographs were taken with the aid of a Dunn Multicolor on Ectachrome color slide film, ASA 100. Color Images 9.8.1–9.8.14 were encoded and produced with the aid of the *Seurat* and *Collage* software systems. *Collage* was designed by Michael Barnsley and Alan Sloan and written by Alan Sloan. *Seurat* was designed by Michael Barnsley and Laurie Reuter and written by Laurie Reuter.